Control of
Animal Cell Proliferation

Volume I

Contributors

Stuart A. Aaronson
Ramesh C. Adlakha
Suzanne K. Beckner
Robert M. Benbow
Daniel F. Bowen-Pope
Alton L. Boynton
R. N. Buick
Jane-Jane Chen
Frederick J. Darfler
Frances M. Davis
Michelle F. Gaudette
Denis Gospodarowicz
Harvey R. Herschman
Pamela J. Hines
Richard Horuk
C. Ronald Kahn
George L. King
Leonard P. Kleine
Katherine S. Koch
Hyam L. Leffert

Daniel H. Levin
Michael C. Lin
Allan Lipton
Irving M. London
E. A. McCulloch
Robert L. Matts
Alan C. Moses
Raymond Petryshyn
Sydne J. Pilistine
Potu N. Rao
R. H. Rixon
Keith C. Robbins
Russell Ross
Bartholomew M. Sefton
Ronald A. Seifert
Masaki Shioda
Charles K. Singleton
N. Shaun B. Thomas
Steven R. Tronick
James F. Whitfield

T. Youdale

Control of Animal Cell Proliferation

Volume I

Edited by

ALTON L. BOYNTON

Cell Physiology Group
Division of Biological Sciences
National Research Council of Canada
Ottawa, Ontario, Canada

HYAM L. LEFFERT

Department of Medicine
Division of Pharmacology
University of California, San Diego
La Jolla, California

1985

ACADEMIC PRESS, INC.

(Harcourt Brace Jovanovich, Publishers)

Orlando San Diego New York London
Toronto Montreal Sydney Tokyo

ACADEMIC PRESS, INC.
Orlando, Florida 32887

United Kingdom Edition published by
ACADEMIC PRESS INC. (LONDON) LTD.
24–28 Oval Road, London NW1 7DX

Library of Congress Cataloging in Publication Data
Main entry under title:

Control of animal cell proliferation.

Includes index.
1. Cell proliferation. 2. Cellular control
mechanisms. 3. Cancer cells. I. Boynton,
Alton L. II. Leffert, H. L.
QH605.C756 1985 591'.08762 85-6061
ISBN 0–12–123061–9 (v. 1 : alk. paper)

PRINTED IN THE UNITED STATES OF AMERICA

85 86 87 88   9 8 7 6 5 4 3 2 1

Contents

I. TRENDS AND ISSUES

1 Oncogenes and Pathways to Malignancy

Stuart A. Aaronson, Steven R. Tronick,
and Keith C. Robbins

2 The Role of Stem Cells in Normal and Malignant Tissue

R. N. Buick and E. A. McCulloch

II. GROWTH FACTORS

3 Epidermal and Fibroblastic Growth Factor

Denis Gospodarowicz

4 Insulin-Like Growth Factors

Alan C. Moses and Sydne J. Pilistine

5 Cyclic AMP Elevators Stimulate the Initiation of DNA Synthesis by Calcium-Deprived Rat Liver Cells

*Alton L. Boynton, Leonard P. Kleine,
and James F. Whitfield*

6 Platelet Growth Factors: Presence and Biological Significance

Allan Lipton

III.   RECEPTORS

7 The EGF Receptor

Harvey R. Herschman

8 Effect of Insulin on Growth *in Vivo* and Cells in Culture

George L. King and C. Ronald Kahn

9 Glucagon Receptors and Their Functions

Suzanne K. Beckner, Richard Horuk,
Frederick J. Darfler, and Michael C. Lin

10 The Platelet-Derived Growth Factor Receptor

Daniel F. Bowen-Pope, Ronald A. Seifert,
and Russell Ross

IV. TRANSDUCTION MECHANISMS

11 The Role of Tyrosine Protein Kinases in the Action of Growth Factors

Bartholomew M. Sefton

12 The Control of Cell Proliferation by Calcium, Ca^{2+}-Calmodulin, and Cyclic AMP

James F. Whitfield, Alton L. Boynton, R. H. Rixon, and T. Youdale

13 Growth Regulation by Sodium Ion Influxes

Hyam L. Leffert and Katherine S. Koch

V.   REGULATION

14 Structural Heterogeneity of Duplex DNA

Charles K. Singleton

15 Initiation of DNA Replication in Eukaryotes

*Robert M. Benbow, Michelle F. Gaudette,
Pamela J. Hines, and Masaki Shioda*

16 Role of Phosphorylation of Nonhistone Proteins in the Regulation of Mitosis

*Ramesh C. Adlakha, Frances M. Davis,
and Potu N. Rao*

17 Translational Regulation of Eukaryotic Protein Synthesis by Phosphorylation of eIF-2α

*Irving M. London, Daniel H. Levin, Robert L. Matts,
N. Shaun B. Thomas, Raymond Petryshyn,
and Jane-Jane Chen*

Contributors

Numbers in parentheses indicate the pages on which the authors' contributions begin.

STUART A. AARONSON (3), Laboratory of Cellular and Molecular Biology, National Cancer Institute, National Institutes of Health, Bethesda, Maryland 20205

RAMESH C. ADLAKHA (485), Department of Chemotherapy Research, The University of Texas M. D. Anderson Hospital and Tumor Institute, Houston, Texas 77030

SUZANNE K. BECKNER (251), Laboratory of Cellular and Developmental Biology, National Institute of Arthritis, Diabetes, and Digestive and Kidney Diseases, National Institutes of Health, Bethesda, Maryland 20205

ROBERT M. BENBOW (449), Department of Biology, The Johns Hopkins University, Baltimore, Maryland 21218, and Department of Zoology, Iowa State University, Ames, Iowa 50011

DANIEL F. BOWEN-POPE (281), Department of Pathology, School of Medicine, University of Washington, Seattle, Washington 98195

ALTON L. BOYNTON[1] (121, 331), Cell Physiology Group, Division of Biological Sciences, National Research Council of Canada, Ottawa, Ontario, Canada K1A OR6

R. N. BUICK (25), Division of Biological Research, Ontario Cancer Institute, Toronto, Ontario, Canada M4X 1K9

JANE-JANE CHEN (515), Harvard-MIT Division of Health Sciences and Technology, and Department of Biology, Massachusetts Institute of Technology, Cambridge, Massachusetts 02139

FREDERICK J. DARFLER (251), Laboratory of Cellular and Developmental Biology, National Institute of Arthritis, Diabetes, and Digestive and Kidney Diseases, National Institutes of Health, Bethesda, Maryland 20205

[1]Present address: Cancer Center of Hawaii, University of Hawaii, Honolulu, Hawaii 96813.

FRANCES M. DAVIS (485), Department of Chemotherapy Research, The University of Texas M. D. Anderson Hospital and Tumor Institute, Houston, Texas 77030

MICHELLE F. GAUDETTE (449), Department of Biology, The Johns Hopkins University, Baltimore, Maryland 21218, and Department of Zoology, Iowa State University, Ames, Iowa 50011

DENIS GOSPODAROWICZ (61), Cancer Research Institute, and Departments of Medicine and Ophthalmology, University of California Medical Center, San Francisco, California 94143

HARVEY R. HERSCHMAN (169), Department of Biological Chemistry, and Laboratory of Biomedical and Environmental Sciences, UCLA Center for Health Sciences, Los Angeles, California 90024

PAMELA J. HINES[2] (449), Department of Biology, The Johns Hopkins University, Baltimore, Maryland 21218, and Department of Zoology, Iowa State University, Ames, Iowa 50011

RICHARD HORUK[3] (251), Laboratory of Cellular and Developmental Biology, National Institute of Arthritis, Diabetes, and Digestive and Kidney Diseases, National Institutes of Health, Bethesda, Maryland 20205

C. RONALD KAHN (201), Research Division, Joslin Diabetes Center, Brigham and Women's Hospital, Harvard Medical School, Boston, Massachusetts 02215

GEORGE L. KING (201), Research Division, Joslin Diabetes Center, Brigham and Women's Hospital, Harvard Medical School, Boston, Massachusetts 02215

LEONARD P. KLEINE (121), Cell Physiology Group, Division of Biological Sciences, National Research Council of Canada, Ottawa, Ontario, Canada K1A OR6

KATHERINE S. KOCH[4] (367), Department of Biology, University of California, San Diego, La Jolla, California 92093

HYAM L. LEFFERT (367), Department of Medicine, Division of Pharmacology, University of California, San Diego, La Jolla, California 92093

DANIEL H. LEVIN (515), Harvard-MIT Division of Health Sciences and Technology, and Department of Biology, Massachusetts Institute of Technology, Cambridge, Massachusetts 02139

MICHAEL C. LIN (251), Laboratory of Cellular and Developmental Biol-

[2]Present address: Zoology Department, University of Washington, Seattle, Washington 98195.

[3]Present address: Department of Medicine, University of California, San Diego, La Jolla, California 92093.

[4]Present address: Department of Medicine, Division of Pharmacology, University of California, San Diego, La Jolla, California 92093.

ogy, National Institute of Arthritis, Diabetes, and Digestive and Kidney Diseases, National Institutes of Health, Bethesda, Maryland 20205

ALLAN LIPTON (151), Division of Oncology, and the Specialized Cancer Research Center, The Milton S. Hershey Medical Center, The Pennsylvania State University, Hershey, Pennsylvania 17033

IRVING M. LONDON (515), Harvard-MIT Division of Health Sciences and Technology, and Department of Biology, Massachusetts Institute of Technology, Cambridge, Massachusetts 02139

E. A. McCULLOCH (25), Division of Biological Research, Ontario Cancer Institute, Toronto, Ontario, Canada M4X 1K9

ROBERT L. MATTS (515), Harvard-MIT Division of Health Sciences and Technology, and Department of Biology, Massachusetts Institute of Technology, Cambridge, Massachusetts 02139

ALAN C. MOSES (91), The Charles A. Dana Research Institute, Department of Medicine, Harvard Medical School, and The Harvard-Thorndike Laboratory of Beth Israel Hospital, Beth Israel Hospital, Boston, Massachustts 02215

RAYMOND PETRYSHYN (515), Harvard-MIT Division of Health Sciences and Technology, and Department of Biology, Massachusetts Institute of Technology, Cambridge, Massachusetts 02139

SYDNE J. PILISTINE[5] (91), The Charles A. Dana Research Institute, Department of Medicine, Harvard Medical School, and The Harvard-Thorndike Laboratory of Beth Israel Hospital, Beth Israel Hospital, Boston, Massachusetts 02215

POTU N. RAO (485), Department of Chemotherapy Research, The University of Texas M. D. Anderson Hospital and Tumor Institute, Houston, Texas 77030

R. H. RIXON (331), Cell Physiology Group, Division of Biological Sciences, National Research Council of Canada, Ottawa, Ontario, Canada K1A OR6

KEITH C. ROBBINS (3), Laboratory of Cellular and Molecular Biology, National Cancer Institute, National Institutes of Health, Bethesda, Maryland 20205

RUSSELL ROSS (281), Department of Pathology, School of Medicine, University of Washington, Seattle, Washington 98195

BARTHOLOMEW M. SEFTON (315), Molecular Biology and Virology Laboratory, Salk Institute, San Diego, California 92138

RONALD A. SEIFERT (281), Department of Pathology, School of Medicine, University of Washington, Seattle, Washington 98195

[5]Present address: Fels Research Institute, Department of Biological Chemistry, Endocrinology Section, Wright State University, Dayton, Ohio 45435.

MASAKI SHIODA (449), Department of Biology, The Johns Hopkins University, Baltimore, Maryland 21218, and Department of Zoology, Iowa State University, Ames, Iowa 50011

CHARLES K. SINGLETON[6] (417), Department of Bacteriology, University of Wisconsin, Madison, Wisconsin 53706

N. SHAUN B. THOMAS (515), Harvard-MIT Division of Health Sciences and Technology, and Department of Biology, Massachusetts Institute of Technology, Cambridge, Massachusetts 02139

STEVEN R. TRONICK (3), Laboratory of Cellular and Molecular Biology, National Cancer Institute, National Institutes of Health, Bethesda, Maryland 20205

JAMES F. WHITFIELD (121, 331), Cell Physiology Group, Division of Biological Sciences, National Research Council of Canada, Ottawa, Ontario, Canada K1A OR6

T. YOUDALE (331), Cell Physiology Group, Division of Biological Sciences, National Research Council of Canada, Ottawa, Ontario, Canada K1A OR6

[6]Present address: Department of Molecular Biology, Vanderbilt University, Nashville, Tennessee 37235.

Preface

How animal cells regulate their proliferative activity and how cells become proliferatively autonomous resulting in malignant behavior are two outstanding questions in cell biology. In humans, the neoplastic process is complicated by the extremely long development time. The question of whether cancer is a disease of proliferation or differentiation or if oncogenes are involved also has not been answered.

Our understanding of mechanisms that control animal cell proliferation is advancing steadily. There are several reasons for this progress. One has been the rapid development of powerful tools of recombinant DNA technology. Another has been the identification, isolation, and physical characterization of many protein growth factors (and their receptors) that stimulate proliferation of different kinds of animal cells. A third reason involves the use (often under chemically defined conditions) and development of cell culture systems—as bioassays for growth-promoting agents—that provide models for physiological processes occurring in the intact animal. These developments have led to an explosive amount of information that has been disseminated in increasingly larger numbers of scientific journals. A central forum for addressing these questions is missing, and it is the purpose of this treatise to address these basic questions.

The chapters in this first volume update areas of animal cell proliferation, beginning with trends and issues (oncogenes and stem cells). Current knowledge of the structure and function of several growth factors (EGF, FGF, insulin, PDGF, IGF, and cAMP) and their receptors (EGF, PDGF, insulin, and glucagon) and of the mechanisms of information transduction (tyrosine protein kinases, calcium, calmodulin, cAMP, and sodium ion fluxes) will be reviewed. Finally, certain aspects of DNA structure, DNA regulation, protein synthesis, and mitosis will be presented.

We hope that by combining these subjects, the field of cell proliferation and growth control will begin to take shape as a unified discipline of biology.

Alton L. Boynton
Hyam L. Leffert

xvii

I

Trends and Issues

1

Oncogenes and Pathways to Malignancy

STUART A. AARONSON, STEVEN R. TRONICK, AND
KEITH C. ROBBINS

Laboratory of Cellular and Molecular Biology
National Cancer Institute
National Institutes of Health
Bethesda, Maryland

Efforts to elucidate the genetic alterations that lead normal cells to become cancer cells have been aided immeasurably by the investigation of acute transforming retroviruses. These agents have substituted viral genes required for replication with discrete segments of host cellular information. When incorporated within the viral genome, these transduced sequences, designated *onc* genes, acquire transforming properties.

In the past few years, evidence has emerged that some of the very same normal cellular genes which have given rise to retroviral *onc* genes can be activated as transforming genes by mechanisms com-

3

pletely independent of retrovirus involvement. This review summarizes the salient features of acute transforming retroviruses as well as evidence that has linked some of their cell-derived transforming genes to cellular genes activated to become oncogenes of human cancer cells.

I. ACUTE TRANSFORMING RETROVIRUSES

Acute transforming retroviruses can cause a variety of malignancies, including sarcomas, carcinomas, and hematopoietic tumors. In tissue culture, these agents rapidly induce growth alterations of susceptible target cells. Such transformants, after propagation in culture, produce tumors in animals, whereas uninfected cells in parallel cultures do not. Thus, the induction of transformed foci by the virus directly reflects its tumor-inducing capability *in vivo* (Weiss *et al.*, 1982; Duesberg, 1983; Aaronson, 1983).

Advances in recombinant DNA technology have made it possible to clone the proviral forms of these RNA viruses (Aaronson *et al.*, 1983). Figure 1 compares the structure of a prototype mouse leukemia virus genome with that of a typical acute transforming retrovirus of mouse origin. The chronic viral genome consists of *gag*, *pol*, and *env* genes, all of which code for proteins required for replication of an infectious virus. The retrovirus genome also contains long terminal repeat sequences, designated LTRs, which possess signals for the initiation, enhancement, and termination of transcription. The extent of each LTR is defined by terminal inverted repeats. Thus, the overall retroviral genomic structure is analogous to that of transposable elements found in prokaryotic and eukaryotic species (Weiss *et al.*, 1982; Temin, 1982).

As shown in Fig. 1, the prototype acute transforming virus has deleted substantial information necessary for replication functions and has substituted, instead, a discrete new segment of information. Specific alteration or deletion of the substitution sequence causes the virus to lose its ability to transform cells or cause tumors. Analogous findings with a number of acute transforming viruses have led to assignment of the viral transforming function(s) to the acquired *onc* sequence (Weiss *et al.*, 1982).

If a retroviral *onc* sequence is used as a DNA probe to analyze normal cellular DNAs, the probe detects homologous sequences in DNAs of a wide range of vertebrate species (Weiss *et al.*, 1982; Duesberg, 1983; Bishop, 1983). These findings, in addition to the fact that *onc* sequences are generally represented as single copy numbers, imply that viral *onc* genes have arisen from normal cellular genes. Evidence that

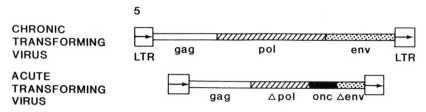

Fig. 1. Comparison of the genomic structures of chronic and acute transforming retroviruses. The structures illustrated were determined by analysis of molecularly cloned DNAs of Moloney murine leukemia virus and BALB murine sarcoma virus (Aaronson, 1983).

these genes, designated proto-oncogenes, are also conserved in evolution further implies that they serve very important functions in normal cellular growth processes.

If any of the thousands of genes that exist in our cells could acquire neoplastic properties when captured by one of these viruses, then we would never expect to observe the same sequence incorporated by more than one of the known two dozen or so acute transforming retrovirus isolates. In fact, several virus isolates from the same or even different species have captured the very same onc sequence (Weiss et al., 1982; Bishop, 1983). These findings have indicated that the number of cellular genes that can acquire transforming properties when incorporated within the retrovirus genome must be rather limited.

II. ASSIGNMENT OF A NORMAL FUNCTION TO A PROTO-ONCOGENE

Investigations aimed at elucidating the functions of the cellular genes that give rise to onc genes have in some cases led to detection of their transforming proteins. With one group of onc genes, the prototype of which is the src gene of Rous sarcoma virus, it has been possible to assign an enzymatic activity to the gene product. These genes code for protein kinases having specificity for tyrosine residues (Collett and Erikson, 1978; Levinson et al., 1978; see Chapter 11). However, identification of the function of proto-oncogenes from which oncogenes have arisen has, until recently, eluded efforts of scientists.

A. The Transforming Gene of Simian Sarcoma Virus

For a number of years, our group has been investigating simian sarcoma virus (SSV), the only acute transforming retrovirus of primate origin. Having developed an appropriate biologic system for SSV characterization, we molecularly cloned the SSV genome, dissected its structure, and identified its cell-derived *onc* gene, v-*sis* (Robbins *et al.*, 1981). More recently, investigators in our laboratory have determined the nucleotide sequence of the complete SSV genome (Devare *et al.*, 1983). These studies revealed an open reading frame within the SSV transforming gene which possessed the coding capacity for a protein of ~26,000 daltons. By preparation of antibodies to small regions of the amino acid sequence predicted for the v-*sis*-coded protein, it was possible to detect the SSV transforming protein in SSV infected cells. Antibody to a peptide derived from the amino terminus of the predicted v-*sis* gene product detected a 28,000-dalton species in SSV transformed but not uninfected cells. Moreover, preincubation of the antibody with unlabeled peptide completely blocked the reaction, establishing this protein, p28^{sis}, as the product of the SSV transforming gene (Robbins *et al.*, 1982a).

B. The Structure of Human Platelet-Derived Growth Factor

Human platelet-derived growth factor (PDGF) is a potent mitogen for connective tissue cells (see Chapter 6). This protein is biologically important because it is thought to be involved in wound healing. PDGF is a heat-stable, cationic protein which is normally stored in the α granules of platelets. Its biologically active forms are disulfide bonded and range in size from 28,000 to 35,000 daltons. Upon reduction, the molecule loses biologic activity, yielding two distinct polypeptides, designated PDGF-1 and -2, each around 18,000 daltons in size (Antoniades and Hunkapiller, 1983).

Antoniades and Hunkapiller (1983) purified human PDGF and reported the N-terminal amino acid sequences of PDGF polypeptides 1 and 2. Upon entering this newly published sequence into his computer bank of predicted protein sequences, Doolittle made the striking observation that 87% of the amino acid sequence determined for PDGF peptide 2 matched, without the introduction of gaps, that predicted for p28^{sis} (Doolittle *et al.*, 1983). Studies by another group working on PDGF structure have led to similar conclusions concerning the relationship of the two proteins (Waterfield *et al.*, 1983). The fact that v-*sis* is of New World primate origin (Robbins *et al.*, 1982b), while PDGF is a human protein, likely explains the very close but not perfect sequence match. Thus, p28^{sis} and PDGF appear to be derived from the same cellular gene.

C. The v-*sis* Gene Product Is Structurally and Immunologically Related to PDGF

Utilizing peptide antibodies against the amino and carboxy termini of the SSV transforming gene product, we were able to demonstrate that the primary translational product, p28^{sis}, of the SSV transforming gene undergoes cleavage to yield 11,000- and 20,000-dalton polypeptides, designated p11^{sis} and p20^{sis}, respectively (Robbins *et al.*, 1983). The latter, which was shown to be derived from the carboxy terminus of p28^{sis}, is very similar in size to the inactive reduced 18,000-dalton form of human PDGF-2. The sequence correspondence between PDGF-2 and the v-*sis* product begins at position 67. If cleavage of p28^{sis} were signaled by the two basic amino acids (Lys and Arg) at positions 65-66, a 160-residue protein approximating p20^{sis} in its theoretical size would result. Thus, p20^{sis} closely corresponds in size and amino acid sequence to that observed for PDGF-2.

Based upon the fact that biologically active PDGF exhibits a dimeric structure involving disulfide bonds, we examined v-*sis* gene products under nonreducing conditions (Fig. 2). Our observations that a protein of 56,000 daltons was detectable with antibodies to both amino and carboxy termini of p28^{sis} was most consistent with the possibility that p56^{sis} was a disulfide-linked dimer of p28^{sis}. Evidence that the carboxy- but not the amino-terminal *sis* antibodies also bound discrete *sis*-related polypeptides ranging in size from 28,000 to 46,000 daltons implied further processing of p56^{sis} to smaller forms, some of which closely resembled the sizes of unreduced PDGF.

Evidence of the close structural and conformational similarity between SSV transforming gene products and PDGF was further strengthened by demonstration of immunologic cross-reactivity. Anti-PDGF serum was shown to recognize the p28^{sis} dimer as well as the other processed unreduced forms of the protein also recognized by the anti-*sis*-C peptide serum. In addition, anti-PDGF recognized a 24,000-dalton protein not detected with either anti-*sis* peptide sera (Fig. 3). Pulse chase analysis using anti-PDGF serum established p56^{sis} as a rapidly formed dimer of p28^{sis}. It was possible to demonstrate further processing in which p24^{sis} emerged as the most stable product of the

SSV transforming gene detectable with the available antisera. Thus, our findings revealing a high degree of relatedness between the transforming gene product of a primate sarcoma virus and a potent growth factor for human fibroblasts, smooth muscle cells, and glial cells suggest that the mechanism by which this *onc* gene transforms cells may involve the constitutive expression of a gene encoding a normal growth factor.

Recent studies have provided an additional link between growth factors and oncogenes. Purification of the cellular receptor for epidermal growth factor (EGF) has led to the elucidation of a portion of its amino acid sequence (see Chapter 7). A computer search has revealed a very close similarity between the EGF receptor and the product of the *erb*-B *onc* gene (Downward *et al.*, 1984). Thus, transformation by an-

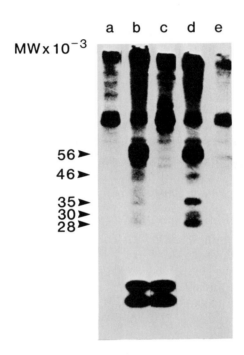

Fig. 2.   Analysis of SSV transforming gene products in nonreducing conditions. Infected cells were metabolically labeled with [^{35}S]amino acids. Cell lysates were immunoprecipitated with preimmune (lane a), anti-*sis*-N (lanes b, c), or anti-*sis*-C (lanes d, e) serum. In some cases antibodies were incubated before immunoprecipitation with 5 μg of *sis*-N (lane c) or *sis*-C (lane e) peptide. Immunoprecipitates were analyzed by SDS–PAGE in the absence of reducing agent.

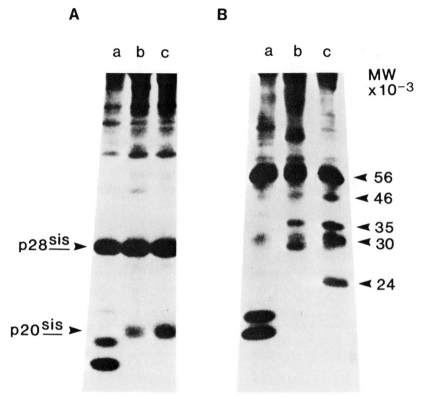

Fig. 3. Structural and immunologic similarities between the products of the SSV transforming gene and PDGF. Lysates of metabolically labeled SSV transformed cells were immunoprecipitated with anti-*sis*-N (lanes a), anti-*sis*-C (lanes b), or anti-PDGF (lanes c) serum, and analyzed by SDS–PAGE in the presence (panel A) or absence (panel B) or reducing agent.

other acute transforming retrovirus may also involve pathways through which growth factors exert their proliferative effects.

III. SPECIFIC PROTO-ONCOGENE REARRANGEMENTS IN HUMAN TUMORS

Since genes related to retroviral *onc* genes are present in human DNA, it became possible to assess the role of proto-oncogenes in the etiology of human cancers. One approach used in this regard has involved the chromosomal mapping of human proto-oncogenes. In human–rodent somatic cell hybrids possessing varying numbers of human chromosomes, it is possible to distinguish rodent and human proto-oncogene sequences by digestion of cellular DNAs with appropriate restriction enzymes. Where possible, cloned DNAs from human proto-oncogene loci have been used as molecular probes to enhance the signal intensity of the human proto-oncogene sequence.

The results of chromosome mapping studies by our laboratory and others are summarized in Table I. It is apparent that proto-oncogenes are widely distributed throughout the human genome. Thus, we have assigned *sis* to chromosome 22 (Swan *et al.*, 1982), *mos* and *myc* to chromosome 8 (Prakash *et al.*, 1982; Taub *et al.*, 1982), and *myb* to chromosome 6 (McBride *et al.*, 1983). Members of the *ras* oncogene family are found on chromosomes 1 (N-*ras*), 6 (K-*ras*), 11 (H-*ras*-1), and 12 (K-*ras*-2) (Dalla Favera *et al.*, 1982b; McBride *et al.*, 1982, 1983). Other groups have made similar observations and have mapped additional proto-oncogenes to chromosomes 3 (*raf*-1), 4 (*raf*-2), 5 (*fms*), 15 (*fes*), 22 (*abl*), and X (H-*ras*-2) (Bonner *et al.*, 1984; Dalla Favera *et al.*, 1982a; deMartinville *et al.*, 1983a,b; Groffen *et al.*, 1983; Hall *et al.*,

TABLE I
Chromosomal Assignments of Human Proto-Oncogenes

Chromosome number	Tumor[a]	Chromosomal aberration	*onc* Gene[b]
1	Neuroblastoma	del 1p	N-*ras*
3	Renal and lung CA, parotid tumors	t(3;14); t(3p); del 3p	*raf*-1
4	—	—	*raf*-2
5	AML	del 5q	*fms*
6	ALL, ovarian CA	6q−; t(6;14)	K-*ras*-1; *myb*
8	Burkitt's lymphoma	t(8;14); t(8;22); t(2;8)	*mos*; *myc*
	AML	t(8;21)	
9	CML	t(9;22)	*abl*
11	Wilms' tumor	11p−	H-*ras*-1
12	CLL	12+	K-*ras*-2
15	APL	t(15;17)	*fes*
20	—	—	*src*
22	CML	t(9;22)	*sis*

[a]Cancers with chromosomal aberrations affecting chromosome to which *onc* gene homologues have been localized.

[b]See Bonner *et al.*, 1984; Dalla Favera *et al.*, 1982a,b; deMartinville *et al.*, 1983a,b; Groffen *et al.*, 1983; Hall *et al.*, 1983; Harper *et al.*, 1983; McBride *et al.*, 1982, 1983; Neel *et al.*, 1982; O'Brien *et al.*, 1983; Prakash *et al.*, 1982; Ryan *et al.*, 1983; Swan *et al.*, 1982; Taub *et al.*, 1982.

1983; Harper et al., 1983; Neel et al., 1982; O'Brien et al., 1983; Roussel et al., 1983; Ryan et al., 1983; Sakaguchi et al., 1983).

Klein (1981) speculated that onc genes might be activated by chromosomal translocations, since specific chromosomal rearrangements were known to occur in certain human tumors. As shown in Table I, chromosomes harboring onc-related genes are involved in several such diseases. It has been possible to show that, in the case of Burkitt's lymphoma, the c-myc locus is translocated from chromosome 8 to chromosome 14 (Dalla Favera et al., 1982b; Taub et al., 1982) into the immunoglobulin (μ chain) α switch region (Taub et al., 1982). Moreover, evidence has been obtained indicating that these tumor cells express increased levels of the myc RNA (Ar-Rushdi et al., 1983), implying that the rearrangement of myc causes the gene to be subject to a new regulatory influence. The high degree of specificity of myc rearrangement in Burkitt's lymphoma supports the concept that it plays a role in the etiology of this tumor.

Human chronic myelogenous leukemia demonstrates a different translocation, termed the Philadelphia chromosome, involving chromosomes 9 and 22 (Table I). Recent studies suggest that a translocation of the proto-oncogene related to the onc gene, abl, from chromosome 9 to chromosome 22 leads to its rearrangement (De Klein et al., 1982) as well as to the expression of an aberrant abl transcript (Canaani et al., 1984). Thus, this approach appears to hold great promise in identifying proto-oncogenes which are targets for genetic alterations associated with chromosomal aberrations.

IV. PROTO-ONCOGENE AMPLIFICATION IN TUMOR CELLS

There are other mechanisms in addition to chromosomal rearrangements that could lead to the altered expression of a proto-oncogene. These include increased gene dosage due to gene amplification, up mutations in regulatory sequences, and DNA rearrangements due to transposition, which place the proto-oncogene under new regulatory controls.

Evidence for the differential expression of oncogenes in human tumor cells was first obtained by probing mRNAs isolated from human cell lines and tumor tissues with cloned DNA probes derived from retroviral onc genes (Eva et al., 1982; Westin et al., 1982). High levels of transcripts related to the v-myc oncogene were found in several tumor cell lines (Eva et al., 1982). In the case of the HL60 leukemia cell line, its high level of expression of *myc* transcripts (Westin *et al.*, 1982) was shown to be associated with a 20- to 30-fold amplification of c-*myc* DNA (Collins and Groudine, 1982; Dalla Favera *et al.*, 1982c). A colon carcinoma, which was found to contain an abundance of c-*myc* transcripts, similarly contained an approximately 16-fold amplification of the c-*myc* gene (Alitalo *et al.*, 1983). More recently, a high percentage of human small-cell lung carcinoma cell lines and also a non-small-cell lung carcinoma line have been shown to contain amplified c-*myc* DNA (Little *et al.*, 1983). Many tumor cells containing amplified genes often display abnormal karyotypic markers such as homogeneously staining regions (HSRs), double minutes (DMs), and abnormally banded regions (ABRs) (Biedler *et al.*, 1973; Biedler and Spengler, 1976; Sandberg *et al.*, 1972). In several instances, amplified *myc* DNA sequences have been found to be present in or associated with such chromosomal abnormalities (Alitalo *et al.*, 1983; Kohl *et al.*, 1983; Little *et al.*, 1983; Nowell *et al.*, 1983; Schwab *et al.*, 1983b).

c-*myc* is not the only oncogene that can be amplified in tumor cells. For example, a sequence partially related to c-*myc*, designated N-*myc*, frequently has been found to be amplified in human neuroblastomas and neuroblastoma-derived cell lines (Kohl *et al.*, 1983; Schwab *et al.*, 1983b). The K-*ras* proto-oncogene has been found to be overexpressed and amplified in a mouse adrenocortical tumor cell line and located within a region of karyotypic abnormality (Schwab *et al.*, 1983a). The *abl* proto-oncogene is amplified about nine times in a cell line derived from a patient with chronic myelogenous leukemia (Collins and Groudine, 1983). It is not known whether proto-oncogene amplification is a primary or secondary event in the chain of events leading to neoplasia. However, the strikingly frequent and consistent amplification of certain oncogenes in specific tumors suggests their relevance to the development of such cancers.

V. DNA INSERTION CAN LEAD TO ONCOGENE ACTIVATION

Chronic leukemia viruses lack a specific *onc* gene sequence (Duesberg, 1983; Weiss *et al.*, 1982). Recent studies have provided evidence that their mechanisms of malignant transformation involve the activation of cellular proto-oncogenes by the nearby integration of regulatory sequences located within the viral LTR (Hayward *et al.*, 1981; Payne *et al.*, 1982).

Recent studies have also provided evidence for oncogene activation in naturally occurring tumors by transposition of endogenous cellular DNA sequences containing regulatory signals. In certain mouse plasmacytomas, the c-mos proto-oncogene was found to be rearranged, accompanied by the abnormal expression of c-mos-related mRNA (Rechavi et al., 1982). It was possible to demonstrate that c-mos in such tumors was activated as an oncogene due to insertion of an endogenous intracisternal A-particle genome within the c-mos proto-oncogene (Canaani et al., 1983). It is not yet resolved whether this insertion occurred by a mechanism involving retrovirus-like integration or by direct transposition.

VI. ONCOGENES DEMONSTRATED BY TRANSFECTION OF HUMAN TUMOR DNA

Another approach toward detecting transforming genes was initiated a few years ago. By the use of a sensitive technique for incorporating exogenous DNA into cells, termed DNA transfection (Graham and Van der Eb, 1973), investigators asked whether DNAs of animal or human tumor cells possessed the capacity to directly confer the neoplastic phenotype to a susceptible assay cell. Such results might indicate the presence of a "transforming gene" in the tumor. In fact, some human tumor DNAs were shown to induce transformed foci in the continuous NIH/3T3 mouse cell line, which was highly susceptible to the uptake and stable incorporation of exogenous DNA (Cooper, 1982; Weinberg, 1982).

In one series of solid tumors examined for transforming DNA sequences (Pulciani et al., 1982), oncogenes were demonstrated in about 15% of the fresh tumors tested (5/28), including two colon carcinomas, carcinomas of the lung and pancreas, and an embryonal rhabdomyosarcoma. Transforming genes were also detected in cell lines established from carcinomas of the colon, lung, gall bladder, urinary bladder, as well as from a fibrosarcoma.

In a series of 22 established human hematopoietic tumor cell lines or primary tumors, 7 were positive in the DNA transfection assay (5/10 from cell lines, 2/12 from tumors) (Eva et al., 1983). The frequency at which transforming genes were detected in hematopoietic tumor cell lines and primary malignancies was similar to that observed with solid tumors. In acute myelogenous leukemia, for example, more than 50% of tumors analyzed have been detected as positive in the NIH/3T3 transfection assay (S. W. Needleman, personal communication). Activation of cellular oncogenes does not appear to be an artifact of tissue culture growth of tumor cells since primary human hematopoietic (Eva et al., 1983) as well as solid tumors (Pulciani et al., 1982) can register as positive in the NIH/3T3 transfection assay.

Many human oncogenes that have been detected by transfection analysis are related to a small group of retroviral oncogenes, designated *ras*. These genes were initially detected as the *onc* genes H- and K-*ras* of the acute transforming viruses, Harvey and Kirsten murine sarcoma viruses (MSV), respectively (Ellis *et al.*, 1981). H- and K-*ras* arise from different cellular genes. However, sequence comparison (Dhar *et al.*, 1982; Tsuchida *et al.*, 1982) and immunologic analysis of their 21,000-dalton products (Andersen *et al.*, 1981) revealed their high degree of homology and, hence, evolutionary relatedness. Recently, Wigler and co-workers have isolated a new oncogene from a human neuroblastoma cell line which is also related to H- and K-*ras* oncogenes (Shimizu *et al.*, 1983b). This oncogene, designated N-*ras*, has not to our knowledge been captured by a retrovirus.

It is striking that a very high fraction of the transforming DNA sequences associated with solid and hematopoeitic tumors analyzed to date are related to members of the *ras* gene family. K-*ras* oncogenes have been detected at high frequency in lung and colon carcinomas (Pulciani *et al.*, 1982). Carcinomas of the digestive tract, including pancreas and gall bladder, as well as genitourinary tract tumors and sarcomas have also been shown to contain *ras* oncogenes. N-*ras* appears to be the most frequently activated *ras* transforming gene in human hematopoietic neoplasms (Eva *et al.*, 1982).

Not only can a variety of tumor types contain the same activated *ras* oncogene, but the same tumor type can contain different activated *ras* oncogenes. Thus, in hematopoietic tumors, we have observed different *ras* oncogenes (K-*ras*, N-*ras*) activated in lymphoid tumors at the same stage of hematopoietic cell differentiation, as well as N-*ras* genes activated in tumors as diverse in origin as acute and chronic myelogenous leukemia (Eva *et al.*, 1982). These findings strongly imply that *ras* oncogenes detected in the NIH/3T3 transfection assay are not specific to a given stage of cell differentiation or tissue type.

Retroviruses that contain *ras*-related *onc* genes are known to possess a wide spectrum of target cells for transformation *in vivo* and *in vitro*. In addition to inducing sarcomas and transforming fibroblasts (Gross, 1970), these viruses are capable of inducing tumors of immature lymphoid cells (Pierce and Aaronson, 1982). They also can stimulate the proliferation of erythroblasts (Hankins and Scolnick, 1981) and monocyte/macrophages (Greenberger, 1979) and can even induce alterations in the growth and differentiation of epithelial cells (Weissman and Aaronson, 1983). Thus, the wide array of tissue types that can be induced to proliferate abnormally by these *onc* genes may help to explain the high frequency of detection of their activated human homologues in diverse human tumors.

It should be noted that not all oncogenes detected by NIH/3T3 transfection analysis are related to known retroviral oncogenes. For example, Cooper and co-workers have identified and molecularly cloned an oncogene, B-*lym*, from a B-cell lymphoma (Goubin *et al.*, 1983). B-*lym* appears to be activated in a large proportion of tumors at a specific stage of B-cell differentiation (Diamond *et al.*, 1983). Studies to date indicate that this oncogene is relatively small in size (<600 bp), and sequence analysis indicates that it possesses distant homology to transferrin (Goubin *et al.*, 1983). These investigators have also detected oncogenes which appear to be specifically activated in tumors at other stages of lymphoid differentiation (Lane *et al.*, 1982) or in mammary carcinomas (Lane *et al.*, 1981). None of these transforming genes appears to possess detectable homology with known retroviral *onc* sequences.

VII. POINT MUTATIONS ARE RESPONSIBLE FOR ACTIVATION OF *ras* ONCOGENES

The availability of molecular clones of the normal and activated alleles of human *ras* proto-oncogenes made it possible to determine the molecular mechanisms responsible for the malignant conversion of these genes. The genetic lesions responsible for activation of a number of *ras* oncogenes have been localized to single base changes in their p21 coding sequences. In the T24/EJ bladder carcinoma oncogene, a transversion of a G to a T causes a valine residue to be incorporated instead of a glycine into the twelfth position of the predicted p21 primary structure (Capon *et al.*, 1983a; Reddy *et al.*, 1982; Tabin *et al.*, 1982; Taparowsky *et al.*, 1982). During our analyses of human cells for transforming DNA sequences, we were able to isolate and molecularly clone an oncogene from a human lung carcinoma, designated Hs242 (Yuasa *et al.*, 1983). This gene was also identified as an activated H-*ras* (human) proto-oncogene, making it possible to compare the mechanisms by which the same human proto-oncogene has been independently activated in human tumor-derived cells.

The Hs242 transforming sequence was isolated and subjected to restriction enzyme analysis in order to compare its physical map with that of the previously reported T24/EJ bladder tumor oncogene (Yuasa *et al.*, 1983) as well as c-H-*ras* cloned from a normal human fetal liver library (Santos *et al.*, 1982). The restriction map of the Hs242 oncogene closely corresponded with both, diverging only outside the region previously shown to be required for the transforming activity of the T24 oncogene.

To map the position of the genetic lesion in Hs242 leading to its malignant activation, recombinants were constructed in which fragments of the Hs242 oncogene were substituted by the homologous sequences of H-*ras* (human) (Fig. 4A). By this analysis the genetic alteration that activated the Hs242 oncogene was localized to a 0.45-kbp (kilobase pair) region that encompassed its second coding exon. Nucleotide sequence analysis of this region revealed that the Hs242 oncogene and H-*ras* (human) differed at a single base within codon 61 (Fig. 4B). The change of an A to a T resulted in the replacement of glutamine by leucine in this codon. Thus, a single amino acid substitution seems sufficient to confer transforming properties on the product of the Hs242 oncogene. These results also established that the site of activation in the Hs242 oncogene was totally different from that of the T24/EJ oncogene (Yuasa *et al.*, 1983).

In subsequent studies, we have assessed the generality of point mutations as the basis for acquisition of malignant properties by *ras* proto-oncogenes by molecularly cloning and analyzing other activated *ras* oncogenes (Table II). Activation of an H-*ras* transforming gene of the Hs0578 human breast carcinosarcoma line has been localized to a point mutation at position 12, changing glycine to aspartic acid in the amino acid sequence (Kraus *et al.*, 1984). Recently, Wigler and co-workers (Taparowsky *et al.*, 1983) reported that the lesion leading to activation of the N-*ras* oncogene in a neuroblastoma line was due to the alteration of codon 61 from CAA to AAA causing the substitution in this case of lysine for glutamine. Another N-*ras* transforming gene, this one isolated from a human lung carcinoma cell line, SW1271, has been shown to result from a single point mutation of an A to a G at position 61 in the coding sequence, resulting in the substitution of arginine for glutamine (Yuasa *et al.*, 1984).

Investigators analyzing the activated forms of the K-*ras* oncogene (Shimizu *et al.*, 1983a; Capon *et al.*, 1983b) have achieved strikingly similar results. In two K-*ras* transforming genes so far analyzed, single point mutations in the twelfth codon have been shown to be responsible for acquisition of malignant properties. Thus, mutations at positions 12 or 61 appear to be the genetic lesions most commonly responsible for activation of *ras* oncogenes under natural conditions in human tumor cells (Table II).

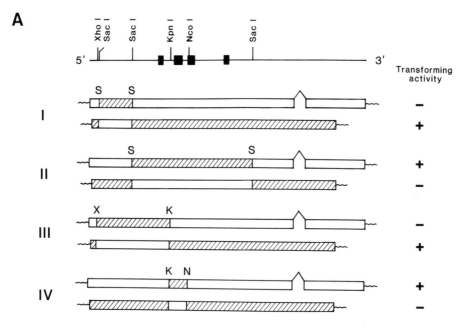

Fig. 4. Mechanism of activation of the Hs242 mammary carcinoma oncogene. A series of recombinants between Hs242 and c-H-*ras* (human) were constructed and tested for their ability to induce transformation of NIH/3T3 cells (A). The region between the Kpn I and Nco I sites of Hs242 conferred transforming activity upon c-H-*ras* (human) and was therefore subjected to nucleotide sequence analysis (B). A transversion of A to T resulted in the change of a Glu to Leu residue in the 61st codon of the p21 amino acid sequence.

TABLE II

Point Mutations That Activate *ras* Oncogenes in Human Tumors

Oncogene	Tumor source	Point mutation	Amino acid change	Codon	References
H-*ras*					
T24	Bladder carcinoma	G → T	Gly → Val	12	Reddy *et al.*, 1982; Tabin *et al.*, 1982; Taparowsky *et al.*, 1982; Capon *et al.*, 1983a
Hs0578	Mammary carcinosarcoma	G → A	Gly → Asp	12	Kraus *et al.*, 1984
Hs242	Lung carcinoma	A → T	Gln → Leu	61	Yuasa *et al.*, 1983
K-*ras*					
Calu-1	Lung carcinoma	G → T	Gly → Cys	12	Shimuzu *et al.*, 1983; Capon *et al.*, 1983b
SW 480	Colon carcinoma	G → T	Gly → Val	12	Capon *et al.*, 1983b
N-*ras*					
SK-N-SH	Neuroblastoma	C → A	Gln → Lys	61	Taparowsky *et al.*, 1983
SW-1271	Lung carcinoma	A → G	Gln → Arg	61	Yuasa *et al.*, 1984

VIII. IMPLICATIONS

The number of genetic alterations required for neoplastic transformation remains to be elucidated. In cell culture systems, transfection of primary rat embryo fibroblasts with two oncogenes, *ras* and *myc* or *ras* and an early gene of the adenovirus genome E1A, has been reported to be sufficient to induce the malignant phenotype, whereas either transforming gene alone is unable to do so (Land *et al.*, 1983; Ruley, 1983). In human cell culture systems, it has been possible to induce neoplastic transformation of human epidermal keratinocytes by the sequential addition of viruses containing different oncogenes (J. S. Rhim, *et al.*, 1985). Such transformants achieved frank malignancy capable of inducing progressively growing squamous cell carcinomas in athymic nude mice, whereas uninfected keratinocytes or keratinocytes infected with only one virus were incapable of inducing tumors. Such findings suggest that the number of discrete steps in the neoplastic pathway may be relatively few, and that there may be complementing groups of oncogenes.

There is a great likelihood that not all of the proto-oncogenes capable of being activated as oncogenes have been discovered. Moreover, genetically predisposing states certainly exist but likely involve recessive alterations whose effects are more subtle than that of the germ line transmission of an activated oncogene. Finally, the few genes that have already been identified as targets of the neoplastic process are highly conserved and in their unaltered state almost certainly are involved in essential growth processes. Thus, any attempt to intervene in such normal functions is likely to be difficult. Nevertheless, the evidence obtained so far strongly argues that the problem of deciphering the neoplastic process is becoming less complex and, thus, more approachable. If so, it is likely that continued investigation of this small group of proto-oncogenes will provide important insights into strategies that may eventually be useful in the diagnosis, treatment, and it is hoped even prevention of human cancer.

REFERENCES

Aaronson, S. A. (1983). Unique aspects of the interactions of retroviruses with vertebrate cells: C. P. Rhodes Memorial Lecture. *Cancer Res.* **43**, 1–5.

Aaronson, S. A., Reddy, E. P., Robbins, K., Devare, S. G., Swan, D. C., Pierce, J. H., and Tronick, S. R. (1983). Retroviruses, *onc* genes and human cancer. In "Human Carcinogenesis" (C. C. Harris and H. N. Autrup, eds.), pp. 609–630. Academic Press, New York.

Alitalo, K., Schwab, M., Lin, C. C., Varmus, H. E., and Bishop, J. M. (1983). Homogeneously staining chromosomal regions contain amplified copies of an abundantly expressed cellular oncogene (c-myc) in malignant neuroendocrine cells from a human colon carcinoma. *Proc. Natl. Acad. Sci. U.S.A.* **80**, 1707–1711.

Andersen, P. R., Devare, S. G., Tronick, S. R., Ellis, R. W., Aaronson, S. A., and Scolnick, E. M. (1981). Generation of BALB-MuSV and Ha-MuSV by type C virus transduction of homologous transforming genes from different species. *Cell (Cambridge, Mass.)* **26**, 129–134.

Antoniades, H. N., and Hunkapiller, M. W. (1983). Human platelet-derived growth factor (PDGF): Amino-terminal amino acid sequence. *Science* **220**, 963–965.

Ar-Rushdi, A., Nishikura, K., Erikson, J., Watt, R., Rovera, G., and Croce, C. M. (1983). Differential expression of the translocated and the untranslocated c-myc oncogene in Burkitt lymphoma. *Science* **222**, 390–393.

Biedler, J. L., and Spengler, B. A. (1976). Metaphase chromosome anomaly: Association with drug resistance and cell-specific products. *Science* **191**, 185–187.

Biedler, J. L., Henson, L., and Spengler, B. A. (1973). Morphology and growth, tumorigenicity, and cytogenetics of human neuroblastoma cells in continuous culture. *Cancer Res.* **33**, 2643–2652.

Bishop, J. M. (1983). Cellular oncogenes and retroviruses. *Annu. Rev. Biochem.* **52**, 301–354.

Bonner, T. I., O'Brien, S. J., Nash, W. G., and Rapp, U. R. (1984). The Human homologs of the *raf* (mil) oncogene are located on human chromosomes 3 and 4. *Science* **223**, 71–74.

Canaani, E., Gale, R. P., Steiner-Saltz, D. Berrebi, A., Aghai, E., and Januszewicz, E. (1984). Altered transcription of an oncogene in chronic myeloid leukaemia. *Lancet* **1**, 593–595.

Canaani, E., Dreazen, O., Klar, A., Rechavi, G., Ram, D., Cohen, J. B., and Givol, D. (1983). Activation of the c-mos oncogene in a mouse plasmacytoma by insertion of an endogenous intracisternal A-particle genome. *Proc. Natl. Acad. Sci. U.S.A.* **80**, 7118–7122.

Capon, D. J., Chen, E. Y., Levinson, A. D., Seeburg, P. H., and Goeddel, D. V. (1983a). Complete nucleotide sequences of the T24 human bladder carcinoma oncogene and its normal homologue. *Nature (London)*, **302**, 33–37.

Capon, D. J., Seeburg, P. H., McGrath, J. P., Hayflick, J. S., Edman, U., Levinson, A. D., and Goeddel, D. V. (1983b). Activation of Ki-ras-2 gene in human colon and lung carcinomas by two different point mutations. *Nature (London)* **304**, 507–513.

Collett, M. S., and Erikson, R. L. (1978). Protein kinase activity associated with the avian sarcoma virus src gene product. *Proc. Natl. Acad. Sci. U.S.A.* **75**, 2021–2024.

Collins, S. J., and Groudine, M. (1982). Amplification of endogenous myc-related DNA sequences in a human myeloid leukemia cell line. *Nature (London)*, **298**, 679–681.

Collins, S. J., and Groudine, M. T. (1983). Rearrangement and amplification of c-abl sequences in the human chronic myelogenous leukemia cell line K-562. *Proc. Natl. Acad. Sci. U.S.A.* **80**, 4813–4817.

Cooper, G. M. (1982). Cellular transforming genes. *Science* **217**, 801–806.

Dalla Favera, R., Bregni, M., Erikson, J., Patterson, D., Gallo, R. C., and Croce, C. M. (1982a). Human c-myc onc gene is located on the region of chromosome 8 that is translocated in Burkitt lymphoma cells. *Proc. Natl. Acad. Sci. U.S.A.* **79**, 7824–7827.

Dalla Favera, R., Gallo, R. C., Giallongo, A., and Croce, C. M. (1982b). Chromosomal localization of the human homolog (c-sis) of the simian sarcoma virus onc gene. *Science* **218**, 686–688.

Dalla Favera, R., Wong-Staal, F., and Gallo, R. C. (1982c). Onc gene amplification in promyelocytic leukemia cell line HL-60 and primary leukemic cells of the same patient. *Nature (London)* **299**, 61–63.

De Klein, A., Van Kessel, A. G., Grosveld, G., Bartram, C. R., Hagemeiger, A., Bootsma, D., Spurr, N. K., Heisterkamp, N., Groffen, N., and Stephenson, J. R. (1982). A cellular oncogene is translocated to the Philadelphia chromosome in chronic myelocytic leukaemia. *Nature (London)* **300**, 765–767.

deMartinville, B., Cunningham, J. M., Murray, M. J., and Francke, U. (1983a). The n-ras oncogene assigned to the short arm of human chromosome 1. *Nucleic Acids Res.* **11**, 5267–5275.

deMartinville, B., Giacalone, J., Shih, C., Weinberg, R. A., and Francke, U. (1983b). Oncogene from human EJ bladder carcinoma is located on the short arm of chromosome 11. *Science* **219**, 498–501.

Devare, S. G., Reddy, E. P., Law, J. D., Robbins, K. C., and Aaronson, S. A. (1983). Nucleotide sequence of the simian sarcoma virus genome: Demonstration that its acquired cellular sequences encode the transforming gene product p28^{sis}. *Proc. Natl. Acad. Sci. U.S.A.* **80**, 731–735.

Dhar, R., Ellis, R. W., Shih, T. Y., Oroszlan, S., Shapiro, B., Maizel, J., Lowy, D., and Scolnick, E. (1982). Nucleotide sequence of the p21 transforming protein of Harvey murine sarcoma virus. *Science* **217**, 934–937.

Diamond, A., Cooper, G. M., Ritz, J., and Lane, M. A. (1983). Identification and molecular cloning of the human B-lym transforming gene activated in Burkitt's lymphomas. *Nature (London)* **305**, 112–116.

Doolittle, R. F., Hunkapiller, M. W., Hood, L. E., Devare, S. G., Robbins, K. C., Aaronson, S. A., and Antoniades, H. N. (1983). Simian sarcoma virus onc gene, v-sis, is derived from the gene (or genes) encoding a platelet-derived growth factor. *Science* **221**, 275–277.

Downward, J., Yarden, Y., Mayes, E., Scrace, G., Totty, P., Stockwell, P., Ullrich, A., Schlessinger, J., and Waterfield, M. D. (1984). Close similarity of epidermal growth factor receptor and v-erb-B oncogene protein sequences. *Nature (London)* **307**, 521–527.

Duesberg, P. H. (1983). Retroviral transforming genes in normal cells (?). *Nature (London)* **304**, 219–226.

Ellis, R., DeFeo, D., Shih, T. Y., Gonda, M. A., Young, H. A., Tsuchida, N., Lowy, D. R., and Scolnick, E. M. (1981). The p21 *src* genes of Harvey and Kirsten sarcoma viruses originate from divergent members of a family of normal vertebrate genes. *Nature (London)* **292**, 506–511.

Eva, A., Robbins, K. C., Andersen, P. R., Srinivasan, A., Tronick, S. R., Reddy, E. P., Ellmore, N. W., Galen, A. T., Lautenberger, J. A., Papas, T. S., Westin, E. H., Wong-Staal, F., Gallo, R. C., and Aaronson, S. A. (1982). Cellular genes analogous to retroviral *onc* genes are transcribed in human tumour cells. *Nature (London)* **295**, 116–119.

Eva, S., Tronick, S. R., Gol, R. A., Pierce, J. H., and Aaronson, S. A. (1983). Transforming genes of human hematopoietic tumors: Frequent detection of *ras*-related oncogenes whose activation appears to be independent of tumor phenotype. *Proc. Natl. Acad. Sci. U.S.A.* **80**, 383–387.

Goubin, G., Goldman, D. S., Luce, J., Neiman, P. E., and Cooper, G. M. (1983). Molecular cloning and nucleotide sequence of a transforming gene detected by transfection of chicken B-cell lymphoma DNA. *Nature (London)* **302**, 114–119.

Graham, F. L., and Van der Eb, A. J. (1973). Transformation of rat cells by DNA of human adenovirus 5. *Virology* **52**, 456–467.

Greenberger, J. S. (1979). Phenotypically distinct target cells for murine sarcoma virus and murine leukemia virus marrow transformation *in vitro*. *JNCI, J. Natl. Cancer Inst.* **62**, 337–344.

Groffen, J., Heisterkamp, N., Spurr, N., Dana, S., Wasmuth, J. J., and Stephenson, J. R. (1983). Chromosomal localization of the human c-*fms* oncogene. *Nucleic Acids Res.* **11**, 6331–6339.

Gross, L. (1970). "Oncogenic Viruses," 2nd ed. Pergamon, Oxford.

Hall, A., Marshall, C. J., Spurr, N. I., and Weiss, R. A. (1983). Identification of transforming gene in two human sarcoma cell lines as a new member of the *ras* gene family located on chromosome 1. *Nature (London)* **303**, 396–400.

Hankins, D. W., and Scolnick, E. M. (1981). Harvey and Kirsten sarcoma viruses promote the growth and differentiation of erythroid precursor cells *in vitro*. *Cell (Cambridge, Mass.)* **26**, 91–97.

Harper, M. E., Franchini, G., Love, J., Simon, M. I., Gallo, R. C., and Wong-Staal, F. (1983). Chromosomal sublocalization of human c-*myb* and c-*fes* cellular *onc* genes. *Nature (London)* **304**, 169–171.

Hayward, W. S., Neel, B. G., and Astrin, S. M. (1981). Activation of a cellular *onc* gene by promoter insertion in ALV-induced lymphoid leukosis. *Nature (London)* **290**, 475–480.

Klein, G. (1981). The role of gene dosage and genetic transposition in carcinogenesis. *Nature (London)* **294**, 290–293.

Kohl, N. E., Kanda, N., Schreck, R. R., Bruns, G., Latt, S., Gilbert, F., and Alt, F. W. (1983). Transposition and amplification of oncogene-related sequences in human neuroblastomas. *Cell (Cambridge, Mass.)* **35**, 359–367.

Kraus, M. H., Yuasa, Y., and Aaronson, S. A. (1984). A position 12-activated H-*ras* oncogene in all HS578T mammary carcinosarcoma cells but not normal mammary cells of the same patient. *Proc. Natl. Acad. Sci. U.S.A.* **81,** 5384–5388.

Land, H., Parada, L. F., and Weinberg, R. A. (1983). Tumorigenic conversion of primary embryo fibroblasts requires at least two cooperating oncogenes. *Nature (London)* **304,** 596–602.

Lane, M. A., Sainten, A., and Cooper, G. M. (1981). Activation of related transforming genes in mouse and human mammary carcinomas. *Proc. Natl. Acad. Sci. U.S.A.* **78,** 5185–5189.

Lane, M. A., Sainten, A., and Cooper, G. M. (1982). Stage-specific transforming genes of human and mouse B- and T-lymphocyte neoplasms. *Cell (Cambridge, Mass.)* **28,** 873–880.

Levinson, A. D., Opperman, H., Levinow, L., Varmus, H. E., and Bishop, J. M. (1978). Evidence that the transforming gene of avian sarcoma virus encodes a protein kinase associated with a phosphoprotein. *Cell (Cambridge, Mass.)* **15,** 561–572.

Little, C. D., Nau, M. M., Carney, D. N., Gazdar, A. F., and Minna J. D. (1983). Amplification and expression of the c-*myc* oncogene in human lung cancer cell lines. *Nature (London)* **306,** 194–196.

McBride, O. W., Swan, D. C., Santos, E., Barbacid, M., Tronick, S. R., and Aaronson, S. A. (1982). Localization of the normal allele of T24 human bladder carcinoma oncogene to chromosome 11. *Nature (London)* **300,** 773–774.

McBride, O. W., Swan, D. C., Tronick, S. R., Gol, R., Klimanis, E., Moore, D. E., and Aaronson, S. A. (1983). Regional chromosomal localization of N-*ras*, K-*ras*-1, K-*ras*-2 and *myb* oncogenes in human cells. *Nucleic Acids Res.* **11,** 8221–8236.

Neel, B. G., Jhanwar, S. C., Chaganti, R. S., and Hayward, W. S. (1982). Two human c-*onc* genes are located on the long arm of chromosome 8. *Proc. Natl. Acad. Sci. U.S.A.* **79,** 7842–7846.

Nowell, P., Finan, J., Dalla Favera, R., Gallo, R. C., Ar-Rushdi, A., Romanczuk, H., Selden, J. R., Emanuel, B. S., Rovera, G., and Croce, C. M. (1983). Association of amplified oncogene c-*myc* with an abnormally banded chromosome 8 in a human leukemia cell line. *Nature (London)* **306,** 494–497.

O'Brien, S. J., Nash, W. G., Goodwin, J. L., Lowy, D. R., and Chang, E. H. (1983). Dispersion of the *ras* family of transforming genes to four different chromosomes in man. *Nature (London)* **302,** 839–842.

Payne, G. S., Bishop, J. M., and Varmus, H. E. (1982). Multiple arrangements of viral DNA and an activated host oncogene in bursal lymphomas. *Nature (London)* **295,** 209–214.

Pierce, J. H., and Aaronson, S. A. (1982). BALB- and Harvey-murine sarcoma virus transformation of a novel lymphoid progenitor cell. *J. Exp. Med.* **156,** 873–887.

Prakash, K., McBride, O. W., Swan, D. C., Devare, S. G., Tronick, S. R., and Aaronson, S. A. (1982). Abelson murine leukemia virus: Structural requirements for transforming gene function. *Proc. Natl. Acad. Sci. U.S.A.* **79,** 5210–5214.

Pulciani, S., Santos, E., Lauver, A. V., Long, L. K., Aaronson, S. A., and Barbacid, M. (1982). Oncogenes in solid tumours. *Nature (London)* **300,** 539–542.

Rechavi, G., Givol, D., and Canaani, E. (1982). Activation of a cellular oncogene by DNA rearrangement: Possible involvement of an IS-like element. *Nature (London)* **300,** 607–610.

Reddy, E. P., Reynolds, R. K., Santos, E., and Barbacid, M. (1982). A point mutation is responsible for the acquisition of transforming properties by the T24 human bladder carcinoma oncogene. *Nature (London)* **300,** 149–152.

Rhim, J. S., Jay, S., Arnstein, P., Price, F. M., Sanford, K. K., and Aaronson, S. A. (1985). Neoplastic transformation of primary human epidermal keratinocytes by the combined action of adeno 12-SV40 virus and Kirsten murine sarcoma virus. *Science,* **227,** 1250–1252.

Robbins, K. C., Devare, S. G., and Aaronson, S. A. (1981). Molecular cloning of integrated simian sarcoma virus: Genome organization of infectious DNA clones. *Proc. Natl. Acad. Sci. U.S.A.* **78,** 2923–2926.

Robbins, K. C., Devare, S. G., Reddy, E. P., and Aaronson, S. A. (1982a). *In vivo* identification of the transforming gene product of simian sarcoma virus. *Science* **218,** 1131–1133.

Robbins, K. C., Hill, R. L., and Aaronson, S. A. (1982b). Primate origin of the cell-derived sequences of simian sarcoma virus. *J. Virol.* **41,** 721–725.

Robbins, K. C., Antoniades, H. N., Devare, S. G., Hunkapiller, M. W., and Aaronson, S. A. (1983). Structural and immunologic similarities between simian sarcoma virus gene product(s) and human platelet-derived growth factor. *Nature (London)* **305,** 605–608.

Roussel, M. F., Sherr, C. J., Barker, P. E., and Ruddle, F. H. (1983). Molecular cloning of the *c-fms* locus and its assignment to human chromosome 5. *J. Virol.* **48,** 770–773.

Ruley, H. E. (1983). Adenovirus early region 1A enables viral and cellular transforming genes to transform primary cells in culture. *Nature (London)* **304,** 602–606.

Ryan, J., Barker, P. E., Shimizu, K., Wigler, M., and Ruddle, F. H. (1983). Chromosomal assignment of a family of human oncogenes. *Proc. Natl. Acad. Sci. U.S.A.* **80,** 4460–4463.

Sakaguchi, A. Y., Naylor, S. L., Shows, T. B., Toole, J. J., McCoy, M., and Weinberg, R. A. (1983). Human C-K1-ras2 proto-oncogene on chromosome 12. *Science* **219,** 1081–1083.

Sandberg, A. A., Sakurai, M., and Holdsworth, R. N. (1972). Chromosomes and causation of human cancer and leukemia: I. DMS chromosomes in a neuroblastoma. *Cancer* **29,** 1671–1678.

Santos, R., Tronick, S. R., Aaronson, S. A., Pulciani, S., and Barbacid M. (1982). T24 human bladder carcinoma oncogene is an activated form of the normal human homologue of BALB- and Harvey-MSV transforming gene. *Nature (London)* **298,** 343–347.

Schwab, M., Alitalo, K., Varmus, H. E., Bishop, J. M., and George, D. (1983a). A cellular oncogene (c-Ki-*ras*) is amplified, overexpressed, and located within karyotypic abnormalities in mouse adrenocortical tumour cells. *Nature (London)* **303,** 496–501.

Schwab, M., Alitalo, K., Klempnauer, K. H., Varmus, H. E., Bishop, J. M., Gilbert, F., Brodeur, G., Goldstein, M., and Trent, J. (1983b). Amplified DNA with limited homology to *myc* cellular oncogenesis shared by human neuroblastoma cell lines and a neuroblastoma tumour. *Nature (London)* **305,** 245–248.

Shimuzu, K., Birnbaum, D., Ruley, M. A., Fasano, O., Suard, Y., Edlund, L., Taparowsky, E., Goldfarb, M., and Wigler M. (1983a). Structure of the Ki-ras gene of the human lung carcinoma cell line, Calu-1. *Nature (London)* **304,** 496–500.

Shimizu, K., Goldfarb, M., Perucho, M., and Wigler, M. (1983b). Isolation and preliminary characterization of the transforming gene of a human neuroblastoma cell line. *Proc. Natl. Acad. Sci. U.S.A.* **80,** 383–387.

Swan, D. C., McBride, O. W., Robbins, K. C., Keithley, D. A., Reddy, E. P., and Aaronson, S. A. (1982). Chromosomal mapping of the simian sarcoma virus *onc* gene analogue in human cells. *Proc. Natl. Acad. Sci. U.S.A.* **79,** 4691–4695.

Tabin, C. J., Brodley, S. M., Bargmann, C. I., Weinberg, R. A., Papageorge, A. G., Scolnick,

E. M., Dhar, R., Long, D. R., and Chang, E. H. (1982). Mechanism of activation of a human oncogene. *Nature (London)* **300**, 143–149.

Taparowsky, E., Suard, Y., Fasano, O., Shimizu, K., Goldfarb, M., and Wigler, M. (1982). Activation of the T24 bladder carcinoma transforming gene is linked to a single amino acid change. *Nature (London)* **300**, 762–765.

Taparowsky, E., Shimizu, K., Goldfarb, M., and Wigler, M. (1983). Structure and activation of the human N-ras gene. *Cell (Cambridge, Mass.)* **34**, 581–586.

Taub, R., Kirsch, I., Morton, C., Lenoir, G., Swan, D., Tronick, S., Aaronson, S. A., and Leder, P. (1982). Translocation of the c-*myc* gene into the immunoglobulin heavy chain locus in human Burkitt lymphoma and murine plasmacytoma cells. *Proc. Natl. Acad. Sci. U.S.A.* **79**, 7837–7841.

Temin, H. (1982). Function of the retrovirus long terminal repeat. *Cell (Cambridge, Mass.)* **28**, 3–5.

Tsuchida, N., Ryder, R., and Ohtsubo, E. (1982). Nucleotide sequence of the oncogene encoding the p21 transforming protein of Kirsten murine sarcoma virus. *Science* **217**, 937–939.

Waterfield, M. D., Scrace, G. T., Whittle, N., Stroobant, P., Johnsson, A., Wasteson, A., Westermark, B., Heldin, C.-H., Huany, J. D., and Deuel, T. F. (1983). Platelet derived growth factor is structurally related to the putative transforming protein p28*sis* of simian sarcoma virus. *Nature (London)* **304**, 35–39.

Weinberg, R. A. (1982). Fewer and fewer oncogenes. *Cell (Cambridge, Mass.)* **30**, 3–4.

Weiss, R. A., Teich, N., Varmus, H., and Coffin, J., eds. (1982). "Molecular Biology of Tumor Viruses, 2nd ed., Cold Spring Harbor, New York.

Weissman, B. E., and Aaronson, S. A. (1983). BALB and Kirsten murine sarcoma viruses alter growth and differentiation of EGF-dependent BALB/c mouse epidermal keratinocyte lines. *Cell (Cambridge, Mass.)* **32**, 599–606.

Westin, E. H., Wong-Staal, F., Gelmann, E. P., Dalla Favera, R. D., Papas, T. S., Lautenberger, J. A., Eva, A., Reddy, E. P., Tronick, S. R., and Aaronson, S. A. (1982). Expression of cellular homologues of retroviral *onc* genes in human hematopoietic cells. *Proc. Natl. Acad. Sci. U.S.A.* **79**, 2490–2494.

Yuasa, Y., Srivastava, S. K., Dunn, C. Y., Rhim, J. A., Reddy, E. P., and Aaronson, S. A. (1983). Acquisition of transforming properties by alternative point mutations within C-Bas/Has human proto-oncogene. *Nature (London)* **303**, 775–779.

Yuasa, Y., Gol, R. A., Chang, A., Chiu, I.-M., Reddy, E. P., Tronick, S. R., and Aaronson, S. A. (1984). Mechanism of activation of an N-*ras* oncogene of SW-1271 human lung carcinoma cells. *Proc. Natl. Acad. Sci. U.S.A.* **81**, 3670–3674.

2

The Role of Stem Cells in
Normal and Malignant Tissue

R. N. BUICK AND E. A. McCULLOCH

Division of Biological Research
Ontario Cancer Institute
Toronto, Ontario, Canada

I. INTRODUCTION

The stem cell concept specifies that a cellular system is maintained by the activities of a small minority of its members; these stem cells have extensive proliferative capacity that permits them not only to maintain their own numbers but also to give rise to descendants with reduced or absent growth potential. Individual stem cells may begin and maintain long-lived cellular clones; indeed, the definition of stem cells is based on this capacity. It also forms part of their attraction for

25

theorists of cellular regulation, since independent clones may readily be considered as units of control within an integrated cellular population.

The purpose of this chapter is to apply the stem cell concept to certain normal and malignant cell populations; particular emphasis will be placed on the hemopoietic system, since it is there that both methodology and theory are most advanced. However, the role of stem cells in cancer generally will be addressed. Stem cells are considered as possible sites of malignant transformation; they are assigned a central role in the generation of tumor cell heterogeneity and proposed as targets for therapeutic intervention. Finally, recent advances in the molecular biology of tumors are considered as these impinge on stem cell theory.

II. STEM CELL RENEWAL IN NORMAL TISSUES

The usual ways of representing cell renewal systems all suggest hierarchical organization. An illustration of a cell renewal system is shown in Fig. 1. Each clone in the cell population is regarded as a cell division hierarchy with the final progeny dominating in a numerical sense. The central stem cell has two critical functions: First, it can act as the initiating cell in a process of cell division and differentiation resulting in the production of large numbers of differentiated descendants, highly specialized for a tissue-specific function. This process is known as clonal expansion. A stem cell of a particular renewal tissue is determined; that is, its potential for provision of cells for differentiation is restricted to that tissue type alone. Second, a stem cell may undergo cell division to produce stem cell daughters which replace the stem

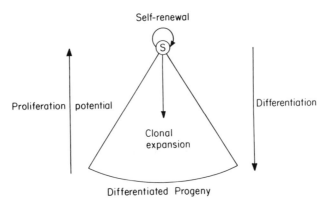

Fig. 1.   A schematic cell renewal hierarchy. From Mackillop *et al.* (1983a).

cells used in clonal expansion. This second process is called self-renewal and is the defining property of a stem cell.

In general terms, control of the cell renewal process is thought to be the consequence of balanced and opposing differentiation and proliferation; with the exception of the stem cell, the clone is self-limiting in size. Within the hierarchy, cells are thought to exhibit a continuous gradation of properties changing unidirectionally. The changes in cellular differentiation and proliferative potential are shown in Fig. 1. At the extremes of the hierarchy, the stem cell is regarded as possessing the maximum proliferative potential, while the terminally differentiated cells are considered to have no capacity for growth. Intermediate cells will have proliferative potential and differentiation states based on their positions in the hierarchy relative to stem and end cells.

The majority of metazoan tissues depend on cell renewal since the lifetimes of the functional cells are less than that of the organism as a whole. Under normal conditions, cell production in these tissues is matched by cell loss, and steady state is maintained. The most commonly studied examples of renewal tissues are gastrointestinal mucosa, epidermis, and hemopoietic tissue. However, all epithelia maintain steady state in this manner although at very different rates that are proportional to cell loss.

Some tissues (parenchymal cells of liver, kidney, and exocrine and endocrine glands) are classified as conditionally renewing. In these tissues, under normal circumstances, cell loss is rare and therefore cell renewal is not required; however, if cell loss is imposed, renewal is triggered (see Chapter 12). For example, if part of the liver is resected, the remaining cells proliferate rapidly to regenerate the original tissue mass. There are other instances in which non-steady-state conditions apply in normal renewal tissues; the proliferation characteristic of non-malignant hyperplasia or metaplasia is distinguished from neoplasia because it is reversible.

The cellular organization of a number of renewal tissues has been studied in detail (Potten et al., 1979; Till and McCulloch, 1980). All appear to have similar general characteristics (Potten et al., 1979). We have chosen to focus on myelopoietic differentiation and renewal to exemplify the normal process since, at this time, cellular and molecular control mechanisms are best understood in this system.

III. MYELOPOIESIS

The myelopoietic system, the source of short-lived blood erythrocytes, granular leukocytes, and platelets, is a typical renewal system. Its organization has been studied extensively (for review, see Till and

McCulloch, 1980); most agree on a description of cellular hierarchies within it and these are often depicted as lineage diagrams similar to that shown in Fig. 2. Myelopoietic stem cells are capable of giving rise to all three major cell classes; their pluripotent nature was first deduced from cytological examination of mice surviving radiation (Barnes *et al.*, 1958). Some hemopoietic cells in these animals contained recognizable chromosomal abnormalities compatible with proliferation. Since each of these was unique, they served as markers of populations with single ancestors. Such clones were found to contain erythrocytes,

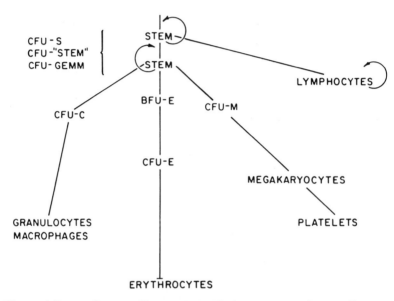

Fig. 2. A lineage diagram of hemopoiesis. The heterogeneity of stem cells is emphasized by showing them twice, both capable of self-renewal. The lineage relationships among CFU-S (Till and McCulloch, 1961), the stem cell colonies of Nakahata and Ogawa (1982b) and Keller and Phillips (1982), and CFU-GEMM (Johnson and Metcalf, 1977; Fauser and Messner, 1979a) are not resolved. CFU-C represents a heterogeneous population of granulocyte–macrophage precursors (Pluznik and Sachs, 1966; Bradley and Metcalf, 1966; Senn *et al.*, 1967; Pike and Robinson, 1970). BFU-E is used to describe early erythropoietic progenitors (Heath *et al.*, 1976) whose maturation stage is related to the time of appearance of multilobulated colonies (Gregory and Eaves, 1978). CFU-E refers to erythropoietic progenitors at a maturation stage just prior to that of identifiable erythroblasts (Stephenson *et al.*, 1971; Tepperman *et al.*, 1974). CFU-M is a megakaryocytic precursor (Nakeff and Daniels-McQueen, 1976; Messner *et al.*, 1982; Burstein *et al.*, 1981). A link is shown between stem cells and lymphocytes (Wu *et al.*, 1968; Abramson *et al.*, 1977; Messner *et al.*, 1981), although details are not indicated. Capacity for self-renewal is indicated by a curved arrow. Stages in lymphoid differentiation are not outlined (Reinherz *et al.*, 1980). From McCulloch (1983).

granulocytes, and megakaryocytes, providing evidence for a common ancestor for these differentiated populations. However, the construction of the lineage diagram shown in Fig. 2 required the development of functional assays for ancestral stem cells and their immediate descendants.

A. Assays Based on Colony Formation

Myelopoietic progenitor cell assays are developmental in nature; that is, a single cell is permitted to disclose its potential for proliferation and differentiation by developing into a recognizable colony. The first such assay was based upon marrow transplantation procedures in mice. Macroscopic nodules were observed in the spleens of heavily irradiated recipients of small numbers of marrow cells (Till and Mc-Culloch, 1961); these proved to be clonal hemopoietic populations (Becker et al., 1963) consisting of the three myelopoietic lineages (Fowler et al., 1967). Observations made using this method, the spleen colony assay, confirmed the existence of pluripotent myelopoietic stem cells; more importantly, the technique provided a way to quantitate stem cells and to determine their properties.

Much progress became possible when Pluznik and Sachs (1965) and Bradley and Metcalf (1966) developed culture methods for colony formation by granulopoietic progenitor cells. Their techniques shared two general features: First, cells were immobilized in agar so that their progeny remained localized as colonies expanded. Second, growth factors were added, either as feeder layers or medium conditioned by cultured cells. Culture assays using these two features have now been developed for many myelopoietic progenitors at various stages of differentiation. Names and references for these are included in Fig. 2 and its caption.

Colony assays have a number of desirable features. They are selective; that is, the circumstances under which colony formation occurs may permit clonal growth from members of a very small subpopulation. Further, assay conditions can be specified that promote colony formation by some progenitors and not others. Colony assays are flexible; populations can be manipulated before assay to determine their properties, or the conditions can be changed during clonal expansion. The methods have enough generality to allow them to be adapted to meet many research requirements. For example, culture techniques first developed for murine cells were found to be applicable, with small changes, to progenitors from other species, including man.

Colony assays have important disadvantages. Perhaps the most significant is inherent in their very nature. In dividing to initiate colony formation, the progenitors themselves are lost; their existence must be deduced from their progeny. This disadvantage is symbolized in the nomenclature. The entities that began colonies are referred to as "colony-forming units" (CFU) rather than as cells.

Their indirect nature has limited the value of colony assays in the study of progenitor phenotypes. Although colony formation has been used extensively as a guide in the purification of progenitors (Francis et al., 1983; Metcalf and MacDonald, 1975; Nicola et al., 1980), their structure remains controversial, and methods such as radioautography which require morphological identification cannot be applied to them; nor is it known whether they express enzymatic or antigenically determined markers. Indeed most biochemical techniques cannot at present be used directly to study stem cells.

Because they are indirect, colony assays can be misleading. Important precautions are needed in developing them and these must be maintained in their applications. For example, if colonies are not of single cell origin, deductions concerning the numbers and properties of CFU can be erroneous. Proof of clonality requires genetic methods that permit cells to be traced to their ancestors. Radiation-induced chromosomal markers, cited earlier for their use in the original demonstration of stem cell pluripotentiality, provide the most satisfactory evidence. Polymorphic x-linked enzyme systems also yield convincing results. These depend on examining cells from females heterozygous for different allelic isoenzymes. Following x inactivation, such individuals are cellular mosaics. Each cell contains one or the other of the two possible forms of the enzyme. Since x inactivation is irreversible in subsequent cell generations, cells with the same isoenzyme share the same progenitor. The isoenzymes of glucose-6-phosphate dehydrogenase (G-6-PD) have been used extensively for this purpose in man (Fialkow, 1982), and phosphoglycerate kinase (PGK-1) (Nielsen and Chapman, 1977) is available for murine cells. Analysis of the cellular distribution of isoenzymes has been valuable in determining the origins of colonies in culture. The presence of two isoenzymes in a single colony is proof that it began with more than one cell. With this technique Singer et al. (1979) showed that colonies in cultures plated at high cell density were not clonal, even though they were well separated in the dishes. Of course, the presence of a single isoenzyme in a colony is not proof of clonality, since a clump consisting of cells with the same isoenzyme would yield the same outcome.

Clumps not only form colonies of multicellular origin. If sufficiently large, they may be mistaken for colonies when cultures are evaluated. It is necessary, therefore, to prove cell proliferation is the basis of colony formation. Repeated examination of cultures may be feasible and convincing. Equally useful and often easier is the determination of the sensitivity of colony formation to ionizing radiation, since survival curves for cell proliferation have characteristic parameters (Whitmore and Till, 1964).

When colony assays are to be used for the quantitation of progenitors, it is helpful if a linear relationship can be demonstrated between cell numbers entered into the procedure and colonies enumerated at the end of it. Then it is reasonable to express the frequency of cells with colony-forming potential as CFU/cell number.

B. Cell Lineages and Differentiation Programs

Colony assays were used in the construction of lineage diagrams of myelopoiesis by ordering progenitors on the basis of their properties. Those with capacity for self-renewal were considered to be stem cells. Their first differentiated descendants were lineage-committed cells, giving rise to colonies without new progenitors (McCulloch et al., 1965b). This order was evident from cellular composition of colonies. Those derived from stem cells were multilineage, while committed progenitors had only one kind of progeny.

These lineage constructs were strengthened and refined by experiments based on the stochastic model of stem cell differentiation (Till et al., 1964). The procedure was to analyze clonal populations (usually colonies) for the distribution of various progenitor classes among them. Marked clone-to-clone variation was encountered and was attributed to randomness, both in selection of self-renewal or commitment and in the lineage emerging after each commitment step. Correlations were then calculated between progenitor pairs as these occurred in many clones. The extent of the correlation was considered to represent the quantitative relationship between the cells in the pair. Thus, progenitors separated by one or very few differentiation steps would have had little opportunity to diverge during colony formation and would be closely correlated. Others, more widely separated in a lineage scheme, would experience more stochastic events and would have a greater opportunity to lose association (Gregory and Henkelman, 1977).

The lineage view of myelopoiesis envisages discrete populations separated by one or more differentiation events. Usually the populations were defined by the colony assays that were used to detect them. However, when cells that all passed the same test of colony formation were characterized by other criteria (for example, size or density), marked heterogeneity was always encountered. Diversity could even be identified within populations of CFU by counting colonies at different times (Gregory and Eaves, 1978). Colony-forming capacity might, then, span several segments in the life history of each cell.

The events that intervene between a single stem cell and each of its many mature descendants might be considered to be differentiation programs (Fig. 3). Segments in each program might be sufficiently small to be specified by a single gene; others, for example, protein glycosylation or phosphorylation, might be posttranslational. Some segments might be included in differentiation programs in response to environmental stimuli. The importance of determination is emphasized in Fig. 3 as a separation between stem cells and progenitors capable only of terminal divisions. Each mitosis is considered to be the origin of a new program. Those occurring before determination might lead to new stem cells or different end cells. Mitoses after determination would initiate programs that could terminate only in cells of a single lineage, but the events leading to each cell in the lineage might differ in number of segments, time in each, and perhaps even in segmental order. For example, it is known that cell divisions may be missed from the erythron under conditions of extreme stimulation (Stohlman, 1970). This view of the progression from a stem cell to a mature cell contains the hypothesis that a number of different programs might lead from a stem cell to its differentiated descendants even if these appeared to be homogeneous (McCulloch, 1984).

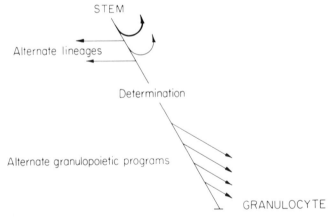

Fig. 3. A diagram of a differentiation program leading from a stem cell to a mature cell. New programs begin at each mitosis. Those occurring before determination may have a number of terminations while those beginning after determination end in cells of the same type.

C. Regulation

Two general classes of myelopoietic regulators have been identified, the one intrinsic to the cells, the other operating from the organ environment required for their growth. Certain genetically determined anemias in mice illustrate this division of control mechanisms. Mice of genotype W/W^v bear a defect intrinsic to stem cells that prevents them from forming colonies when injected into irradiated normal ($\pm/\pm$) recipients (McCulloch et al., 1964). In contrast, mice of genotype Sl/Sl^d, although phenotypically very similar to W/W^v mice, have an environmental lesion; their stem cells form colonies normally when injected into $\pm/\pm$ recipients, but normal cells do not grow or form colonies when injected into irradiated Sl/Sl^d recipients (McCulloch et al., 1965a).

Earlier, reference was made to the stochastic model of the regulation of stem cell differentiation. This was proposed as an explanation for intrinsic mechanisms, controlling which of one or more alternative steps would be incorporated into each separate differentiation program. Its central premise was that these segments would be assembled at random, governed by fixed probabilities. It was first proposed in the context of the major contrasting fates of stem cells. These might either undergo self-renewal ("birth") or determination ("death"). The model has now been expanded to include the genetically determined segments of differentiation programs that specify which lineage will develop following determination (Nakahata et al., 1982). Indeed, alternate assemblies of segments might be made randomly, restricted only by mechanisms that ensure that functionally effective programs will be constructed from the available genetic information.

Intrinsically acting stochastic mechanisms are selective in nature; that is, alternatives are available but no inductive or instructional message is postulated. "Selection" is not even a consequence of permissive environmental influences, but rather derives from genetic information that either sets probabilities or limits the variety of programs that may be assembled. Extrinsically acting regulatory mechanisms have been more controversial; some investigators have postulated that these are inductive or even instructional. An example of this view is provided by the "hemopoietic-inductive-microenvironment" (HIM) model proposed by Trentin (1970). Spleen colonies were classified on the basis of their cellular composition as determined from microscopic examination of sections. Often this procedure disclosed only a single lineage within a colony; with increasing time of colony expansion, the number of multilineage colonies increased. These observations were attributed to environmental influences; specifically, that marrow and spleen contained microenvironments which specified by induction the differentiation patterns of stem cells within them. Evidence against the HIM model was obtained as cell culture methods became available. Colonies, apparently homogeneous by histological criteria, were found to contain progenitors of other lineages when examined using colony assays, and multilineage colonies could be obtained in culture without the presence of an organized microenvironment (Johnson and Metcalf, 1977; Fauser and Messner, 1979a; Nakahata and Ogawa, 1982b). Although most commentators now reject the details of Trentin's model (Ogawa et al., 1983; Till, 1982), the concept of induction remains attractive as an explanation of the activity of stimulators required for myelopoietic colony formation in culture.

D. Growth Factors

There is now an extensive literature on factors that affect colony formation in culture (Metcalf, 1982; Metcalf and Nicola, 1984; Stanley and Heard, 1977; Sachs, 1982; Nakahata and Ogawa, 1982a). Two classes with different specificities have been described. First, bioactive molecules derived from lectin-stimulated lymphocytes of certain cell lines have stage specificity. They promote colony formation from primitive cells, often pluripotent progenitors. Conditioned media with activity of this class and purified molecules from them have been described by investigators, who have used different descriptive names. Iscove coined the term "burst-promoting activity" (BPA) because his assay depended on the stimulation of erythroid bursts in cultures with reduced serum content (Iscove et al., 1982). Johnson and Metcalf (1977) used media conditioned by mouse spleen cells in the presence of pokeweed mitogen in their assay for multilineage colony formation. They call the active factor CSF mix. Ihle uses the term interleukin 3 (Ihle et al., 1983). Insufficient evidence is available at this time to define the molecular similarities and differences between the active species from each source.

To various degrees stimulators of the second class have lineage specificity. The best known of these is the hormone erythropoietin (epo). It is an acidic glycoprotein with molecular weight of ~40,000 (Miyake et al., 1977; Lee-Huang, 1980) and monoclonal antibody has been prepared agianst iι (Weiss et al., 1982). In spite of this detailed information, the mechanisms of its action are not known. However, its cellular targets have been identified. These are erythropoietic-committed progenitors well removed in differentiation from stem cells. In normal erythropoiesis, the hormone is required both for their expansion and the synthesis of hemoglobin by their progeny (Eaves and Eaves, 1984). Convincing evidence for erythropoietin activity outside of the erythron is not available.

Lineage-specific stimulatory factors obtained from culture supernatants played an important role in the development of colony assays. A major goal of purification has been to obtain preparations with sufficient specificity for analytical studies and determinations of structure–function relationships. Mouse fibroblasts (L cells) proved a convenient source of a glycoprotein, CSF-1, that is needed for the formation of macrophage colonies (Stanley et al., 1975). However, progenitor specificity is seen clearly only when colony formation is the criterion of activity. Binding has been demonstrated to mature granulopoietic cells (Shadduck et al., 1983), but a regulatory role has not been identified in this context. Different stimulators have been obtained from a variety of sources, including media conditioned by human leukocytes (LCM) (Iscove et al., 1971) and extracts of mouse lung (MLE) (Burgess et al., 1977). These characteristically are required for the formation of colonies of granulocytes and/or macrophages. Other molecules have been obtained with specificity for neutrophilic colony formation only (Nicola and Metcalf, 1981).

Usually specificity was obtained by purifying a factor from a crude preparation using as assay end points colonies of defined cellular composition. However, as highly purified preparations became available, specificity appeared to be less than anticipated. Significant stimulation of early cells and even of erythropoiesis has been obtained with preparations considered initially to be specific for granulocytes/macrophages or neutrophils (Johnson, 1984).

Two issues arise from the study of factors required for differentiation in culture. The first concerns the mechanism of stimulation. In general, active molecules are produced by cells different from, but coexisting with their targets; therefore the effect is to establish cell-to-cell communications (Price and McCulloch, 1978). Receptors for CSF-1 have been identified, suggesting that the stimulator may act by mechanisms similar to those established for growth factors in other systems (Cohen et al., 1982). Receptors have not as yet been demonstrated for molecules with other specificities. However, it does not appear necessary to postulate "inductive" or "instructive" mechanisms. Rather, conditioned media or glycoproteins purified from them may act in culture to select for certain progenitors or to be permissive for their growth.

Second, although hemopoietic growth factors have been shown to be effective stimulators in culture, their role in vivo is not well established. Epo has been shown to stimulate erythropoiesis in plethoric mice (Filmanowicz and Gurney, 1961), but in the cow erythropoietic colony formation may normally be independent of the hormone (Kaaya et al., 1982). Other factors, both stage and lineage specific, have not been shown to be rate-limiting under physiological conditions. Indeed, failure to establish specificity for purified preparations argues against an important function for them except in cell culture. A resolution of this issue may require more knowledge of their structure and function, particularly at the molecular level.

IV. STEM CELLS IN TUMORS

Although this review has emphasized myelopoiesis as an example of a normal cell renewal system, similar discussions could have been constructed about other tissues (Fialkow, 1976; Potten et al., 1979; Pierce et al., 1978). For the purpose of the subsequent treatment of stem cells in tumors, it is important to stress the generality of stem cell regulation in all renewal tissues.

Human tumors arise primarily in tissues which normally (or conditionally) proliferate by cell renewal mechanisms; further, tumors are tissue specific. One can speculate, therefore, that the origin of tumors lies in a cell (or group of cells) which is determined with respect to tissue-specific differentiation and which proliferates under normal conditions. There are two general theories which accommodate these facts. First, it has long been suggested that a carcinogenic event can occur in a differentiated cell of a particular tissue, rendering that cell proliferative, although retaining its ability to organize tissue-specific differentiation. The acquisition of proliferative features in a differentiated cell necessitates proposing "dedifferentiation." The second class of theories would propose that tumors arise from carcinogenic events occurring in the stem cells of a particular tissue. The properties of stem cells mean that it is not necessary to assume dedifferentiation as a mechanism of tumor induction. Rather, it is proposed that the changes associated with the carcinogenic insult cause a defect in the control of the normal stem cell functions, cell renewal, and differentiation (Mackillop et al., 1983a).

A large body of information supports the validity of stem cell mechanisms to the growth of human tumors (Selby et al., 1983a). First, tissue-specific differentiation is a distinguishing feature of many human tumors, and evidence exists to support the ability of human tumor cells to differentiate in vivo and in vitro (Pierce et al., 1978; Flynn et al., 1983; Dexter and Hager, 1980). In many instances an inverse relationship has been seen between indices of cell differentiation (tumor grade) and proliferation (labeling index, mitotic index) (Steel, 1977; Meyer et al., 1977; Selby et al., 1983b). This is consistent with a cellular organization governed by a differentiation hierarchy.

Second, radiation therapy experience suggests that in many human tumors only a small proportion of tumor cells have regenerative capacity. These conclusions are based on the knowledge of radiation sensitivity of human cells, and the fact that small radiation doses can achieve permanent local control of large skin, breast, and cervical tumors (Bush and Hill, 1975; Tepper, 1981). It has been calculated that such results may be explained if the proportion of stem cells in such tumors was approximately 1 in 1000, and only these needed to be sterilized to achieve cure.

Third, a major source of information on tumor tissue organization has derived from the use of developmental assays similar to those described earlier in the context of myelopoiesis (Buick, 1980). Tissue culture or animal transplantation procedures have been devised to identify cells with high proliferative capacity by their colony-forming ability. In a general sense, such procedures reinforce the view of human tumors as "caricatures" (Pierce et al., 1978) of cell renewal in their normal tissue equivalents. Colony-forming cells have typically been found to be present in tumor cell suspensions at frequencies of 0.001–1%. This estimate, of course, may be low because of cell trauma during preparation of cell suspensions or inadequate culture conditions (Selby et al., 1983a).

Fourth, analysis of clonal markers, particularly in hemopoietic malignancies (Blackstock and Garson, 1974; Adamson et al., 1976, 1978) (see later), has supported the notion of stem cell origin and organization of tumors. Similar genetic information favoring clonality is accumulating for epithelial malignancies, although considerably greater technical difficulties are associated with data collection.

Fifth, a large degree of heterogeneity in tumor cell populations has been seen when individual cells are analyzed for cell proliferative or cell differentiation features (Mackillop and Buick, 1981; Mackillop et al., 1983b). In certain cases, it has been possible to show that proliferative activity and expression of differentiated phenotypes are restricted to separate subpopulations of cells. These findings are consistent with the existence of a cell differentiation hierarchy within tumors.

Human tumor cells are heterogeneous with respect to proliferative potential and cell differentiation, perhaps because of the relative position in the cell renewal hierarchy of the clone from which they are

Fig. 4. Schematic representation of interclonal heterogeneity within a tumor. A transformed stem cell (T_s) is shown to have capacity to initiate a clonal hierarchy with renewal and differentiation capability. Through processes related to genetic instability, new stem cell lines (n in number) are initiated from the original T_s. These stem cells are also capable of renewal and differentiation, but likely with probabilities different from T_s.

derived (Mackillop et al., 1983a). In addition to this complexity, human tumors are frequently considered to have undergone clonal progression by the time of clinical presentation. This process is thought to occur as a result of the genetic instability acquired by the cells at the time of carcinogenesis (Nowell, 1976). Thus, although tumors probably arise in single stem cells, during growth new genetically discrete subclones are produced. Some of these may have sufficient growth advantage to overgrow others and become dominant in the tumor. A diagrammatic representation of clonal heterogeneity is shown in Fig. 4. This process may be particularly important in explaining the emergence of new drug-resistant clones during chemotherapy (Goldie and Coltman, 1983). Thus, the heterogeneity imposed by cell renewal and differentiation within a tumor cell population may be increased by several orders of magnitude because multiple related clones are present in a tumor at the time of diagnosis (Buick, 1984b).

A. Clonal Assays Applied to Human Tumors

Attempts to quantitate clonogenic cells from human tumors have drawn heavily on the experience gained through study of normal and malignant human hemopoiesis and experimental animal tumors (Ogawa et al., 1973; Park et al., 1971; Moore et al., 1974). A number of reports have discussed specific alterations to normal hemopoietic colony-forming assays which allow selective growth of malignant cells (Buick et al., 1977; Park et al., 1977; Dicke et al., 1976; Izaguirre et al., 1981). Salmon and his collaborators have developed assays using diluted agar in an enriched medium to support the growth of colony-forming cells in hemopoietic and solid tumors, human myeloma (Hamburger and Salmon, 1977; Durie et al., 1983), ovarian cancer (Hamburger et al., 1978), and melanoma (Meyskens et al., 1981). With the use of similar techniques, a variety of additional solid tumors have been studied, e.g., small cell carcinoma of the lung (Carney et al., 1980), neuroblastoma (von Hoff et al., 1980), colon cancer (Laboisse et al., 1981), and transitional cell carcinoma of the bladder (Buick, et al., 1979b). Concurrently, Steel and his associates have developed clonic assays from human tumor xenografts or directly from tumors (Courtenay et al., 1976, 1978; Courtenay and Mills, 1978). This method uses replenishable liquid medium over agar. Tumor cells are admixed with rat red cells in the agar and reduced oxygen tension is used during incubation. In parallel studies, this latter procedure gives higher plating efficiency for several tumor types when compared to the assay of Hamburger and Salmon (Tveit et al., 1981).

While colony-forming assays for human hemopoietic tumor cells have invariably required the addition of specific conditioning factors (usually supplied by conditioned medium), such factors appear to be unnecessary for growth of colonies from most solid tumors. There are, however, recent reports of manipulation of the growth of clonogenic cells from solid tumors with hormonal additives of physiological relevance, including epidermal growth factor (Hamburger et al., 1981; Meyskens and Salmon, 1981). These finding indicate ways in which improved plating efficiencies may be achieved.

Even though plating efficiency is probably not now optimal, the properties of clonogenic cells are understood to some degree, using the assumption that those cells which do grow are representative of the total with growth potential. The proportion of clonogenic cells found to be in active cell cycle [by tritiated thymidine ([^{3}H]TdR) suicide assays] has been found to be very high (40–80%) in myeloma, acute myeloblastic leukemia (AML), and ovarian carcinoma (Hamburger et al., 1978; Shimizu et al., 1982; Minden et al., 1978). In addition, the growth of colonies from a number of epithelial tumors seems to be affected by factors released from coexisting macrophage-like cells (Hamburger and White, 1981; Buick et al., 1980).

The physical properties of clonogenic tumor cells are also known to some degree. They can be enriched in some tumor populations using physical fractionation procedures (Mackillop and Buick, 1981; Mackillop et al., 1982). Within a particular tumor type, the physical characteristics of the clonogenic tumor cells are similar in different patients.

B. The Relationship between Clonogenic Cells and Stem Cells

It has been recognized that assessment of colony formation in culture by individual tumor cells (hemopoietic or epithelial) does not necessarily quantitate tumor stem cells (Buick et al., 1980). Normal hemopoiesis provides a precedent for colony formation from cells without renewal capacity. In the absence of an in vivo assay for tumor regeneration (as has been possible in transplantable animal tumor models), indirect methods have been used to probe the relationship between clonogenic tumor cells and tumor stem cells.

For example, it has been found that a proportion of tumor colonies derived from small-cell carcinomas of the lung have the proliferative capacity to form tumors in immune-privileged sites in nude mice (Carney et al., 1981). Colony-replating assays to assess the property of self-renewal have also proved useful in this regard. It has been shown that a small and variable proportion of colonies derived from AML (Buick et al., 1979a), ovarian carcinoma (Buick and Mackillop, 1981), and melanoma (Thomson et al., 1982) have the capacity to regrow when disaggregated and replated. Furthermore, the quantitation of this property has led to the conclusion that the capacity for self-renewal may be a biological property of importance in determining prognosis in AML (McCulloch et al., 1982). Recently, colony size has been proposed as a means of determining which clonogenic cells have stem cell properties (Mackillop et al., 1983a). It has been noted that in ovarian carcinoma the clonogenic cells with the greatest proliferative capacity, as judged by colony size, are likely to display stem cell properties (Buick, 1984a).

V. THE CLONAL HEMOPATHIES

Certain hemopoietic diseases are excellent examples of neoplasms originating in stem cells. With the use of both the analysis of the isoenzymes of G-6-PD (Fialkow, 1980) and chromosomal abnormalities (Nowell and Hungerford, 1961; Blackstock and Garson, 1974), hemopoietic populations in affected individuals have been shown to belong to a single clone. Coexisting normal hemopoietic progenitors could be demonstrated only by using culture assays for progenitors (Adamson et al., 1976), and in some instances, these decreased or disappeared with time (Adamson et al., 1978). Chronic myeloblastic leukemia (CML), AML, polycythemia rubra vera (P vera), and idiopathic myelofibrosis (IMF) are members of this group, for which the designation "clonal hemopathy" has been suggested (for a recent review, see Adamson, 1984).

Clonal Composition

In at least some patients with each disease the clonal population contains the three myelopoietic lineages, indicating origin in pluripotent stem cells. However, variation in clonal composition among the diseases has also been noted.

In P vera and IMF only erythropoiesis, granulopoiesis, and platelet formation have been observed. However, in CML both T and B lymphocytes have been identified as belonging to malignant clones (Fialkow, 1982). In contrast, AML shows patient-to-patient variation. While abnormal blast cells are always found, in some instances only granulopoietic differentiation can be identified, while in others erythroblasts and stromal cells are present in the clone (Fialkow et al., 1981; Singer et al., 1984). These findings might be interpreted as evidence for malignant transformation occurring in stem cells at various levels of differentiation. Thus, CML might start in one or more abnormal stem cell capable of both myeloid and lymphoid maturation, while, in contrast, P vera and IMF would derive from myeloid-committed stem cells. From this point of view, it would be necessary to postulate AML beginning either in granulopoietic-committed progenitors or in early myeloid stem cells.

It is not necessary to accept without reservation this attractive interpretation of cellular variation in the clonal hemopathies. It is possible that changes in stem cells occurring at transformation or immediately following it may have altered their differentiation potential. Indeed, in general, negative findings (for example, the absence of a lineage) should not be considered as strong evidence for limited potential in the clonal ancestor. The problems of interpretation encountered by Trentin (1970), described earlier, are examples of the difficulty. Further, the observation that CML or P vera may terminate with the production of the blast population characteristic of AML argues against making a firm distinction between them.

The clinical diagnosis in the clonal hemopathies is based on the cellular composition of the abnormal clone in each patient. For example, P vera patients have large excesses of red cells, while in CML marked granular leukocytosis is usually associated with anemia. The mechanism underlying such deviations from normal is not well understood. However, changes in the relative frequency of commitment toward either erythropoiesis or granulopoiesis have not been detected in either disorder (Eaves and Eaves, 1984; Adamson, 1984). It appears probable that in each instance differentiation programs normally expressed rarely, if at all, become predominant. If, as postulated earlier, programmatic components can be assembled in a variety of ways, part of the defect in each of the clonal hemopathies may affect mechanisms that limit programmatic variation or select for efficient (i.e., normal) program assemblies.

VI. LYMPHOPOIESIS AND ACUTE LYMPHOBLASTIC LEUKEMIA

Malignancies of the lymphoid system, particularly those consisting of B cells, have long been recognized as clonal because they produce homogeneous immunoglobulins (Ig). Moreover, the normal lymphopoietic system, like myelopoiesis, contains cells at various levels of differentiation. For immunoglobulin producing B cells, stages are defined in terms of the presence and location of Ig molecules. Thus, pre-B cells have cytoplasmic μ chains and Ig is found at the surfaces of B cells and in the cytoplasm of mature plasma cells (Bhan *et al.*, 1981) (Fig. 5). Earlier cells in the system do not produce Ig, and it is now known that Ig genes must undergo rearrangement prior to the synthesis of their products (Tonegawa, 1983). In some patients, acute lymphoblastic leukemia is characterized by blast cells that cannot be identified with the B or T lymphocytic lineages [non B, non T acute lymphoblastic leukemia (ALL)]. However, in many such blast populations, gene rearrange-

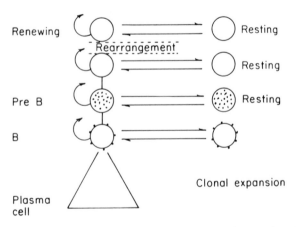

Fig. 5. A diagram of B-cell differentiation. Lineage specificity is found following Ig gene rearrangement (see text). From McCulloch (1983).

ments can be detected using Southern blotting techniques (Korsmeyer *et al.*, 1983) which may be aberrant and incompatible with gene expression. It is reasonable to suppose that the malignant transformation occurred in an earlier cell which subsequently was unable to complete functional Ig gene rearrangement. Such a cell would be an undifferentiated stem cell, and at least some cases of ALL would begin at a differentiation level comparable to that of the clonal hemopathies.

The relationships between myelopoietic and lymphoid stem cells are controversial. There is good evidence that a single cell may have both myelopoietic and lymphoid progeny (Wu *et al.*, 1968; Abramson *et al.*, 1977). But it is not certain whether this cell is different from the myelopoietic stem cell detected by colony assays in culture. B and T lymphocytes might share a common lymphoid-restricted stem cell (Messner *et al.*, 1981), but such a progenitor has not been experimentally detected. Indeed, the difficulty in dissecting stem cell heterogeneity arises, at least in part, from the lack of colony assays for hemopoietic stem cells with fully defined differentiation capacities beyond myelopoiesis. The cellular compositions of the abnormal clones in the hemopathies, described earlier, support the concept of stem cell heterogeneity, but this evidence is flawed by its source in transformed cells.

Regardless, myelopoiesis and lymphopoiesis differ in the effect of differentiation on proliferative capacity. Lymphoid maturation may proceed without loss of self-renewal. Thus, the stem cell concept in lymphopoiesis resembles that applicable to conditionally renewing tissues, since even mature B cells may be considered to be stem cells capable of extensive clonal expansion following exposure to antigen. Yet, as emphasized earlier, tumors of conditionally renewing tissues are organized as cellular hierarchies that include proliferatively inert end cells. Colony assays for the blast cells in ALL provide evidence for both birth and death in that disease (Izaguirre, 1984). The progenitors of ALL blast colonies are capable of self-renewal, but the majority of the cells within the colonies do not grow on replating (Izaguirre *et al.*, 1981). Thus, one consequence of malignant transformation in conditionally renewing tissues is the possibility that cells may lose proliferative capacity. Indeed, without such a "death probability," tumor growth would be so rapid that cancer would be an acutely fatal illness.

VII. HETEROGENEITY

The marked heterogeneity present in solid tumors has been emphasized and attributed both to continuing differentiation in malignant clones (Mackillop *et al.*, 1983a) and clonal progression with time (Nowell, 1976). In hemopoietic malignancies, the extent and significance of heterogeneity is controversial. Particularly in lymphatic leukemia, the malignant cells appear uniform morphologically, and, with the use of immunologically defined phenotypic markers, they can be related to a recognized stage in normal differentiation. These findings have led to the suggestion that differentiation is "blocked" or "frozen" in leukemia (Greaves, 1982a; Seligmann *et al.*, 1981). This view of leukemic growth would limit heterogeneity since differentiation, the process through which variation is generated, would not go to conclusion. Implicit in the model is that differentiation programs are followed faithfully until their premature termination (Greaves, 1982a).

The findings in AML clones are open to a different interpretation. These have been shown to contain at least completely differentiating granulocytes, and in some patients all the myeloid lineages are represented (see previous discussion). Thus, heterogeneity need not be less than that seen in normal hemopoietic clones. However, the blast cells characteristic of the disease are dominant numerically and, in each patient, appear morphologically homogeneous. A "frozen differentiation" model might propose that these are early cells, lacking diversity because they are unable to complete differentiation. Recently, it has been possible to test this explanation for the source of AML blasts using an assay that permits the formation of colonies of blast cells morphologically similar to those found circulating in AML patients (Buick *et al.*, 1977; Minden *et al.*, 1979) (Fig. 6). Analysis of such colonies showed that they contained a small subpopulation of cells capable of colony formation on replating while the majority were unable to grow (Buick *et al.*, 1979a). This finding suggested that the blast population might be organized as a lineage similar to those found in normal myelopoiesis; that is, blast progenitor cells might either undergo renewal or lose proliferative capacity after a determination-like step. These and further studies led to the proposal of a model of blast cells that postulated that these followed differentiation programs assembled abnormally, but using normal components (McCulloch *et al.*, 1978; McCullock and Till, 1981).

Support for this view was obtained as more immunological reagents (particularly monoclonal antibodies) became available. The phenotypes of blasts in both AML and ALL were found to be very complex. In ~30% of patients, leukemic cells were identified that expressed simultaneously markers usually associated with a single differentiation lineage ("lineage infidelity") (Smith *et al.*, 1983). Similar phenotypes have not been encountered in normal hemopoiesis, even after extensive

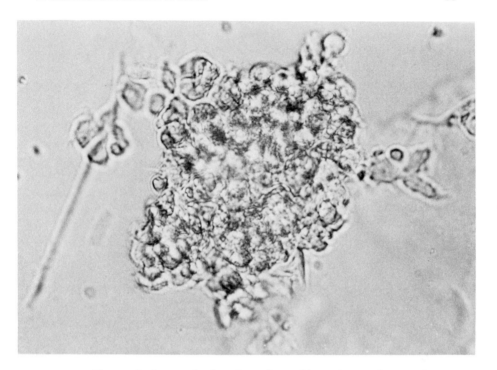

Fig. 6. A photograph of a colony of AML blast cells in culture.

searches, although their existence in leukemia has been confirmed (Bettelheim *et al.*, 1982; Jani *et al.*, 1983). Taken together, the cell culture findings and the phenotypic evidence provide a strong case for considering that AML blasts are very heterogeneous, both within a single patient and between patients (Griffin *et al.*, 1983; McCulloch, 1983). Therefore, AML conforms to the usual behavior of tumors; that is, heterogeneity is increased in comparison with the corresponding normal tissue. A similar conclusion may be reached for ALL as the blast cells are studied in more detail.

VIII. MOLECULAR CONSIDERATIONS

The explosive recent developments in molecular biology will certainly prove highly relevant to the cellular organization in normal and malignant tissues. At this time only general observations can be made.

Most agree that the normal genome contains proto-oncogenes or cellular oncogenes (c-*oncs*) with considerable homology to the transforming oncogenes of retroviruses (v-*oncs*) (Temin, 1984; see Chapter 1). Malignant transformation is associated with the activation of one or more c-*oncs*. Several mechanisms have been identified as possible sources of such activation, including point mutation, translocation, and gene amplification (for review, see Land *et al.*, 1983a). While the importance of oncogenes was first appreciated in the context of viral transformation, oncogene activation has been shown to occur during chemical carcinogenesis (Shih *et al.*, 1979) and may play an essential role in the pathogenesis of most malignancies.

If, as suggested in this review, stem cells are central to the cellular organization of most cancers, c-*oncs* should be active at the stem cell stage in differentiation. Indeed, it has been suggested that proto-oncogenes might function normally in development (Müller *et al.*, 1983) and during liver regeneration (Goyette *et al.*, 1983). Several potential sites of oncogene action emerge from consideration of stem cell physiology. For example, self-renewal has been cited as the defining property of stem cells. There is indirect evidence that the information needed for self-renewal is encoded specifically in genes in addition to those required for cell proliferation generally (McCulloch *et al.*, 1984; Boettiger and Dexter, 1984). Pathological activation of self-renewal genes might then convey growth advantages on the stem cells in which they functioned normally.

Growth factors provide another potential link between activated c-*oncs* and stem cells. Recently, evidence has been presented that the gene product of the *sis* oncogene has extensive homology with platelet-derived growth factor (Waterfield *et al.*, 1983; see Chapters 6, 7, and 11) and that expression of c-*myc* in lymphoid cells can be regulated by mitogenic growth factors (Kelly *et al.*, 1983). If c-*oncs* sometimes code for growth factors, their activation might lead to factor excess or production of stimulators in stem cells that normally served only as their targets. The resulting autostimulation might contribute to the behavior of malignant clones. *src* gene products are known to be protein kinases (Dasidar and Fox, 1983) with specificity for tyrosine residues. Stanley and Jubinsky (1984) have suggested that phosphorylation of tyrosines in receptor proteins might change the effect of factor binding and lead to abnormal proliferation.

Malignant transformation is usually considered to require at least two separate genetic events (Knudson, 1978). This view is easily compatible with the observed phenotypes in malignancy generally and the clonal hemopathies in particular. For example, in the development of AML, a hemopoietic stem cell must become dominant in the population, and genes coding for stage- and differentiation-specific antigens appear to be expressed abnormally. While these multiple effects might be traced to a single pleiotropic gene, it seems more likely that more than one locus would be involved. Thus, a few proto-oncogenes, activated together or in sequence, might use a limited number of mechanisms to alter growth. Such events might be carcinogenic only if they occurred in cells that normally possessed capacity for self-renewal. Tissue specificity would derive from the differentiation state of the target stem cells. The recent findings with respect to complementation groups of oncogenes in transformation of normal fibroblasts (Ruley, 1983; Land et al., 1983b) are also compatible with a multistage model of carcinogenesis. A relationship has been postulated between stem cell growth control and oncogene complementation groups (Buick and Pollak, 1984).

Taken together, these speculations might explain how a small repertoire of mechanisms would yield the wide variety of malignant diseases encountered in nature. At a minimum, present information does not suggest that recent developments in molecular biology will prove incompatible with a stem cell model of cancer.

IX. TREATMENT GOALS

Knowledge of the clonal nature and cellular organization of tumors permits the establishment of rational treatment goals. If the purpose is to cure the disease, it is necessary to destroy the malignant clone. The target need not be all the abnormal cells; it suffices if all the stem cells are destroyed. This goal may be achieved only if normal function is not compromised too severely, and therefore may be applicable as a rationale for the surgical excision of solid tumors.

Hemopoietic tumors, however, must be regarded in a different light. In AML, cure is a reasonable objective if normal stem cells are present or can be supplied by marrow transplantation. This modality, following ablative chemotherapy and radiotherapy, is now showing real promise in the treatment of CML (Messner, 1984a). Alternatively, if distress is related to a large mass of differentiated cells, the goal of treatment may be control of disease by reducing cell number. Phlebotomy for the management of P vera or plasmapheresis in CML are examples of this approach. Attempts may also be made to alter differentiation patterns. Small doses of cytoreductive drugs are used in the treatment of CML with this objective; the abnormal clone remains (Whang et al., 1963), but erythropoietic differentiation programs become commoner, with alleviation of the anemia and reduction in leukocytosis. Given the life-threatening toxicity associated with attempts to cure leukemia, control is an attractive option. Unfortunately, although differentiation can be induced in certain cell lines in culture (Ferrero and Rovera, 1984), it has yet to be achieved clinically. An example of ongoing efforts is the recent failure to control promyelocytic leukemia with retinoic acid, although the agent led to differentiation of the patient's blast cells in culture (Flynn et al., 1983). A better understanding of molecular mechanisms as these affect stem cells may suggest more successful strategies.

X. CONCLUSIONS

This review describes and endorses a stem cell model for cellular organization in normal cell renewal systems and the tumors that derive from them. The major feature of the model is that both normal and malignant populations are considered to consist of independent clones, each maintained by a self-renewing stem cell; the majority of the cells in such clones, however, are either proliferatively inert or capable only of terminal divisions. The stem cell model contrasts with others that postulate dedifferentiation as part of transformation, and it assumes that all or most cells in tumors can maintain progressive growth.

A stem cell model of cancer has many implications. It assumes that carcinogenic events occur in stem cells. In molecular terms, activated oncogenes may be expressed effectively only at this early stage in differentiation. The tissue specificity of each tumor is a consequence of the pretransformation determination of stem cells committed to a particular specialized organ system.

Many characteristics of established tumors may reflect stem cell properties. Particularly the heterogeneity of malignant populations derives from the capacity of stem cells to produce differentiated progeny. Indeed, the variety of such descendants may be greater in malignant than in normal clones.

Maintenance in tumor populations by a small number of stem cells casts light on the results of treatment using cytoreductive modalities. Doses of radiation or chemotherapy may be effective in controlling growth or in curing cancers, although they might not be expected to inactivate all the cells in a tumor. Particularly in man, where tumors may contain $>10^{10}$ cells, reduction in cell viability by 10 decades would almost certainly be associated with unacceptable toxicity. The favorable results obtained with tolerable doses may reflect the need to inactivate only tumor stem cells.

   Concepts of stem cell function have been developed by cell biologists often using indirect development methods. Molecular biologists have made their remarkable advances using homogeneous cell lines as sources of nucleic acids or targets for transfection. Clinicians have improved the results of treatment using clinical trials, usually based on previous experience or available agents rather than biological considerations. The opportunity is before us now to conduct unified cancer research. For example, gene expression can be detected directly using appropriate probes; in situ hybridization allows resolution at the level of single cells. Gene products can be identified with monoclonal antibodies. Clinicians can use laboratory methods to relate cellular heterogeneity to treatment outcome. Interactions such as these may be guided by the stem cell model and may provide its ultimate test.

ACKNOWLEDGMENTS

   This work was supported by grants from the Medical Research Council of Canada, the Ontario Cancer Treatment and Research Foundation and the National Cancer Institute of Canada.

REFERENCES

Abramson, S., Miller, R. G., and Phillips, R. A. (1977). The identification in adult bone marrow of pluripotent and restricted stem cells of the myeloid and lymphoid systems. *J. Exp. Med.* **145,** 1567–1579.

Adamson, J. W. (1984). Analysis of haematopoiesis: The use of cell markers and in vitro culture techniques in studies of clonal haemopathies in man. *Clin. Haematol.* **13,** 489–502.

Adamson, J. W., Fialkow, P. J., Murphy, S., Prchal, J. F., and Steinmann, L. (1976). Polycythemia vera: Stem cell and probable clonal origin of the disease. *N. Engl. J. Med.* **295,** 913–916.

Adamson, J. W., Singer, J. W., Murphy, S., Steinmann, L., and Fialkow, P. J. (1978). Polycythemia vera (PV): Further in vitro studies of marrow regulation. *Blood* **52,** Suppl. 1, 199.

Barnes, D. W. H., Ford, C. E., Gray, S. M., and Loutit, J. F. (1958). Spontaneous and induced changes in cell populations in heavily irradiated mice. *Proc. U.N. Int. Conf. Peaceful Uses At. Energy, 2nd, 1958,* Vol. 23, pp. 10–16.

Becker, A. J., McCulloch, E. A., and Till, J. E. (1963). Cytological demonstration of the clonal nature of spleen colonies derived from transplanted mouse marrow cells. *Nature (London),* **197,** 452–454.

Bettelheim, P., Paietta, E., Majdic, O., Gadner, H., Schwarzmeier, J., and Knapp, W. (1982). Expression of a myeloid marker on TdT-positive acute lymphocytic leukemic cells: Evidence by double-fluorescence staining. *Blood* **60,** 1392–1396.

Bhan, A. K., Nadler, L. M., Stashenko, P., McCluskey, R. T., and Schlossman, S. F. (1981). Stages of B cell differentiation in human lymphoid tissue. *J. Exp. Med.* **154,** 737–749.

Blackstock, A. M., and Garson, O. M. (1974). Direct evidence for involvement of erythroid cells in acute myeloblastic leukemia. *Lancet* **2**, 1178–1179.

Boettiger, D., and Dexter, M. (1984). Long-term marrow cultures in analysis of viral leukaemogenesis. *Clin. Haematol.* **13**, 349–370.

Bradley, T. R., and Metcalf, D. (1966). The growth of mouse bone marrow cells *in vitro Aust. J. Exp. Biol. Med. Sci.* **44**, 287–299.

Buick, R. N. (1980). *In vitro* clonogenicity of primary human tumor cells: Quantitation and relationship to tumor stem cells. In "Cloning of Human Tumor Stem Cells" (S. E. Salmon, ed.), pp. 15–20. Alan R. Liss, Inc., New York.

Buick, R. N. (1984a). The cell renewal hierarchy in ovarian cancer. In "Human Tumor Cloning" (S. E. Salmon and J. M. Trent, eds.) pp. 3–13. Grune and Stratton, New York.

Buick, R. N. (1984b). Cell heterogeneity in human ovarian carcinoma. *J. Cell Physiol.* *Suppl.* **3**, 117–122.

Buick, R. N., and Mackillop, W. J. (1981). Measurement of self-renewal in culture of clonogenic cells from human ovarian carcinoma. *Br. J. Cancer* **44**, 349–355.

Buick, R. N., and Pollak, M. (1984). Perspectives on human tumor clonogenic cells, stem cells and oncogenes. *Cancer Res.* **44**, 4909–4918.

Buick, R. N., Till, J. E., and McCulloch, E. A. (1977). Colony assay for proliferative blast cells circulating in myeloblastic leukaemia. *Lancet* **1**, 862–863.

Buick, R. N., Minden, M. D., and McCulloch, E. A. (1979a). Self-renewal in culture of proliferative blast progenitor cells in acute myeloblastic leukemia. *Blood* **54**, 95–104.

Buick, R. N., Stanisic, T. H., Fry, S. E., Salmon, S. E., Trent, J. M., and Krasovich, P. (1979b). Development of an agar–methylcellulose clonogenic assay for cells in transitional cell carcinoma of the human bladder. *Cancer Res.* **39**, 5051–5056.

Buick, R. N., Fry, S. E., and Salmon, S. E. (1980). Effect of host–cell interactions on clonogenic carcinoma cells in human maligmant effusions. *Br. J. Cancer* **41**, 695–704.

Burgess, A. W., Camakaris, J., and Metcalf, D. (1977). Purification and properties of colony-stimulating factor from mouse-lung conditioned medium. *J. Biol. Chem.* **252**, 1998–2003.

Burstein, S. A., Adamson, J. W., Erb, S. K., and Harker, L. A. (1981). Megakaryocytopoiesis in the mouse: Response to varying platelet demand. *J. Cell. Physiol.* **109**, 333–341.

Bush, R. S., and Hill, R. P. (1975). Biologic discussions augmenting radiation effects and model systems. *Laryngoscope* **85**, 1119–1133.

Carney, D. N., Gazdar, A. F., and Minna, J. D. (1980). Positive correlation between histological tumor involvement and generation of tumor cell colonies in agarose in specimens taken directly from patients with small-cell carcinoma of the lung. *Cancer Res.* **40**, 1820–1822.

Carney, D. N., Gazdar, A. F., Bunn, P. A., Jr., and Guccion, J. G. (1981). Demonstration of the stem cell nature of clonogenic tumor cells from lung cancer patients. *Stem Cells* **1**, 149–164.

Cohen, S., Ushiro, H., Stoscheck, C., and Chinkers, M. (1982). A native 170,000 epidermal growth factor receptor–kinase complex from shed plasma membrane vesicles. *J. Biol. Chem.* **257**, 1523–1531.

Courtenay, V. D., and Mills, J. (1978). An *in vitro* colony assay for human tumours grown in immune-suppressed mice and treated *in vivo* with cytotoxic agents. *Br. J. Cancer* **37**, 261–268.

Courtenay, V. D., Smith, I. E., Peckham, M. J., and Steel, G. G. (1976). *In vitro* and *in vivo* radiosensitivity of human tumor cells obtained from a pancreatic carcinoma xenograft. *Nature (London)* **263**, 771–773.

Courtenay, V. D., Selby, P. J., Smith, I. E., Mills, J., and Peckham, M. J. (1978). Growth of human tumour cell colonies from biopsies using two soft agar techniques. *Br. J. Cancer* **38**, 77–81.

Dasidar, P. G., and Fox, C. F. (1983). Epidermal growth factor receptor dependent phosphorylation of a $M_r = 34,000$ protein substrate for pp60[src]. *J. Biol. Chem.* **258**, 2041–2044.

Dexter, D. L., and Hager, J. C. (1980). Maturation-induction of tumor cells using a human colon carcinoma model. *Cancer* **45**, 1178–1184.

Dicke, K. A., Spitzer, G., and Ahearn, M. J. (1976). Colony formation *in vitro* by leukaemic cells in acute myelogenous leukaemia with phytohaemagglutinin as stimulator factor. *Nature (London)* **259**, 129–130.

Durie, B. G. M., Young, L. A., and Salmon, S. E. (1983). Human myeloma *in vitro* colony growth; interrelationships between drug sensitivity, cell kinetics, and patient survival duration. *Blood* **61**, 929–934.

Eaves, A. C., and Eaves, C. J. (1984). Erythropoiesis in culture. *Clin. Haematol.* **13**, 371–391.

Fauser, A. A., and Messner, H. A. (1979a). Identification of megakaryocytes, macrophages and eosinophils in colonies of human bone marrow containing neutrophilic granulocytes and erythroblasts. *Blood* **53**, 1023–1027.

Fauser, A. A., and Messner, H. A. (1979b). Fetal hemoglobin in mixed hemopoietic colonies (CFU-GEMM), erythroid bursts (BFU-E) and erythroid colonies (CFU-E): Assessment by radioimmune assay and immunofluorescence. *Blood* **54**, 1384–1394.

Ferrero, D., and Rovera, G. (1984). Human leukemic cell lines. *Clin. Haematol.* **13**, 461–487.

Fialkow, P. J. (1976). Clonal origin of human tumors. *Biochim. Biophys Acta* **458**, 283–321.

Fialkow, P. J. (1980). Clonal and stem cell origin of blood cell neoplasms. *Contemp. Hematol./Oncol.* **1**, 1–46.

Fialkow, P. J. (1982). Cell lineages in hematopoietic neoplasia studied with glucose-6-phosphate dehydrogenase cell markers. *In* "Cellular and Molecular Biology of Hemopoietic Stem Cell Differentiation" (T. W. Mak and E. A. McCulloch, eds.), pp. 37–43. Alan R. Liss, Inc., New York.

Fialkow, P. J., Singer, J. W., Adamson, J. W., Vaidya, K., Dow, L. W., Ochs, J., and Moohr, J. W. (1981). Acute non-lymphocytic leukemia: Heterogeneity of stem cell origin. *Blood* **57**, 1068–1073.

Filmanowicz, E., and Gurney, C. G. (1961). Studies on erythropoiesis: XVI. Response to a single dose of erythropoietin in polycythemic mouse. *J. Lab. Clin. Med.* **57**, 65–72.

Flynn, P. J., Miler, W. J., Weisdorf, D. J., Arthur, D. C., Brunning, R., and Branda, R. F. (1983). Retinoic acid treatment of acute promyelocytic leukemia: *In vitro* and *in vivo* observations. *Blood* **62**, 1211–1217.

Fowler, J. H., Wu, A. M., Till, J. E., McCulloch, E. A., and Siminovitch, L. (1967). The cellular composition of hemopoietic spleen colonies. *J. Cell Physiol.* **69**, 65–72.

Francis, G. E., Wing, M. A., Berney, J. J., and Guimaraes, J. E. T. (1983). CFU_{gemm}, CFU_{gm}, CFU_{mk}: Analysis by equilibrium density centrifugation. *Exp. Hematol.* **11**, 481–489.

Goldie, J. H., and Coltman, A. J. (1983). Quantitative model for multiple levels of drug resistance in clinical tumors. *Cancer Treat. Rep.* **67**, 923–931.

Goyette, M., Petropoulos, G. J., Shank, P. R., and Fausto, N. (1983). Expression of a cellular oncogene during liver regeneration. *Science* **219**, 510–512.

Greaves, M. F. (1982a). "Target" cells, cellular phenotypes, and lineage fidelity in human leukaemia. *J. Cell. Physiol.* **111**, Suppl. 1, 113–125.

Greaves, M. F. (1982b). Leukaemogenesis and differentiation: A commentary on recent progress and ideas. *Cancer Surv.* **1**(2), 189–204.

Gregory, C. J., and Eaves, A. C. (1978). Three stages of erythropoietic progenitor cell differentiation distinguished by a number of physical and biological properties. *Blood* **51**, 527–537.

Gregory, C. J., and Henkelman, R. M. (1977). Relationships between early hemopoietic progenitor cells determined by correlation analysis of their numbers in individual spleen colonies. *In* "Experimental Hematology Today" (S. J. Baum and G. D. Ledney, eds.), pp. 93–101. Springer-Verlag, Berlin and New York.

Griffin, J. D., Larcom, P., and Schlossman, S. F. (1983). Use of surface markers to identify a subset of acute meylomonocytic leukemia cells with progenitor cell properties. *Blood* **62**, 1300–1303.

Hamburger, A. W., and Salmon, S. E. (1977). Primary bioassay of human myeloma stem cells. *J. Clin. Invest.* **60**, 846–854.

Hamburger, A. W., and White, C. P. (1981). Interaction between macrophages and human tumor clonogenic cells. *Stem Cells* **1**, 209–223.

Hamburger, A. W., Salmon, S. E., Kim, M. B., Soehnler, B. J., and Alberts, D. J. (1978). Direct cloning of human ovarian carcinoma cells in agar. *Cancer Res.* **38**, 3438–3444.

Hamburger, A. W., White, C. P., and Brown, R. W. (1981). Effect of epidermal growth factor on proliferation of human tumor cells in soft agar. *JNCI, J. Natl. Cancer Inst.* **67**, 825–831.

Heath, D. S., Axelrad, A. A., McLeod, D. L., and Shreeve, M. (1976). Separation of the erythropoietin-responsive progenitors BFU-E and CFU-E in mouse bone marrow by unit gravity sedimentation. *Blood* **47**, 777–792.

Ihle, J. N., Keller, J., Oroszlan, S., Henderson, L. E., Copeland, T. D., Fitch, F., Prystowsky, M. B., Goldwasser, E., Schrader, J. W., Plaszynski, E., Dy, M., and Lebel, B. (1983). Biological properties of homogeneous interleukin 3: I. Demonstration of WEHI-3 growth factor activity, mast cell growth factor activity, P cell-stimulating factor activity, colony-stimulating factor activity, and histamine-producing cell-stimulating factor activity. *J. Immunol.* **131**, 282–287.

Iscove, N. N., Senn, J. S., Till, J. E., and McCulloch, E. A. (1971). Colony formation by normal and leukemic human marrow cells in culture: Effect of conditioned medium from human leukocytes. *Blood* **37**, 1–5.

Iscove, N. N., Roitsch, C. A., Williams, N., and Guilbert, L. J. (1982). Molecules stimulating early red cell, granulocyte, macrophage, and megakaryocyte precursors in culture: Similarity in size, hydrophobicity, and charge. *J. Cell. Physiol.* **111**, Suppl. 1, 65–78.

Izaguirre, C. A. (1984). Colony formation by lymphoid cells. *Clin. Haematol.* **13**, 405–422.

Izaguirre, C. A., Curtis, J. E., Messner, H. A., and McCulloch, E. A. (1981). A colony assay for blast cell progenitors in non-B non-T (common) acute lymphoblastic leukemia. *Blood* **57**, 823–829.

Jani, P., Verbi, W., Greaves, M. F., Bevan, D., and Bollum, F. (1983). Terminal deoxynucleotidyl transferase in acute myeloid leukaemia. *Leuk. Res.* **7**, 17–29.

Johnson, G. R. (1984). Haemopoietic multipotential stem cell in culture. *Clin. Haematol.* **13**, 309–328.

Johnson, G. R., and Metcalf, D. (1977). Pure and mixed erythroid colony formation *in vitro* stimulated by spleen conditioned medium with no detectable erythropoietin. *Proc. Natl. Acad. Sci. U.S.A.* **74**, 3879–3882.

Kaaya, G. P., Jamal, N., Maxie, M. G., and Messner, H. A. (1982). Characteristics of erythroid colonies formed by bovine marrow progenitor cells in methylcellulose cultures. *Res. Vet. Sci.* **32**, 213–220.

Keller, G. M., and Phillips, R. A. (1982). Detection *in vitro* of a unique, multipotent hemopoietic progenitor. *J. Cell. Physiol.* **111**, Suppl. 1, 31–36.

Kelly, K., Cochran, B. H., Stiles, C. D., and Leder, P. (1983). Cell-specific regulation of the c-myc gene by lymphocyte mitogens and platelet-derived growth factor. *Cell (Cambridge, Mass.)* **35**, 603–610.

Knudson, A. G., Jr. (1978). Retinoblastoma: A prototypic hereditary neoplasm. *Semin. Oncol.* **5**, 57–60.

Korsmeyer, S. J., Arnold, A., Bakhshi, A., Ravetch, J. V., Siebenlist, U., Hieter, P. A., Sharrow, S. O., LeBien, T. W., Kersey, J. H., Poplack, D. G., Leder, P., and Waldmann, T. A. (1983). Immunoglobulin gene rearrangement and cell surface antigen expression in acute lymphocytic leukemias of T-cell and B-cell precursor origins. *J. Clin. Invest.* **71**, 301–313.

Laboisse, C. L., Augeron, C., and Potet, F. (1981). Growth and differentiation of human gastrointestinal adenocarcinoma stem cells in soft agarose. *Cancer Res.* **41**, 310–315.

Land, H., Parada, L. F., and Weinberg, R. A. (1983a). Cellular oncogenes and multistep carcinogenesis. *Science* **222**, 771–778.

Land, H., Parada, L. F., and Weinberg, R. A. (1983b). Tumorigenic conversion of primary embryo fibroblasts requires at least two co-operating oncogenes. *Nature (London)* **304**, 596–602.

Lee-Huang, S. (1980). A new preparative method of isolation of human erythropoietin with hydrophobic interaction chromatography. *Blood* **56**, 620–624.

McCulloch, E. A. (1983). Stem cells in normal and leukemic hemopoiesis. *Blood* **62**, 1–13.

McCulloch, E. A. (1984). Diversity in normal and leukemic hemopoiesis. *In* "Aplastic Anemia: Stem Cell Biology and Advances in Treatment" (R. K. Humphries and N. S. Young, eds.), pp. 59–70. Alan R. Liss, Inc., New York.

McCulloch, E. A., and Till, J. E. (1981). Blast cells in acute myeloblastic leukemia: A model. *Blood Cells* **7**, 63–77.

McCulloch, E. A., Siminovitch, L., and Till, J. E. (1964). Spleen colony formation in anemic mice of genotype W/W^v. *Science* **144**, 844–846.

McCulloch, E. A., Siminovitch, L., Till, J. E., Russell, E. S., and Bernstein, S. E. (1965a). The cellular basis of the genetically determined hemopoietic defect in anemic mice of genotype Sl/Sl^d. *Blood* **26**, 399–410.

McCulloch, E. A., Till J. E., and Siminovitch, L. (1965b). The role of independent and dependent stem cells in the control of hemopoietic and immunologic responses. *In* "Methodological Approaches to the Study of Leukemias" (V. Defendi, ed.), pp. 61–68. Wistar Inst. Press, Philadelphia, Pennsylvania.

McCulloch, E. A., Buick, R. N., Minden, M. D., and Izaguirre, C. A. (1978). Differentiation programs underlying cellular heterogeneity in the myeloblastic leukemias of man. *In* "Hematopoietic Cell Differentiation" (D. W. Golde, M. J. Cline, D. Metcalf, and C. F. Fox, eds.), pp. 317–333. Academic Press, New York.

McCulloch, E. A., Curtis, J. E., Messner, H. A., Senn, J. S., and Germanson, T. P. (1982).

The contribution of blast cell properties to outcome variation in acute myeloblastic leukaemia. *Blood* **59,** 601–609.

McCulloch, E. A., Motoji, T., Smith, L. J., and Curtis, J. E. (1984). Hemopoietic stem cells: Their roles in human leukemia and certain continuous cell lines. *J. Cell Physiol. Suppl.* **3,** 13–20.

Mackillop, W. J., and Buick, R. N. (1981). Cellular heterogeneity in human ovarian carcinoma studied by density gradient fractionation. *Stem Cells* **1,** 355–366.

Mackillop, W. J., Stewart, S. S., and Buick, R. N. (1982). Density/volume analysis in the study of cellular heterogeneity in human ovarian carcinoma. *Br. J. Cancer* **45,** 812–820.

Mackillop, W. J., Ciampi, A., Till, J. E., and Buick, R. N. (1983a). A stem cell model of human tumor growth: Implications for tumor cell clonogenic assays. *JNCI, J. Natl. Cancer Inst.* **70,** 9–16.

Mackillop, W. J., Trent, J. M., Stewart, S. S., and Buick, R. N. (1983b). Tumor progression studied by analysis of cellular features of serial ascitic ovarian carcinoma tumors. *Cancer Res.* **43,** 874–878.

Messner, H. A. (1984a). Bone marrow transplantation for hematologic malignancies other than acute leukemia. *In* "Clinical Bone Marrow Transplantation" (K. G. Blume and L. D. Petz, eds.), pp. 299–312. Churchill-Livingston, Inc., New York.

Messner, H. A. (1984b). Human stem cells in culture. *Clin. Haematol.* **13,** 393–404.

Messner, H. A., Izaguirre, C. A., and Jamal, N. (1981). Identification of T-lymphocytes in human mixed hemopoietic colonies. *Blood* **58,** 402–405.

Messner, H. A., Jamal, N., and Izaguirre, C. (1982). The growth of large megakaryocyte colonies from human bone marrow. *J. Cell Physiol.* **111,** Suppl. 1, 45–51.

Metcalf, D. (1982). Regulatory control of the proliferation and differentiation of normal and leukemia cells. *Natl. Cancer Inst. Monogr.* **60,** 123–131.

Metcalf, D., and MacDonald, H. R. (1975). Heterogeneity of *in vitro* colony and cluster forming cells in mouse marrow. Segregation by velocity sedimentation. *J. Cell. Physiol.* **85,** 643–654.

Metcalf, D., and Nicola, N. A. (1984). The regulatory factors controlling murine erythropoiesis *in vitro*. *In* "Aplastic Anemia: Stem Cell Biology and Advances in Treatment" (R. K. Humphries and N. Young, eds.), pp. 93–105. Alan R. Liss, Inc., New York.

Meyer, J. S., Rao, B. R., Stevens, S. C., and White, W. L. (1977). Low incidence of estrogen receptor in breast cancers with rapid rates of cellular replication. *Cancer* **40,** 2290–2298.

Meyskens, F. L., Jr., and Salmon, S. E. (1981). Modulation of clonogenic human melanoma cells by follicle-stimulating hormone, melatonin, and nerve growth factor. *Br. J. Cancer* **43,** 111–115.

Meyskens, F. L., Jr., Soehnlen, B. J., Saxe, D. F., Casey, W. J., and Salmon, S. E. (1981). *In vitro* clonal assay for human metastatic melanoma cells. *Stem Cells* **1,** 61–72.

Minden, M. D., Till, J. E., and McCulloch, E. A. (1978). Proliferative state of blast cell progenitors in acute myeloblastic leukemia (AML). *Blood* **52,** 592–600.

Minden, M. D., Buick, R. N., and McCulloch, E. A. (1979). Separation of blast cell and T-lymphocyte progenitors in the blood of patients with acute myeloblastic leukemia. *Blood* **54,** 186–195.

Miyake, T., Kung, C. K. H., and Goldwasser, E. (1977). The purification of human erythropoietin. *J. Biol. Chem.* **252,** 5558–5564.

Moore, M. A. J., Spitzer, G., Williams, N., Metcalf, D., and Buckley, J. (1974). Agar culture studies in 127 cases of untreated acute leukemia: The prognostic value of reclassification of leukemia according to *in vitro* growth characteristics. *Blood* **44**, 1–18.

Müller, R., Slamon, D. J., and Tremblay, J. M. (1983). Differential expression of cellular oncogenes during pre- and post-natal development of the mouse. *Nature (London)* **299**, 640–644.

Nakahata, T., and Ogawa, M. (1982a). Clonal origin of murine hemopoietic colonies with apparent restriction to granulocyte-macrophage-megakaryocyte (GMM) differentiation. *J. Cell. Physiol.* **111**, 239–246.

Nakahata, T., and Ogawa, M. (1982b). Identification in culture of a class of hemopoietic colony-forming units with extensive capability to self-renew and generate multipotential hemopoietic colonies. *Proc. Natl. Acad. Sci. U.S.A.* **79**, 3843–3847.

Nakahata, T., Gross, A. J., and Ogawa, M. (1982). A stochastic model of self-renewal and commitment to differentiation of the primitive hemopoietic stem cells in culture. *J. Cell. Physiol.* **113**, 455–458.

Nakeff, A., and Daniels-McQueen, S. (1976). In vitro colony assay for a new class of megakaryocyte precursor: Colony-forming unit mega-(CFU-M). *Proc. Soc. Exp. Biol. Med.* **151**, 587–590.

Nicola, N. A., and Metcalf, D. (1981). Biochemical properties of differentiation factors for murine myelomonocytic leukemic cells in organ-conditioned media: Separation from colony-stimulating factors. *J. Cell. Physiol.* **109**, 253–264.

Nicola, N. A., Burgess, A. W., Staber, F. G., Johnson, G. R., Metcalf, D., and Battye, F. L. (1980). Differential expression of granulocyte–macrophage progenitor cells. *J. Cell. Physiol.* **103**, 217–237.

Nielsen, J. T., and Chapman, V. M. (1977). Electrophoretic variation for x-chromosome-linked phosphoglycerate kinase (PGK-1) in the mouse. *Genetics* **87**, 319–325.

Nowell, P. C. (1976). Clonal evolution of tumor cell populations. *Science* **194**, 23–28.

Nowell, P. C., and Hungerford, D. A. (1961). Chromosome studies in human leukemia: II. Chronic granulocytic leukemia. *JNCI, J. Natl. Cancer Inst.* **27**, 1013–1035.

Ogawa, M., Bergsagel, D. E., and McCulloch, E. A. (1973). Chemotherapy of mouse myeloma: Quantitative cell cultures predictive of response *in vivo*. *Blood* **41**, 7–15.

Ogawa, M., Porter, P. N., and Nakahata, T. (1983). Renewal and commitment to differentiation of hemopoietic stem cells (an interpretive review). *Blood* **61**, 823–829.

Park, C. H., Bergsagel, D. E., and McCulloch, E. A. (1971). Mouse myeloma tumor stem cells: A primary cell culture assay. *JNCI, J. Natl. Cancer Inst.* **46**, 411–422.

Park, C. H., Savin, M. A., Hoogstraten, B., Amare, M., and Hathaway, P. (1977). Improved growth of *in vitro* colonies in human acute leukemia with the feeding culture method. *Cancer Res.* **37**, 4595–4601.

Pierce, G. B., Shikes, R., and Fink, L. M. (1978). Cancer: A problem of developmental biology. In "Foundations of Developmental Biology Series," (C. L. Markert, ed.). Prentice-Hall, Englewood Cliffs, New Jersey.

Pike, B. L., and Robinson, W. A. (1970). Human bone marrow colony growth in agar-gel. *J. Cell. Physiol.* **76**, 77–84.

Pluznik, D. H., and Sachs, L. (1965). The cloning of normal "mast" cells in tissue culture. *J. Cell. Comp. Physiol.* **66**, 319–324.

Pluznik, D. H., and Sachs, L. (1966). The induction of clones of normal mast cells by a substance from conditioned medium. *Exp. Cell Res.* **43**, 553–563.

Potten, C. S., Schofield, R., and Lajtha, L. G. (1979). A comparison of cell renewal in bone marrow, testis and three regions of surface epithelium. *Biochim. Biophys. Acta* **560**, 281–299.

Price, G. B., and McCulloch, E. A. (1978). Cell surfaces and the regulation of hemopoiesis. *Semin. Hematol.* **15**, 283–300.

Reinherz, E. L., Kung, P. C., Goldstein, G., Levey, R. H., and Schlossman, S. F. (1980). Discrete stages of human intrathymic differentiation: Analysis of normal thymocytes and leukemic lymphoblasts of T lineage. *Proc. Natl. Acad. Sci. U.S.A.* **77**, 1588–1592.

Ruley, H. E. (1983). Adenovirus early region 1A enables viral and cellular tranforming genes to transform primary cells in culture. *Nature (London)* **304**, 302–306.

Sachs, L. (1982). Control of growth and normal differentiation in leukemic cells: Regulation of the developmental program and restoration of the normal phenotype in myeloid leukemia. *J. Cell. Physiol.* **111**, Suppl. 1, 151–164.

Selby, P., Buick, R. N., and Tannock, I. (1983a). A critical appraisal of the human tumor stem cell assay. *N. Engl. J. Med.* **308**, 129–134.

Selby, P., Bizzari, J.-P., and Buick, R. N. (1983b). Therapeutic implications of a stem cell model for human breast cancer: A hypothesis. *Cancer Treat. Rep.* **67**, 659–663.

Seligmann, M., Vogler, L. B., Preud'homme, J. L., Guglielmi, P., and Brouet, J. C. (1981). Immunological phenotypes of human leukemias of the B-cell lineage. *Blood Cells* **7**, 237–246.

Senn, J. S., McCulloch, E. A., and Till, J. E. (1967). Comparison of the colony-forming ability of normal and leukemic human marrow in cell culture. *Lancet* **2**, 597–598.

Shadduck, R. K., Pigoli, G., Caramatti, C., Degliantoni, G., Rizzoli, V., Porcellini, A., Waheed, A., and Shiffer, L. (1983). Identification of hemopoietic cells responsive to colony-stimulating factors by autoradiography. *Blood* **62**, 1197–1202.

Shih, C., Shilo, B. Z., Goldfarb, M. P., Dannenberg, A., and Weinberg, R. A. (1979). Passage of phenotypes of chemically transformed cells via transfection of DNA and chromatin. *Proc. Natl. Acad. Sci. U.S.A.* **76**, 5714–5718.

Shimizu, T., Motoji, T., Oshimi, K., and Mizogushi, H. (1982). Proliferative state and radiosensitivity of human myeloma stem cells. *Br. J. Cancer* **45**, 679–683.

Singer, J. W., Fialkow, P. J., Dow, L. W., Ernst, C., and Steinmann, L. (1979). Unicellular or multicellular origin of human granulocyte–macrophage colonies in vitro. *Blood* **54**, 1395–1399.

Singer, J. W., Keating, A., Cuttner, J., Gown, A. M., Jacobson, R., Killen, P. D., Moohr, J. W., Najfeld, V., Powell, J., Sanders, J., Striker, G. E., and Fialkow, P. J. (1984). Evidence for a stem cell common to hematopoiesis and its in vitro microenvironment: Studies of patients with clonal hematopoietic neoplasia. *Leuk. Res.* **8**, 535–545.

Smith, L. J., Curtis, J. E., Messner, H. A., Senn, J. S., Furthmayr, H., and McCulloch, E. A. (1983). Lineage infidelity in acute leukemia. *Blood* **61**, 1138–1145.

Stanley, E. R., and Heard, P. M. (1977). Factors regulating macrophage production and growth: Purification and some properties of the colony-stimulating factor from medium conditioned by mouse L. cells. *J. Biol. Chem.* **252**, 4305–4312.

Stanley, E. R., and Jubinsky, P. T. (1984). Factors affecting the growth and differentiation of hemopoietic cells in culture. *Clin. Haematol.* **13**, 329–348.

Stanley, E. R., Hansen, G., Woodcock, J., and Metcalf, D. (1975). Colony stimulating factor and the regulation of granulopoiesis and macrophage production. *Fed. Proc., Fed. Am. Soc. Exp. Biol.* **34**, 2272–2278.

Steel, G. G. (1977). Growth and survival of tumour stem cells. In "Growth Kinetics of Tumours," pp. 217–267. Oxford Univ. Press (Clarendon), London and New York.

Stephenson, J. R., Axelrad, A. A., McLeod, D. C., and Shreeve, M. M. (1971). Induction of colonies of hemoglobin-synthesizing cells by erythropoietin *in vitro. Proc. Natl. Acad. Sci. U.S.A.* **68**, 1542–1546.

Stohlman, F., Jr. (1970). Kinetics of erythropoiesis. *In* "Regulation of Hematopoiesis" (A. S. Gordon, ed.), pp. 317–326. Appleton-Century-Crofts, New York.

Temin, H. M. (1984). Do we understand the genetic mechanisms of oncogenesis? *J. Cell Physiol. Suppl.* **3**, 1–11.

Tepper, J. (1981). Clonogenic potential of human tumors—a hypothesis. *Acta Radiol. Oncol. Radiat. Phys. Biol.* **20**, 283–288.

Tepperman, A. D., Curtis, J. E., and McCulloch, E. A. (1974). Erythropoietic colonies in cultures of human marrow. *Blood* **44**, 659–669.

Thomson, S. P., and Meyskens, F. L., Jr. (1982). Method for measurement of self-renewal capacity of clonogenic cells from biopsies of metastatic human malignant melanoma. *Cancer Res.* **42**, 4606–4613.

Till, J. E. (1982). Stem cells in differentiation and neoplasia. *J. Cell. Physiol.* **111**, Suppl. 1, 2–11.

Till, J. E., and McCulloch, E. A. (1961). A direct measurement of the radiation sensitivity of normal mouse bone marrow cells. *Radiat. Res.* **14**, 213–222.

Till, J. E., and McCulloch, E. A. (1980). Hemopoietic stem cell differentiation. *Biochim. Biophys. Acta* **605**, 431–459.

Till, J. E., McCulloch, E. A., and Siminovitch, L. (1964). A stochastic model of stem cell proliferation, based on the growth of spleen colony forming cells. *Proc. Natl. Acad. Sci. U.S.A.* **51**, 29–36.

Tonegawa, S. (1983). Somatic generation of antibody diversity. *Nature (London)* **302**, 575–581.

Trentin, J. J. (1970). Influence of haematopoietic organ stroma (hematopoietic inductive microenvironments) on stem cell differentiation. *In* "Regulation of Hematopoiesis" (A. S. Gordon, ed.), pp. 161–186. Appleton, Century, Crofts, New York.

Tveit, K. M., Endresen, L., Rugstad, H. E., Fodstad, O., and Pihl, A. (1981). Comparison of two soft-agar methods of assaying chemosensitivity of human tumours *in vitro:* Malignant melanomas. *Br. J. Cancer* **44**, 539–545.

von Hoff, D. D., Casper, J., Bradley, E., Trent, J. M., Hodach, A., Reichert, C., Makuch, R., and Altman, A. (1980). Direct cloning of human neuroblastoma cells in soft agar culture. *Cancer Res.* **40**, 3591–3597.

Waterfield, M. D., Scrace, G. T., Whittle, N., Stroobant, P., Johnson, A., Wasteson, A., Westermark, B., Herdin, C. H., Huang, J. S., and Deuel, T. F. (1983). Platelet-derived growth factor is structurally related to the putative transforming protein p28[sis] of simian sarcoma virus. *Nature (London)*, **304**, 35–39.

Weiss, T. L., Kavinsky, C. J., and Goldwasser, E. (1982). Characterization of a monoclonal antibody to human erythropoietin. *Proc. Natl. Acad. Sci. U.S.A.* **79**, 5465–5469.

Whang, J., Frei, E., Tjio, J. H., Carbone, P. P., and Brecher, G. (1963). The distribution of the Philadelphia chromosome in patients with chronic myelogenous leukemia. *Blood* **22**, 664–673.

Whitmore, G. F., and Till, J. E. (1964). Quantitation of cellular radiobiological responses. *Annu. Rev. Nucl. Sci.* **14**, 347–374.

Wu, A. M., Till, J. E., Siminovitch, L., and McCulloch, E. A. (1968). Cytological evidence for a relationship between normal hematopoietic colony-forming cells and cells of the lymphoid system. *J. Exp. Med.* **127**, 455–467.

II

Growth Factors

3

Epidermal and Fibroblast Growth Factor

DENIS GOSPODAROWICZ

Cancer Research Institute
and
Departments of Medicine and Ophthalmology
University of California Medical Center
San Francisco, California

I. INTRODUCTION

Interest in factors involved in growth control has led during the past 20 years to the newly and actively developing field of growth factors. Until 1970 identified growth factors were in limited number. Those were colony-stimulating factor and erythropoietin for the hematopoietic system, nerve growth factor (NGF) for the peripheric nervous system, and epidermal growth factor (EGF) for basal epithelial cell layers. Since then, numerous growth factors acting at various parts of the cell cycle and on various cell types have been identified. Among these is fibroblast growth factor (FGF), present in brain and pituitary tissue. Although its role in vivo is still to be determined, it is likely that

61

it will play an important role in wound-healing processes as reflected by its ability to stimulate vascular endothelial and smooth muscle cell proliferation both *in vivo* and *in vitro*. In view of its mitogenic effect on a large number of tissue derived from the primary and secondary mesenchyme as well as from neuroectoderm, it could also play a role in developmental processes. The tissue range sensitivity of FGF complements and sometimes overlaps with that of EGF. In the present review, we will outline recent *in vivo* and *in vitro* studies which could eventually help to elucidate the function as well as mechanisms of action of both mitogens.

II. EPIDERMAL GROWTH FACTOR

A. Characterization and Structure

Epidermal growth factor is a single-chain polypeptide composed of 53 amino acids with an isoelectric point of 4.6 and a molecular weight of 6045 (Carpenter and Cohen, 1979). It has been isolated from mouse submaxillary gland (Cohen, 1959; 1962) and is present in that organ as a high-molecular-weight dimer complex in which one EGF peptide chain is associated with a specific arginine esteropeptidase: the EGF binding protein. This binding protein is involved in the production of active EGF from a precursor molecule with an MW of 9000 (Frey *et al.*, 1979). It is likely that EGF is derived from an even larger precursor as indicated by the nucleotide sequence of the EGF gene which codes for a 130-kilodalton protein (Gray *et al.*, 1983; Scott *et al.*, 1983). On the basis of a radioligand assay, a factor with all the biological properties of EGF has been isolated from human urine (hEGF). The similarity between mouse EGF (mEGF) and hEGF has been demonstrated by Gregory (1975), who reported that the amino sequence of human urogastrone has a 70% homology with that of mEGF. Since both urogastrone or hEGF has the same ability to compete with ^{125}I-labeled mEGF in radioreceptor assays (Hollenberg and Gregory, 1976), it is likely that the two factors are identical. Their identity is further supported by their ability to elicit nearly identical biological responses in humans or mice. Both urogastrone and mEGF can promote early eyelid opening in mice and block the release of HCl from the gastric mucosa (Gregory, 1975). The inhibition of HCl release correlates with morphological changes of the apical pole of parietal cells and reflects a general effect of EGF on cytoskeletal function (Gonzalez *et al.*, 1981).

B. Biological Effect of EGF both in Vivo and in Vitro

EGF has been shown to be a potent mitogen for a variety of cultured cells of ectodermal, mesodermal, and endodermal origin (Carpenter and Cohen, 1979). Among the cells of ectodermal origin which respond to it are a variety of keratinocytes derived from skin, conjunctival, or pharyngeal tissue (Sun and Green, 1977). In all cases, EGF markedly stimulates their proliferation and leads to enhanced keratinization and squame production. It also delays the ultimate senescence of the cells, thereby increasing their culture lifetime (Rheinwald and Green, 1977). Owing to these effects, culture of epidermal keratinocytes, particularly those of the human, has been greatly improved (Green et al., 1979). Single cultured cells can now generate in vitro stratified colonies that ultimately fuse and form an epithelium that is a reasonable approximation of the epidermis. Among the cells of mesodermal origin which respond to EGF in culture are granulosa cells, corneal endothelial cells, vascular smooth muscle, chondrocytes, and fibroblasts (Gospodarowicz et al., 1978a). This last cell type has been used extensively to study the sequence of events that are part of the mitogenic response produced by EGF (for a recent review, see Carpenter and Cohen, 1979). Among the cell types of endodermal origin known to be affected by EGF are liver cells (Richman et al., 1976; Bucher, 1978; Koch and Leffert, 1979; Leffert and Koch, 1982; McGowan et al., 1981) and thyroid cells (Rogers and Dumont, 1982; Westermark et al., 1983).

Although EGF has a very limited ability to support the growth of normal cells in the anchorage-independent configuration, having in that regard no transforming activity, it has been shown by Roberts et al. (1982) to be a strong potentiator of one of the class of transforming growth factors (βTGF). This class of TGF requires the presence of EGF in order to promote the growth in soft agar of normal fibroblasts (Roberts et al., 1982).

The biological effect of EGF both in vivo and in organ culture is to promote the proliferation of the basal cell layer of various epithelia of ectodermal origin (Cohen, 1965; Cohen and Taylor, 1974). This effect has been observed in fetal, neonatal, and adult tissues (Cohen and Elliott, 1963) and requires proper cell attachment to basement membrane (Cohen, 1965). In the case of chick skin, for example, when the isolated epidermis is maintained under organ culture conditions, an EGF effect can be observed, provided that the basal cell layer rests on collagen, killed dermis, or a Millipore filter, but not when it rests on plastic. Similar observations were made using corneal rather than skin epithelium as a target tissue (Gospodarowicz et al., 1977a; 1978e).

The ability of EGF to trigger epithelial cell proliferation in the neo-nate and adult animals has raised the possibility that EGF could be a fetal growth hormone responsible for the proliferation and develop-ment of specific epithelial territories in the embryo. This possibility has been tested in three different organs, the lung (Sundell et al., 1975; Catterton et al., 1979), the skin (Thorburn et al., 1981), and the second-ary palate (Hassell, 1975; Hassell and Pratt, 1977).

Studies done by Sundell et al. (1975, 1980) have shown that the constant infusion of EGF into fetal lambs for 3–5 days stimulates epi-thelial growth in many sites, including upper and lower airways. In addition, EGF appears to afford protection against the development of hyaline membrane disease when given in utero at 123–130 days of gestation. This suggests that cell differentiation is stimulated in addi-tion to cell growth (Sundell et al., 1980). A similar conclusion was reached by Catterton et al. (1979), who have observed that injection of EGF into 24-day-old rabbit fetuses also induces accelerated maturation of the lung.

Thorburn et al. (1981) have studied the possible role of EGF in a number of developmental processes associated with maturation of ovine fetus. EGF was infused over the period 110–125 days of gesta-tion, before the initiation of many of the maturational events related to postnatal survival of the fetus. The infusion of EGF induced gross mor-phological changes in all fetuses. The most striking changes were a marked increase in skin wrinkling and a shedding of wool fibers. Edema of the skin and appendages (ears and scrotum) was also evident. A highly significant increase was observed in the weights of the adrenal and thyroid glands and the liver. A marked decrease in thymus weight was also seen. No significant changes were noted in total body weight, crown–rump length, and hind or forelimb length. Likewise, no changes were seen in the weight of the pancreas, heart, lungs, brain, or pituitary gland. Radiographic measurement of fetal bones revealed retarded growth of the tibial and radial diaphyses of long-term EGF-infused fetuses (9–10 days) relative to controls. Softening of the skull was noticed in experimental animals, and this may relate to the ability of EGF to induce bone resorption (Raisz et al., 1980).

Since EGF can inhibit the degeneration of the medial edge palate epithelium and promote the hypertrophy and keratinization of these cells (Hassell, 1975; Hassell and Pratt, 1977; Pratt et al., 1979), it sug-gests that the palate might be dependent upon EGF for some aspects of its growth and differentiation. The response to EGF can be prevented by the addition of dibutyryl-cAMP (Hassell and Pratt, 1977). It has also been shown that cessation of DNA synthesis and cell death within the medial palatal epithelium can be inhibited in organ culture by the addition of EGF (Hassell, 1975; Pratt *et al.*, 1978). EGF, when administered after cells have lost their ability to make DNA, acts as a survival agent and prevents cell death within this region. In contrast, if present in the culture medium before cells have lost their ability to make DNA, EGF will both induce hyperplasia and prevent cell death within the medial epithelium (Hassell, 1975; Pratt *et al.*, 1978). If EGF is added to organ culture of palatal shelves which are in their final stage of differentiation (day 13 or 14 of gestation in mice), while it can still affect the morphology of the isolated epithelium, it will not stimulate DNA synthesis or prevent cell death (Tyler and Pratt, 1980). Therefore, depending on the temporal sequence of epithelial cell differentiation in the palate, some aspects of palatal shelf growth and differentiation may depend on EGF.

The possibility exists that the fetal form of EGF is different from the adult form. Nexo *et al.* (1980), using both EGF radioreceptors and radioimmunoassays, have analyzed the appearance of an EGF-like substance as a function of the age of the mouse embryo. Although little or no EGF was found prior to day 11, a dramatic increase occurred on day 12 and continued to rise through day 14 (Nexo *et al.*, 1980; Pratt, 1980). No corresponding increase occurred in the maternal serum during this time, although levels were considerably higher than those found in nonpregnant serum. This increase in EGF-like substance closely paralleled the increase in specific binding of EGF to membranes of crude embryonic homogenates, which increased 10-fold from day 12 to 14 (Nexo *et al.*, 1980). The source and composition of the EGF present in the embryo is uncertain. Although the maternal level of EGF could contribute to the fetal level through transplacental transport, this is unlikely because maternal and fetal EGF differ in that the fetal form does not cross-react to the same extent as maternal EGF with anti-EGF antibodies. This suggests that the fetal EGF could be distinct from adult EGF antigenically and, although related to it, may not be the same molecule (Nexo *et al.*, 1980).

If one were to reflect on the *in vivo* effect of EGF, although it has been shown to affect multiple developmental processes, its effects are either teratogenic (cleft palate, softening of the skull) or acceleration of processes which would have taken place anyway (tooth eruption, eyelid opening). As such and in contrast with NGF, no defined role of EGF in various developmental processes *in vivo* has yet been established. In particular, and in contrast with NGF, the use of EGF antibodies *in vivo* has been ineffective to modify normal development. The role of EGF *in vivo* therefore remains elusive. This could reflect either (1) that the fetal form of EGF differed from the adult one, (2) that all tissues which are the target of EGF could produce it (autocrine control of growth), or (3) that depletion of the circulating form of EGF is not possible. Circumstantial evidence exists for it since sialectomy (salivary gland removal), although resulting in a strong decrease in EGF blood content for 3 weeks, is ineffectual on a long-term basis. The blood content of EGF is always coming back to its normal level, presumably because tissues other than the salivary gland can release EGF into the bloodstream.

C. The EGF Receptor

The interaction of EGF with the cell surface and the subsequent fate of the EGF–receptor complexes have been extensively studied using cultured cells (Carpenter et al., 1975; Carpenter and Cohen, 1979; Hollenberg and Gregory, 1976; see Chapter 7). The failure of EGF to be denatured when linked to radioactive, fluorescent, photoaffinity, or particulate probes has made it possible to study its interaction with cultured cells as well as the fate of the cell surface EGF–receptor sites once EGF interacts with them (review by Haigler, 1982). After the initial binding events, an increased fluidity of the cell surface (Linsley et al., 1979), as reflected by increased lateral mobility of concanavalin A (Con A) receptor ligand complexes, as well as a transient increase in pinocytotic activity have been observed (Haigler et al., 1979b). This increase in pinocytotic activity correlates with rapid morphological changes. Within less than 5 min following the addition of EGF to the cells, extensive ruffling of the cell membrane and extension of filopodia were observed. After 15 min the ruffling activity decreases. The cells retract from the surface of the tissue culture substrate, become more rounded, and apparently pile up to form multilayered colonies. The molecular mechanisms responsible for these EGF effects have recently been elucidated by Schlessinger and Geiger (1981), who observed that upon addition of EGF to the cells, the actin-associated microfilament system underwent striking changes, as visualized by labeling for both actin and α-actinin. These changes included the loss of tightly organized filament bundles and the simultaneous increase in diffuse labeling for these proteins, in the cytoplasm, in membrane ruffles, and in retraction fibers. The effects of EGF on cell morphology and organization of the cytoskeleton are specific to this hormone and transient. The first effects are seen after incubation of the cells for 30–45 min in the presence of the hormone, and maximal effect occurs after 2 hr and then declines. After about 8 hr of incubation, the effect is no longer apparent (Schlessinger and Geiger, 1981).

The binding characteristics of the EGF receptor are affected by other growth factors. Studies examining the effect of platelet-derived growth factor (PDGF) showed that it decreases the cellular binding of labeled EGF (Wrann et al., 1980; Heldin et al., 1982). This effect is due to an alteration in EGF receptor affinity with no change in receptor number (Bowen-Pope et al., 1983). Likewise, receptor binding of fibroblast-derived growth factor (Rozengurt et al., 1982) or vasopressin (Rozengurt et al., 1981) results in a decreased affinity in receptors for EGF. Recently, Assoian et al. (1984) have shown that when βTGF binds to its receptor, it results in an increase in the number of EGF receptors. This suggests that delayed biological effects of βTGF (which resemble those of EGF) may be indirect consequences of its ability to regulate EGF receptors and therefore amplify EGF-induced responses.

The binding of EGF to its membrane receptor activates a cyclic nucleotide-independent protein kinase (Carpenter et al., 1979) specific for tyrosine residues (Ushiro and Cohen, 1980). This kinase was later shown to be the receptor itself (Buhrow et al., 1982), which is composed of three major domains: an EGF binding domain which lies external to the plasma membrane; a transmembrane domain; and a cytoplasmic kinase domain having both kinase activity and an autophosphorylation site. Structural studies done by Downward et al. (1984) have demonstrated that 74 out of 83 residues present in the transmembrane and cytoplasmic domain of the EGF receptor are identical to those of the transforming protein encoded by the v-erb-B oncogene of avian erythroblastosis virus (AEV). In view of the limited size of the v-erb-B oncogene, it is not expected that it would code for the binding domain of the EGF receptor. These results suggest that v-erb-B may be derived from cellular sequences which encode for a truncated version of the EGF receptors with tyrosine kinase activity associated with it.

The fate of EGF cell surface receptor sites following the binding of EGF to them has been studied with a wide variety of biochemical and morphological techniques. Incubation of affinity-labeled cells at 37°C resulted in a time-dependent loss of radioactivity from the EGF–receptor covalent complex (M_r 170,000) (Das and Fox, 1978). Ninety percent of the radioactivity lost from the band of M_r 170,000 during a 1 hr incubation at 37°C appeared in three bands of M_r 62,000, 47,000, and 37,000. The cross-linked EGF–receptor complex on intact cells was accessible to the action of trypsin at 4°C and cofractionated with the plasmalemmal fraction. The proteolytic processing products of receptors were inaccessible to trypsin and banded with the lysosomal fraction upon subcellular fractionation. The rate of internalization and proteolytic processing of radiolabeled receptors was the same as the rate of reduction of binding activity induced by EGF. A study of the relationship between EGF-induced receptor internalization and processing, and stimulation of DNA synthesis, showed that both these processes were half-maximally stimulated at ~0.1 nM EGF, a concentration at which only 10% of the receptor sites are occupied. These data indicate that at concentrations of EGF which are subsaturating for binding but optimal for biological activity, there is a slow, continuous process of receptor internalization and degradation which could be limiting for EGF-induced mitogenesis.

Schlessinger and his colleagues (1978), using fluorescent analogs of EGF, have visualized the binding, aggregation, and internalization of EGF in confluent monolayers of BALB/c 3T3 cells. The cells labeled with the fluorescent analogues were visualized with a sensitive video intensification microscopic system that allowed direct observation of the location of the fluorescent hormone on the surface and within the living fibroblasts. It was observed that EGF initially bound diffusely to the cell surface and at 4°C remained dispersed. Within a few minutes at 23°C or 37°C, the hormone–receptor complexes aggregated into patches that could be readily removed by trypsin but not by excess native hormone. The hormone–receptor complexes, which were initially mobile in the plane of the membrane, became later immobilized, as the consequence of receptor aggregation or internalization. Within ~30 min at 37°C, much of the labeled hormone was found within the cell in endocytotic vesicles that moved randomly in the cytoplasm. Ferritin-tagged EGF, which retains substantial binding affinity for cell receptors and is biologically active, has also been used to study the fate of EGF bound to the cell surface of a carcinoma cell line (A-431) (Haigler et al., 1979a). EGF:ferritin is initially localized exclusively on the plasma membrane, and when unfixed cells were warmed to 37°C, the conjugate rapidly redistributed in the plane of the membrane so that nearly 50% was located within 45 sec in microclusters. The redistribution was not dependent on metabolic energy, since it occurred in glucose-free media containing sodium azide and dinitrophenol. Time-course studies revealed that the clusters of EGF:ferritin on the cell surface were rapidly pinocytosed into vesicles. The internalization occurred at both coated and noncoated regions of the membrane, and the formation of endocytic vesicles required metabolic energy, since no EGF:ferritin-containing vesicles were observed when cells were incubated with conjugate in glucose-free media containing sodium azide and dinitrophenol (Haigler et al., 1979a).

EGF:ferritin-containing vesicles apparently serve to shuttle the hormone to multivesicular bodies (secondary lysosomes). The rapidity (10–15 min) with which ^{125}I-labeled tyrosine is released into the media subsequent to the binding of ^{125}I-labeled EGF to cell surface receptors suggests that lysosomal degradation begins when EGF:ferritin starts to accumulate in multivesicular bodies.

At the same time that the EGF–receptor complexes are endocytosed by an energy-dependent process (Gorden et al., 1978; Haigler et al., 1979a), an array of cellular metabolic alterations can be observed. These include early effects such as stimulation of nutrient uptake (Carpenter and Cohen, 1979; Haigler et al., 1979b) and ion fluxes (Rozengurt and Heppel, 1975), enhanced phosphorylation of endogenous membrane proteins (Carpenter et al., 1979; King et al., 1980), changes in cell morphology (Chinkers et al., 1979), and cytoskeletal organization (Schlessinger and Geiger, 1981). Delayed effects of EGF include the activation of cytoplasmic enzymes and the stimulation of protein and DNA synthesis (Carpenter and Cohen, 1979). The question remains open of when and how EGF generates biological effects and, in particular, of what the initial events are which are important in inducing DNA synthesis.

The importance of EGF–receptor aggregation for the induction of the EGF mitogenic response is indicated by the use of an EGF derivative obtained by cyanogen bromide cleavage (CNBr) of the hormone (Schecter et al., 1979). Cyanogen bromide-cleaved EGF (CNBr-EGF) binds with reduced affinity to EGF receptors and is as potent as EGF in activating both early and late cell responses to the hormone. The early events include phosphorylation of membrane proteins, the stimulation of Na^+,K^+-ATPase as reflected by the ouabain-sensitive uptake of ^{86}Rb (see Chapter 13), and changes in the organization of microfilaments and in cell morphology, while late events include the activation of the enzyme ornithine decarboxylase (Schreiber et al., 1981; Yarden et al., 1982). In contrast, CNBr-EGF, but not EGF, fails to induce receptor clustering and DNA synthesis. These observations therefore tend to demonstrate that the early or late events triggered by CNBr-EGF are apparently not sufficient signals for induction of DNA synthesis if clustering of EGF receptor molecules does not occur. In contrast, cross-linking of CNBr-EGF with bivalent anti-EGF antibodies restores both cluster formation and stimulation of DNA synthesis (Schechter et al., 1979; Schlessinger, 1980), thus suggesting that initial receptor clustering is an important event in the stimulation of DNA synthesis. Similar observations were made by Schreiber et al. (1983), who used monoclonal antibodies (2G2-IgM and TL5-IgG) directed against the EGF receptor and binding to different parts of the EGF receptor. 2G2-IgM and its monovalent Fab fragment (2G2-Fab) compete for the binding of EGF to the EGF receptor. While 2G2-IgM can both activate the EGF-sensitive kinase and induce receptor clustering and DNA synthesis, its Fab fragment, while activating the kinase, fails to induce receptor clustering and DNA synthesis. Cross-linking of 2G2-Fab fragments on the cell surface with anti-mouse IgG antibodies results in EGF receptor clustering and restores its capacity to induce DNA synthesis. The second antibody denoted TL5-IgG binds to EGF receptor in a domain which differs from that of EGF and does not compete with EGF for binding to the receptor molecule. This antibody fails to induce receptor clustering and possesses no intrinsic "EGF-like" activity. However, the cross-linking of TL5-IgG on the cell surface of fibroblasts with anti-mouse IgG antibodies leads to clustering of EGF receptor molecules and stimulation of DNA synthesis. One could therefore conclude that EGF–receptor clustering and its subsequent internalization induced by monoclonal antibodies of either polyvalent (2G2-IgM) or cross-linked monovalent (2G2-Fab fragments or TL5-IgG) ligands bound to the EGF receptor appear to constitute sufficient triggering for the induction of DNA synthesis, even when mediated via domains distinct from the EGF binding domain on the EGF receptor molecule (for another view, see Chapter 13).

Now that the path of the EGF–receptor complex has been followed as far as the point of degradation, the question arises of which second messenger(s) responsible for mitogenesis could be generated and of whether, indeed, a second messenger(s) is even required. Is there a membrane-based receptor-transducer mechanism acting near the beginning of the pathway, or is there a lysosomal "activation mechanism" generating active fragments acting near the end of the internalization pathway? The demonstration that primary alkylamines such as methylamine or antibiotics such as bacitracin, which greatly reduced cluster formation, do not inhibit the mitogenic activity of EGF (Maxfield *et al.*, 1979) favors a membrane-based receptor-transducer mechanism acting near the beginning of the pathway. This further suggests that the rapid internalization of occupied receptors via coated or uncoated pits may be a mechanism which limits the response to hormones and that the primary site of action of EGF is at the cell surface. However, methylamine could also act intracellularly and does not block EGF internalization totally (King *et al.*, 1981). The recent observation that neither leupeptin nor antipain, two inhibitors of lysosomal enzymes capable of blocking the intracellular degradation of EGF by 95%, significantly affects the mitogenic response speaks against the theory that a lysosomal activation mechanism could generate active fragments from the EGF–receptor complex (Savion et al., 1980). It therefore seems more likely that lysosomal degradation, rather than generating active fragments from the EGF–receptor complex, in fact represents one means by which the cells destroy the internalized EGF–receptor complex so that each EGF molecule acts only once. Interestingly enough, it has recently been demonstrated that in cells exposed to either leupeptin or chloroquine, both intracellular accumulation of intact EGF–receptor complexes and their increased accumulation in the nuclear fraction can be observed (Savion et al., 1981; Johnson et al., 1980). The significance of the accumulation of EGF–receptor complexes in nuclei is still unclear, but could represent one significant step in the mitogenic activity of EGF–receptor complexes (see Chapter 7).

D. EGF Receptor and Oncogenes

The agents inducing cell transformation are multiple, but they may all play upon a common genetic substrate within the cells. Recent evidence suggests that such substrates could be cellular oncogenes (c-oncogenes), copies of which are found in the oncogenes of retroviruses (v-oncogenes) (Bishop, 1983; see Chapter 7). Until recently, the role of cellular oncogenes in normal cells was unknown. Evidence has been mounting that they could, at least in part, regulate cell division at two different levels: either by coding for growth factors (see Chapter 10) or by coding for growth factor receptors (see Chapter 7).

Direct evidence that viruses or c-oncogenes, once activated, could produce growth factors has been provided by the studies of Doolittle et al. (1983) and Waterfield et al. (1983). Both groups have shown that the transforming protein of simian sarcoma virus (SSV) has a close structural and functional relationship to PDGF, supporting the hypothesis that autocrine growth factor production may be involved in abnormal growth control and neoplasia. Others have shown that infecting cells with retroviruses, such as Snyder Theilin feline sarcoma virus (FSV), led to the production of TGF, which can then autostimulate the cells to transform (Marquardt et al., 1984). Those factors have in recent studies been shown to have a strong homology with mEGF and hEGF (33.3 and 43.8%, respectively), and promote cell transformation by interacting both with EGF receptors as well as with their own receptors.

Direct evidence that oncogenes could also produce receptors for growth factors such as EGF has been provided by the studies of Downward et al. (1984), demonstrating that six peptides derived from hEGF receptor very closely match a part of the deduced sequence of the v-erb-

B transforming protein of AEV. The AEV progenitor may have acquired the cellular gene sequences of a truncated EGF receptor (or closely related protein) lacking the external EGF-binding domain, but retaining the transmembrane domain and a domain involved in stimulating cell proliferation. Transformation of cells by AEV may therefore result, in part, from the inappropriate acquisition of a truncated EGF receptor form of the c-erb-B gene. These observations together with those presented above illustrate two distinct but related mechanisms for subversion of normal growth regulation. In the case of SSV or FSV, the oncogene encodes a growth factor which can act as a mitogen for target cells having PDGF or EGF receptors. AEV, on the other hand, appears to utilize a different mechanism where a part of a growth factor receptor which is thought to be involved in transducing the EGF signal may be expressed in transformed cells. The absence of EGF binding might cause continuous generation of a signal equivalent to that produced by EGF, causing cells to proliferate rapidly. However, how this could result in the block in differentiation observed in AEV-infected hemopoietic cells is unclear. The structural relationship between v-erb-B transforming protein and EGF receptor suggests that other oncogenes from this subset of retroviruses could be derived in part from sequences which encode proteins having a less direct functional relationship to EGF and other receptors.

III. FIBROBLAST GROWTH FACTOR

A. Characterization

The observation that pituitary extracts and partially purified pituitary hormone preparations are mitogenic for various cultured cell types (Armelin, 1973; Holley and Kiernan, 1968; Clark et al., 1972; Corvol et al., 1972) led to the purification from pituitary tissue of FGF (Gospodarowicz, 1974, 1975; Gospodarowicz et al., 1978b). This peptide has a molecular weight of 15,000 and an isoelectric point of 9.6 (Gospodarowicz, 1974, 1975; Gospodarowicz, et al., 1978b, 1984). During the course of the purification of pituitary FGF, it was realized that brain tissue contains a similar activity (Gospodarowicz, 1974; Gospodarowicz et al., 1978b).

Both pituitary and brain FGF have many common chemical characteristics. They are basic polypeptides (pI = 9.6) with similar molecular weights under reducing conditions. The growth-promoting activity is destroyed by proteolytic digestion and by acid (below pH 4.0) or heat

(65°C, 5 min) treatment (Gospodarowicz, 1975, 1984a; Gospodarowicz et al., 1978b). Both mutagens are active at concentrations ranging from 2×10^{-13} to 2×10^{-11} M with similar dose–response curves (Gospodarowicz et al., 1978b, 1984, 1985).

B. Biological Effect of FGF both *in Vivo* and *in Vitro*

In vivo, brain FGF has been shown to stimulate the proliferation of blastemal cells in frogs and *Triturus virudescens* and could therefore contribute to the regeneration process in lower vertebrates by eliciting the initial recruitment of the cells in the area of amputation injury (Gospodarowicz et al., 1975, 1978d; Mescher and Gospodarowicz, 1979; Gospodarowicz and Mescher, 1981; Mescher, 1983). Both pituitary and brain FGF have been shown to be potent angiogenic factors in vivo. At concentrations as low as 0.3 ng/day, they induce massive capillary formation in either the rabbit cornea (Gospodarowicz et al., 1979c) or the hamster cheek pouch (A. Schreiber and D. Gospodarowicz, unpublished observations). In agreement with those observations, brain FGF also increases formation of granulation tissue in vivo (Buntrock et al., 1982a,b). This increase is associated with stimulation of the biosynthetic functions of fibroblasts and myofibroblasts. Finally, FGF in vivo has been shown to induce articular cartilage regeneration (Jentzsch et al., 1980).

The biological effects of FGF in vitro can be seen with respect to cell transformation, migration, proliferation, differentiation, and senescence. FGF has both acute and long-term effects on the morphology and growth pattern of responsive cells. Thus, human skin fibroblasts maintained in serum-supplemented medium together with FGF become extremely elongated with long slender projections which are retraction fibrils resulting from an increase in locomotory activity (Gospodarowicz and Moran, 1975). The same is true for vascular endothelial and smooth muscle cells which in the presence of FGF become bipolar (Gospodarowicz, 1984b). Even more striking is the effect of FGF on cell morphology when the growth factor is added to confluent and resting monolayers of BALB/c3T3 cells. Such cells, which in confluent culture exposed to serum-supplemented medium show a regular cobblestone pattern, become spindle shaped and grow in an irregular crisscross pattern when exposed to FGF for 6–12 hr (Gospodarowicz and Moran, 1974). Large membrane ruffles, often associated with macropinocytotic vesicles, can be seen on most of the cells. Morphologically, the FGF-treated cells look "transformed," since a reduced cell–substratum adhesion, growth in crisscross pattern, and increased membrane ruffling are features typical of transformation. The similarity in phenotype of transformed and FGF-treated cells is consistent with the hypothesis that transformation and FGF might at least partially have a common metabolic pathway. Comparison of the locomotor activity of arterial smooth muscle cells and endothelial cells in media supplemented with plasma and FGF, respectively, has given circumstantial evidence that FGF stimulates cell migration (Gospodarowicz, 1984b).

Both pituitary and brain FGF have been shown to be potent mitogens for mesoderm-derived cells (Gospodarowicz, et al., 1978b,c,d, 1983; Gospodarowicz, 1979). Until now, pituitary and brain FGF have mostly been used in vitro to develop new cell strains (Gospodarowicz and Zetter, 1977; Gospodarowicz, 1979), particularly from the vascular and corneal endothelia (Gospodarowicz et al., 1976, 1977a, 1978a,b, 1979b, 1980, 1981a,b). FGF can be observed to be mitogenic both for cells seeded at clonal density and for low-density cultures (Gospodarowicz and Zetter, 1977). Although the response of the cells can vary widely depending on the culture conditions (types of serum and media used, Gospodarowicz and Lui, 1981), addition of FGF to the cultures of most mesoderm-derived cells results in a greatly reduced average doubling time, which in the case of vascular endothelial cells, for example, can decrease from 72 to 18 hr (Gospodarowicz et al., 1980; Duthu and Smith, 1980). This results primarily from a shortening of the G_1 phase of the cell cycle (Gospodarowicz et al., 1980, 1981a,b).

FGF stabilizes the phenotypic expression of cultured cells (Vlodavsky and Gospodarowicz, 1979; Vlodavsky et al., 1979). This is a particularly interesting characteristic of FGF, since it has made possible the long-term culturing of various cell types which otherwise would lose their normal phenotypes in culture when passaged repeatedly at low cell density. This biological effect of FGF has been best studied using vascular or corneal endothelial cell cultures cloned and maintained in the presence of FGF and then deprived of it for various time periods (Vlodavsky and Gospodarowicz, 1979; Vlodavsky et al., 1979; Greenburg et al., 1980; Gospodarowicz et al., 1981b; Tseng et al., 1982; Gospodarowicz 1983a,b, 1984a). It has been related to the ability of FGF to control the synthesis of various basement lamina components (Gospodarowicz et al., 1978e, 1979a,b; Gospodarowicz and Tauber, 1980) which would in turn affect cell polarity and gene expression (Gospodarowicz and Greenburg, 1981; Gospodarowicz et al., 1982b). In the case of sheep preadipocyte fibroblasts, it has been reported that FGF promotes their differentiation into adipocytes (Broad and Ham, 1983). A similar observation has been made by Serrero and Khoo (1982) using the 1246 cell line.

FGF can also significantly delay the ultimate senescence of cultured cells. In the case of granulosa cells, addition of FGF to either clonal or mass cultures of granulosa cells can extent their life span in culture from 10 to 60 generations (Gospodarowicz and Bialecki, 1978). Adrenal cortex cell lines cloned in the presence of FGF show a similar dependence on this growth factor during their limited *in vitro* life span. Its removal from the culture medium results not only in a greatly extended doubling time but also in rapid cell senescence (Simonian and Gill, 1979; Simonian et *al.*, 1979). In the case of vascular (Gospodarowicz et *al.*, 1978b; Gospodarowicz, 1979; Duthu and Smith, 1980) and corneal endothelial cells (Gospodarowicz et *al.*, 1981b), FGF has been shown to extend the life span of the cultures, the effect being best observed with corneal endothelial cells which, when maintained in the absence of FGF, have a life span of 20–30 generations, while in its presence they can proliferate for 200 generations (Gospodarowicz et *al.*, 1981b; Giguere et *al.*, 1982).

In summary, the following effects of FGF have been observed with various cell types: It increases the rate of cellular proliferation; it stimulates cell migration; and it stabilizes their phenotypic expression and extends the culture life span.

C. Mechanism of Action

With the use of cultures of vascular endothelial cells, the early effect of FGF on membrane motility has been studied (Gospodarowicz et *al.*, 1979b). The interaction between various lectins and glycoproteins that are capable of lateral diffusion in the plane of the membrane can lead to a selective redistribution of lectin–receptor complexes in a process that results in "patch" and subsequent "cap formation" (segregation of the cross-linked patches to one pole of the cell by an active, microfilament-dependent process). Long-range receptor movement involved in patching and capping can be directly observed by looking at the distribution of fluorescein-labeled lectins such as Con A (Nicholson, 1976). Short-range lateral movements like those involved in the lectin-mediated receptor clustering and cell agglutination can be indirectly quantified by measuring the ability of cells to bind to nylon fibers coated with different densities of lectin molecules (Vlodavsky and Sachs, 1975). This binding requires a short-range lateral mobility of the appropriate receptors to allow their alignment with the lectin molecules on the fiber. By using both of these techniques, it was observed that both the short- and long-range types of Con A receptor mobility were increased in bovine vascular endothelial cells preincubated with FGF. A 6- to 9-

hr preincubation with FGF gave a nearly maximal increase (two- to fourfold) in the binding of cells to Con A molecules on the fibers, and a significant effect was obtained at 3 hr. A longer preincubation period (12–24 hr) with FGF is required to induce in a large proportion (60–80%) of the cells the ability to form caps upon addition of Con A. Such changes were not obtained when cells were preincubated with EGF, which does not bind and has no mitogenic effect on bovine vascular endothelial cells (Gospodarowicz et al., 1978a,b).

The early effect of FGF on increased membrane fluidity and increased ruffling activity correlate, as has been shown in the case of EGF (Schlessinger and Geiger, 1981) or PDGF (Westermark et al., 1982), with rapid changes in the dynamic structure of the actin cytoskeleton (see Chapter 13). By using the indirect immunofluorescence technique, it has been observed that FGF causes profound, acute changes in the distribution of actin. Within 1 min, a patchy fluorescence appears on the upper surface, probably corresponding to small ruffles. After 5–15 min, a lattice of coarse filaments is seen, and in many cells an intensive circular actin staining is visible, corresponding to the circular ruffles observed by phase contrast microscopy (Gospodarowicz, 1984b).

Those events that include sequential changes, including stimulation of cellular transport systems, polyribosome formation, protein synthesis, ribosomal and tRNA synthesis, and eventually DNA synthesis followed by cell division, inducing a "pleiotypic" and mitogenic response, have been analyzed using confluent cultures of BALB/c 3T3 cells. The addition of FGF to confluent and resting cultures of BALB/c 3T3 cells leads to an increased uptake of uridine, total amino acids, and thymidine (Rudland et al., 1974). The time course of the increase in net incorporation into the cell was almost identical for both FGF and serum additions. Amino acid uptake increased immediately, followed by a 15-min lag time for uridine, and approximately a 3- to 5-hr lag time for thymidine. The rate of protein synthesis increased (about 300%) within 5 hr, that of mRNA (about 15–20%) within 5 hr, and DNA (60- to 100-fold) within 24 hr. These increases are almost identical for cultures activated with FGF or serum, both in the kinetics and extent of the increases. Finally, cell division occurred in both cases 25–30 hr later. The dose responses of varying concentrations of FGF for the stimulation of protein synthesis at 6 hr, ribosomal RNA synthesis at 14 hr, and the eventual induction of DNA synthesis are roughly parallel. The abilities of FGF and serum to induce polysome reformation measured 6 hr after additions are also virtually identical. The mRNA location in resting BALB/c 3T3 cells is ~28% in polysomes and 72% in the 30 S–80 S messenger ribonucleoprotein complex. Addition of either FGF or fresh serum results in greater than 85% of the mRNA being located in the polysomal region of the sucrose gradient. This increase in polysome formation correlates with an increase in protein synthesis.

Based on data indicating that specific proteins need to be made for the rate of DNA synthesis to increase in response to growth stimulation, Nilsen-Hamilton et al. (1980, 1981, 1982) have identified proteins that are specifically synthesized in growth-arrested BALB/c 3T3 in response to FGF and serum. The most dramatic specific increases in labeling with [^{35}S]methionine were of secreted rather than intracellular proteins. The secreted levels of these proteins begin to rise within 2 hr of adding FGF to the growth-arrested 3T3 cells (Nilsen-Hamilton et al., 1980). The secreted levels of one of these proteins with a molecular weight of 39,000, also called a "major excreted protein" (MEP) (Gottesman, 1978), increase about eightfold in response to FGF (Nilsen-Hamilton et al., 1981). FGF also increases the secreted levels of a set of five proteins that share the property of being "superinduced" by cycloheximide; that is, their secreted levels are increased two- to fivefold by concentrations of cycloheximide that inhibit total protein synthesis by about 85% when this agent was present during induction, but not present during labeling with [^{35}S]methionine (Nilsen-Hamilton et al., 1982). Cycloheximide acts in synergism with FGF. These cellular proteins have been named "superinducible proteins," or SIPs. They have molecular weights ranging from 12,000(SIP 12) to 62,000(SIP 62), and they may be the same proteins that increase intracellularly in response to stimulation by PDGF (Pledger et al., 1981). By contrast, when cycloheximide is present during the induction period, it prevents FGF from raising the secreted levels of MEP.

The secreted levels of MEP and the SIPs are also regulated differently in that NH$_4$Cl, which inhibits lysosomal proteolysis, increases the secreted level of MEP, but not the SIPs. Finally, the ability of FGF to raise the secreted levels of MEP and the SIPs depends on whether the BALB/c 3T3 cells are sparse and growing or confluent and quiescent. For growing BALB/c 3T3 cells, FGF is much more effective at raising the secreted level of MEP than for quiescent 3T3 cells, the opposite being true for the SIPs (Nilsen-Hamilton et al., 1982).

As with PDGF, FGF has been shown by Jimenez de Asua and Rudland to be a competence factor for BALB/c as well as for Swiss 3T3 cell lines (Jimenez de Asua et al., 1977a,b; Rudland and Jimenez de Asua, 1979; Jimenez de Asua, 1980; see Chapter 6). This was later confirmed by Stiles et al. (1979).

Among the plasma components which could be held responsible for the progression of cells committed to divide by brief exposure to FGF or

PDGF are the somatomedins or IGF (Stiles et $al.$, 1979; see Chapter 4). However, these factors are not the only plasma factors required, since when competent cells are transferred to media containing only somatomedin C but no plasma, cells do not progress through their G_0/ G_1 phase. Recent studies using normal diploid cells maintained in serum-free medium have demonstrated that high-density lipoproteins (HDL) together with transferrin could act as progression factors (Gospodarowicz and Tauber, 1980; Longenecker et $al.$, 1984; Gospodarowicz et $al.$, 1982b). Insulin-like growth factors, except in the case of granulosa and adrenal cortex cells, were not required and had only a minimal effect on other cell types. These include vascular aortic smooth muscle and endothelial cells, corneal endothelial, lens epithelial, kidney epithelial cells, and a number of tumor or established cell lines (Gospodarowicz, 1983b). The ability of HDL to support cell cycle progression is not only due to its ability to donate cholesterol (Longenecker et $al.$, 1984). HDL effects are more likely to be exerted through the reverse process (cholesterol egress). This would activate the cellular 3-hydroxy-3-methylglutaryl (HMG) CoA reductase (Cohen et $al.$, 1982a,b). This cellular enzyme is responsible for the synthesis of mevalonic acid, which is later utilized for the synthesis of dolichol, ubiquinone, isopentenyl adenyl adenine, and cholesterol (reviewed in Brown and Goldstein, 1980). Of these compounds, the nonsterol products are probably required for cell proliferation (Quesney-Huneeus et $al.$, 1980).

FGF or PDGF-modulated competence appears to require the expression of modulated gene products. FGF or PDGF-inducible gene sequences account for 0.1–0.3% of the total genes transcribed in 3T3 cells. The sequences characterized thus far correspond to low-abundance mRNAs (70–100 copies per cell) in quiescent cells; following PDGF treatment, the abundance level increases to 700–3000 mRNA copies per cell (Cochran et $al.$, 1983). This increase in mRNA content occurs within the same time frame and with the same dose–response characteristics as induction of "competence" by those growth factors. Two-dimensional gel electrophoresis of 3T3 cells labeled in $vivo$ with [^{35}S]methionine reveals rare FGF or PDGF-inducible proteins which are the presumptive translation products of these growth factor-regulated mRNAs (Cochran et $al.$, 1983). Variants of the 3T3 cell line that express these proteins constitutively can proliferate in the absence of growth factors (Pledger et $al.$, 1981).

FGF as well as PDGF have been shown recently to regulate cellular oncogenes. There is ample evidence that perturbations of cellular oncogenes contribute to specific tumorigenesis (Cooper, 1982; Weinberg,

1982). The c-myc gene, identified originally as the cellular homolog of the transforming determinant carried by avian myelocytomatosis virus, is altered in association with a broad spectrum of neoplasms. Activation of c-myc by avian leukosis virus has been implicated in the genesis of bursal lymphomas (Hayward et al., 1981), and elevations in c-myc expression occur in at least some isolates of various tumor types (Eva et al., 1982). It has been suggested that an altered c-myc gene operationally encodes an immortalization function, expression of which is contingent upon coordinate expression of an altered c-ras gene (Land et al., 1983). Immortalization may result from the ability of cells to traverse specific restriction points in the cell cycle (Pledger et al., 1978; Pardee, 1974) that allow growth in vitro. Therefore, it is reasonable to suggest that c-myc is involved in the progression of many or all cells through the cell cycle.

In agreement with those proposals, it has recently been demonstrated that c-myc is an inducible gene that is regulated by specific growth signals in a cell cycle-dependent manner (Kelly et al., 1983). Specifically, agents that initiate the first phase of a proliferative response in lymphocytes (lipopolysaccharide or Con A) and fibroblasts (PDGF and FGF) induce c-myc mRNA (Kelly et al., 1983). Within 1–3 hr after the addition of these mitogens to the appropriate cells, c-myc mRNA concentration is increased between 10- and 40-fold. This induction of c-myc mRNA concentration occurs in the presence of cycloheximide and, therefore, does not require the synthesis of new protein species (Kelly et al., 1983). Possibly other oncogenes might exhibit cell cycle-associated roles and be induced by specific growth factors. The results relating oncogenes and growth factors suggest a rational mechanism by which these genes may contribute to the growth of normal or cancer cells. Little is presently known about the precise function of oncogenes (Bishop, 1983). The next steps in understanding will therefore require a progression from molecular biology to the more difficult domain of cell biology.

IV. CONCLUSION

Our basic knowledge of the way in which growth factors induce cells to divide has been improved greatly due to the results from various laboratories which, using EGF as a probe, have investigated the ways in which cells can process the information provided by the interaction of the mitogen with cell surface receptor sites (Carpenter and Cohen, 1979; see Chapter 7). In fact, it is with EGF that the pathway of internalization and degradation of mitogen–receptor complexes was first explored and is now best understood. One should note, however, that the physiological meaning of every step is not entirely clear. The EGF-enhanced protein kinase (S. Cohen *et al.*, 1980; see Chapter 11) which is associated with the EGF receptor could be crucial to our understanding of the control of cell proliferation, if not of cell transformation as well. Indeed, the association of both virus- and EGF-induced cell growth with the synthesis or activation of a membrane protein kinase is intriguing, and more so if one considers that part of the EGF receptor can be coded for by an oncogene. Evidence from Schlessinger (1980) and his collaborators suggests that microaggregation followed by internalization of EGF–receptor complexes are important steps which convey to the cells the signal(s) to divide. It is therefore possible that synthesis of a truncated EGF receptor made of the transmembrane and cytoplasmic domain induced by AEV infection of various cell types could lead to a permanent signal for those cells to divide (Downward *et al.*, 1984). Another way growth factors could be linked to oncogenes is through their regulation of various oncogenes which have been shown to be permanently transcribed in human tumor cells. Such examples are provided by the ability of both FGF and PDGF to activate the c-myc gene (Kelly *et al.*, 1983). What role those oncogenes could play in normal cell proliferation as well as differentiation is actually unclear (Bishop, 1983), but will likely be the object of intense investigation.

Also important for our understanding of growth control has been the finding that growth factors could act at various points in the cell cycle (see Chapter 13). This has led to the development of the competence versus progression theory. While cells required only brief exposure to growth factors such as FGF or PDGF in order to be sensitive to divide, they will not do so unless progression factors such as HDL or insulin-like growth factors are present. This observation helps to explain why multiple factors are required in order to promote cell proliferation, and why some growth factors such as IGF or somatomedin C could be inactive in some systems while being active in others.

Where research is progressing the slowest is in the *in vivo* effect of both EGF and FGF. This is not due, however, to a lack of *in vivo* effect of both growth factors. It rather reflects the extremely limited number of physiologists being actively engaged in that area of research. This contrasts with the huge number of molecular biologists presently working with growth factors. One can only hope that this situation would change, since a growth factor active *in vitro*, but without an established role *in vivo*, would be of extremely limited value. In that context, one cannot help to think about the events which followed the discovery of

NGF. Within 2 years of its identification, its purification was done, its role *in vivo* was established, and use of NGF antibodies further helped to demonstrate its crucial role in the establishment of the peripheral nervous system. Only later did its structure, nature of its receptor, and mechanism of action start to be investigated. In contrast with EGF, we know its structure, the nature of its receptor, and its relationship with oncogenes, yet 20 years after its discovery, its role *in vivo* has not been established. Nor has it been established for any growth factors other than NGF (for another view, see Leffert and Koch, 1982). The *in vivo* effect of growth factors is therefore an area of research which should strongly be encouraged to develop.

REFERENCES

Armelin, H. A. (1973). Pituitary extracts and steroid hormones in the control of 3T3 cell growth. *Proc. Natl. Acad. Sci. U.S.A.* **70,** 2702–2706.

Assoian, R. K., Frolik, C. A., Roberts, A. B., Miller, D. M., and Sporn, M. B. (1984). Transforming growth factors B control receptor levels for epidermal growth factors in NRK fibroblasts. *Cell (Cambridge, Mass.)* **36,** 35–41.

Bishop, J. M. (1983). Cellular oncogenes and retroviruses. *Annu. Rev. Biochem.* **52,** 301–354.

Bowen-Pope, D., Dicorleto, P., and Ross, R. (1983). Interactions between receptors for platelet-derived growth factor and epidermal growth factor. *J. Cell. Physiol.* **96,** 679–683.

Broad, T. E., and Ham, R. G. (1983). Growth and adipose differentiation of sheep pre-adipocyte fibroblasts in serum-free medium. *Eur. J. Biochem.* **135,** 33–39.

Brown, M., and Goldstein, J. (1980). Multivalent feedback regulation of HMG CoA reductase, a control mechanism coordinating isoprenoid synthesis on cell growth. *J. Lipid Res.* **21,** 505–517.

Bucher, N. L. R. (1978). Hormonal factors and liver growth. *Adv. Enzyme Regul.* **16,** 205–215.

Buhrow, S. A., Cohen, S., and Staros, J. V. (1982). Affinity labeling of the protein kinase associated with the epidermal growth factor receptor in membrane vesicles from A-431 cells. *J. Biol. Chem.* **257,** 4019–4022.

Buntrock, P., Jentzsch, K. D., and Heder, G. (1982a). Stimulation of wound healing using brain extract with FGF activity: I. Quantitative and biochemical studies into formation of granulation tissue. *Exp. Pathol.* **21,** 46–53.

Buntrock, P., Jentzsch, K. D., and Heder, G. (1982b). Stimulation of wound healing using brain extract with FGF activity: II. Histological and morphometric examination of cells and capillaries. *Exp. Pathol.* **21,** 62–67.

Carpenter, G., and Cohen, S. (1979). Epidermal growth factor. *Annu. Rev. Biochem.* **48,** 193–216.

Carpenter, G., Lembach, K. J., Morrison, M., and Cohen, S. (1975). Characterization of the binding of ^{125}I-labeled epidermal growth factor to human fibroblasts. *J. Biol. Chem.* **250,** 4297–4304.

Carpenter, G., King, L., and Cohen, S. (1979). Epidermal growth factor–receptor-protein kinase interactions. *J. Biol. Chem.* **255,** 4834–4842.

Catterton, W. Z., Escobedo, M. B., Sexson, W. R., Gray, M. E., Sundell, H. W., and Stahlman, M. T. (1979). The effects of epidermal growth factor on lung maturation in fetal rabbit. *Pediatr. Res.* **13**, 104–108.

Chinkers, M., McKanna, J. A., and Cohen, S. (1979). Rapid induction of morphological changes in human carcinoma cells A-431 by epidermal growth factor. *J. Cell Biol.* **83**, 260–265.

Clark, J. F., Jones, K. L., Gospodarowicz, D., and Sato, G. H. (1972). Hormone dependent growth response of a newly established ovarian cell line. *Nature (London), New Biol.* **236**, 180–182.

Cochran, B. H., Reffel, A. C., and Stiles, C. D. (1983). Molecular cloning of gene sequences regulated by platelet-derived growth factor. *Cell (Cambridge, Mass.)* **33**, 939–947.

Cohen, D. C., Massoglia, S., and Gospodarowicz, D. (1982a). Correlation between two effects of high density lipoproteins on vascular endothelial cells: The induction of 3-hydroxy-3-methylglutaryl coenzyme A reductase activity and the support of cellular proliferation. *J. Biol. Chem.* **257**, 9429–9437.

Cohen, D. C., Massoglia, S., and Gospodarowicz, D. (1982b). Feedback regulation of 3-hydroxy-3-methylglutaryl coenzyme A reductase in vascular endothelial cells: Separate sterol and non-sterol components. *J. Biol. Chem.* **257**, 11106–11112.

Cohen, S. (1959). Purification and metabolic effects of nerve growth-promoting protein from snake venom. *J. Biol. Chem.* **234**, 1129–1137.

Cohen, S. (1962). Isolation of a mouse submaxillary gland protein accelerating incisor eruption and eyelid opening in the newborn animal. *J. Biol. Chem.* **237**, 1562–1568.

Cohen, S. (1965). The stimulation of epidermal proliferation by a specific protein (EGF). *Dev. Biol.* **12**, 394–407.

Cohen, S., and Elliott, G. A. (1963). The stimulation of epidermal keratinization by a protein isolated from the submaxillary gland of the mouse. *J. Invest. Dermatol.* **40**, 1–5.

Cohen, S., and Taylor, J. M. (1974). Epidermal growth factor: Chemical and biological characterization. *Recent Prog. Horm. Res.* **30**, 533–550.

Cohen, S., Carpenter, G., and King, L. (1980). Epidermal growth factor–receptor-protein kinase interactions. *J. Biol. Chem.* **255**, 4834–4842.

Cooper, G. M. (1982). Cellular transforming genes. *Science* **217**, 801–806.

Corvol, M. T., Malemud, C. J., and Sokoloff, L. (1972). A pituitary growth promoting factor for articular chondrocytes in monolayer culture. *Endocrinology* **90**, 262–271.

Das, M., and Fox, C. R. (1978). Molecular mechanism of mitogen action. Processing of receptor induced by epidermal growth factor. *Proc. Natl. Acad. Sci. U.S.A.* **75**, 2644–2648.

Doolittle, R. F., Hunkapiller, M. W., Hood, E., Devare, S. G., Robbins, K. C., Aaronson, S. A., and Antoniades, H. N. (1983). Simian sarcoma virus onc C gene v-sis is derived from the genes encoding a platelet-derived growth factor. *Science* **221**, 275–278.

Downward, J., Yarden, Y., Mayes, E., Scrace, G., Totly, N., Stockwell, P., Ullrich, A., Schlessinger, J., and Waterfield, M. D. (1984). Close similarity of epidermal growth factor receptor and v-erb-B oncogenes protein sequences. *Nature (London)* **307**, 521–527.

Duthu, G. S., and J. R. Smith. 1980. In vitro proliferation and life-span of bovine aorta endothelial cells: Effect of culture conditions and fibroblast growth factor. *J. Cell. Physiol.* **103**, 385–392.

Eva, A., Robbins, K. C., Andersen, P. R., Srinivasan, A., Tronick, S. R., Reddy, E. P., Nelson, W. E., Galen, A. T., Lautenberger, J. A., Papas, T. S., Westin, E. H., Wong-Staal, F., Gallo, R. C., and Aaronson, S. A. (1982). Cellular genes analogous to retroviral oncogenes are transcribed in human tumor cells. *Nature (London)* **295**, 116–119.

Frey, P., Forand, R., Maciag, T., and Shooter, E. M. (1979). The biosynthetic precursor of epidermal growth factor and the mechanism of its processing. *Proc. Natl. Acad. Sci. U.S.A.* **76**, 6294–6298.

Giguere, L., Cheng, J., and Gospodarowicz, D. (1982). Factors involved in the control of proliferation of bovine corneal endothelial cells maintained in serum-free medium. *J. Cell. Physiol.* **110**, 72–80.

Gonzalez, A., Garrido, J., and Vial, J. D. (1981). Epidermal growth factor inhibits cytoskeleton related changes in the surface of parietal cells. *J. Cell Biol.* **88**, 108–118.

Gorden, P., Carpentier, J. L., Cohen, S., and Orci, L. (1978). Epidermal growth factor: Morphological demonstration of binding, internalization, and lysosomal association in human fibroblasts. *Proc. Natl. Acad. Sci. U.S.A.* **75**, 5025–5029.

Gospodarowicz, D. (1974). Localization of a fibroblast growth factor and its effect alone and with hydrocortisone on 3T3 cell growth. *Nature (London)* **249**, 123–127.

Gospodarowicz, D. (1975). Purification of a fibroblast growth factor from bovine pituitary. *J. Biol. Chem.* **250**, 2515–2519.

Gospodarowicz, D. (1979). Fibroblast and epidermal growth factors. Their uses *in vivo* and *in vitro* in studies on cell functions and cell transplantation. *Mol. Cell. Biochem.* **25**, 79–110.

Gospodarowicz, D. (1983a). Control of endothelial cell proliferation and repair. *In* "Biochemical Interaction at the Endothelium" (A. Cryer, ed.), pp. 366–396. Elsevier, North-Holland, New York.

Gospodarowicz, D. (1983b). The control of mammalian cell proliferation by growth factors, basement lamina, and lipoproteins. *J. Invest. Dermatol.* **81**, 405–505.

Gospodarowicz, D. (1984a). Purification of pituitary and brain fibroblast growth factors and their use in cell culture. *In* "Methods in Molecular and Cell Biology" (D. Barnes, D. Sirbasku, and G. Sato, eds.), Vol. II., pp. 167–198. A. R. Liss, Inc., New York.

Gospodarowicz, D. (1984b). Fibroblast growth factor. *In* "Hormonal Proteins and Peptides" (C. H. Li, ed.), Vol. 12., pp. 205–230. Academic Press, New York.

Gospodarowicz, D., and Bialecki, H. (1978). The effects of the epidermal and fibroblast growth factor on the replicative lifespan of bovine granulosa cells in culture. *Endocrinology* **103**, 854–858.

Gospodarowicz, D., and Greenburg, G. (1981). Growth control of mammalian cells. Growth factors and extracellular matrix. *In* "The Biology of Normal Human Growth" (M. Ritzen, A. Aperia, K. Hall, A. Larsson, A. Zetterberg, and R. Zetterström, eds.), pp. 1–21. Raven Press, New York.

Gospodarowicz, D., and Lui, G.-M. (1981). Effect of substrata and fibroblast growth factor on the proliferation *in vitro* of bovine aortic endothelial cells. *J. Cell. Physiol.* **109**, 69–81.

Gospodarowicz, D., and Mescher, A. L. (1981). Fibroblast growth factor and vertebrate regeneration. *Adv. Neurol.* **29**, 149–171.

Gospodarowicz, D., and Moran, J. S. (1974). Effect of a fibroblast growth factor, insulin, dexamethasone and serum on the morphology of BALB/c 3T3 cells. *Proc. Natl. Acad. Sci. U.S.A.* **71**, 4648–4652.

Gospodarowicz, D., and Moran, J. S. (1975). Mitogenic effect of fibroblast factor on early passage cultures of human and murine fibroblasts. *J. Cell Biol.* **66**, 451–556.

Gospodarowicz, D., and Tauber, J.-P. (1980). Growth factors and extracellular matrix. *Endocr. Rev.* **1**, 201–227.

Gospodarowicz, D., and Zetter, B. (1977). The use of fibroblast and epidermal growth factors to lower the serum requirement for growth of normal diploid cells in early passage: A new method for cloning. *Dev. Biol. Stand.* **37,** 109–130.

Gospodarowicz, D., Rudland, P., Lindstrom, J., and Benirschke, K. (1975). Fibroblast growth factor: Localization, purification, mode of action, and physiological significance. Nobel Symposium on Growth Factors. *Adv. Metab. Disorders* **8,** 301–335.

Gospodarowicz, D., Moran, J. S., Braun, D., and Birdwell, C. R. (1976).Clonal growth of bovine endothelial cells in tissue culture: Fibroblast growth factor as a survival agent. *Proc. Natl. Acad. Sci. U.S.A.* **73,** 4120–4124.

Gospodarowicz, D., Mescher, A. L., Brown, K., and Birdwell, C. R. (1977a). The role of fibroblastic growth factor and epidermal growth factor in the proliferative response of the corneal and lens epithelium. *Exp. Eye Res.* **25,** 631–649.

Gospodarowicz, D., Moran, J. S., and Braun, D. (1977b). Control of proliferation of bovine vascular endothelial cells. *J. Cell. Physiol.* **91,** 377–388.

Gospodarowicz, D., Brown, K. S., Birdwell, C. R., and Zetter, B. (1978a). Control of proliferation of human vascular endothelial cells of human origin: I. Characterization of the response of human umbilical vein endothelial cells to fibroblast growth factor, epidermal growth factor, and thrombin. *J. Cell Biol.* **77,** 774–788.

Gospodarowicz, D., Greenburg, G., Bialecki, H., and Zetter, B. (1978b). factors involved in the modulation of cell proliferation *in vivo* and *in vitro:* The role of fibroblast and epidermal growth factors in the proliferative response of mammalian cells. *In Vitro* **14,** 85–118.

Gospodarowicz, D., Mescher, A. L., and Birdwell, C. R. (1978c). Control of cellular proliferation by the fibroblast and epidermal growth factors. *Nat. Cancer Inst. Monogr.* **48,** 109–130.

Gospodarowicz, D., Mescher, A. L., and Moran, J. S. (1978d). Cellular specificities of fibroblast growth factor and epidermal growth factor. *Symp. Soc. Dev. Biol.* **35,** 33–61.

Gospodarowicz, D., Greenburg, G., and Birdwell, C. R. (1978e). Determination of cellular shape by the extracellular matrix and its correlation with the control of cellular growth. *Cancer Res.* **38,** 4155–4171.

Gospodarowicz, D., Bialecki, H., and Greenburg, G. (1978f). Purification of the fibroblast growth factor activity from bovine brain. *J. Biol. Chem.* **253,** 3736–3743.

Gospodarowicz, D., Vlodavsky, I., and Savion, N. (1980). The extracellular matrix and the control of proliferation of vascular endothelial and vascular smooth muscle cells. *J. Supramol. Struct.* **13,** 339–372.

Gospodarowicz, D., Vlodavsky, I., and Savion, N. (1981a). The extracellular matrix and the control of proliferation of vascular endothelial and vascular smooth muscle cells. *Prog. Clin. Biol. Res.* **66A,** 53–86.

Gospodarowicz, D., Vlodavsky, I., and Savion, N. (1981b). The role of fibroblast growth factor and the extracellular matrix in the control of proliferation and differentiation of corneal endothelial cells. *Vision Res.* **21,** 87–103.

Gospodarowicz, D., Lui, G.-M., and Cheng, J. C. (1982a). Purification in high yield of brain fibroblast growth factor by preparative isoelectric focusing at pH 9.6. *J. Biol. Chem.* **257,** 12,266–12,276.

Gospodarowicz, D., Vlodavsky, I., Greenburg, G., Alvarado, J., Johnson, L. K., and Moran, J. (1979a). Cellular shape is determined by the extracellular matrix and is responsible for the control of cellular growth and function. *Cold Spring Harbor Conf. Cell Proliferation* **9,** 561–592.

Gospodarowicz, D., Vlodavsky, I., Greenburg, G., Alvarado, J., Johnson, L. K., and Moran,

J. (1979b). Studies on atherogenesis and corneal transplantation using cultured vascular and corneal endothelia. *Recent Prog. Horm. Res.* **35**, 375–448.

Gospodarowicz, D., Bialecki, H., and Thakral, T. K. (1979c). The angiogenic activity of the fibroblast and epidermal growth factor. *Exp. Eye Res.* **28**, 501–514.

Gospodarowicz, D., Cohen, D. C., and Fujii, D. K. (1982b). Regulation of cell growth by the basal lamina and plasma factors: Relevance to embryonic control of cell proliferation and differentiation. *Cold Spring Harbor Conf. Cell Proliferation* **9**, 95–124.

Gospodarowicz, D., Cheng, J., and Lirette, M. (1983). Bovine brain and pituitary fibroblast growth factors: Comparison of their abilities to support the proliferation of human and bovine vascular endothelial cells. *J. Cell Biol.* **97**, 1677–1685.

Gospodarowicz, D., Cheng, J., Lui, G.-M., Baird, A., and Bohlen, P. (1984). Isolation by reparin-sepharose affinity chromatography of brain fibroblast growth factor: Identity with pituitary fibroblast growth factor. *Proc. Natl. Acad. Sci. U.S.A.* **81**, 6963–6967.

Gospodarowicz, D., Massoglia, S., Cheng, J., Lui, G.-M., and Bohlen, P. (1985). Isolation of bovine pituitary fibroblast growth factor purified by fast protein liquid chromatography (FPLC): Partial chemical and biological characterization. *J. Cell Physiol.* **122**, 323–332.

Gottesman, M. M. (1978). Transformation dependent secretion of low molecular weight protein by murine fibroblasts. *Proc. Natl. Acad. Sci. U.S.A.* **75**, 2767–2771.

Gray, A., Dull, T. J., and Ullrich, A. (1983). Nucleotide sequence of epidermal growth factor cDNA predicts a 128,000-molecular weight protein precursor. *Nature (London)* **303**, 722–725.

Green, H., Kehinde, O., and Thomas, J. (1979). Growth of cultured human epidermal cells into multiple epithelia suitable for grafting. *Proc. Natl. Acad. Sci. U.S.A.* **76**, 5665–5668.

Greenburg, G., Vlodavsky, I., Foidart, J.-M., and Gospodarowicz, D. (1980). Conditioned medium from endothelial cell cultures can restore the normal phenotypic expression of vascular endothelium maintained *in vitro* in the absence of fibroblast growth factor. *J. Cell. Physiol.* **103**, 333–347.

Gregory, H. (1975). Isolation and structure of urogastrone and its relationship to epidermal growth factor. *Nature (London)* **257**, 325–327.

Haigler, H. T. (1982). Epidermal growth factor: Cellular binding and consequences. *In* "Growth and Maturation Factors" (G. Guroff, ed.), Vol. I, pp. 119–147. Wiley, New York.

Haigler, H. T., McKanna, J. A., and Cohen, S. (1979a). Direct visualization of the binding and internalization of epidermal growth factor in human carcinoma cells A-431. *J. Cell Biol.* **81**, 382–395.

Haigler, H. T., McKanna, J. A., and Cohen, S. (1979b). Rapid stimulation of pinocytosis in human carcinoma A431 by epidermal growth factor. *J. Cell Biol.* **83**, 82–90.

Hassell, J. R. (1975). The development of the rat palatal shelves *in vitro*: A structural analysis of the inhibition of epithelial cell death and palatal fusion by EGF. *Dev. Biol.* **45**, 90–102.

Hassell, J. R., and Pratt, R. M. (1977). Elevated level of cAMP altered the effect of epidermal growth factor *in vitro* on programmed cell death in the secondary palate epithelium. *Exp. Cell Res.* **106**, 155–172.

Hayward, W. S., Neel, B. G., and Astrin, S. M. (1981). Activation of cellular oncogenes by promoter insertion in ALV-induced lymphoid leudosis. *Nature (London)* **296**, 475–479.

Heldin, C.-H., Wasteson, A., and Westermark, B. (1982). Interaction of platelet-derived growth factor with its fibroblast receptor: Demonstration of ligand degradation and receptor modulation. *J. Biol. Chem.* **257**, 4216–4221.

Hollenberg, M. D., and Gregory, H. (1976). Human urogastrone and mouse epidermal growth factor share a common receptor site in cultured human fibroblasts. *Life Sci.* **20**, 267–274.

Holley, R. W., and Kiernan, J. A. (1968). Contact inhibition of cell division in 3T3 cells. *Proc. Natl. Acad. Sci. U.S.A.* **70**, 300–304.

Jentzsch, K. D., Wellmitz, G., Heder, G., Petzold, E., Buntrock, P., and Gehme, P. (1980). A bovine brain fraction with fibroblast growth factor activity inducing articular cartilage regeneration *in vivo*. *Acta Biol. Med. Ger.* **39**, 967–971.

Jimenez de Asua, L. (1980). An order sequence of temporal steps regulate the rate of initiation of DNA synthesis in cultured mouse cells. *In* "Control Mechanisms in Animal Cells" (L. Jimenez de Asua, R. Levi-Montalcini, R. Sheilds, and S. Iacobelli, eds.), pp. 173–197. Raven Press, New York.

Jimenez de Asua, L., O'Farrell, M. K., Clinigan, D., and Rudlannd, P. S. (1977a). Temporal sequence of hormonal interactions during the prereplicative phase of quiescent cultured 3T3 fibroblasts. *Proc. Natl. Acad. Sci. U.S.A.* **74**, 3845–3849.

Jimenez de Asua, L., O'Farrell, M. K., Bennett, D., Clinigan, D., and Rudland, P. S. (1977b). Interaction of two hormones and their effect on observed rate of initiation of DNA synthesis in 3T3 cells. *Nature (London)* **265**, 151–153.

Johnson, L. K., Vlodavsky, I., Baxter, J. D., and Gospodarowicz, D. (1980). Epidermal growth factor and expression of specific genes. Effects on cultured rat pituitary cells are dissociable from the mitogenic response. *Nature (London)* **287**, 340–343.

Kelly, K., Cochran, B. H., Stiles, C. D., and Leder, P. (1983). Cell specific regulation of the c-*myc* by lymphocyte mitogens and platelet-derived growth factor. *Cell (Cambridge, Mass.)* **55**, 603–610.

King, A. C., Hernaez-Davis, L., and Cuatrecasas, P. (1981). Lysosomotropic amines inhibit mitogenesis induced by growth factors. *Proc. Natl. Acad. Sci. U.S.A.* **78**, 717–721.

King, L. E., Jr., Carpenter, G., and Cohen, S. (1980). Characterization by electrophoresis of epidermal growth factor stimulated phosphorylation using A-431 membranes. *Biochemistry* **19**, 1524–1528.

Koch, K. S., and Leffert, H. L. (1979). Increased sodium ion influx is necessary to initiate rat hepatocyte proliferation. *Cell (Cambridge, Mass.)* **18**, 153–163.

Land, H., Parada, L. F., and Weinberg, R. A. (1983). Tumorigenic conversion of primary embryo fibroblasts requires at least two cooperating oncogenes. *Nature (London)* **304**, 596–602.

Leffert, H. L., and Koch, K. S. (1982). Hepatocyte growth regulation by hormones in chemically defined medium: A two-signal hypothesis. *Cold Spring Harbor Symp. Cell Proliferation* **9**, 597.

Linsley, P. S., Blifeld, C., Wrann, M., and Fox, C. F. (1979). Direct linkage of epidermal growth factor to its receptor. *Nature (London)* **278**, 745–768.

Longenecker, J. P., Kilty, L. A., and Johnson, L. K. (1984). Glucocorticoid inhibition of vascular smooth muscle proliferation: Influence of homologous extracellular matrix and serum proteins. *J. Cell Biol.* **98**, 534–540.

McGowan, J. A., Strain, A. J., and Bucher, N. L. R. (1981). DNA synthesis in primary cultures of adult rat hepatocytes in a defined medium: Effect of epidermal growth factor, insulin, glucagon, and cyclic AMP. *J. Cell. Physiol.* **108**, 353–364.

Marquardt, H., Hunkapiller, M. W., Hood, L. E., and Todaro, G. J. (1984). Rat transforming growth factor type I: Structure and relationship to epidermal growth factor. *Science* **223**, 1079–1082.

Maxfield, F. R., Davies, P. J. A., Klempner, L., Willingham, M. C., and Pastan, I. (1979). Epidermal growth factor stimulation of DNA synthesis is potentiated by compounds that inhibit its clustering in coated pits. *Proc. Natl. Acad. Sci. U.S.A.* **76**, 5731–5735.

Mescher, A. L. (1983). Growth factors from nerves and their roles during limb regeneration. In "Limb Development and Regeneration," Part A, pp. 501–512. Alan R. Liss, Inc., New York.

Mescher, A. L., and Gospodarowicz, D. (1979). Mitogenic effect of a growth factor derived from myelin on denervated regenerates of newt forelimbs. *J. Exp. Zool.* **207**, 497–503.

Nexo, E., Hollenberg, M. D., Figueroa A., and Pratt, R. M. (1980). Detection of epidermal growth factor: Urogastrone and its receptor during fetal mouse development. *Proc. Natl. Acad. Sci. U.S.A.* **77**, 2782–2785.

Nicholson, G. L. (1976). Transmembrane control of the receptors on normal and tumor cells: II. Surface changes associated with transformation and malignancy. *Biochim. Biophys. Acta* **458**, 1–72.

Nilsen-Hamilton, M., Shapiro, J. M., Massoglia, S. L., and Hamilton, R. T. (1980). Selective stimulation by mitogens of incorporation of ^{35}S-methionine into a family of proteins released into the medium by 3T3 cells. *Cell (Cambridge, Mass.)* **20**, 19–28.

Nilsen-Hamilton, M., Hamilton, R. T., Allen, W. R., and Massoglia, S. L. (1981). Stimulation of the release of two glycoproteins from mouse 3T3 cells by growth factors and by agents that increase intralysosomal pH. *Biochem. Biophys. Res. Commun.* **101**, 411–417.

Nilsen-Hamilton, M., Hamilton, R. T., and Adams, G. A. (1982). *Biochem. Biophys. Res. Commun* **101**, 158–166.

Pardee, A. B. (1974). A restriction point for control of normal animal cell proliferation. *Proc. Natl. Acad. Sci. U.S.A.* **71**, 1286–1290.

Pledger, W. J., Stiles, C. D., Antoniades, H. N., and Scher, C. D. (1978). An ordered sequence of events is required before BALB/c-3T3 cells become committed to DNA synthesis. *Proc. Natl. Acad. Sci. U.S.A.* **75**, 2839–2843.

Pledger, W. J., Hart, D. A., Locatell, K. L., and Scher, C. D. (1981). Platelet-derived growth factor modulated protein constitutive synthesis by a transformed cell line. *Proc. Natl. Acad. Sci. U.S.A.* **78**, 4358–4361.

Pratt, R. M. (1980). Involvement of hormones and growth factors in the development of secondary palate. In "Development in Mammals" (M. Johnson, ed.), pp. 203–231. Elsevier/North-Holland, New York.

Pratt, R. M., Figueroa, A. A., Nexo, E., and Hollenberg, M. D. (1978). Involvement of EGF during secondary palatal development. *J. Cell Biol.* **79**, 24a.

Pratt, R. M., Figueroa, A. A., Greene, R. M., and Salomon, D. S. (1979). Involvement of glucocorticoids and epidermal growth factor in secondary palate development. In "Abnormal Embryogenesis: Cellular and Molecular Aspects" (T. V. N. Persaud, ed.), pp. 161–176. MTP Press, Ltd., Lancaster.

Quesney Huneeus, V., Wiley, M., and Siperstein, M. (1980). Isopentenyladenine as a mediator of mevalonate-regulated DNA replication. *Proc. Natl. Acad. Sci. U.S.A.* **77**, 5842–5846.

Raisz, L. G., Simmons, H. A., Sandberg, A. L., and Canalis, E. (1980). Direct stimulation of bvc resorption by epidermal growth factor. *Endocrinology* **107**, 270–293.

Rheinwald, J. G., and Green, H. (1977). Epidermal growth factor and the multiplication of cultured human epidermal keratinocytes. *Nature (London)* **265**, 421–424.

Richman, R. A., Claus, T. H., Pilkis, S. J., and Friedman, D. L. (1976). Hormonal stimulation of DNA synthesis in primary cultures of adult rat hepatocytes. *Proc. Natl. Acad. Sci. U.S.A.* **60,** 1620–1624.

Roberts, A. B., Anzano, M. A., Lamb, L. C., Smith, J. M., and Sporn, M. B. (1981). New class of transforming growth factors potentiated by epidermal growth factor: Isolation from neoplastic tissues. *Proc. Natl. Acad. Sci. U.S.A.* **78,** 5339–5343.

Roberts, A. B., Anzano, M. A., Lamb, L. C., Smith, J. M., Frolik, C. A., Marquardt, H., Todaro, G. J., and Sporn, M. B. (1982). Isolation from murine sarcoma cells of novel transforming growth factors potentiated by EGF. *Nature (London)* **295,** 417–419.

Roger, P. P., and Dumont, J. E. (1982). Epidermal growth factor controls the proliferation and the expression of differentiation in canine thyroid cells in primary culture. *FEBS Lett.* **144,** 209–212.

Rozengurt, E., and Heppel, L. A. (1975). Serum rapidly stimulates ouabain-sensitive ^{86}Rb$^+$ influx in quiescent 3T3 cells. *Proc. Natl. Acad. Sci. U.S.A.* **72,** 4492–4495.

Rozengurt, E., Brown, K., and Pettican, P. (1981). Vasopressin inhibition of epidermal growth factor binding to cultured mouse cells. *J. Biol. Chem.* **256,** 716–722.

Rozengurt, E., Collins, M., Brown, K., and Pettican, P. (1982). Inhibition of epidermal growth factor binding to mouse cultured cells by fibroblast-derived growth factor. *J. Biol. Chem.* **257,** 3680–3686.

Rudland, P. S., and Jimenez de Asua, L. (1979). Action of growth factors in the cell cycle. *Biochim. Biophys. Acta* **560,** 91–133.

Rudland, P. S., Seifert, W., and Gospodarowicz, D. (1974). Growth control in cultured mouse fibroblasts: Induction of the pleiotypic and mitogenic responses by a purified growth factor. *Proc. Natl. Acad. Sci. U.S.A.* **71,** 2600–2604.

Savion, N., Vlodavsky, I., and Gospodarowicz, D. (1980). Role of the degradation process in the mitogenic effect of epidermal growth factor. *Proc. Natl. Acad. Sci. U.S.A.* **77,** 1466–1470.

Savion, N., Vlodavsky, I., and Gospodarowicz, D. (1981). Nuclear accumulation of epidermal growth factor in cultured bovine corneal endothelial and granulosa cells. *J. Biol. Chem.* **256,** 1149–1154.

Schecter, Y., Hernaez, L., Schlessinger, J., and Cuatrecasas, P. (1979). Epidermal growth factor: Biological activity requires persistent occupation of high-affinity cell surface receptors. *Nature (London)* **278,** 835–838.

Schlessinger, J. (1980). The mechanism and role of hormone induced clustering of membrane receptor. *Trends Biol. Sci.* **5,** 210–214.

Schlessinger, J., and Geiger, B. (1981). Epidermal growth factor induces redistribution of actin and actinin in human epidermal carcinoma cells. *Exp. Cell Res.* **134,** 273–279.

Schlessinger, J., Schechter, Y., Willingham, M. C., and Pastan, I. (1978). Direct visualization of binding, aggregation, and internalization of insulin and epidermal growth factor on living fibroblastic cells. *Proc. Natl. Acad. Sci. U.S.A.* **75,** 2659–2663.

Schreiber, A. B., Yarden, Y., and Schlessinger, J. (1981). A non-mitogenic analogue of epidermal growth factor enhances the phosphorylation of endogenous membrane protein. *Biochem. Biophys. Res. Commun.* **101,** 517–523.

Schreiber, A. B., Libermann, T. A., Lax, I., Yarden, Y., and Schlessinger, J. (1983). Biological role of epidermal growth factor receptor clustering investigation with monoclonal antibodies. *J. Biol. Chem.* **258,** 846–853.

Scott, J., Urdea, M., Quiroga, M., Sanchez-Pescador, R., Fong, N., Selby, M., Rutter, W. J., and Bell, G. (1983). Structure of a mouse submaxillary messenger RNA encoding epidermal growth factor and seven related proteins. *Science* **221,** 236–240.

Serrero, G., and Ghoo, J. C. (1982). An *in vitro* model to study adipose differentiation in serum-free medium. *Anal. Biochem.* **120,** 351–359.

Simonian, M. H., and Gill, G. N. (1979). Regulation of deoxyribonucleic acid synthesis in bovine adrenocortical cells in culture. *Endocrinology* **104**, 588–595.

Simonian, M. H., Hornsby, P. J., Ill. C. R., O'Hare, M. J., and Gill, G. (1979). Characterization of cultured bovine adrenal cortical cells and derived clonal lines: Regulation of steroidogenesis and culture life span. *Endocrinology* **105**, 99–108.

Stiles, C. D., Capne, G. T., Scher, C. D., Antoniades, H. M., Van Wyk, J. J., and Pledger, W. J. (1979). Dual control of cell growth by somatomedins and "competence factors." *Proc. Natl. Acad. Sci. U.S.A.* **76**, 1279–1283.

Sun, T. T., and Green, H. (1977). Cultured epithelial cells of cornea, conjunctiva and skin: Absence of marked intrinsic divergence of their differentiated states. *Nature (London)* **269**, 489–492.

Sundell, H. W., Serenius, R. S., Barthe, P., Friedman, Z., Kanarek, K. S., Escobedo, M. B., Orth, D. N., and Stahlman, M. T. (1975). The effect of EGF on fetal lamb lung maturation. *Pediatr. Res.* **9**, 371.

Sundell, H. W., Gray, M. E., Serenius, R. S., Escobedo, M. B., and Stahlman, M. T. (1980). Maturation in fetal lamb. *Am. J. Pathol.* **100**, 707–726.

Thorburn, G. D., Waters, M. J., Young, I. R., Dolling, M., Buntine, D., and Hopkins, P. S. (1981). Epidermal growth factor: A critical factor in fetal maturation. *Ciba Found. Symp.* **86**, 172–198.

Tseng, S., Savion, N., Stern, R., and Gospodarowicz, D. (1982). Fibroblast growth factor modulates synthesis of collagen in cultured vascular endothelial cells. *Eur. J. Biochem.* **122**, 355–360.

Tyler, M. S., and Pratt, R. M. (1980). Effect of EGF on secondary palatal epithelium *in vitro*: Tissue isolation and recombination studies. *J. Embryol. Exp. Morphol.* **58**, 93–106.

Ushiro, H., and Cohen, S. (1980). Identification of phosphotyrosine as a product of epidermal growth factor-activated protein kinase in A-431 cell membrane. *J. Biol. Chem.* **255**, 8363–8365.

Vlodavsky, I., and Gospodarowicz, D. (1979). Structural and functional alterations in the surface of vascular endothelial cells associated with the formation of a confluent cell monolayer and with the withdrawal of fibroblast growth factor. *J. Supramol. Struct.* **12**, 73–114.

Vlodavsky, I., and Sachs, L. (1975). Restriction of receptor mobility and the agglutination of cells by concanavalin A. *Exp. Cell Res.* **96**, 202–214.

Vlodavsky, I., Johnson, L. K., Greenburg, G., and Gospodarowicz, D. (1979). Vascular endothelial cells maintained in the absence of fibroblast growth factor undergo structural and functional alterations that are incompatible with their *in vivo* differentiated properties. *J. Cell Biol.* **83**, 468–486.

Waterfield, M. D., Scrace, G. T., Whittle, N., Stroobaut, P., Johnsson, A., Wasteson, A., Westermark, B., Heldin, C. H., Huang, J. S., and Deuel, T. (1983). Platelet derived growth factor is structurally related to the putative transforming protein p28[sis] of simian sarcoma virus. *Nature (London)* **304**, 35–39.

Weinberg, R. A. (1982). Oncogenesis of spontaneous and chemically induced tumors. *Adv. Cancer Res.* **36**, 149–163.

Westermark, B., Heldin, C. H., Ek, B., Johnson, A., Mellström, K., Nister, M., and Wasteson, A. (1982). Biochemistry and biology of platelet-derived growth factor. *In* "Growth and Maturation Factors" (G. Guroff, ed.), Vol. I, pp. 75–112. Wiley, New York.

Westermark, K., Karlsson, F. A., and Westermark, B. (1983). Epidermal growth factor modulates thyroid growth and function in culture. *Endocrinology* **112**, 1680–1686.

Wrann, M., Fox, C., and Ross, R. (1980). Modulation of epidermal growth factor receptors on 3T3 cells by platelet-derived growth factor. *Science* **210,** 1363–1365.

Yarden, Y., Schreiber, A. B., and Schlessinger, J. (1982). A non-mitogenic analogue of epidermal growth factor induces early responses indicated by epidermal growth factor. *J. Cell Biol.* **92,** 688–692.

4

Insulin-Like Growth Factors

ALAN C. MOSES AND SYDNE J. PILISTINE[1]

The Charles A. Dana Research Institute
Department of Medicine, Harvard Medical School
and
The Harvard-Thorndike Laboratory of Beth Israel Hospital
Beth Israel Hospital
Boston, Massachusetts

I. INTRODUCTION

Over the past 25 years our understanding of the insulin-like growth factors (IGFs) or somatomedins has progressed from the concept of an ill-defined growth hormone-dependent biological activity to the isolation and characterization from two species of several polypeptides with insulin-like biological activity. These growth factors have now assumed a prominent role in both clinical endocrinology and cell biology. Many early observations concerning the biological activity, receptor interaction, and clinical utility of these factors are being

[1]Present address: Fels Research Institute, Department of Biological Chemistry, Endocrinology Section, Wright State University, Dayton, Ohio 45435.

91

reinterpreted since the recognition that the spectrum of biological activity assumed to be the result of one peptide is in fact the result of two physically distinct peptides that are under different regulatory controls. This chapter will review the biochemistry, spectrum of biological activity, physiology, and measurement of the insulin-like growth factors.

II. BACKGROUND

The broad interest in insulin-like growth factors or somatomedins as anabolic hormones stems from three divergent areas of research. In the late 1950s Daughaday and his co-workers (Salmon and Daughaday, 1957) demonstrated that neither serum from hypophysectomized (hypox) rats nor growth hormone (GH) added to serum from hypox rats stimulated the incorporation of [^{35}S]sulfate into cartilage *in vitro*. Rather, they demonstrated that GH treatment of hypophysectomized rats induced a substance, termed sulfation factor, that did stimulate [^{35}S]sulfate incorporation. Sulfation factor was later renamed somatomedin in the early 1970s (Daughaday *et al.*, 1972) when it was realized that more than one polypeptide with multiple biological effects *in vitro* accounted for this biological activity.

At about the same time, Froesch and his co-workers (Oelz *et al.*, 1972) were studying the insulin-like biological activity in human plasma that could not be inhibited by anti-insulin antibodies. They used either glucose oxidation or lipogenesis in isolated rat fat pads as a bioassay for this material. They soon recognized that the low-molecular-weight nonsuppressible insulin-like activity (NSILAs) of human plasma also could stimulate sulfate incorporation into cartilage; that is, it had somatomedin-like biological activity (Zingg and Froesch, 1973).

Meanwhile, also at the same time, Temin and his co-workers were interested in the macromolecular constituents of serum that promoted the growth of chicken cells *in vitro* (Temin, 1971). They initially purified a substance from calf serum (multiplication stimulating activity, MSA) that stimulated the replication of chick embryo fibroblasts (Pierson and Temin, 1972). They later turned their attention to serum-free media conditioned by the BRL-3A cell line derived from a normal Buffalo rat liver (Dulak and Temin, 1973). It was soon realized that the MSA activity in the rat liver conditioned media had both insulin-like activity and sulfation factor activity in a rat cartilage sulfation factor assay (Smith and Temin, 1974). Hence, three separate areas of investigation resulted in the isolation of polypeptides that had a common and broad spectrum of biological actions in many different cells and tissues.

Despite the early and intense interest in NSILA and sulfation factor, it took nearly 25 years to purify sufficient homogeneous material to elucidate the structure of these factors and to begin to understand the nature of the cellular interaction, the spectrum of biological activities, and the physiologic control of the IGFs.

III. IGF STRUCTURE

A. Amino Acid Sequence and Three-Dimensional Structure

IGFs are single-chain polypeptides that are homologous to proinsulin and are predicted to have three-dimensional conformations similar to each other and to proinsulin (see Chapter 8). The amino acid sequences of the human IGFs, IGF I (Rinderknecht and Humbel, 1978a), SMC (Klapper et al., 1983), and IGF II (Rinderknecht and Humbel, 1978b), and rat MSA III-2 (Marquardt et al., 1981) have been completed (Fig. 1).

Rinderknecht and Humbel (1976, 1978a,b) were the first to isolate sufficient purified insulin-like growth factor from human plasma to

CHAIN	POLYPEPTIDE	AMINO ACID
B		1 2 3 4 5 6 7 8 9 10 11 12 13 14 15 16 17 18 19 20 21 22 23 24 25 26 27 28 29 30
	HUMAN PROINSULIN	Phe Val Asn Gln His Leu Cys Gly Ser His Leu Val Glu Ala Leu Tyr Leu Val Cys Gly Glu Arg Gly Phe Phe Tyr Thr Pro Lys Thr →
	HUMAN IGF I	Gly Pro Glu Thr Leu Cys Gly Ala Glu Leu Val Asp Ala Leu Gln Phe Val Cys Gly Asp Arg Gly Phe Tyr Phe Asn Lys Pro Thr→
	HUMAN IGF II	Ala Tyr Arg Pro Ser Glu Thr Leu Cys Gly Gly Glu Leu Val Asp Thr Leu Gln Phe Val Cys Gly Asp Arg Gly Phe Tyr Phe Ser Arg Pro Ala→
	RAT IGF II	Ala Tyr Arg Pro Ser Glu Thr Leu Cys Gly Gly Glu Leu Val Asp Thr Leu Gln Phe Val Cys Ser Asp Arg Gly Phe Tyr Phe Ser Arg Pro Ser→
C	HUMAN PROINSULIN	Arg Arg Glu Ala Glu Asp Leu Gln Val Gly Gln Val Glu Leu Gly Gly Gly Pro Gly Ala Gly Ser Leu Gln Pro Leu Ala Leu Glu Gly Ser Leu Gln Lys Arg -
	HUMAN IGF I	Gly Tyr Gly Ser Ser Ser Arg Arg Ala Pro Gln Thr →
	HUMAN IGF II	Ser - - Arg Val Ser Arg Arg Ser Arg - - →
	RAT IGF II	Gly - - Arg Ala Asn Arg Arg Ser Arg - - →
A	HUMAN PROINSULIN	Gly Ile Val Glu Gln Cys Cys Thr Ser Ile Cys Ser Leu Tyr Gln Leu Glu Asn Tyr Cys Asn→
	HUMAN IGF I	Gly Ile Val Asp Glu Cys Cys Phe Arg Ser Cys Asp Leu Arg Arg Leu Glu Met Tyr Cys Ala →
	HUMAN IGF II	Gly Ile Val Glu Glu Cys Cys Phe Arg Ser Cys Asp Leu Ala Leu Leu Glu Thr Tyr Cys Ala →
	RAT IGF II	Gly Ile Val Glu Glu Cys Cys Phe Arg Ser Cys Asp Leu Ala Leu Leu Glu Thr Tyr Cys Ala →
D	HUMAN PROINSULIN	- - - - - - - -
	HUMAN IGF I	Pro Leu Lys Pro Ala Lys Ser Ala
	HUMAN IGF II	Thr - - Pro Ala Lys Ser Glu
	RAT IGF II	Thr - - Pro Ala Lys Ser Glu

Fig. 1. Amino acid sequence homologies of IGFs and proinsulin. The amino acid sequences are aligned based on the sequence of human proinsulin. The amino acid residues of proinsulin that are underlined are conserved in the IGFs. The amino acid residues of rat IGF II that are underlined are those that *differ* from residues in human IGF II. Sequence data are from the following sources: proinsulin (Steiner et al., 1974); IGF I (Rinderknecht and Humbel, 1978a); IGF II (Rinderknecht and Humbel, 1978b); rat IGF II (MSA III-2) (Marquardt et al., 1981).

TABLE I

Characteristics of Insulin-Like Growth Factors Whose Amino Acid
Sequences Have Been Determined

Species	pI	MW	Source	MW of native receptor complex
IGF I[a]/Som C[b]	Basic	7649	Human plasma	360,000
IGF II[c]	Neutral	7471	Human plasma	240,000
MSA III-2[d] (rat IGF II)	Neutral	7484	Buffalo rat liver cell conditioned media	240,000

[a]Rinderknecht and Humbel, 1978a.
[b]Klapper et al., 1983.
[c]Rinderknecht and Humbel, 1978b.
[d]Marquardt et al., 1981.

determine their amino acid sequences. They found that what they
thought to be one biological activity at the start of the purification was
in fact the result of two homologous polypeptides with remarkable
structural similarities to proinsulin (Rinderknecht and Humbel, 1976).
They named these polypeptides insulin-like growth factor I and insulin-like growth factor II. IGF I contains 76 amino acids and has a
molecular weight (MW) of 7649 and an isoelectric point of 8.6 (Table I)
(Rinderknecht and Humbel, 1978a). IGF II contains 67 amino acids, has
an MW of 7468, and has a neutral pI (6.5) (Table I) (Rinderknecht and
Humbel, 1978b). IGF I and II exhibit 62% sequence homology with each
other and 45% homology with human proinsulin. Each retains a distinct C-peptide region analogous but not homologous to that of the
proinsulin molecule. Each contains a short extension (six or eight amino acids) at the carboxy terminus of the molecule, the so-called D
region.

A larger species of IGF II also has been identified in human spinal
fluid (Haselbacher and Humbel, 1982). "Big IGF II" has a molecular
weight of 9000 and comprises approximately half of the IGF II found in
spinal fluid and a much smaller percentage of that found in serum. This
big IGF II may represent a precursor of the smaller IGF II. Its amino acid
sequence has not been published but may provide insight into posttranslational processing of IGF II.

The relationship of IGF I and IGF II to larger proteins with insulin-
like biological activity [NSILA-p (Zapf et al., 1978a), NSILP (Poffenbarger, 1975)] is not known, although NSILA-p cannot be cleaved to
form the smaller IGFs.

Concurrently with the isolation of IGF I, somatomedin C was purified
from outdated human plasma by Van Wyk and his colleagues (Svoboda

TABLE II

IGF Equivalents[a]

Human
Somatomedin C = IGF I = basic somatomedin = ILA (p*I* 8.2)
IGF II = ILA (p*I* 6.5)
Rat
Rat somatomedin ≃ human IGF I
MSA (III-2) ≃ human IGF II

[a]The equivalence of the human IGF peptides has been determined on the basis of identical amino acid sequence or the inability to distinguish two IGFs in at least two radioligand assays, one of which is a radioimmunoassay.

et al., 1980). Recently, published data on the sequence of somatomedin C confirm that it is identical to IGF I (Klapper *et al.*, 1983). Current terminology favors the designation of IGF I/somatomedin C when referring to this polypeptide (Tables I and II).

MSA appears to be the rat homologue of IGF II. At least seven different peptides with MSA biological activity and at least one common antigenic site have been identified by polyacrylamide gel electrophoresis of serum-free media conditioned by the Buffalo rat liver 3A (BRL-3A) cell line (Moses *et al.*, 1980a,b). One of these polypeptides, probably that designated MSA III-2, has been sequenced (Marquardt *et al.*, 1981) (Fig. 1). It contains 67 amino acids and has a molecular weight of 7484 (Table I). It differs from human IGF II in only five amino acid residues, three of which are in the C-peptide region that is highly variant between IGF I, IGF II, and proinsulin (Fig. 1, Table II). Using rabbit antisera generated against MSA polypeptides of MW 8700, Acquaviva *et al.* (1982) were able to specifically precipitate an apparent precursor of MSA (pre-pro-MSA?) of 21,600 MW following *in vitro* translation of BRL-3A messenger RNA. The steps and the control of posttranslational processing of MSA species are not known, although the relative biological activity of these peptides suggests that the smallest of these peptides (MSA III-2, MW 7100) is the most potent mitogen (Moses *et al.*, 1980a).

Partial sequence analysis of "rat basic SM" (Rubin *et al.*, 1982) shows that this IGF is the rat analog of human IGF I/SM-C (Table II).

Some controversy exists concerning the relationship of somatomedin A to the other insulin-like growth factors purified from outdated human plasma. According to original descriptions (Fryklund *et al.*, 1975), it is unclear whether somatomedin A is more closely related to IGF I or IGF II. More recent data suggests that somatomedin A may be a deamidated derivative of IGF I (Enberg *et al.*, 1984), although the procedure utilized to purify this material differed from that described by Fryklund (1975). Spencer et al. (1983) have isolated somatomedin A by yet another procedure and claim that this peptide is identical to IGF II. Operationally, it seems appropriate to recognize two major classes of insulin-like growth factors or somatomedins, IGF I/somatomedin C and IGF II. Other investigators have purified insulin-like growth factors from human plasma, and despite a variety of names, these peptides can be considered either IGF I (Bhaumick et al., 1981a), IGF II (Posner et al., 1978), or a mixture of the two (Table II). Kato et al. (1981) have identified a growth factor isolated from fetal bovine cartilage that is strikingly similar to the IGFs in its range of biological activity.

Somatomedin B also has been purified from human plasma (Fryklund et al., 1975). This polypeptide now is recognized not to be a somatomedin, but rather to have the sequence of a protease inhibitor. More recently, its biological activity for human glial cells has been determined to be the result of contamination by epidermal growth factor and not an inherent property of the somatomedin B molecule (Heldin et al., 1981). Somatomedin B should no longer be considered a member of the insulin-like growth factor family.

Using the X-ray crystallographic data of the three-dimensional structure of insulin as a model, Blundell et al. (1978, 1983) have predicted the tertiary structures of IGF I and IGF II. In these models, the hydrophobic core of insulin, consisting of the A and B chains, is preserved in the IGFs, as is a modified form of the C-peptide region of proinsulin. The proposed arrangements of the C-peptide regions of the IGFs, the COOH-terminal extension of the "A" chain (the D region) in the IGFs, and different amino acid residues at the antigenic sites (of insulin), account for the inability of anti-insulin antibodies to recognize the IGFs and for the ability of IGF I and II to interact with the insulin receptor. The C-peptide region of proinsulin, unlike that of the IGFs, is flanked by two basic amino acids (arginine) which appear to act as a signal directing the cleavage of the C peptide during the conversion of proinsulin to insulin (Steiner et al., 1974). The COOH-terminal extension of the IGF I molecule (D region) appears to be responsible for much of the growth-promoting activity of IGF I as assessed by the biological activity of insulin-IGF hybrids (King et al., 1982).

The elucidation of the amino acid sequence of IGF I has resulted in the solid phase synthesis of this peptide (Li et al., 1983). Recently, the genes for both IGF I (Jansen et al., 1983) and IGF II (Bell et al., 1984; Dull et al., 1984) have been isolated from human liver cDNA libraries and have been sequenced. The DNA sequences reveal that both the IGF I and the IGF II coding regions are flanked by regions capable of encoding amino- and carboxy-terminal extensions of these peptides. Similar data has been published for the gene sequence of MSA (Dull et al., 1984; Whitfield et al., 1984). These data are compatible with direct evidence from in vitro translation of BRL 3A messenger RNA that IGF II (MSA) is synthesized as a precursor molecule (Acquaviva et al., 1982; Yang et al., 1985). The location of the IGF I gene has been assigned to chromosome 12 (Brissenden et al., 1984; Tricoli et al., 1984) and the location of the IGF II gene to chromosome 11 (Brissenden et al., 1984; Tricoli et al., 1984).

B. Immunologic Comparisons of the Different IGFs

Antisera to the various IGFs have been produced in several laboratories (Table III). As expected from their identical amino acid sequences, IGF I and somatomedin C are indistinguishable by all antisera raised

TABLE III
Reactivity of IGFs in Radioligand Assays

Radioimmunoassays	Potency in descending order	Reference
IGF I/Som C		
Rabbit anti-Som C	Som C = IGF I > Som A > IGF II > MSA	Furlanetto et al., 1977; Van Wyk et al., 1980
Mouse monoclonal anti-IGF I	IGF I > IGF II	Laubli et al., 1982
Rabbit anti-IGF I D-peptide	IGF I > IGF II = MSA	Hintz et al., 1980b
Rabbit anti-basic Sm	IGF I = Basic Sm >> IGF II >> MSA	Bala and Bhaumick, 1979
Mouse monoclonal anti-IGF I	Som C ~ IGF I >> IGF II	Baxter et al., 1982
Rabbit anti-IGF I	IGF I = Som C > Som A > IGF II	Zapf et al., 1981
IGF II		
Rabbit anti-IGF II	IGF II >> IGF I = Som C = Som A	Zapf et al., 1981
Rabbit anti-MSA	MSA III-2 > MSA II > IGF II > IGF I	Moses et al., 1980b
Rabbit anti-IGF II C-peptide	IGF II	Hintz and Liu, 1982
Radioreceptor assays		
Human placental membranes (IGF I)	IGF I > IGF II > MSA >> insulin	Van Wyk et al., 1980
Rat placental membranes (IGF II)	IGF II > MSA III-2 > IGF I	Pilistine et al., 1983
Rat liver membranes (IGF II, MSA)	IGF II > MSA > IGF I	Rechler et al., 1980
Human serum carrier protein	IGF II > MSA > IGF I	Zapf et al., 1977

against either the intact molecule (Van Wyk *et al.*, 1980) or selected fragments of the molecule (Hintz *et al.*, 1980a,b). The considerable differences in antigenic cross-reactivity of MSA and IGF II, despite a difference in only five amino acid residues, suggest that the C-peptide region is a major antigenic site in these molecules. The mouse anti-human somatomedin C antibody raised by Baxter (Baxter *et al.*, 1982) is also highly species specific. SM-A, on the other hand, may contain antigenic sites common to both IGF I and IGF II (Hall *et al.*, 1979). Insulin is recognized poorly or not at all by all IGF antisera reported to date.

C. Purification of IGFs

The major limitation to defining the structure, biological activity, and physiological significance of the IGFs has been the scarcity of homogeneous IGF peptides. With the exception of serum-free media conditioned by the BRL 3A rat liver cell line, no tissue source has been found that is richer in its content of IGF(s) than human or animal sera (D'Ercole *et al.*, 1984). Until recently, procedures for purifying IGFs from serum sources have been laborious, costly, and have yielded a very low percentage of the biological activity present in the starting material (Rinderknecht and Humbel, 1976). The availability of high-performance liquid chromatography (HPLC) (Klapper *et al.*, 1983) and the use of specific affinity reagents (IGF carrier proteins or monoclonal antibodies) recently have reduced the problems of IGF purification (Laubli *et al.*, 1982). In the rat, sera can be enriched in its content of IGF I by implanting serum donor animals with the growth hormone and prolactin secreting pituitary tumor, MStT/W15 (Chochinov *et al.*, 1977a). Since IGF I and IGF II both have similar molecular weights and compete for binding to serum carrier proteins, cell membrane receptors, and some IGF-directed antibodies, separation of these two peptides requires a procedure that takes advantage of their different isoelectric points, 8.2 for IGF I and 6.5 for IGF II (Table II).

MSA derived from BRL-3A conditioned media remains the most plentiful (1 µg/ml of serum-free media) and the most easily purified of the IGFs. The purification of MSA can be accomplished by as few as three steps (Moses *et al.*, 1980a; Marquardt *et al.*, 1981). This material does not appear to be contaminated by rat IGF I.

The recent publication of the DNA sequences coding for IGF I (Jansen *et al.*, 1983) and IGF II (Bell *et al.*, 1984; Dull *et al.*, 1984) predicts an increased availability of these polypeptides from recombinant DNA technology and gene expression in bacterial or mammalian systems. One biosynthetically modified IGF I, which has a tryptophan for methionine substitution at position 59, already has been demonstrated to be equivalent in two radioligand assays to IGF I purified from human plasma (Schalch et al., 1984).

IV. BIOLOGICAL ACTIVITY OF INSULIN-LIKE GROWTH FACTORS

Somatomedins were first recognized on the basis of their ability to stimulate [^{35}S]sulfate into the macromolecules of cartilage pieces in vitro (Salmon and Daughaday, 1957). Since that time, the spectrum of their biological activity has been expanded enormously (Table IV). There has been no convincing evidence to date that IGF biological activity in vivo is restricted to specific tissues or to specific cell types (Table V). It was soon recognized that these factors stimulated general protein synthesis and DNA synthesis in cartilage (Salmon and Duvall, 1970). Further, with the recognition of their nonsuppressible insulin-like activity, it was recognized that the IGFs had acute metabolic actions related to the cellular uptake of nutrients as well as to the synthesis of macromolecules. The range of the biological activities of the IGFs are given in Table IV. These activities include (1) the stimulation of the acute cellular uptake of cations, glucose, xylose, and amino acids, (2) the specific stimulation of a variety of enzymes including ornithine decarboxylase and tyrosine aminotransferase, and (3) the stimulation of synthesis of glycosaminoglycans, protein, and DNA. The cell types and tissues that respond to the IGFs have been expanded from cartilage (Salmon and Daughaday, 1957) to fat cells (Zapf et al., 1978b) to liver cells (Koch et al., 1982) to ovarian granulosa cells (Veldhuis and Hammond, 1979) and to skin fibroblasts (Rechler et al., 1978) (Table V). Recently, it has been recognized that the insulin-like growth factors act synergistically in their biological effects on macromolecular synthesis with a variety of other growth factors and hormones including platelet-derived growth factor (PDGF) (Stiles et al., 1979), epidermal growth factor (EGF) (Stiles et al., 1979; Kato et al., 1983), triiodothyronine (Froesch et al., 1976), glucocorticoids (Conover et al., 1983), and constituents of hypopituitary platelet-poor plasma (Wharton et al., 1981), and glucagon (Koch et al., 1982) (Table VI).

The mechanism(s) by which the IGFs elicit specific biological responses remains unclear. It is generally accepted that a prerequisite for their biological activity is their interaction with specific receptors on the surface membrane of cells in responsive target tissues. An understanding of IGF–receptor interaction is complicated by the presence of at least three types of receptors that could mediate the biological re-

TABLE IV

IGF Biological Activity *in Vitro*

Activity	Reference
I. Transport	
Ion flux	
Chick embryo fibroblasts (MSA)	Smith, 1977
Glucose uptake	
Rat adipocytes (IGF I and II)	Zapf et al., 1978b
Human skin fibroblasts (ILA)[a]	Harley et al., 1980
Amino acid transport	
Human Skin Fibroblast[c]	
ILA	Harley et al., 1980
MSA	Knight et al., 1981
Som C/IGF I	Kaplowitz et al., 1984
Rat soleus muscle (IGF I, MSA)	Yu and Czech, 1984
II. Macromolecular synthesis	
RNA	
Rat cartilage (Sm)[b]	Salmon and DuVall, 1970
Chondrosarcoma cells (MSA)	Stevens et al., 1981
Protein (generalized effect)	
Rat cartilage (Sm)	Salmon and DuVall, 1970
Protein (enzyme inductions)	
Chick embryo fibroblasts (IGF)	Haselbacher and Humbel, 1976
HTC hepatoma cells (MSA)	Heaton et al., 1980
Lipid	
Rat adipocytes	Zapf et al., 1978a
DNA (general)	
Rat calvaria (IGF I)	Canalis, 1980
Human skin fibroblasts (MSA, Som A)	Rechler et al., 1978
DNA (specific)	
Chicken oviduct (MSA)	Evans et al., 1981
III. Cell division	
Chick embryo fibroblasts (Sm)	Cohen and Nissley, 1976
F9 embryonal carcinoma cells (MSA)	Nagarajan et al., 1982
L-6 myoblasts (MSA, Som A)	Ewton and Florini, 1980
Human skin fibroblasts (IGF I/Som C)	Stiles et al., 1979
Rat hepatocytes (IGF I and II)	Koch et al., 1982
IV. Cell differentiation	
Chick embryo cells (IGF I and II)	Schmid et al., 1983

[a]ILA, Insulin-like activity.
[b]Sm, Somatomedin activity.

TABLE V

Some Tissues and Cell Lines That Respond to IGF Biological Activity

Tissue or cell line	Reference
Rat adipocytes	Zapf et al., 1978b
Human fibroblasts	Harley et al., 1980; Clemmons and Shaw, 1983
Endothelial cells	King et al., 1983
Chondrosarcoma cells	Stevens et al., 1981
GH₃ pituitary cell line	Hayashi and Sato, 1976
Fetal rat brain	Lenoir and Honegger 1983
Lymphoblasts	Schimpff et al., 1983
F9 embryonal carcinoma cells	Nagarajan et al., 1982
Sertoli cells	Borland et al., 1984
Granulosa cells	Veldhuis and Hammond, 1979
Pancreatic islet cells	Rabinovitch et al., 1982
Fetal mouse mesenchymal cultures	Kaplowitz et al., 1982
Rat calvaria	Canalis, 1980
HeLa cells	Salmon and Hosse, 1971
Lens (frog and rabbit)	Rothstein et al., 1980; Reddan and Wilson-Dziedzic, 1983

sponses to these peptides. These receptors include the classical insulin receptor, a type I IGF receptor that is structurally homologous to the insulin receptor (Bhaumick et al., 1981b), and a type II IGF receptor (Kasuga et al., 1981) (Table I and see Chapter 8). The variety of receptors with which the IGFs can interact may, in part, explain the different rank order of IGF biological activity in different tissues and for different biological effects (Table VII). All three receptors have been found in many cells and tissues. While in most tissues the IGFs appear to initiate their growth-promoting effects through their own receptors and in-

TABLE VI

Factors That Potentiate the Biological Activity of Insulin-Like Growth Factors

Factor	Reference
Platelet-derived growth factor (PDGF)	Stiles et al., 1979
Epidermal growth factor (EGF)	Kato et al., 1983; Koch et al., 1982
Fibroblast growth factor (FGF)	Kato et al., 1983
Cohn Fraction VI	Cohen and Nissley, 1979
Glucocorticoids	Conover et al., 1983
Thyroxine	Froesch et al., 1976
Platelet-poor plasma	Wharton et al., 1981
Glucagon	Leffert and Koch, 1982

TABLE VII
Relative Biological Potencies of IGFs

Tissue	Effect	Potency in descending order	Reference
Rat adipocytes	Glucose oxidation	Insulin >> IGF II > IGF I	Zapf et al., 1978a
	Lipogenesis	Insulin >> IGF II > IGF I = MSA	Zapf et al., 1978a
Chick embryo fibroblasts	[³H]Thymidine incorporation	IGF I ~ IGF II > MSA = Sm A >> insulin	Rinderknecht and Humbel, 1976; Nissley and Rechler, 1978
Rat cartilage	[³⁵S]Sulfate	IGF I = Sm A > IGF II	Zapf et al., 1978a

sulin-like effects through the insulin receptor, there are exceptions to this rule (Stevens et al., 1981; Koch et al., 1982; Massague et al., 1982). King et al., (1980) have shown that Fab fragments of polyclonal antibodies directed against the insulin receptor block the insulin-like biological effects of the IGFs. Further, these reagents demonstrate that the growth effects of these polypeptides in normal rat hepatocytes and in human fibroblasts are mediated through an IGF receptor. However, in the H35 rat hepatoma cell line, insulin appears to stimulate its growth effects through its own receptor (Koch et al., 1982; Massague et al., 1982). Further, in the rat Swarm chondrosarcoma cell line, insulin stimulates macromolecular synthesis at low concentrations through its own receptor (Stevens et al., 1981). Even in human skin fibroblasts, insulin may exert some of its effects on cell replication through its own as well as the type I IGF receptor (Moses et al., 1985). Classical insulin-like effects mediated through an IGF receptor(s) have been suggested in mouse soleus muscle (Poggi et al., 1979). Both IGF I and IGF II (MSA) appear to stimulate AIB uptake in certain cell lines through a type I IGF receptor (Yu and Czech, 1984).

Recent data suggest that IGF I stimulates the autophosphorylation of its own receptor (Jacobs et al., 1983) in a manner analogous to insulin-stimulated autophosphorylation of its receptor (Kasuga et al., 1983). IGF II does not appear to stimulate the autophosphorylation of the type II IGF receptor (M. Czech, personal communication). The role of receptor phosphorylation in mediating the biological response to insulin or the IGFs remains unclear.

Attempts to identify a "second messenger(s)" that mediates IGF biological responses have so far been unsuccessful. As noted above, IGF I stimulates the autophosphorylation of its receptor. IGF I also stimulates phosphorylation of ribosomal protein 6 S (Haselbacher et al., 1979). Cyclic AMP does not appear to mediate the biological activity of the

IGFs. In fact, experiments with several tissues suggest that IGFs inhibit the generation of cAMP induced by a number of other stimuli, including glucagon and prostaglandin E_1 (Zederman et al., 1980; Anderson et al., 1979; Koch et al., 1982). One report suggests that cGMP may be involved in the biological activity of IGFs in chondrocytes, but this remains unproved (Stuart et al., 1982).

The ability of the IGFs to initiate and support a full cycle of DNA synthesis and cell division is dependent on the interaction of the IGFs with a variety of other growth factors (both peptide and nonpeptide) and on the passage number and cell density of the cells being stimulated (Clemmons and Van Wyk, 1981). Stiles et al., (1979) have suggested that IGFs stimulate cell multiplication only in cells that are made competent by other growth factors. They have proposed that PDGF is an example of a competence factor that recruits cells from a resting phase (G_0) and allows them to progress through the G_1 and S phases of the cell cycle in the presence of sufficient concentrations of IGF. At least one effect of PDGF is to induce the expression of the c-myc gene in responsive cells (Kelly et al., 1983; Armelin et al., 1984). PDGF also may act as a competence factor by increasing the density of cell surface IGF receptors (Clemmons et al., 1980a).

The growth promoting effects of IGFs are not dependent on competence factors in all cell types. For example, chick embryo fibroblasts maintained in serum-free medium synthesize DNA and divide in response to IGFs in the absence of any other growth factor (Rechler et al., 1976). It is not clear whether cells of embryonic origin synthesize their own competence factors or are independent from competence factors. Other cells are capable of dividing in serum-free medium without added competence or progression factors. In the BRL-3A2 cell line, a subclone of the cell line (BRL-3A) that makes MSA, continued cell division occurs in serum-free medium despite the inability of this cell line to produce MSA (Nissley et al., 1977).

The growth-promoting effects of the IGFs are additive or synergistic with a wide variety of other factors, including glucocorticoids (Conover et al., 1983), hypopituitary platelet-poor plasma (Stiles et al., 1979), and low-molecular-weight components of Cohn Fraction VI derived from human plasma (Cohen and Nissley, 1979) (Table VI). Glucocorticoids markedly potentiate the growth-promoting effects of IGFs on human skin fibroblasts. They appear to have differential effects on DNA synthesis stimulated by insulin or by IGF I (Conover et al., 1985). The IGFs also stimulate the differentiation of chick embryo myoblasts into functioning myotubes (Schmid et al., 1983). Since the growth response of a given tissue (or cell line) is usually the result of the cumulative effects of a variety of factors, it is important to carefully define a system when testing the biological activity of specific growth factor(s).

It has become apparent that interpreting the biological effects of IGFs on cells in culture is complicated by the ability of many cells to synthesize and secrete IGFs themselves. This appears to be particularly true for cells of embryonic origin including rat embryo fibroblasts (Adams et al., 1983a) and fetal rat hepatocytes (Rechler et al., 1979). IGF production by fibroblasts in culture is dependent on the age (passage) of the culture and the cell density as well as the availability of adequate nutrients and other growth factors (Clemmons and Shaw, 1983). IGF I production by human skin fibroblasts as well as serum levels of IGF I are dependent on growth hormone (Clemmons et al., 1981a). Fibroblasts derived from rats show a developmental pattern of IGF secretion that mirrors the developmental pattern of IGF levels in rat serum (Adams et al., 1983b). Fetal rat fibroblasts secrete primarily IGF II that is responsive to stimulation by ovine placental lactogen, while adult rat fibroblasts secrete primarily IGF I that is responsive both to growth hormone and to placental lactogen (Adams et al., 1983b). Transformed cell lines may be particular candidates for the synthesis of IGFs (or other growth factors) (DeLarco and Todaro, 1978; Roberts et al., 1983).

The in vivo biological activity of the IGFs is just beginning to be explored. It has been demonstrated clearly that homogeneous IGF I is capable of stimulating an increase in body weight and an increase in tibial epiphyseal width in hypophysectomized rats (Schoenle et al., 1982). IGF II is less active (Schoenle et al., 1983). Neither IGF is capable of stimulating an increase in the production of a large molecular size IGF carrier protein (Zapf et al., 1983). The acute administration of somatomedin A or IGF I to intact rats results in a decrease in blood glucose concentration (Fryklund et al., 1975). As noted elsewhere, this effect may be blocked by the simultaneous administration of IGF carrier protein with the IGF (Meuli et al., 1978). An impure preparation of IGF (mixture of IGF I and IGF II) blunts spontaneous GH secretion in rats following intracerebroventricular administration (Tannenbaum et al., 1983). Insufficient quantities of highly purified IGFs have been available to test other or long-term effects of the IGFs in vivo.

V. CIRCULATING AND SECRETED FORMS OF THE IGFS

Early work on the biological activity of the IGFs recognized a discrepancy between the molecular size of the IGFs in native human (or animal) plasma and in the test tube following purification (Oelz et al.,

1972; Poffenbarger, 1975; Zapf et al., 1978a). Part of this size discrepancy can be accounted for by large molecular weight nonsuppressible insulin-like activity, NSILP, or NSILAp. However, Hintz and Liu (1977) and Megyesi et al., (1975a) demonstrated a reversible association of small molecular weight (5,000–10,000) insulin-like activity with a large molecular size carrier protein. The interaction of IGFs with specific carrier proteins and the structure and potential function of these carrier proteins have become an area of intensive investigation.

Two IGF carrier proteins of different molecular size (160K and 40–50K) in human serum were first described by Zapf et al., (1975). Moses et al., (1976) then described that in rat serum the 160K carrier protein was, like somatomedin itself, growth hormone dependent. This work was confirmed by other investigators (Kaufmann et al., 1978) and extended to human sera (White et al., 1981; Hintz and Liu, 1977). All of the homogeneous IGFs isolated to date interact specifically with one or more IGF carrier proteins, although their relative affinity for a specific carrier protein varies (Moses et al., 1979, 1983; Binoux et al., 1982). The percentage of "free" IGFs in human sera (not bound to carrier protein) is exceedingly low as assessed by molecular exclusion gel chromatography of sera at neutral pH and by ultracentrifugation (Daughaday et al., 1982).

All tissues and cells in vitro that secrete IGFs also secrete IGF carrier proteins. The spectrum of these tissues includes fetal rat liver explants (Rechler et al., 1979), perfused adult rat liver (Schalch et al., 1979), human skin fibroblasts (Clemmons et al., 1981a), rat pituitary explants (Binoux et al., 1981), rat embryo fibroblasts (Adams et al., 1983a), and several cloned liver cell lines (Moses et al., 1979, 1983; Morris et al., 1981). Human amniotic fluid also contains a high-affinity IGF carrier protein (Chochinov et al., 1977b; Drop et al., 1979). At least two liver cell lines, the rat BRL-3A2 line (Moses et al., 1979) and the human HEP G2 cell line (Moses et al., 1983), secrete IGF carrier protein without IGF, suggesting that the IGF and carrier protein are separate gene products that associate after intracellular processing. Acquaviva et al., (1982) have identified a 34K IGF carrier protein following in vitro translation of BRL-3A mRNA.

IGF carrier proteins have not yet been purified from human or rat sera. Knauer et al., (1981) have isolated an apparently homogeneous IGF carrier protein from serum-free media conditioned by the BRL-3A cell line, but chemical characterization of this protein is not complete.

Both gel chromatography at neutral pH and ion exchange chromatography of rat or human sera reveal a large carrier protein–IGF complex of M_r 160,000, a carrier protein of 40,000–60,000 and very small amounts of "free" IGF (Moses et al., 1979; Furlanetto, 1980;

White *et al.*, 1981; Daughaday *et al.*, 1982). D'Ercole and Wilkins (1984) have suggested even more molecular size heterogeneity of rat serum IGF carrier–proteins based on SDS polyacrylamide gel electrophoresis of radiolabelled somatomedin C, covalently cross-linked to carrier proteins by disuccinimidyl suberate. The 160K IGF-carrier protein complex can be dissociated to yield an acid-labile binding component of M_r 40,000–60,000 (Furlanetto, 1980). There also appears to be another acid-stable protein (M_r 40,000–60,000) capable of binding IGFs, which is not part of the acid-labile complex (Furlanetto, 1980). The formation of the 160K carrier protein complex is GH dependent. In the presence of low GH levels, the large molecular weight complex (and its acid-stable breakdown products) is decreased or absent from human and rat serum and only the lower molecular weight, acid-stable carrier protein is present (White *et al.*, 1981; Cohen and Nissley, 1976). While treatment of hypophysectomized rats and hypopituitary humans with GH restores the M_r 160K binding activity in serum to normal, treatment with IGF I or IGF II does not (Zapf *et al.*, 1983). Only the 40–60K IGF carrier protein is present in fetal rat serum (White *et al.*, 1982).

Two functions have been proposed for IGF carrier proteins. These proteins prolong the serum half-life of the IGFs, since in the presence of the M_r 160K carrier proteins, IGFs have a significantly longer half-life than when they are bound to the low-molecular-weight species (Cohen and Nissley, 1976). For this reason, GH may exert its effect on the serum levels of the IGFs partly or wholly by inducing the synthesis and/or secretion of the large molecular weight carrier protein.

The second function, proposed by Zapf and co-workers (1979), may be to prevent uncontrolled expression of the insulin-like (hypoglycemic) effects of the IGFs *in vivo*. There is evidence to support such a role, since carrier proteins inhibit the ability of IGF to stimulate glucose uptake and utilization in the perfused rat heart (Meuli *et al.*, 1978). Further, partially purified carrier proteins inhibit binding of IGFs to adipose cells (Zapf *et al.*, 1979) and chick embryo fibroblasts (Knauer and Smith, 1980).

VI. MEASUREMENT OF INSULIN-LIKE GROWTH FACTORS

The concentrations of various IGFs can be estimated by biological assays, radioreceptor assays, and specific radioimmunoassays. The biological assays share the advantage of measuring net IGF biological activity. This is the sum of both stimulatory and inhibitory factors (Phillips *et al.*, 1979). For this reason, these assays do not measure the specific content of the IGFs in a biological fluid. The original bioassay, that is, sulfate incorporation in hypophysectomized rat costal cartilage, is still being used (Phillips and Weiss, 1982). Other sulfate incorporation bioassays have been devised, including a porcine cartilage (Spencer and Taylor, 1978) and chick embryonic pelvic leaflet assay (Herington et al., 1976). Other laboratories have continued to rely on the insulin-like effects of the IGFs to measure IGF levels under different conditions (Franklin et al., 1976).

Many laboratories have evaluated the growth-stimulating effect of the IGFs by looking at the incorporation of [³H]thymidine into DNA (Cohen et al., 1975). These assays have the advantage of being relatively easy to perform if tissue culture facilities are available. They too have the disadvantage of looking at the sum of stimulatory and inhibitory influences on DNA synthesis. They also have the disadvantage of being dependent on factors other than the IGFs for maximal DNA stimulation. Nonetheless, these assays have been very helpful in defining the additive or synergistic factors that complement IGF activity (Cohen and Nissley, 1979; Clemmons and Shaw, 1983; Conover et al., 1983).

A variety of relatively specific radioreceptor assays have been described for the measurement of the different IGFs (Table III). IGF receptors are plentiful on a variety of tissues, including human placenta (Hintz et al., 1972), rat placenta (Daughaday et al., 1981a), rat liver membranes (Megyesi et al., 1975b), and rat adipocytes (Zapf et al., 1978b). In addition, the IGF serum carrier proteins serve as a convenient source for binding of radiolabeled IGFs (Zapf et al., 1977). These assays have the disadvantage of not being totally specific for the IGFs being measured, since IGF I and IGF II both interact with the type I and type II receptors (Rechler et al., 1980). Generally, IGF I is 10% as potent as IGF II in competing for the IGF II receptor and IGF II is 10% as potent as IGF I in competing for the IGF I receptor. The issue is made even more complex by the fact that most tissues studied contain both IGF I and IGF II receptors. Nonetheless, these easily performed assays may provide valuable information about the relative content of IGFs in biological samples from animals or the tissue culture laboratory.

Many laboratories have generated either polyclonal or monoclonal antibodies directed against the IGFs (Furlanetto et al., 1977; Bala and Bhaumick, 1979; Hall et al., 1979; Hintz et al., 1980a,b; Baxter et al., 1982) (Table III). These reagents are more specific than either radioreceptor assays or biological assays. Antibodies have been generated against the intact IGF molecules (Furlanetto et al., 1977) and against synthetic fragments of these molecules (Hintz et al., 1980a,b; Hintz and Liu, 1982). Both the radioreceptor and radioimmunoassays are dependent upon highly purified preparations of the IGFs for accurate IGF measurement. The IGF radioimmunoassays do not necessarily reflect the biologically available IGFs, since IGF carrier proteins may block some (or all) of the IGF biological activity in serum (Zapf et al., 1979; Knauer and Smith, 1980). In any assay system, the ability of IGF carrier proteins present in whole sera or tissue culture media to interfere with the measurement of the IGFs must be taken into account. Various methods, including acid–ethanol extraction (Daughaday et al., 1980) or molecular sizing chromatography at acid pH (Zapf et al., 1981), have been devised to remove contaminating IGF carrier proteins from the sample to be assayed.

VII. PHYSIOLOGICAL CONTROL OF IGF SECRETION

Evidence from many laboratories supports the original observation that sulfation factor is a growth hormone-dependent serum factor. However, the regulation of serum levels or cellular secretion of IGF I and IGF II differs. Much of the confusion that developed around the regulation of these polypeptides was the result of imprecise biological or radioreceptor assays.

In vivo, the serum levels of radioimmunoassayable IGF I are strongly dependent on growth hormone (Furlanetto et al., 1977; Zapf et al., 1981). IGF I levels are markedly reduced in states of growth hormone deficiency and increased in states of growth hormone excess (acromegaly). Adequate nutritional intake also is essential in maintaining serum levels of IGF I/somatomedin C (Clemmons et al., 1981c; Prewitt et al., 1982; Isley et al., 1983). Poorly controlled diabetes mellitus may result in decreased serum levels of IGF I that return toward normal with restoration of good blood glucose control (Tamborlane et al., 1981). Under certain conditions, estradiol (Clemmons et al., 1980b), prolactin (Clemmons et al., 1981b), and placental lactogen (Hurley et al., 1977; Daughaday et al., 1979; Pilistine et al., 1984) also may regulate the serum levels of IGF I.

IGF II is less growth hormone dependent than IGF I. IGF II levels are decreased moderately with growth hormone deficiency, but are not elevated in acromegaly (Zapf et al., 1981). The roles of nutrition and diabetes mellitus on serum levels of IGF II have not been clearly defined. In man, IGF II levels as measured by a radioreceptor assay may be elevated in some cases of nonislet cell tumor hypoglycemia (Megyesi et al., 1974; Daughaday et al., 1981b; Gordon et al., 1981), but this has not been confirmed in IGF II radioimmunoassay (Zapf et al., 1981). IGF II

(MSA) levels are much higher in fetal rat sera than in maternal rat sera (Moses et al., 1980c), and this has raised the possibility that IGF II may be an important fetal growth factor.

With increasing availability of specific radioimmunoassays for IGFs, additional physiological control mechanisms will likely become apparent.

VIII. UNRESOLVED QUESTIONS

Despite the explosion of information on the structure, cellular interaction, and biological activity of IGFs over the past 25 years, much remains to be learned. It is still unclear how important the IGFs are as stimulators of cellular growth and differentiation. Further, it is unclear what the primary source of the IGFs is. Are IGFs hormones that are secreted from a central secretory organ to be transported in the blood to target tissues, or are IGFs autocrine or paracrine factors that are secreted locally and act locally? If the latter is correct, do serum levels of IGFs reflect their biological activity and biological availability? How do the IGFs produce their biological effects? Are both the type I and type II IGF receptors important for the growth-promoting effects of the IGFs? How many specific IGF carrier proteins are there and what is their relationship to IGF serum half-life and IGF biological activity? What are the primary IGF gene products? Is posttranslational processing of the IGFs important for their biological activity and clearance? How are IGF receptors regulated and how do these receptors interact with the insulin receptor and the biological responses induced by insulin (see Chapter 8)? What is the significance of the structural relationship between proinsulin and insulin-like growth factors? Are insulin-like growth factors important for fetal growth and development?

These are but a few of the unanswered questions in this exciting area of cell biology and clinical medicine.

ACKNOWLEDGMENTS

The authors wish to acknowledge the many investigators who have contributed significantly to our understanding of insulin-like growth factors and whose work was not cited in the present review due to space limitations.

Terri Wiseman worked tirelessly and patiently in preparing this manuscript.

REFERENCES

Acquaviva, A. M., Bruni, C. B., Nissley, S. P., and Rechler, M. M. (1982). Cell-free synthesis of rat insulin-like growth factor II. *Diabetes* **31**, 656–658.

Adams, S. O., Nissley, S. P., Greenstein, L. A., Yang, Y. W. H., and Rechler, M. M. (1983a). Synthesis of multiplication-stimulating activity (rat insulin-like growth factor II) by rat embryo fibroblasts. *Endocrinology* **112**, 979–987.

Adams, S. O., Nissley, S. P., Handwerger, S., and Rechler, M. M. (1983b). Developmental patterns of insulin-like growth factor I and II synthesis and regulation in rat fibroblasts. *Nature (London)* **302**, 150–153.

Anderson, W. B., Wilson, J., Rechler, M. M., and Nissley, S. P. (1979). Effect of multiplication stimulating activity (MSA) on intracellular cAMP levels and adenylate cyclase activity in chick embryo fibroblasts. *Exp. Cell Res.* **120**, 47–53.

Armelin, H. A., Armelin, M. C., Kelly, K., Stewart, T., Leder, P., Cochran, B. H., and Stiles, C. D. (1984). Functional role for c-myc in mitogenic response to platelet-derived growth factor. *Nature* **310**, 655–660.

Bala, R. M., and Bhaumick, B. (1979). Radioimmunoassay of a basic somatomedin: Comparison of various assay techniques and somatomedin levels in various sera. *J. Clin. Endocrinol. Metab.* **49**, 770–777.

Baxter, R. C., Axiak, S., and Raison, R. L. (1982). Monoclonal antibody against human somatomedin-C/insulin-like growth factor I. *J. Clin. Endocrinol. Metab.* **54**, 474–476.

Bell, G. I., Merryweaher, J. P., Sanchez-Pescador, R., Stempien, M. M., Priestley, L., Scott, J., and Rall, L. B. (1984). Sequence of a cDNA clone encoding human preproinsulin-like growth factor II. *Nature (London)* **310**, 775–777.

Bhaumick, B., Goren, H. J., and Bala, R. M. (1981a). Further characterization of human basic somatomedin: Comparison with insulin-like growth factors I and II. *Horm. Metab. Res.* **13**, 515–518.

Bhaumick, B., Bala, R. M., and Hollenberg, M. D. (1981b). Somatomedin receptor of human placenta: Solubilization, photolabelling, partial purification and comparison with insulin receptor. *Proc. Natl. Acad. Sci. U.S.A.* **78**(7), 4279–4283.

Binoux, M., Hossenlopp, P., Lassarre, C., and Hardouin, N. (1981). Production of insulin-like growth factors and their carrier by rat pituitary gland and brain explants in culture. *FEBS Lett.* **124**, 178–184.

Binoux, M., Hardouin, S., Lassarre, C., and Hossenlopp, P. (1982). Evidence for production by the liver of two IGF binding proteins with similar molecular weights but different affinities for IGF I and IGF II: Their relations with serum and cerebrospinal fluid IGF binding proteins. *J. Clin. Endocrinol. Metab.* **55**, 600–602.

Blundell, T. L., Bedarker, S., Rinderknecht, E., and Humbel, R. E. (1978). Insulin-like growth factor: A model for tertiary structure accounting for immunoreactivity and receptor binding. *Proc. Natl. Acad. Sci. U.S.A.* **75**(1), 180–184.

Blundell, T. L., Bedarkar, S., and Humbel, R. E. (1983). Tertiary structures, receptor binding and antigenicity of insulin-like growth factors. *Fed. Proc., Fed. Am. Soc. Exp. Biol.* **42**, 2592–2597.

Borland, K., Mita, M., Oppenheimer, C. L., Blinderman, L. A., Massague, P. F., Hall, P. F., and Czech, M. P. (1984). The actions of insulin-like growth factors I and II on cultured Sertoli cells. *Endocrinology* **114**, 240–246.

Brissenden, J. E., Ullrich, A., and Francke, U. (1984). Human chromosomal mapping of genes for insulin-like growth factors I and II and epidermal growth factor. *Nature (London)* **310**, 781–784.

Canalis, E. (1980). Efect of insulin-like growth factor I on DNA and protein synthesis in cultured rat calvaria. *J. Clin. Invest.* **66,** 709–719.

Chochinov, R. H., Mariz, I. K., and Daughaday, W. H. (1977a). Isolation of a somatomedin from plasma of rats bearing growth hormone secreting tumors. *Endocrinology* **100,** 549–556.

Chochinov, R. H., Mariz, I. K., Hajek, A. S., and Daughaday, W. H. (1977b). Characterization of a protein in midterm human amniotic fluid which reacts in the somatomedin-C radioimmunoassay. *J. Clin. Endocrinol. Metab.* **44**(5), 902–908.

Clemmons, D. R., and Shaw, D. S. (1983). Variables controlling somatomedin production by cultured human fibroblasts. *J. Cell. Physiol.* **115,** 137–142.

Clemmons, D. R., and Van Wyk, J. J. (1981). Somatomedin-C and platelet-derived growth factor stimulate human fibroblast replication. *J. Cell. Physiol.* **106,** 361–367.

Clemmons, D. R., Van Wyk, J. J., and Pledger, W. J. (1980a). Sequential addition of platelet factor and plasma to BALB/c 3T3 fibroblast cultures stimulates somatomedin-C binding early in cell cycle. *Proc. Natl. Acad. Sci. U.S.A.* **77**(11), 6644–6648.

Clemmons, D. R., Underwood, L. E., Ridgway, E. C., Kliman, B., Kjellberg, R. N., and Van Wyk, J. J. (1980b). Estradiol treatment of acromegaly. Reduction of immunoreactive somatomedin-C and improvement in metabolic status. *Am. J. Med.* **69,** 571–575.

Clemmons, D. R., Underwood, L. E., and Van Wyk, J. J. (1981a). Hormonal control of immunoreactive somatomedin production by cultured human fibroblasts. *J. Clin. Invest.* **67**(1), 10–19.

Clemmons, D. R., Underwood, L. E., Ridgway, E. C., Climan, B., and Van Wyk, J. J. (1981b). Hyperprolactinemia is associated with increased immunoreactive somatomedin C in hypopituitarism. *J. Clin. Endocrinol. Metab.* **52,** 731–735.

Clemmons, D. R., Klibanski, A., Underwood, L. E., McArthur, J. W., Ridgway, E. C., Beitins, I. Z., and Van Wyk, J. J. (1981c). Reduction of plasma immunoreactive somatomedin C during fasting in humans. *J. Clin. Endocrinol. Metab.* **53,** 1247–1250.

Cohen, K. L., and Nissley, S. P. (1976). Serum half-life of somatomedin activity: Evidence for growth hormone dependence. *Acta Endocrinol (Copenhagen)* **83**(2), 243–258.

Cohen, K. L., and Nissley, S. P. (1979). Cohn Fraction VI enhances the growth stimulating effect of multiplication stimulating activity (MSA) on chick embryo fibroblasts in culture. *Horm. Metab. Res.* **12,** 164–168.

Cohen, K. L., Short, P. A., and Nissley, S. P. (1975). Growth hormone-dependent serum stimulation of DNA synthesis in chick embryo fibroblasts in culture. *Endocrinology* **96,** 193–198.

Conover, C. A., Dollar, L. A., Hintz, R. L., and Rosenfeld, R. G. (1983). Insulin-like growth factor I/somatomedin-C (IGF-I/SM-C) and glucocorticoids synergistically regulate mitosis in competent human fibroblasts. *J. Cell. Physiol.* **116,** 191–197.

Conover, C. A., Hintz, R. A., and Rosenfeld, R. G. (1985). Comparative effects of somatomedin C and insulin on the metabolism and growth of cultured human fibroblasts. *J. Cell. Physiol.* **122,** 133–141.

Daughaday, W. H., Hall, K., Raben, M. S., Salmon, W. D., Van den Brande, J. L., and Van Wyk, J. J. (1972). Somatomedin: Proposed designation for sulfation factor. *Nature (London)* **235,** 107.

Daughaday, W. H., Trivedi, B., and Kapadia, M. (1979). The effect of hypophysectomy on rat chorionic somatomammotropin as measured by prolactin and growth hormone radioreceptor assays: Possible significance in maintenance of somatomedin generation. *Endocrinology* **105,** 210–214.

Daughaday, W. H., Mariz, I. K., and Blethen, S. L. (1980). Inhibition of access of bound somatomedin to membrane receptor and immunobinding sites: A comparison of radioreceptor and radioimmunoassay of somatomedin in native and acid–ethanol-extracted serum. *J. Clin. Endocrinol. Metab.* **51**(4), 781–787.

Daughaday, W. H., Mariz, I. K., and Trivedi, B. (1981a). A preferential binding site for insulin-like growth factor II in human and rat placental membranes. *J. Clin. Endocrinol. Metab.* **53**(2), 282–288.

Daughaday, W. H., Trivedi, B., and Kapadia, M. (1981b). Measurement of insulin-like growth factor II by a specific radioreceptor assay in serum of normal individuals, patients with abnormal growth hormone secretion and patients with tumor-associated hypoglycemia. *J. Clin. Endocrinol. Metab.* **53**, 289–294.

Daughaday, W. H., Ward, A. P., Goldberg, A. C., Trivedi, B., and Kapadia, M. (1982). Characterization of somatomedin binding in human serum by ultracentrifugation and gel filtration. *J. Clin. Endocrinol. Metab.* **55**, 916–921.

DeLarco, J. E., and Todaro, G. J. (1978). A human fibrosarcoma cell line producing multiplication stimulating activity (MSA)-related peptides. *Nature (London)* **272**, 356–358.

D'Ercole, A. J., and Wilkins, J. R. (1984). Affinity-labeled somatomedin C-binding proteins in rat sera. *Endocrinol.* **114**, 1141–1144.

D'Ercole, A. J., Stiles, A. D., and Underwood, L. E. (1984). Tissue concentrations of somatomedin C: Further evidence for multiple sites of synthesis and paracrine or autocrine mechanisms of action. *Proc. Natl. Acad. Sci. U.S.A.* **81**, 935–939.

Drop, S. L. S., Valiquette, G., Guyda, H. J., Corvol, M. T., and Posner, B. I. (1979). Partial purification and characterization of a binding protein for insulin-like activities in human amniotic fluid: A possible inhibitor of insulin-like activity. *Acta Endocrinol. (Copenhagen)* **90**(3), 505–518.

Dulak, N. C., and Temin, H. M. (1973). A partially purified polypeptide fraction from rat liver cell conditioned medium with multiplication stimulating activity (MSA) for embryo fibroblasts. *J. Cell. Physiol.* **81**, 153–160.

Dull, T. J., Gray, A., Hayflick, J. S., and Ullrich, A. (1984). Insulin-like growth factor II precursor gene organization in relation to insulin gene family. *Nature (London)* **310**, 777–781.

Enberg, G., Carlquist, M., Jornvall, H., and Hall, K. (1984). The characterization of somatomedin A, isolated by microcomputer-controlled chromatography, reveals an apparent identity to insulin-like growth factor: 1. *Eur. J. Biochem.* **143**, 117–124.

Evans, M. I., Haser, L. J., and McKnight, G. S. (1981). A somatomedin-like peptide hormone is required during the estrogen-mediated induction of ovalbumin gene transcription. *Cell (Cambridge, Mass.)* **25**, 187–193.

Ewton, D. Z., and Florini, J. R. (1980). Relative effects of the somatomedins MSA and growth hormone on myoblasts and myotubes in culture. *Endocrinology* **106**, 577–584.

Franklin, R. C., Rennie, G. C., Burger, H. G., and Cameron, D. P. (1976). A bioassay for NSILA-S in individual serum samples and its relationship to somatotropin. *J. Clin. Endocrinol. Metab.* **43**, 1164–1169.

Froesch, E. R., Zapf, J., Audhya, T. K., Ben-Porath, E., Segen, B. J., and Gibson, K. D. (1976). Nonsuppressible insulin-like activity and thyroid hormones: Major pituitary-dependent sulfation factors for chick embryo cartilage. *Proc. Natl. Acad. Sci. U.S.A.* **73**, 2904–2908.

Fryklund, L., Skottner, A., Sievertsson, H., and Hall, K. (1975). Somatomedins A and B. Isolation, chemistry and in vivo effects. *Int. Congr. Ser.—Excerpta Med.* **381**, 156–168.

Furlanetto, R. W. (1980). The somatomedin C binding protein: Evidence for a hetero-logous subunit structure. *J. Clin. Endocrinol. Metab.* **51**(1), 12–19.

Furlanetto, R. W., Underwood, L. E., Van Wyk, J. J., and D'Ercole, A. J. (1977). Estimation of somatomedin-C levels in normals and patients with pituitary disease by radioim-munoassay. *J. Clin. Invest.* **60**, 648–657.

Gordon, P., Hendricks, C. M., Kahn, C. R., Megyesi, K., and Roth, J. (1981). Hypoglycemia associated with non-islet cell tumor and insulin-like growth factors. *N. Engl. J. Med.* **305**(24), 1452–1455.

Hall, K., Brant, J., Enberg, G., and Fryklund, L. (1979). Somatomedin A by radioim-munoassay and radioreceptor assay and its relation to carrier protein. *In* "Somatomedins and Growth" (J. Giordano, J. J. Van Wyk, and F. Minuto, eds.), pp. 119–122. Academic Press, London.

Harley, C. B., Goldstein, S., Posner, B., and Guyda, H. (1980). Insulin-like peptides stimulate metabolism but not proliferation of human fibroblasts. *Am. J. Physiol.* **239**, E125–E131.

Haselbacher, G. K., and Humbel, R. E. (1976). Stimulation of ornithine decarboxylase activity in chick fibroblasts by nonsuppressible insulin-like activity, insulin and serum. *J. Cell. Physiol.* **88**, 239–246.

Haselbacher, G. K., and Humbel, R. E. (1982). Evidence for two species of insulin-like growth factor II (IGF-II and "big" IGF-II) in human spinal fluid. *Endocrinology* **110**(6), 1822–1824.

Haselbacher, G. K., Humbel, R. E., and Thomas, G. (1979). Insulin-like growth factor: Insulin or serum increase phosphorylation of ribosomal protein S6 during transi-tion of stationary chick embryo fibroblasts into early G_1 phase of the cell cycle. *FEBS Lett.* **100**, 185–190.

Hayashi, I., and Sato, G. H. (1976). Replacement of serum by hormones permits growth of cells in a defined medium. *Nature (London)* **259**, 132–134.

Heaton, J. H., Schilling, E. E., Gelehrter, T. D., Rechler, M. M., Spencer, C. J., and Nissley, S. P. (1980). Induction of tyrosine aminotransferase and amino acid transport in rat hepatoma cells by insulin and the insulin-like growth factor, multiplication stim-ulating activity. *Biochim. Biophys. Acta* **632**(2), 192–203.

Heldin, C. H., Fryklund, L., and Westermark, B. (1981). Somatomedin B: Mitogenic activity derived from contaminant epidermal growth factor. *Science* **213**, 1122–1123.

Herington, A. C., Phillips, L. S., and Daughaday, W. H. (1976). Factors governing the stimulation of embryonic chick cartilage by somatomedin. *Acta Endocrinol. (Copenhagen)* **83**, 259–268.

Hintz, R. L., and Liu, F. (1977). Demonstration of specific plasma protein binding sites for somatomedin. *J. Clin. Endocrinol. Metab.* **45**, 988–995.

Hintz, R. L., and Liu, F. (1982). A radioimmunoassay for insulin-like growth factor II specific for the C-peptide region. *J. Clin. Endocrinol. Metab.* **54**, 442–446.

Hintz, R. L., Clemmons, D. R., Underwood, L. E., and Van Wyk, J. J. (1972). Competitive binding of somatomedin to the insulin receptors of adipocytes, chondrocytes, and liver membranes. *Proc. Natl. Acad. Sci. U.S.A.* **69**, 2351–2353.

Hintz, R. L., Liu, F., Marshall, L. B., and Chang, D. (1980a). Interaction of somatomedin-C with an antibody directed against the synthetic C-peptide region of insulin-like growth factor-I. *J. Clin Endocrinol. Metab.* **50**, 405–407.

Hintz, R. L., Liu, F., and Rinderknecht, E. (1980b). Somatomedin-C shares the carboxy-terminal antigenic determinants with insulin-like growth factor-I. *J. Clin. Endo-crinol. Metab.* **51**, 672–673.

Hurley, T. W., D'Ercole, A. J., Handwerger, S., Underwood, L. E., Furlanetto, R. W., and

Fellows, R. E. (1977). Ovine placental lactogen induces somatomedin: A possible role in fetal growth. *Endocrinology* **101**, 1635–1638.

Isley, W. L., Underwood, L. E., and Clemmons, D. R. (1983). Dietary components that regulate serum somatomedin-C concentrations in humans. *J. Clin. Invest.* **71**, 175–182.

Jacobs, S., Kull, F. C., Jr., Earp, H. S., Svoboda, M. E., Van Wyk, J. J., and Cuatrecasas, P. (1983). Somatomedin C stimulates the phosphorylation of the B-subunit of its own receptor. *J. Biol. Chem.* **258**, 9581–9584.

Jansen, M., van Schaik, F. M. A., Ricker, A. T., Bullock, B., Woods, D. E., Gabbay, K. H., Nussbaum, A. L., Sussenbach, J. S., and Van den Brande, J. L. (1983), Sequence of cDNA encoding human insulin-like growth factor I precursor. *Nature (London)* **306**, 609–611.

Kaplowitz, P. B., D'Ercole, A. J., and Underwood, L. E. (1982). Stimulation of embryonic mouse limb bud mesenchymal cell growth by peptide growth factors. *J. Cell. Physiol.* **112**, 353–359.

Kaplowitz, P. B., D'Ercole, A. J., Underwood, L. E., and Van Wyk, J. J. (1984). Stimulation by somatomedin-C of aminoisobutyric acid uptake in human fibroblasts: A possible test for cellular responsiveness to somatomedin. *J. Clin. Endocrinol. Metab.* **58**, 176–181.

Kasuga, M., Van Obberghen, E. E., Nissley, S. P., and Rechler, M. M. (1981). Demonstration of two subtypes of insulin-like growth factor receptors by affinity cross-linking. *J. Biol. Chem.* **256**, 5305–5308.

Kasuga, M., Van Obberghen, E. E., Nissley, S. P., and Rechler, M. M. (1981). Demonstration of two subtypes of insulin-like growth factor receptors by affinity cross-linking. *J. Biol. Chem.* **256**, 5305–5308.

Kato, Y., Nomura, Y., Tsuji, M., Kinoshita, M., Ohmae, H., and Suzuki, F. (1981). Somatomedin-like peptide(s) isolated from fetal bovine cartilage (cartilage-derived factor): Isolation and some properties. *Proc. Natl. Acad. Sci. U.S.A.* **78**, 6831–6835.

Kato, Y., Hiraki, Y., Inoue, H., Kinoshita, M., Yutani, Y., and Suzuki, F. (1983). Differential and synergistic actions of somatomedin-like growth factors, fibroblast growth factor and epidermal growth factor in rabbit costal chondrocytes. *Eur. J. Biochem.* **129**, 685–690.

Kaufmann, U., Zapf, J., and Froesch, E. R. (1978). Growth-hormone dependence of non-suppressible insulin-like activity (NSILA) and of NSILA-carrier protein in rats. *Acta Endocrinol. (Copenhagen)* **87**, 716–727.

Kelly, K., Cochran, B. H., Stiles, C. D., and Leder, P. (1983). Cell-specific regulation of the c-myc gene by lymphocyte mitogens and platelet-derived growth factor. *Cell (Cambridge, Mass.)* **35**, 603–610.

King, G. L., Kahn, C. R., Rechler, M. M., and Nissley, S. P. (1980). Direct demonstration of separate receptors for growth and metabolic activities of insulin and MSA (an insulin-like growth factor) using antibodies to the insulin receptor. *J. Clin. Invest.* **66**(1), 130–140.

King, G. L., Kahn, C. R., Samuels, B., Danho, W., Bullesbach, E. E., and Gattner, H. G. (1982). Synthesis and characterization of molecular hybrids of insulin and insulin-like growth factor I. *J. Biol. Chem.* **257**(18), 10869–10873.

King, G. L., Goodman, A. D., Buzney, S. M., Moses, A. C., and Kahn, C. R. (1985). Specific binding and growth effect of insulin-like growth factors in retinal and aortic endothelium and vascular supporting cells. *J. Clin. Invest.* **75**, 1028–1036.

Klapper, D. G., Svoboda, M. E., and Van Wyk, J. J. (1983). Sequence analysis of somatomedin-C; confirmation of identity with insulin-like growth factor I. *Endocrinology* **112**(6), 2215–2217.

Knauer, D. J., and Smith, G. L. (1980). Inhibition of biological activity of multiplication stimulating activity by binding to its carrier protein. *Proc. Natl. Acad. Sci. U.S.A.* **77**, 7252–7256.

Knauer, D. J., Wagner, F. W., and Smith, G. L. (1981). Purification and characterization of multiplication-stimulating activity (MSA) carrier protein. *J. Supramol. Struct.* **15**, 177–191.

Knight, A. B., Rechler, M. M., Romanus, J. A., Van Obberghen-Schilling, E. E., and Nissley, S. P. (1981). Stimulation of glucose incorporation and amino acid transport by insulin and an insulin-like growth factor in fibroblasts with defective insulin receptors cultured from a patient with leprechaunism. *Proc. Natl. Acad. Sci. U.S.A.* **78**, 2554–2558.

Koch, K. S., Shapiro, P., Skelly, H., and Leffert, H. L. (1982). Rat hepatocyte proliferation is stimulated by insulin-like peptides in defined medium. *Biochem. Biophys. Res. Commun.* **109**, 1054–1060.

Laubli, U. K., Baier, W., Binz, H., Celio, M. R., and Humbel, R. E. (1982). Monoclonal antibodies directed to human insulin-like growth factor I (IGF-I): Use for radioimmunoassay and immunopurification of IGF. *FEBS Lett.* **149**, 109–112.

Leffert, H. L., and Koch, K. S. (1982). Hepatocyte growth regulation by hormones in chemically defined media: A two-signal hypothesis. *Cold Spring Harbor Conf. Cell Proliferation* **9**, 597–613.

Lenoir, D., and Honegger, P. (1983). Insulin-like growth factor I (IGF I) stimulates DNA synthesis in fetal rat brain cell cultures. *Dev. Brain Res.* **7**, 205–213.

Li, C. H., Yamashiro, D., Gospodarowicz, D., Kaplan, S. L., and Van Vliet, G. (1983). Total synthesis of insulin-like growth factor I (somatomedin C). *Proc. Natl. Acad. Sci. U.S.A.* **80**, 2216–2220.

Marquardt, H., Todaro, G. J., Henderson, L. E., and Oroszlan, S. (1981). Purification and primary structure of a polypeptide with multiplication-stimulation activity from rat liver cell cultures. *J. Biol. Chem.* **256**, 6859–6865.

Massague, J., Blinderman, L. A., and Czech, M. P. (1982). The high affinity insulin receptor mediates growth stimulation in rat hepatoma cells. *J. Biol. Chem.* **257**, 13958–13963.

Megyesi, K., Kahn, C. R., Roth, J., and Gorden, P. (1974). Hypoglycemia in association with extrapancreatic tumors: Demonstration of elevated plasma NSILAs by a new radioreceptor assay. *J. Clin. Endocrinol. Metab.* **38**, 931–934.

Megyesi, K., Kahn, R., Roth, J., and Gordon, P. (1975a). Circulating NSILAs in man: Basal and stimulated levels and binding to plasma compounds. *J. Clin. Endocrinol. Metab.* **41**, 475.

Megyesi, K., Kahn, C. R., and Roth, J. (1975b). The NSILA's receptor in liver plasma membranes. *J. Biol. Chem.* **250**, 8990–8996.

Meuli, C., Zapf, J., and Froesch, E. R. (1978). Nonsuppressible insulin-like (NSILA) activity-carrier protein abolishes the action of NSILAs on perfused rat heart. *Diabetologia* **14**(4), 255–259.

Morris, D. H., Schalch, D. S., and Monty-Miles, B. (1981). Identification of a somatomedin-binding protein produced by a rat hepatoma cell line. *FEBS Lett.* **127**, 221–224.

Moses, A. C., Nissley, S. P., Cohen, K. L., and Rechler, M. M. (1976). Specific binding of a somatomedin-like polypeptide in rat serum depends on growth hormone. *Nature (London)* **263**, 137–140.

Moses, A. C., Nissley, S. P., Passamani, J., and White, R. M. (1979). Further characterization of growth hormone-dependent somatomedin (SM) binding proteins in rat serum and demonstration of SM-binding protein produced by rat liver cells in culture. *Endocrinology* **104**(2), 536–546.

Moses, A. C., Nissley, S. P., Short, P. A., Rechler, M. M., and Podskalny, J. M. (1980a). Purification and characterization of multiplication-stimulating activity: Insulin-like growth factors purified from rat liver cell-conditioned medium. *Eur. J. Biochem.* **103**, 387–400.

Moses, A. C., Nissley, S. P., Short, P. A., and Rechler, M. M. (1980b). Immunological cross-reactivity of multiplication-stimulation activity polypeptides. *Eur. J. Biochem.* **103**, 401–408.

Moses, A. C., Nissley, S. P., Short, P. A., Rechler, M. M., White, R. M., Knight, A. B., and Higa, O. Z. (1980c). Increased levels of multiplication stimulating activity (MSA), an insulin-like growth factor, in fetal rat serum. *Proc. Natl. Acad. Sci. U.S.A.* **77**(6), 3649–3653.

Moses, A. C., Freinkel, A. J., Knowles, B. B., and Aden, D. P. (1983). Demonstration that a human hepatoma cell line produces a specific insulin-like growth factor carrier protein. *J. Clin. Endocrinol. Metab.* **56**, 1003–1008.

Moses, A. C., Usher, P., and Flier, J. S. (1985). Monoclonal antibody to the IGF-I receptor reveals which receptors mediate IGF and insulin stimulated DNA synthesis in human skin fibroblasts. Abstract submitted to American Diabetes Association Annual Meeting, manuscript in preparation.

Nagarajan, L., Nissley, S. P., Rechler, M. M., and Anderson, W. B. (1982). Multiplication stimulating activity stimulates the multiplication of F9 embryonal carcinoma cells. *Endocrinology* **110**, 1231–1237.

Nissley, S. P., and Rechler, M. M. (1978). Multiplication stimulating activity (MSA): A somatomedin-like polypeptide from cultured rat liver cells. *Natl. Cancer Inst. Monogr.* **48**, 167–177.

Nissley, S. P., Short, P. A., Rechler, M. M., and Podskalny, J. M. (1977). Proliferation of Buffalo rat liver cells in serum-free medium does not depend upon multiplication-stimulating activity (MSA). *Cell (Cambridge, Mass.)* **11**, 441–446.

Oelz, O., Froesch, E. R., Buenzli, H. F., Humbel, R. E., and Ritschard, W. J. (1972). Antibody-suppressible and non-suppressible insulin-like activities. In "Handbook of Physiology" (D. F. Steiner and N. Freinkel, eds.), Sect. 7, Vol. I, pp. 685–702. Am. Physiol. Soc., Washington, D.C.

Phillips, L. S., and Weiss, L. J. (1982). Circulating growth factor studies in growth plate versus resting cartilage *in vitro:* Tissue responsiveness. *Endocrinology* **111**, 1255–1262.

Phillips, L. S., Belosky, D. C., Young, H. S., and Reichard, L. A. (1979). Nutrition and somatomedin. VI. Somatomedin activity and somatomedin inhibitory activity in sera from normal and diabetic rats. *Endocrinology* **104**, 1519–1524.

Pierson, R. W., and Temin, H. M. (1972). Partial purification from calf serum of a fraction with multiplication stimulating activity for chicken fibroblasts in cell culture and with non-suppressible insulin-like activity. *J. Cell. Physiol.* **79**, 319–330.

Pilistine, S. J., Moses, A. C., and Munro, H. N. (1984). Insulin-like growth factor receptors in rat placental membranes. *Endocrinology* **115**, 1060–1065.

Pilistene, S. J., Moses, A. C., and Munro, H. N. (1984). Placental lactogen administration reverses the effect of low-protein diet on maternal and fetal serum somatomedin levels in the pregnant rat. *Proc. Natl. Acad. Sci.* **81**, 5853–5857.

Poffenbarger, P. L. (1975). The purification and partial characterization of an insulin-like protein from human serum. *J. Clin. Invest.* **56**, 1455–1463.

Poggi, C., LeMarchand-Brustel, Y., Zapf, J., Froesch, E. R., and Freychet, P. (1979). Effects and binding of insulin-like growth factor I in the isolated soleus muscle of lean and obese mice: Comparison with insulin. *Endocrinology* **105**, 723–730.

Posner, B. I., Guyda, H. J., Corvol, M. T., Rappaport, R., Harley, C., and Goldstein, S. (1978). Partial purification, characterization, and assay of a slightly acidic insulin-like peptide (ILAs) from human plasma. *J. Clin. Endocrinol. Metab.* **47**, 1240–1250.

Prewitt, T. E., D'Ercole, A. J., Switzer, B. R., and Van Wyk, J. J. (1982). Relationship of serum immunoreactive somatomedin-C to dietary protein and energy in growing rats. *J. Nutr.* **112**, 144–150.

Rabinovitch, A., Quigley, C., Russell, T., Patel, Y., and Mintz, D. H. (1982). Insulin and multiplication stimulating activity (an insulin-like growth factor) stimulate islet B-cell replication in neonatal rat pancreatic monolayer cultures. *Diabetes* **31**, 160–164.

Rechler, M. M., Podskalny, J. M., and Nissley, S. P. (1976). Interaction of multiplication stimulating activity with chick embryo fibroblasts demonstrates a growth receptor. *Nature (London)* **259**, 134–136.

Rechler, M. M., Fryklund, L., Nissley, S. P., Hall, K., Podskalny, J. M., Skottner, A., and Moses, A. C. (1978). Purified human somatomedin A and rat multiplication stimulating activity, mitogens for cultured fibroblasts that cross-react with the same growth receptors. *Eur. J. Biochem.* **82**, 5–12.

Rechler, M. M., Eisen, H. J., Higa, O. Z., Nissley, S. P., Moses, A. C., Schilling, E. E., Fennoy, I., Bruni, C. B., Phillips, L. S., and Baird, K. L. (1979). Characterization of a somatomedin (insulin-like growth factor) synthesized by fetal rat liver organ cultures. *J. Biol. Chem.* **254**, 7942–7950.

Rechler, M. M., Zapf, J., Nissley, S. P., Froesch, E. R., Moses, A. C., Podskalny, J. M., Schilling, E. E., and Humbel, R. E. (1980). Interactions of insulin-like growth factors I and II and multiplication stimulating activity with receptors and serum carrier proteins. *Endocrinology* **107**, 1451–1459.

Reddan, J. R., and Wilson-Dziedzic, D. (1983). Insulin growth factor and epidermal growth factor trigger mitosis in lenses cultured in a serum-free medium. *Invest. Ophthalmol. Visual Sci.* **24**, 409–416.

Rinderknecht, E., and Humbel, R. E. (1976). Polypeptides with non-suppressible insulin-like and cell-growth promoting activities in human serum: Isolation, chemical characterization and some biological properties of forms I and II. *Proc. Natl. Acad. Sci. U.S.A.* **73**, 2365–2369.

Rinderknecht, E., and Humbel, R. E. (1978a). The amino acid sequence of human insulin-like growth factor I and its structural homology with proinsulin. *J. Biol. Chem.* **253**, 2769–2776.

Rinderknecht, E., and Humbel, R. E. (1978b). Primary structure of human insulin-like growth factor II. *FEBS Lett.* **89**, 283–286.

Roberts, A. B., Frolik, C. A., Anzano, M. A., and Sporn, M. B. (1983). Transforming growth factors from neoplastic and nonneoplastic tissues. *FED. Proc., Fed. Am. Soc. Exp. Biol.* **42**, 2621–2626.

Rothstein, H., Van Wyk, J. J., Hayden, J. H., Gordon, S. R., and Weinsieder, A. (1980). Somatomedin C: Restoration *in vivo* of cycle traverse in G_0/G_1 blocked cells of hypophysectomized animals. *Science* **208**, 410–412.

Rubin, J. S., Mariz, I., Jacobs, J. W., Daughaday, W. H., and Bradshaw, R. A. (1982).

Isolation and partial sequence analysis of rat basic somatomedin. *Endocrinology* **110**, 734–740.

Salmon, W. D., Jr., and Daughaday, W. H. (1957). A hormonally controlled serum factor which stimulates sulfate incorporation by cartilage *in vitro. J. Lab. Clin. Med.* **49**, 825–836.

Salmon, W. D., Jr., and DuVall, M. R. (1970). A serum fraction with "sulfation factor activity" stimulates *in vitro* incorporation of leucine and sulfate into protein–polysaccharide complexes, uridine into RNA, and thymidine into DNA of costal cartilage from hypophysectomized rats. *Endocrinology* **86**, 721–727.

Salmon, W. D., and Hosse, B. R. (1971). Stimulation of HeLa cell growth by a serum fraction with sulfation factor activity. *Proc. Soc. Exp. Biol. Med.* **136**, 805–808.

Schalch, D. S., Heinrich, U. E., Draznin, B., Johnson, C. J., and Miller, L. L. (1979). Role of the liver in regulating somatomedin activity: Hormonal effects on the synthesis and release of insulin-like growth factor and its carrier protein by the isolated perfused rat liver. *Endocrinology* **104**, 1143–1151.

Schalch, D., Reismann, D., Emler, C., Humbel, R., Li, C. H., Peters, M., and Lau, E. (1984). Insulin-like growth factor I/somatomedin C (IGF-I/Sm-C): Comparison of natural, solid phase synthetic, and recombinant DNA analog peptides in two radioligand assays. *J. Clin. Endocrinol. Metab.* **115**, 2490–2492.

Schimpff, R. M., Repellin, A. M., Salvatoni, A., Thieriot-Prevost, G., and Chatelain, P. (1983). Effect of purified somatomedins on thymidine incorporation into lectin-activated human lymphocytes. *Acta Endocrinol. (Copenhagen)* **102**, 21–26.

Schmid, C., Steiner, T., and Froesch, E. R. (1983). Preferential enhancement of myoblast differentiation by insulin-like growth factors (IGF I and IGF II) in primary cultures of chicken embryonic cells. *FEBS Lett.* **161**, 117–121.

Schoenle, E., Zapf, J., Humbel, R. E., and Froesch, E. R. (1982). Insulin-like growth factor I stimulates growth in hypophysectomized rats. *Nature (London)* **296**, 252–253.

Schoenle, E., Zapf, J., and Froesch, E. R. (1983). Long-term *in vivo* effects of insulin-like growth factors (IGF) I and II on growth indices: Direct evidence in favor of the somatomedin hypothesis. *In* "Insulin-Like Growth Factor Somatomedins" (E. M. Spencer, ed.), pp. 51–55. de Gruyter, Berlin.

Smith, G. L. (1977). Increased ouabain-sensitive [86]Rubidium uptake after mitogenic stimulation of quiescent chicken embryo fibroblasts with purified multiplication-stimulating activity. *J. Cell Biol.* **73**, 761–767.

Smith, G. L., and Temin, H. M. (1974). Purified multiplication-stimulating activity from rat liver cell conditioned medium: Comparison of biological activities with calf serum, insulin, and somatomedin. *J. Cell. Physiol.* **84**, 181–192.

Spencer, E. M., Ross, M., and Smith, B. (1983). The identify of human insulin-like growth factors I and II with somatomedins C and A and homology with rat IGF I and II. *In*, "Insulin-like Growth Factors, Somatomedins." (E. M. Spencer, ed.) pp. 81–96. de Gruyter, Berlin, 1983.

Spencer, G. S. G., and Taylor, A. M. (1978). A rapid simplified bioassay for somatomedin. *J. Endocrinol.* **78**, 83–88.

Steiner, D. F., Kemmler, W., Tager, H. S., and Peterson, J. D. (1974). Proteolytic processing in the biosynthesis of insulin and other proteins. *Fed. Proc., Fed. Am. Soc. Exp. Biol.* **33**, 2105–2115.

Stevens, R. L., Nissley, S. P., Kimura, J. H., Rechler, M. M., Caplan, A. I., and Hascall, V. C. (1981). Effects of insulin and multiplication stimulating activity on proteoglycan biosynthesis in chondrocytes from the Swarm rat chondrosarcoma. *J. Biol. Chem.* **256**(4), 2045–2052.

Stiles, C. D., Capone, G. T., Scher, C. D., Antoniades, H. N., Van Wyk, J. J., and Pledger, W. J. (1979). Dual control of cell growth by somatomedins and platelet-derived growth factor. Proc. Natl. Acad. Sci. U.S.A. **76**(3), 1279–1283.

Stuart, C. A., Vesely, D. L., Provow, S. A., and Furlanetto, R. W. (1982). Cyclic nucleotides and somatomedin action in cartilage. Endocrinology **111**, 553–558.

Svoboda, M. E., Van Wyk, J. J., Klapper, D. G., Fellows, R. E., Grissom, F. E., and Schlueter, R. J. (1980). Purification of somatomedin-C from human plasma: Chemical and biological properties, partial sequence analysis and relationship to other somatomedins. Biochemistry **19**, 790–797.

Tamborlane, W. V., Hintz, R. L., Bergman, M., Genel, M., Felig, P., and Sherwin, R. S. (1981). Insulin-infusion pump treatment of diabetes. N. Engl. J. Med. **305**, 303–307.

Tannenbaum, G. S., Guyda, H. J., and Posner, B. I. (1983). Insulin-like growth factors: A role in growth hormone negative feedback and body weight regulation via brain. Science **220**, 77–79.

Temin, H. M. (1971). Stimulation by serum of multiplication of stationary chicken cells. J. Cell. Physiol. **78**, 161–170.

Tricoli, J. V., Rall, L. B., Scott, J., Bell, G. I., and Shows, T. B. (1984). Localization of insulin-like growth factor genes to human chromosomes 11 and 12. Nature (London) **310**, 784–786.

Van Wyk, J. J., Svoboda, M. E., and Underwood, L. E. (1980). Evidence from radioligand assays that somatomedin-C and insulin-like growth factor-I are similar to each other and different from other somatomedins. J. Clin. Endocrinol. Metab. **50**, 206–208.

Veldhuis, J. D., and Hammond, J. M. (1979). Multiplication stimulating activity regulates ornithine decarboxylase in isolated porcine granulosa cells in vitro. Endocr. Res. Commun. **6**, 299–309.

Wharton, W., Van Wyk, J. J., and Pledger, W. J. (1981). Inhibition of BALB/c-3T3 cells in late G_1: Commitment to DNA synthesis controlled by somatomedin C. J. Cell. Physiol. **107**, 31–39.

White, R. M., Nissley, S. P., Moses, A. C., Rechler, M. M., Johnsonbaugh, R. E. (1981). The growth hormone dependence of a somatomedin-binding protein in human serum. J. Clin. Endocrinol. Metab. **53**, 49–57.

White, R. M., Nissley, S. P., and Short, P. A. (1982). Developmental pattern of a serum binding protein for multiplication stimulating activity in the rat. J. Clin. Invest. **69**, 1239–1252.

Whitfield, H. J., Bruni, C. B., Frunzio, R., Terrell, J. E., Nissley, S. P., and Rechler, M. M. (1984). Isolation of a cDNA clone encoding rat insulin-like growth factor-II precursor. Nature (London) **312**, 277–280.

Yang, Y. W.-H., Romanus, J. A., Liu, T.-H., Nissley, S. P., and Rechler, M. M. (1985). Biosynthesis of rat insulin-like growth factor II: I. Immunochemical demonstration of a ~20-kilodalton biosynthetic precursor of rat insulin-like growth factor II in metabolically labeled BRL-3A rat liver cells. J. Biol. Chem. **260**, 2570–2577.

Yu, K. T., and Czech, M. P. (1984). The type I insulin-like growth factor receptor mediates the rapid effects of multiplication-stimulating activity on membrane transport systems in rat soleus muscle. J. Biol. Chem. **259**, 3090–3095.

Zapf, J., Waldvogel, M., and Froesch, E. R. (1975). Binding of nonsuppressible insulin-like activity to human serum: Evidence for a carrier protein. Arch. Biochem. Biophys. **168**, 638–645.

Zapf, J., Kaufmann, U., Eigenmann, E. J., and Froesch, E. R. (1977). Determination of nonsuppressible insulin-like activity in human serum by a sensitive protein-binding assay. Clin. Chem. (Winston-Salem, N.C.) **23**, 677–682.

Zapf, J., Jagars, G., Sand, I., and Froesch, E. R. (1978a). Evidence for the existence in human serum of large molecular weight nonsuppressible insulin-like activity (NSILA) different from the small molecular weight forms. *FEBS Lett.* **90**, 135–140.

Zapf, J., Schoenle, E., and Froesch, E. R. (1978b). Insulin-like growth factors I and II: Some biological actions and receptor binding characteristics of two purified constituents of nonsuppressible insulin-like activity of human serum. *Eur. J. Biochem.* **87**, 285–296.

Zapf, J., Schoenle, E., Jagars, G., Grunwald, J., and Froesch, E. R. (1979). Inhibition of the action of nonsuppressible insulin-like activity on isolated rat fat cells by binding to its carrier protein. *J. Clin. Invest.* **63**, 1077–1084.

Zapf, J., Walter, H., and Froesch, E. R. (1981). Radioimmunological determination of insulin-like growth factors I and II in normal subjects and in patients with growth disorders and extrapancreatic tumor hypoglycemia. *J. Clin. Invest.* **68**, 1321–1330.

Zapf, J., Schoenle, E., and Froesch, E. R. (1983). ^{125}I-IGF binding patterns of sera from hypophysectomized rats after long-term treatment with IGF-I, IGF-II and growth hormone (GH): Evidence for effects of GH not mediated by IGF. In "Insulin-Like Growth Factors Somatomedins" (E. M. Spencer, ed.), pp. 57–61. de Gruyter, Berlin.

Zederman, R., Grebing, C., Hall, K., and Low, H. (1980). Effect of somatomedin A and insulin on cyclic AMP generation in isolated rat hepatocytes. *Horm. Metab. Res.* **12**, 251–256.

Zingg, A. E., and Froesch, E. R. (1973). Effects of partially purified preparations with nonsuppressible insulin-like activity (NSILAs) on sulfate incorporation into rat and chicken cartilage. *Diabetologia* **9**, 472–476.

5

Cyclic AMP Elevators Stimulate the Initiation of DNA Synthesis by Calcium-Deprived Rat Liver Cells

ALTON L. BOYNTON,[1] LEONARD P. KLEINE,
AND JAMES F. WHITFIELD

Cell Phsiology Group
Division of Biological Sciences
National Research Council of Canada
Ottawa, Ontario, Canada

[1]Present address: Cancer Center of Hawaii, University of Hawaii, Honolulu, Hawaii 96813.

121

CONTROL OF ANIMAL CELL PROLIFERATION,
VOLUME I

I. INTRODUCTION

Cyclic AMP has long been known to be the second messenger in many hormone actions (see Chapter 9). More recently, it has become clear that this cyclic nucleotide, together with Ca^{2+} and calcium's intracellular receptor and multipurpose enzyme activator calmodulin, functions briefly in late prereplicative development to trigger events leading to the initiation of DNA synthesis (Boynton and Whitfield, 1983; see Chapter 12). In this chapter, we explore the various routes by which cyclic AMP, Ca^{2+}, Ca^{2+}-calmodulin (calmodulin's active form), and the Ca^{2+}- and cyclic AMP-dependent protein kinases trigger the initiation of DNA synthesis.

II. MODEL SYSTEM: THE Ca^{2+}-DEPRIVED T51B RAT LIVER CELL

The original clone of T51B cells was isolated in this laboratory (Swierenga *et al.*, 1978b) from a culture of T51 rat liver cells originally isolated in the laboratory of E. Farber (University of Toronto). They are probably of epithelial origin because they contain cytokeratin (Swierenga *et al.*, 1981) and have glucagon receptors (i.e., glucagon stimulates them to synthesize cyclic AMP) which are found on hepatocytes rather than the nonparenchymal liver cells (Christoffersen and Berg, 1974; Barrazzone *et al.*, 1980). T51B cells are not neoplastic because they do not form tumors in athymic nude mice or have such neoplastic properties as the ability to proliferate in soft agar, independence from exogenous growth factors, resistance to proliferative inhibition by high culture density, or the ability to proliferate in severely Ca^{2+}-deficient medium (Boynton and Whitfield, 1980b; Boynton *et al.*, 1984).

The extracellular Ca^{2+} requirement for the proliferation of non-neoplastic cells can be demonstrated in various ways (Boynton *et al.*, 1982, 1984). For example, in our T51B liver cell model system, simply dropping the Ca^{2+} concentration in the 90% BME (Eagle's basal medium)-10% BCS (bovine calf serum) medium (Boynton *et al.*, 1981b; Borle and Briggs, 1968) from 1.8 mM to 0.02–0.03 mM greatly reduces the ability of the cells to form colonies (Fig. 1). Cells maintained in this calcium-deficient medium for 2 weeks must remain viable and attached to the surface of the petri dish because re-adding Ca^{2+} to a final concentration of 1.25 mM causes colonies to appear during the next few days (Fig. 1). We can find out where in their cell cycle these Ca^{2+}-

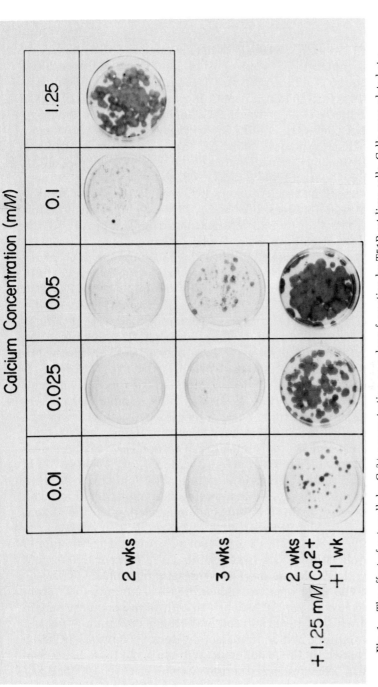

Fig. 1. The effect of extracellular Ca^{2+} concentration on colony formation by T51B rat liver cells. Cells were plated at a density of 38/cm² in medium consisting of 90% (v/v) Eagle's basal medium (BME) and 10% (v/v) bovine calf serum. Twenty-four hours later the medium was replaced with fresh medium containing only 0.015 mM calcium to which various substandard amounts of Ca^{2+} (i.e., 0.01, 0.025, 0.05, and 0.1 mM) were added as previously described by Boynton *et al.* (1982). All cultures were incubated for 2 weeks. Some cultures were then fixed and stained, some were given 1.25 mM Ca^{2+} and incubated for one more week, and the rest were left in their various low Ca^{2+} media and incubated an additional week before fixing and staining. From Boynton *et al.* (1982).

deprived T51B cells are arrested by determining the distribution of cellular DNA contents using propidium iodide staining and flow cytometry. About 55% of the cells in T51B cultures growing randomly in 90% BME-10% BCS medium containing 1.8 mM Ca^{2+} are in G_0/G_1 (i.e., they have the $2n$ DNA content), 36% are in S phase (i.e., they have $> 2n < 4n$ DNA contents), and 9% are in G_2/M [i.e., they have the $4n$ DNA content (Fig. 2A)]. Ca^{2+} deprivation for 24–48 hr in medium containing only 0.025 mM Ca^{2+} increases the fraction of G_0/G_1 cells to 84% and reduces the fraction of S-phase cells to 11% and the fraction of G_2/M cells to 5% (Fig. 2B). To determine more accurately where the Ca^{2+}-deprived cells are blocked in G_0/G_1, we need only raise the external Ca^{2+} concentration to 1.25 mM and autoradiographically follow the ensuing flow of cells into the S phase. This tells us that Ca^{2+}-deprived T51B cells are blocked in at least two regions of their G_1 phase because the added Ca^{2+} causes some, but not all, of the arrested cells to initiate DNA synthesis almost immediately (within 1–2 hr; Fig. 2C). This "rapidly responding" (G_1/S blocked) fraction represents about 40–50% of the arrested G_0/G_1 cells which all enter the S phase by 4–6 hr as seen by the lowering of the rate of entry into S phase at this time. The remaining 40–50% of the arrested G_0/G_1 cells are most likely blocked in a G_0 state like confluent or serum-deprived cells. These "slowly responding," or G_0, Ca^{2+}-deprived cells enter the S phase between 8 and 12 hr after Ca^{2+} addition (Fig. 2C). Many other nonneoplastic cells (e.g., BALB/c 3T3 mouse cells, C3H10T$\frac{1}{2}$ mouse cells, mouse skin primary cultures, NRK rat cells, rat muscle and rat thymic lymphoblast primary cultures, and WI-38 human cells) respond to Ca^{2+} removal and readdition much like T51B cells (Boynton *et al.*, 1976, 1977a,b; Swierenga *et al.*, 1978a; Durkin *et al.*, 1981; Whitfield *et al.*, 1976). It should be noted that no rapidly responding subpopulation of arrested cells accumulates when T51B cultures are incubated in Ca^{2+}-deficient BME medium supplemented with serum (BCS or FBS) which has been treated with the nonspecific Chelex 100 resin instead of the highly specific Ca^{2+}-chelating EGTA to reduce its Ca^{2+} content to 0.02 mM (A. L. Boynton, L. P. Kleine, and J. F. Whitfield, unpublished observations). Chelex 100 resin removes something in addition to Ca^{2+} from serum which is needed for Ca^{2+}-deprived cells to reach and/or remain in late G_1 and which can be replaced by William's medium E, but not BME. Thus, a rapidly responding fraction does accumulate in T51B cultures incubated in Ca^{2+}-deficient William's medium E supplemented with Chelex resin-treated serum (Swierenga *et al.*, 1978b; Whitfield *et al.*, 1980; and more recent unpublished observations).

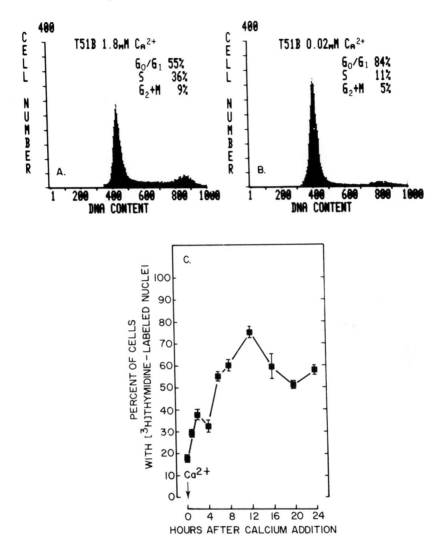

Fig. 2. The distribution of the flow cytofluorometrically determined DNA contents (hence the positions in the cell cycle) of the cells in T51B cultures which had been incubated for 4 days in either (A) 90% BME-10% BCS medium containing 1.8 mM Ca^{2+} or in (B) 90% BME-10% BCS medium containing 0.02 mM Ca^{2+} (Boynton et al., 1984). (C) The "rapid" and "slow" DNA synthetic responses of cells to Ca^{2+} addition to cultures of Ca^{2+}-deprived (0.02 mM for 48 hr) T51B cells. From Swierenga et al. (1980). Points are the means ±SEM of values from four separate cultures.

III. MECHANISM OF INDUCTION OF DNA SYNTHESIS
BY Ca^{2+}

Ca^{2+}, like cyclic AMP, triggers or is otherwise necessary for many cellular functions. The steep Ca^{2+} gradient across the plasma membrane ($10^{-3}M$ outside, about $10^{-7}M$ inside), which the cell spends considerable energy to maintain, can be used to transmit signals and trigger various activities. Thus, internal Ca^{2+} surges contract muscles (Ebashi, 1980), release neurotransmitters (Greengard, 1978), trigger secretion (Rubin, 1982), activate eggs (Gwatkin, 1977), and may mediate the stimulation of quiescent BALB/c 3T3 mouse cells by the platelet-derived growth factor (PDGF) (Tucker *et al.*, 1983).

A. Ca^{2+}-Calmodulin

Investigation of any Ca^{2+}-dependent function inevitably leads to an assessment of Ca^{2+}-calmodulin's role in that function. Ca^{2+}-calmodulin is a multipurpose Ca^{2+}-receptor protein which activates various enzymes such as myosin light-chain kinase, glycogen synthetase, NAD kinase, adenylate cyclase, and cyclic AMP phosphodiesterase (Cheung, 1980).

The first indication of an involvement of Ca^{2+}-calmodulin in the events leading to the initiation of chromosome replication was reported by Boynton *et al.* (1980). According to them, adding trifluoperazine (TFP; an inhibitor of Ca^{2+}-calmodulin action) 15 min before Ca^{2+} prevents the rapidly responding cells in Ca^{2+}-deprived T51B cultures from responding to Ca^{2+} addition, but these cells promptly initiate DNA synthesis when Ca^{2+}-calmodulin is added to a final concentration of 10^{-6} M 4 hr after TFP (Fig. 3A). Unfortunately, the meaning of inhibition by TFP is now rather uncertain because the drug also inhibits Ca^{2+}/phospholipid-dependent protein kinase (protein kinase C) and phospholipase A$_2$ (Lapetina, 1982b; Schatzmann *et al.*, 1981; Walenga *et al.*, 1981), both of which may be involved in the Ca^{2+}-triggered DNA synthetic response (see Sections III, B and IV, C; Boynton and Whitfield, 1981; Boynton *et al.*, 1983). However, a critical role for Ca^{2+}-calmodulin in the G$_1$/S transition is indicated by the facts that Ca^{2+}-calmodulin overrides TFP blockage of the rapid response of Ca^{2+}-deprived T51B cells to Ca^{2+} (Boynton *et al.*, 1980), that Ca^{2+}-calmodulin triggers a prompt DNA synthetic response from cells in Ca^{2+}-deprived BALB/c 3T3 mouse cell cultures (A. L. Boynton, L. P. Kleine, and J. F. Whitfield, unpublished observations) and in T51B rat liver cell cultures (Fig. 3B; Boynton *et al.*, 1980) which have not been

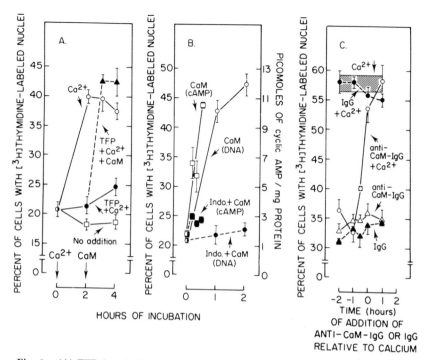

Fig. 3. (A) TFP (10^{-5} M) prevents the DNA synthetic response of Ca^{2+}-deprived T51B cells to Ca^{2+} (1.25 mM) addition, but these TFP-treated cells can respond to a later addition of Ca^{2+}-calmodulin (10^{-6}M). Adapted from Boynton et al. (1980). (B) Ca^{2+}-calmodulin (10^{-6} M) triggers a cyclic AMP surge and DNA synthesis in Ca^{2+}-deprived T51B cells, and these responses are blocked by the cyclooxygenase inhibitor indomethacin (10^{-6} M). From Adv. Cyclic Nucleo. Res., A. L. Boynton and J. F. Whitfield, Copyright 1981. Raven Press, New York. (C) Anti-calmodulin IgG, but not preimmune IgG blocks the DNA synthetic response of Ca^{2+}-deprived T51B cells to Ca^{2+} addition. From Jones et al. (1982). Points in A, B, and C are means ±SEM of values from four separate cultures.

treated with TFP, and, most significantly, that exposing Ca^{2+}-deprived T51B rat liver cell cultures to polyclonal anti-calmodulin IgG (but not preimmune IgG) for 30–60 min before Ca^{2+} addition prevents Ca^{2+} from stimulating the rapidly responding cells to initiate DNA synthesis (Fig. 3C; Jones et al., 1982). Since it takes 30–60 min of pretreatment with anti-calmodulin IgG to make the cell unable to respond to Ca^{2+}, it seems unlikely that the IgG acts by binding directly to Ca^{2+}-calmodulin exposed on the cell surface, but it may need that amount of time to somehow enter the cell and inactivate internal replication-triggering Ca^{2+}-calmodulin complexes.

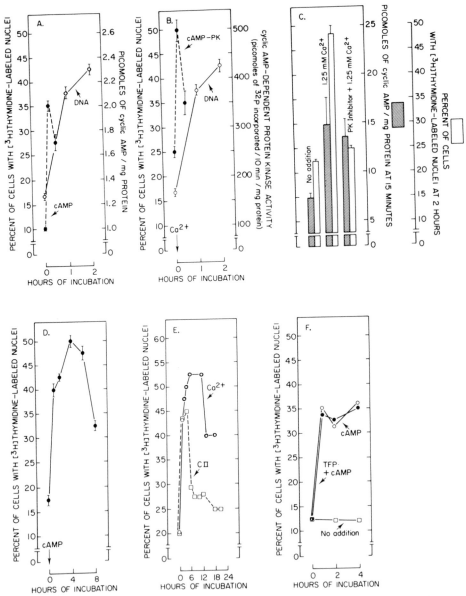

Fig. 4. (A) Adding Ca^{2+} to Ca^{2+}-deprived T51B cultures causes a transient cyclic AMP surge and the prompt initiation of DNA synthesis by the cells of the rapidly responding subpopulation. From Boynton and Whitfield (1979). (B) The Ca^{2+}-induced burst of cyclic AMP-dependent protein kinase activity and the prompt initiation of DNA synthesis by the cells of the rapidly responding subpopulation of Ca^{2+}-deprived T51B cultures. From Boynton *et al.* (1981b). (C) The ability of the cyclic AMP protein kinase

If Ca^{2+} and/or Ca^{2+}-calmodulin function is required for the G_1/S transition of T51B cells, then we must find out what this function is. The most obvious candidates are activation of adenylate cyclase, cyclic nucleotide phosphodiesterase, and/or phospholipase A_2 and C. Activation of adenylate cyclase by Ca^{2+}-calmodulin raises the cyclic AMP level while activation of cyclic nucleotide phosphodiesterase degrades cyclic AMP. Phospholipase activation by Ca^{2+} (but not Ca^{2+}-calmodulin) releases arachidonic acid from membrane phospholipids and triggers an arachidonate cascade during which arachidonic acid is converted to prostaglandins that stimulate cyclic AMP synthesis and consequently activate cyclic AMP-dependent protein kinases (cAMP-PK). Thus, we have two basic pathways leading to adenylate cyclase stimulation, a cyclic AMP surge and a burst of cAMP-PK activity which both involve Ca^{2+} and/or Ca^{2+}-calmodulin function.

B. Cyclic AMP and Cyclic AMP-Dependent Protein Kinases

The simple DNA synthesis-triggering addition of Ca^{2+} (1.25 mM) to Ca^{2+}-deprived T51B cells also causes a pre-DNA synthesis cyclic AMP surge which could be due to stimulation of adenylate cyclase or inhibition of cyclic nucleotide phosphodiesterase (Fig. 4A; Boynton and Whitfield, 1979). The subsequent fall in the intracellular cyclic AMP level which is as rapid as the rise must be due to increased cyclic nucleotide phosphodiesterase activity and/or efflux of cyclic AMP from the cell (Barber and Butcher, 1983). Stimulation of adenylate cyclase by various hormones (see Section V), inhibition of cyclic nucleotide phosphodiesterase by 3-isobutyl-1-methylxanthine, or addition of cyclic AMP itself all stimulate the rapidly responding (but not the slowly responding) cells in Ca^{2+}-deprived T51B liver cells (Figs. 3D, 6B,C, 9A, 10; Boynton and Whitfield, 1979).

Cyclic AMP has only two intracellular receptors, the well-known

---

inhibitor (100 µg/ml) to prevent Ca^{2+} (1.25 mM) from stimulating DNA synthesis without affecting the ability of Ca^{2+} to stimulate cyclic AMP accumulation in Ca^{2+}-deprived cultures. From Boynton et al. (1981b). (D) Cyclic AMP (10^{-5} M)-induced DNA synthesis in the rapidly responding cells in Ca^{2+}-deprived T51B cultures. From Boynton and Whitfield (1979). (E) Catalytic subunits (C subunits; 10 phosphorylating units/ml) from cyclic AMP-dependent protein kinases stimulate the initiation of DNA synthesis in the rapidly responding but not the slowly responding subpopulations of Ca^{2+}-deprived T51B cultures, while Ca^{2+} (1.25 mM) stimulates both subpopulations. (F) The inability of TFP (10^{-5} M) to prevent cyclic AMP (10^{-5} M) from triggering DNA synthesis in the rapidly responding cells in Ca^{2+}-deprived T51B cultures. The points in A–F are the means ±SEM of values from four separate cultures.

types I and II cyclic AMP-dependent protein kinase holoenzymes (cAMP-PK) which consist of different regulatory subunits R^I or R^{II} and the same catalytic or C subunit (i.e., $R_2^I C_2$ and $R_2^{II} C_2$). Binding of cyclic AMP to the holoenzyme's R receptor subunits dissociates it into cyclic AMP·R and C subunits. The liberated C subunits then catalyze the phosphorylation of serine or threonine residues of various proteins.

Cyclic AMP-dependent protein kinase activity increases briefly after Ca^{2+} (1.25 mM) addition to Ca^{2+}-deprived T51B cultures and before the rapidly responding cells initiate DNA synthesis (Fig. 4B; Boynton and Whitfield, 1981; Boynton et al., 1981b). If this burst of protein kinase activity were linked to the initiation of DNA synthesis, then inhibiting either it or the cyclic AMP surge which triggers it should prevent the DNA synthetic response. The necessity for the cyclic AMP surge is indicated by the inhibition of the rapid DNA synthetic response of Ca^{2+}-deprived T51B cells to Ca^{2+} by imidazole, a cyclic nucleotide phosphodiesterase stimulator, which prevents the cyclic AMP surge (Boynton and Whitfield, 1979). The importance of the burst of cAMP-PK activity is indicated by the prevention of the DNA synthetic response, but not the cyclic AMP surge of the rapidly responding cells in Ca^{2+}-deprived T51B cultures to Ca^{2+} by the cAMP-PK inhibitor (PKInh), which specifically inhibits the activity of the C subunits of cAMP-PK (Fig. 4C; Boynton et al., 1981b).

According to the above evidence, cyclic AMP accumulation and a burst of cAMP-PK activity are involved in the Ca^{2+}-induced initiation of DNA synthesis by the rapidly responding cells in Ca^{2+}-deprived T51B cultures. However, the cyclic AMP surge and the burst of cAMP-PK activity may have occurred only in the slowly responding cells, only in the rapidly responding cells, or in the entire cell population. Perhaps the most convincing evidence that cyclic AMP mediates calcium's action on the rapidly responding subpopulation is the fact that cyclic AMP itself (10^{-8}–10^{-5} M) stimulates only these cells and not the slowly responding cells to initiate DNA synthesis (Fig. 4D; Boynton and Whitfield, 1979). Cyclic AMP also greatly increases cAMP-PK activity, and both this and the initiation of DNA synthesis are inhibited by the C subunit inhibitor PKInh. Moreover, adding C subunits (5–100 phosphorylating units/ml) to Ca^{2+}-deprived T51B cultures stimulates the rapidly responding cells to synthesize DNA, a response which is inhibited by adding PKInh before the C subunits (Fig. 4E; Boynton et al., 1981b).

The notion that the cAMP-PK stimulated step follows the Ca^{2+}/Ca^{2+}-calmodulin stimulated step can be confirmed by first treating Ca^{2+}-deprived T51B cells with an inhibitory concentration (10^{-5}

M) of TFP and then a stimulatory concentration (10^{-5} M) of cyclic AMP. Cyclic AMP, unlike Ca^{2+}, can stimulate the TFP-treated rapidly responding cells to make DNA, which is consistent with Ca^{2+}-calmodulin function (or perhaps protein kinase C and/or phospholipase A_2 activity) somehow being responsible for the cyclic AMP surge (Fig. 4F). This is supported further by the fact that adding Ca^{2+}-calmodulin to Ca^{2+}-deprived T51B cultures first stimulates a large accumulation of cyclic AMP and shortly thereafter the initiation of DNA synthesis by the rapidly responding cells (Fig. 3B; Boynton and Whitfield, 1981). However, these results do not indicate the mechanism by which Ca^{2+} and/or Ca^{2+}-calmodulin stimulates adenylate cyclase.

C. Arachidonic Acid Cascade

Because calcium's action (i.e., increased intracellular cyclic AMP, cAMP-PK activity, and subsequent DNA synthesis) on Ca^{2+}-deprived T51B cells seems to be mediated by cyclic AMP, we must find out how Ca^{2+} stimulates adenylate cyclase. It might activate adenylate cyclase indirectly via Ca^{2+}-calmodulin or it might directly activate phospholipases and trigger an arachidonate cascade with its adenylate cyclase-stimulating prostaglandin products. We might be able to answer this question by using various agents which affect the arachidonate cascade and prostaglandin production.

Phospholipase A_2 (PLA_2), a Ca^{2+}-dependent enzyme, releases fatty acids such as arachidonic acid from membrane phospholipids. Adding PLA_2 stimulates the rapidly responding cells in Ca^{2+}-deprived T51B cultures to initiate DNA synthesis (Fig. 5A). This response is blocked by pretreating the cultures with indomethacin (Fig. 5A), a cyclooxygenase inhibitor which blocks adenylate cyclase stimulation by preventing prostaglandin production. PLA_2 could release arachidonic acid either from phospholipids in the medium's serum component or from the cell's plasma membrane. Arachidonic acid itself (10^{-9}–10^{-6} moles/liter) causes a large cyclic AMP surge which is followed by the initiation of DNA synthesis by the rapidly responding cells (Fig. 5B,C; Boynton and Whitfield, 1980a). This DNA synthetic response to arachidonic acid is inhibited by indomethacin (Fig. 5B; Boynton and Whitfield, 1980a) which also inhibits the cyclic AMP and DNA synthetic responses to Ca^{2+} (A. L. Boynton, L. P. Kleine, and J. F. Whitfield, unpublished observations) and Ca^{2+}-calmodulin (Fig. 3B; Boynton and Whitfield, 1981). Moreover, at least one arachidonate product, prostaglandin E_1, stimulates the rapidly responding cells in Ca^{2+}-deprived T51B cell cultures to initiate DNA synthesis (Fig. 5D).

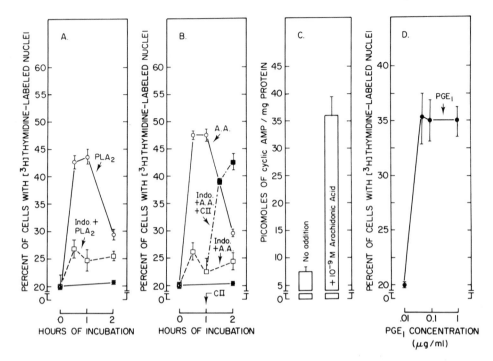

Fig. 5. Evidence supporting the involvement of an arachidonic acid cascade in the initiation of DNA synthesis by the rapidly responding cells in Ca^{2+}-deprived T51B cultures. (A) Phospholipase A_2 (PLA$_2$; 5 units/ml) stimulates DNA synthesis, and its action is blocked by adding indomethacin (10^{-6} M). (B) Arachidonic acid triggers a DNA synthetic response which is blocked by adding indomethacin (10^{-6} M) 15 min before arachidonic acid. Indomethacin's blocking action is rapidly reversed by adding catalytic subunits (10 phosphorylating units/ml) from cyclic AMP-dependent protein kinases. Adapted from Boynton and Whitfield (1980a). (C) Arachidonic acid stimulates cyclic AMP accumulation. (D) PGE$_1$ stimulates DNA synthesis. The points in A–D are the means ±SEM of values from four separate cultures.

Thus, one or more products of the arachidonate cascade could be responsible for the cyclic AMP surge, the burst of cAMP-PK activity, and the prompt DNA synthetic response triggered by Ca^{2+} or Ca^{2+}-calmodulin in Ca^{2+}-deprived T51B cultures. Of course, this will have to be proved by measurements of phospholipid turnover, arachidonic acid release, and prostaglandin synthesis after addition of Ca^{2+}, Ca^{2+}-calmodulin, and cyclic AMP-elevating agonists to Ca^{2+}-deprived T51B cells.

An alternative to a Ca^{2+}-triggered, PLA$_2$-induced arachidonic acid cascade is a Ca^{2+}-triggered phospholipase C-induced cascade, one product of which is diacylglycerol. The accumulating diacylglycerol could serve one or more of three functions. First, it could set off an arachidonate cascade by being a substrate for diacylglyceride lipase which releases arachidonic acid from diacylglycerol; second, it might be converted by diglyceride kinase to phosphatidic acid, a known calcium ionophore (Lapetina, 1982a); or third, it might stimulate membrane Ca^{2+}/phospholipid-dependent protein kinase C by increasing the enzyme's affinity for Ca^{2+} (Takai et al., 1979). As we shall see in Section IV,C, protein kinase C may play an important role in prereplicative development because its activity in nonneoplastic T51B cells, like the proliferative activity of these nonneoplastic cells, is reduced by Ca^{2+} deprivation, while neither it nor proliferative activity is affected by Ca^{2+} deprivation of neoplastic T51B cells (Boynton et al., 1983).

It should be noted that phospholipase function may be controlled by a modulator protein called lipomodulin (Hirata, 1981). This is a 40,000-dalton protein inhibitor of phospholipase activity which is inactivated when phosphorylated by a serine protein kinase and activated by dephosphorylation (Hirata, 1981). Thus, Ca^{2+} may trigger an arachidonate cascade by stimulating the phosphorylation of lipomodulin by some Ca^{2+}-dependent protein kinase such as protein kinase C.

IV. MECHANISM OF INDUCTION OF DNA SYNTHESIS BY TUMOR PROMOTERS

Neoplastic cells of mesenchymal origin and parenchymal (as opposed to lining; Chapter 12, this volume) epithelial cells need from 25- to 100-fold less extracellular calcium to proliferate than the corresponding normal cells (Boynton and Whitfield, 1976; Boynton et al., 1981a; Swierenga et al., 1981). Thus, various neoplastically transformed (by aflatoxin B_1, N-methyl-N'-nitro-N-nitrosoguanidine, or 4-nitroquinoline-1-oxide) T51B rat liver cells proliferate rapidly and indefinitely in medium containing 0.02 mM Ca^{2+} while nonneoplastic T51B cells need at least 0.5 mM Ca^{2+} to proliferate maximally and indefinitely.

A. Ca^{2+}-Calmodulin

Tumor-promoting agents [e.g., cyclamates, 2,4-dichlorophenoxyacetic acid ("2,4-D"), dichlorodiphenyltrichloroethane (DDT), deoxycholic acid, epidermal growth factor, lithocholic acid, phenobarbital, saccharin, and 12-O-tetradecanoylphorbol-13-acetate (TPA)] cause

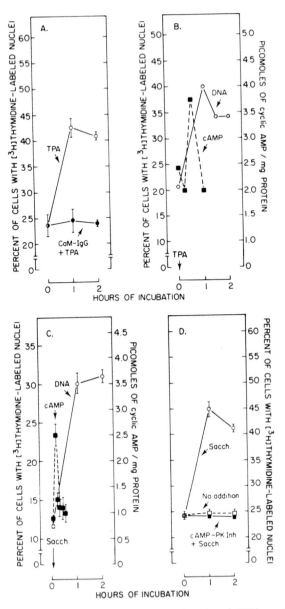

Fig. 6. (A) The inhibition of TPA (50 ng/ml)-induced DNA synthesis by anti-calmodulin IgG (20 μg/ml) in cultures of Ca^{2+}-deprived T51B cells. From Jones *et al.* (1982). (B,C) Stimulation of cyclic AMP accumulation (determined at 30 min after addition of stimulator) by TPA (50 ng/ml) and saccharin (50 μg/ml) in Ca^{2+}-deprived T51B cells. From Boynton *et al.* (1982). (D) The inhibition of saccharin-induced DNA synthesis by the inhibitor (100 μg/ml) of cyclic AMP-dependent protein kinase C subunit activity in cultures of Ca^{2+}-deprived T51B cultures (Boynton *et al.*, 1982). The points in A–D are means ±SEM of values from four separate cultures.

nonneoplastic Ca^{2+}-deprived T51B cells to behave like neoplastic cells by initiating DNA synthesis and forming colonies in Ca^{2+}-deficient medium. The DNA synthetic response of the rapidly responding cells in Ca^{2+}-deprived T51B cultures to TPA, like their response to Ca^{2+}, is probably mediated at some point by Ca^{2+}-calmodulin because adding TFP or polyclonal anti-calmodulin antibody to the cultures before TPA blocks the response to the promoter (Fig. 6A). The small amount of Ca^{2+} in the Ca^{2+}-deficient medium seems to be needed for a TPA-induced Ca^{2+} flux which raises the internal Ca^{2+} level (A. L. Boynton, L. P. Kleine, and J. F. Whitfield, unpublished observations) and may be a product of and/or the trigger of phospholipase activation and an arachidonic acid cascade. Indeed, TPA stimulates arachidonic acid release and prostaglandin synthesis in C3H10T$\frac{1}{2}$ mouse fibroblasts (Mufson et al., 1979), and phenylglyoxal, a PLA_2 inhibitor (Shier and Durkin, 1982), blocks the responses of the rapidly responding cells in Ca^{2+}-deprived T51B cultures to TPA (A. L. Boynton, L. P. Kleine, and J. F. Whitfield, unpublished observations).

B. Cyclic AMP and Cyclic AMP-Dependent Protein Kinases

Since a cyclic AMP surge is needed for the initiation of DNA synthesis by the rapidly responding cells in Ca^{2+}-deprived T51B cultures, TPA and saccharin should cause such a surge. Indeed, both tumor promoters cause a cyclic AMP surge which is followed almost immediately by the initiation of DNA synthesis (Fig. 6B,C). It may not be surprising that TPA can elicit such a response, since it can replace Ca^{2+} and stimulates a Ca^{2+}-dependent adenylate cyclase in crude homogenates of rat pituitary cells (Brostrom et al., 1982). Further support for an involvement of cyclic AMP and cAMP-PK activity in both TPA and saccharin actions is the fact that PKInh prevents these agents from stimulating the rapidly responding cells in Ca^{2+}-deprived T51B cultures (Fig. 6D).

C. Ca^{2+}/Phospholipid-Dependent Protein Kinase

The ability of TPA to stimulate DNA synthesis and proliferation of nonneoplastic cells in Ca^{2+}-deficient medium may be linked to the recently discovered Ca^{2+}/phospholipid-dependent protein kinase. This ubiquitous protein kinase is apparently stimulated in vivo by having its affinity for Ca^{2+} and phosphatidylserine raised by the diacylglycerol generated by an increase in phosphatidylinositol turnover.

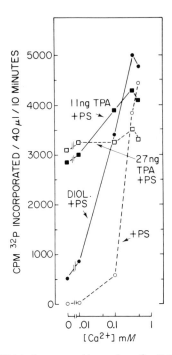

Fig. 7. The ability of TPA to bypass and/or replace the Ca^{2+} and diacylglycerol (e.g., diolein) normally needed for maximum activity of protein kinase C. The "0" Ca^{2+} condition was obtained with excess (i.e., 0.3 mM) EGTA. Details of the protein kinase C assay can be found in Boynton *et al.* (1983).

over. The TPA receptor has recently been found to copurify with protein kinase C (Niedel *et al.*, 1983), which suggests that this protein kinase may actually be part of TPA's receptor in the plasma membrane. Furthermore, TPA (in an *in vitro* assay) can replace diacylglycerol and Ca^{2+} and thus stimulate the enzyme by increasing its affinity for Ca^{2+} and phosphatidylserine (Fig. 7; Castagna *et al.*, 1982; Niedel *et al.*, 1983; Ashendel *et al.*, 1983). This plus the fact that TPA also enables normal T51B cells to proliferate in Ca^{2+}-deficient medium suggest that a persistence of protein kinase C activity is the basis of the preneoplastic or neoplastic cell's ability to continue proliferating after being transferred to a severely Ca^{2+}-deficient medium. Indeed, nonneoplastic T51B cells have three- to four-fold less EDTA/EGTA extractable protein kinase C activity when in Ca^{2+}-deficient (0.02 mM) medium than when they are in medium containing 1.25–1.8 mM Ca^{2+}. However, preneoplastic and neoplastic T51B cells have a normal or even supranormal level of EDTA/EGTA extractable protein kinase C activity when incubated in this Ca^{2+}-deficient medium (Boynton et al., 1983). It is interesting to note that treatment of carcinogen (MNNG)-initiated cultures of T51B cells with the tumor promoter saccharin for 21 consecutive days enables a significant number of them to form colonies in Ca^{2+}-deficient medium. The cells from several of these clones can be cultured indefinitely in calcium-deficient medium where they maintain high levels of EDTA/EGTA-extractable protein kinase C activity. Thus, protein kinase C may be one component of the normally Ca^{2+}-dependent prereplicative control mechanism, but its relation to cyclic AMP and cAMP-PK activity is simply not yet understood. Further support for this is our recent finding that the EDTA/EGTA-extractable protein kinase C activity transiently surges (i.e., 0–20 min, 1–2 hr, and again at 8–10 hr) only during the two Ca^{2+}-dependent stages of the prereplicative development of the cells in serum-stimulated confluent T51B cultures (Fig. 8).

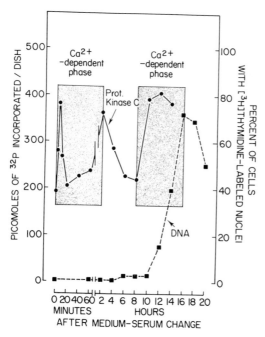

Fig. 8. Protein kinase C activity rises transiently during the Ca^{2+}-dependent stages of the type 2 (see Chapter 12) prereplicative period of serum-stimulated confluent cultures of T51B cells. Details of the experimental procedures and the measurements of EDTA/EGTA-extractable enzyme activity can be found in Boynton et al. (1983, 1984). The protein kinase C data are from one of six such experiments. The points are the means ±SEM of the values from three separate cultures.

V. INDUCTION OF DNA SYNTHESIS BY OTHER CYCLIC AMP ELEVATORS

The evidence just reviewed strongly suggests that a cyclic AMP surge and a burst of cAMP-PK activity are required for events in the late G_1 phase of the growth-division cycle. This is supported by the fact that various kinds of cells ranging from free-living organisms such as yeast and diatoms to cells in mammalian tissues generate a transient cyclic AMP surge as they near the end of the G_1 phase of their growth-division cycle (Boynton and Whitfield, 1983). Moreover, a wide variety of cells need exogenous cyclic AMP or a cyclic AMP elevator such as cholera toxin to proliferate maximally *in vitro* (Boynton and Whitfield, 1983).

Glucagon, one of the classic liver adenylate cyclase stimulators (see Chapter 9), stimulates cyclic AMP accumulation and the initiation of DNA synthesis in the rapidly responding cells in Ca^{2+}-deprived T51B cultures (Fig. 9A,B; Boynton and Whitfield, 1983). This DNA synthetic response is inhibited by further dropping the Ca^{2+} concentration in the already low (0.025 mM) Ca^{2+} medium by adding more EGTA [Fig. 9B; EGTA chelates calcium and strips calcium from cell surface binding sites (Damluji and Riley, 1979)] or by adding $LaCl_3$ [Fig. 9B; $LaCl_3$ displaces calcium from the cell surface and blocks transmembrane calcium fluxes (Langer and Frank, 1972; Martin and Richardson, 1979)]. Ca^{2+}-calmodulin or protein kinase C may also be involved in this DNA synthetic response to glucagon because it was inhibited by adding TFP to the Ca^{2+}-deprived T51B cultures before glucagon (Fig. 9C). However, we do not know whether EGTA, $LaCl_3$, and TFP prevent glucagon from causing a cyclic AMP surge in the Ca^{2+}-deprived T51B cells. Certainly the hormone does not need Ca^{2+} and Ca^{2+}-calmodulin to stimulate adenylate cyclase in cell-free preparations (Rodbell, 1983), which suggests that the role of Ca^{2+} may not be related to cyclic AMP accumulation. It may be that treating calcium-deprived cells with EGTA and/or $LaCl_3$ alters membranes in such a way that glucagon can no longer stimulate adenylate cyclase. Alternatively, and more likely, certain Ca^{2+}/Ca^{2+}-calmodulin-dependent processes may be required along with those triggered by cyclic AMP-dependent protein kinase activity to trigger DNA replication.

Other small peptides and hormones also cause a cyclic AMP surge and stimulate the initiation of DNA synthesis in the rapidly responding cells of Ca^{2+}-deprived T51B cultures. These cyclic AMP elevators include calcitonin, cholera toxin, EGF, and parathyroid hormone (Fig. 10; Swierenga *et al.*, 1980; also A. L. Boynton, L. P. Kleine, and J. F. Whitfield, unpublished observations). Treatment of calcitonin-treated cells

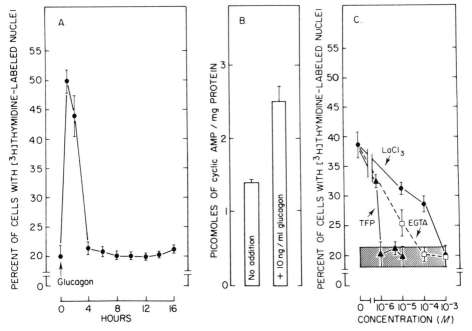

Fig. 9. (A) Glucagon (10 ng/ml) stimulates the initiation of DNA synthesis by the rapidly responding cells in Ca^{2+}-deprived T51B cultures, and (B) this DNA synthetic response is preceded by a rise in cyclic AMP. (C) The DNA synthetic response is blocked by removing the residual Ca^{2+} from the already low calcium (0.02 mM) medium–serum mixture with the highly specific Ca^{2+} chelator EGTA (10^{-5} M) by incubating the cultures with $LaCl_3$ (10^{-5} M) which inhibits transmembrane Ca^{2+} fluxes and displaces Ca^{2+} from the cell surface binding sites, or by incubating the cultures with the Ca^{2+}-calmodulin antagonist and protein kinase C inhibitor TFP (10^{-5} M). The points in A–C are the means ±SEM of values from four separate cultures.

with the cyclic nucleotide phosphodiesterase-activating imidazole inhibits both the cyclic AMP surge and the initiation of DNA synthesis (Fig. 10B). Significantly, human mammary T47D tumor cells can be stimulated to initiate a growth-division cycle by calcitonin, which in this case also stimulates cyclic AMP synthesis and specifically activates type II cyclic AMP-dependent protein kinase (Ng *et al.*, 1983). Cyclic GMP stimulates cyclic AMP accumulation and DNA synthesis of the rapidly responding cells (Boynton and Whitfield, 1983), as does the cyclic nucleotide phosphodiesterase-inhibiting 3-isobutyl-1-methyl xanthine (Boynton and Whitfield, 1979, 1983). We do not know whether these cyclic AMP elevators also trigger an arachidonate cascade and prostaglandin synthesis; nor do we know if their stimulation of DNA

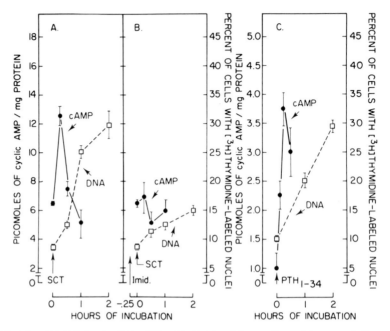

Fig. 10. (A) Calcitonin (1.0 milliunit/ml) stimulates cyclic AMP accumulation and the initiation of DNA synthesis by the rapidly responding cells of Ca^{2+}-deprived T51B cultures, and (B) these responses are prevented by the cyclic nucleotide phosphodiesterase stimulator imidazole (5 mM). (C) Parathyroid hormone (1.0 μg/ml) stimulates cyclic AMP accumulation and the initiation of DNA synthesis by the rapidly responding cells of Ca^{2+}-deprived cultures of T51B cells. The points are means ±SEM of values from four separate cultures.

synthesis depends on Ca^{2+}-calmodulin function. However, we do know that the DNA synthetic response of the rapidly responding cells in Ca^{2+}-deprived T51B cultures to EGF is inhibited by polyclonal anti-calmodulin IgG (Boynton *et al.*, 1982), which suggests that this small peptide requires Ca^{2+}-calmodulin at some point to stimulate DNA synthesis. Regardless of the mechanisms of action of these several cyclic AMP elevators, it is clear that an increase in the intracellular cyclic AMP level starts DNA replication in the rapidly responding calcium-deprived T51B cells.

It should be noted that primary cultures of rat thymic lymphocytes contain a subpopulation of cycling lymphoblasts with short G_1 phases which respond to Ca^{2+} deprivation and readdition in exactly the same manner as T51B cells (i.e., with a rapid increase in DNA synthesis preceded by a cyclic AMP surge; Boynton and Whitfield, 1979, 1983; Whitfield *et al.*, 1976, 1980). The cyclic AMP-elevating agents which

TABLE I

Agents Which Stimulate Cyclic AMP Accumulation and
the Initiation of DNA Synthesis by Rapidly
Responding Lymphoblasts in Primary Cultures of
Rat Thymic Lymphocytes

Acetylcholine	Cyclic GMP
Ca^{2+}	Parathyroid Hormone
Calcitonin	PGA_1
Concanavalin A	PGE_1
Epinephrine	

stimulate DNA synthesis by these thymic lymphoblasts are listed in
Table I.

In the context of interacting Ca^{2+}-triggered phospholipase–arachidonate cascade and cyclic AMP surge, it is interesting that Ca^{2+}-de-
prived thymic lymphoblasts can be stimulated to replicate their DNA
by a Ca^{2+}-dependent, cyclic AMP-mediated action of bradykinin, a
vasoactive nonapeptide which is generated in all injured tissues (Table
I; Perris and Whitfield, 1969; Whitfield et al., 1970). This mitogenic
wound hormone (Perris and Whitfield, 1969) also triggers prereplicative development of serum-deprived human foreskin fibroblasts
(Owen and Villereal, 1982) by a mechanism which also includes a burst
of cyclic AMP production (Bareis et al., 1983). Bareis et al. (1983) have
proposed that this nonapeptide mitogen primarily stimulates Ca^{2+} influx (possibly by stimulating phospholipid methyltransferase activity),
which in turn triggers an arachidonate cascade and the generation of
cyclic AMP-elevating prostaglandins either directly by activating PLA_2
or less directly through the generation of diacylglycerol by phospholipase C.

VI. STIMULATION OF CELL SURFACE PROTEIN KINASES

A. Ca^{2+}

Many of these cyclic AMP-elevating agonists and protein kinases are
too large to readily cross the plasma membrane (i.e., Ca^{2+}-calmodulin,
16 kilodaltons; C subunits, 39 kilodaltons; cAMP-PK holoenzyme, 185
kilodaltons, while others such as EGF, glucagon, and TPA interact with
specific receptors on the cell surface (Carpenter et al., 1978; Niedel et
al., 1983; Rodbell, 1983). Thus, it is not unreasonable to expect that one
of the initial events that occur when any one of these agents interacts

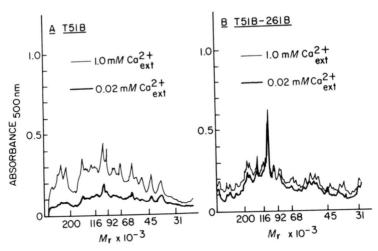

Fig. 11. Ca^{2+} affects the phosphorylation of surface proteins (A) and proliferation (Boynton et al., 1983a,b) of nonneoplastic T51B rat liver cells, but it does not affect the phosphorylation of surface proteins (B) or the proliferation (Boynton et al., 1983) of the neoplastic T51B-261B cells. These are densitometric scans of autoradiographs of ^{32}P-labeled (from $[\gamma-^{32}P]ATP$) surface proteins prepared according to Boynton et al. (1982). Culture techniques may be found in Boynton et al. (1982, 1983).

with calcium-deprived T51B cells is the phosphorylation of certain critical membrane proteins or the activation of cell surface protein kinases (cyclic AMP-dependent and independent) which phosphorylate such proliferation-related membrane proteins. Incubating actively cycling T51B cells for 5 min in $[\gamma-^{32}P]ATP$-containing phosphorylation buffer labels several polypeptides (Fig. 11A). Incubating T51B cells in Ca^{2+}-deficient medium (0.025 mM) for 24–48 hr traps them in the G_1 phase and inhibits or reduces the phosphorylation of most of these polypeptides (Figs. 11A and 12; Boynton et al., 1982; Kleine et al., 1984). A requirement for the proliferative activity is suggested by the fact that re-adding Ca^{2+} (1.25 mM) causes a rapid (within 5 min) phosphorylation of several of the polypeptides, the most prominent of which have molecular weights of 37,000, 42,000, 56,000, 96,000, 110,000, and 182,000 (Fig. 12).

There are four main reasons to believe that these proteins and the protein kinases which catalyze their phosphorylation are on the cell surface.

1. The concentration of $[\gamma-^{32}P]ATP$ in the phosphorylation buffer is in the micromolar range, while the internal ATP concentration is in the millimolar range. If $[\gamma-^{32}P]ATP$ must enter the cell to label these pro-

teins, it would be diluted instantly about 1000-fold in the large endogenous unlabeled ATP pool. In fact, there is no detectable phosphorylation when we add a 100-fold excess of unlabeled ATP to the [γ-^{32}P]ATP-containing phosphorylation buffer.

2. Postphosphorylation treatment of whole cells (which are still at-

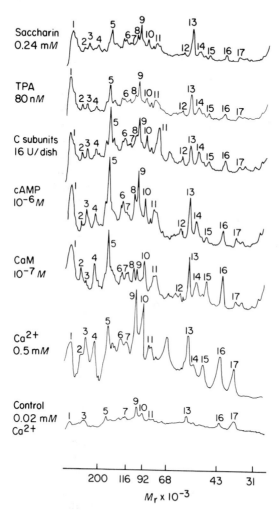

Fig. 12. Densitometric scans of autoradiographs of cell surface phosphoproteins labeled in the presence of exogenous [^{32}P]ATP by adding Ca^{2+}, Ca^{2+}-calmodulin, cyclic AMP, cyclic AMP-dependent protein kinase catalytic subunits, TPA, and saccharin to Ca^{2+}-deprived T51B cultures. Experimental details are described in Boynton *et al.* (1982) and Kleine *et al.* (1984).

tached to the petri dish surface) with dilute (0.001%) trypsin solution removed all traces of phosphorylation (Fig. 12). Since this very brief trypsin treatment does not generally destroy cellular protein and since the protein bands on SDS–polyacrylamide gels from trypsin-treated and untreated cells are essentially the same, the enzyme probably does not enter the cell.

3. If [γ-^{32}P]ATP enters cells, or if ^{32}P$_i$ from externally degraded [γ-^{32}P]ATP enters the cells and is incorporated into endogenous ATP pools, then some or all of the trypsin-sensitive phosphoproteins must come from inside the cell and the trypsin used to destroy them must enter the cell. If this is the case, then intracellular proteins labeled with ^{32}P$_i$ instead of [γ-^{32}P]ATP should also be destroyed by our brief trypsin treatment. This possibility can be tested by labeling internal cellular proteins with ^{32}P$_i$ for 1 hr and then exposing the cells to 0.001% trypsin for 5 min, removing the cells from the petri dish surface, and finally resolving the labeled proteins. Trypsin does not affect the resulting radioactive phosphoproteins and therefore does not enter the cell (A. L. Boynton, L. P. Kleine, and J. F. Whitfield, unpublished observations). Finally, the surface protein kinases seem unable to use ^{32}P$_i$ from externally degraded [γ-^{32}P]ATP to label surface proteins because a 5-min exposure to exogenous ^{32}P$_i$ does not label surface proteins (A. L. Boynton, L. P. Kleine, and J. F. Whitfield, unpublished observations).

4. The protein phosphorylation may also occur only in damaged or broken cells. However, more than 99.5% of the cells are undamaged and viable under the conditions used for surface phosphorylation because they can be stained with fluorescein diacetate.

B. Cyclic AMP and Cyclic AMP Elevators

Cyclic AMP and cAMP-PK C subunits stimulate the phosphorylation of apparently the same proteins on the surfaces of Ca^{2+}-deprived T51B cells as Ca^{2+}. However, they also stimulate the phosphorylation of a protein having a molecular weight of 112,000 (Fig. 12).

Because cyclic AMP almost certainly stimulates cell surface phosphorylations through cAMP-PK, exogenous cyclic AMP may bind preferentially to one or both of the two R (R^I and R^{II}) subunits of cAMP-PK holoenzymes on the cell surface. The photoaffinity probe azido-[^{32}P]-cyclic AMP combines with three trypsin-sensitive receptors on the surface of Ca^{2+}-deprived T51B cells. These three receptor proteins have molecular weights of 48,000, 54,000, and 56,000, and must be, respectively, the R^I, R^{II}, and phosphorylated R^{II} subunits because they comigrate with the authentic purified proteins on SDS–polyacrylamide gels. We do not know if activation of one or both of the cAMP-PKs is required to trigger the DNA synthetic response of the rapidly responding Ca^{2+}-deprived cells. These answers will come eventually from experiments using anti-R^I and anti-R^{II} antibodies.

Other agents that increase cyclic AMP, such as Ca^{2+}-calmodulin and the tumor promoters saccharin and TPA, stimulate the phosphorylation of apparently the same phosphoproteins as cyclic AMP and cAMP-PK C subunits. It may be that all agents which increase the cyclic AMP level and trigger DNA synthesis in the rapidly responding Ca^{2+}-deprived T51B cells act directly or indirectly by phosphorylating the same surface protein(s). However, it should be noted that not all surface phosphorylations are due to cAMP-PK activities because the PKInh did not inhibit all of them (A. L. Boynton, L. P. Kleine, and J. F. Whitfield, unpublished observations). Thus, the G_1/S transition in Ca^{2+}-deprived cells may be the result of a combination of cyclic AMP-dependent and independent (e.g., protein kinase C) protein kinase activity.

C. Neoplastic Cells

T51B-261B cells are neoplastic derivatives of T51B cells which, unlike T51B cells, proliferate in Ca^{2+}-deficient medium. They also have a surface phosphorylation pattern which is essentially unaffected by incubation in Ca^{2+}-deficient medium (Fig. 11B). Thus, the protein kinases (cyclic AMP dependent and independent) responsible for these cell surface phosphorylations must somehow have been permanently deregulated or sensitized to external Ca^{2+} in order to function in Ca^{2+}-deficient medium.

VII. SUMMARY AND CONCLUSIONS

The results discussed above support the contention that a transient cyclic AMP surge and a burst of cyclic AMP-dependent protein kinase activity are required for the G_1/S transition because a wide variety of agents which increase intracellular cyclic AMP levels promote the initiation of chromosome replication. They also suggest at least two routes to the triggering cyclic AMP surge. First, Ca^{2+}-calmodulin could stimulate adenylate cyclase directly. Second, adenylate cyclase-stimulating prostaglandins could be generated from the arachidonic acid released from membrane phospholipids by increased phospholipase activity. Cyclic AMP activation of cyclic AMP-dependent protein kinases followed by phosphorylation of certain proteins on the cell surface by these enzymes would then trigger events leading to the initiation of chromosome replication.

Neoplastic transformation which deregulates cell proliferation alters the requirement for extracellular Ca^{2+}. Preneoplastic and neoplastic derivates of T51B cells are able to proliferate in medium containing 25- to 100-fold less extracellular calcium. This reduction of the extracellular Ca^{2+} requirement may be the work of the recently discovered Ca^{2+}/phospholipid-dependent protein kinase C. Thus, nonneoplastic T51B cells reduce their protein kinase C activity and stop proliferating when incubated in Ca^{2+}-deficient medium. By contrast, preneoplastic and neoplastic derivatives of these cells maintain a high or even higher protein kinase C activity and continue proliferating when incubated in Ca^{2+}-deficient medium. This fact coupled with the observations that the catalytic subunit of the cyclic AMP-dependent protein kinases causes Ca^{2+}-deprived T51B cells to initiate DNA synthesis and form colonies like their neoplastic counterparts, and the fact that the transforming gene products of a variety of oncogenic viruses are either protein kinases or closely associated with protein kinase activity, strongly suggest a role for persistent and/or inappropriate protein kinase activity in neoplastic development (see Chapter 1).

ACKNOWLEDGMENTS

We wish to thank R. Tremblay and R. Isaacs for technical assistance and D. Gillan for preparing illustrations. This work was supported in part by Grant no. CA 28340 from the National Cancer Institute, National Institutes of Health.

REFERENCES

Ashendel, C. L., Staller, J. M., and Boutwell, R. K. (1983). Stabilization, purification, and reconstitution of a phorbol ester receptor from the particulate protein fraction of mouse brain. *Cancer Res.* **43**, 4327–4332.

Barber, R., and Butcher, R. W. (1983). The egress of cyclic AMP from metazoan cells. *Adv. Cyclic Nucleotide Res.* **15**, 119–138.

Bareis, D. L., Manganiello, V. C., Hirata, F., Vaughan, M., and Axelrod, J. (1983). Bradykinin stimulates phospholipid methylation, calcium influx, prostaglandin formation, and AMP accumulation in human fibroblasts. *Proc. Natl. Acad. Sci. U.S.A.* **80**, 2514–2518.

Barrazzone, P., Gorden, P., Carpentier, J. L., Orci, L., Freychet, P., and Canivet, B. (1980). Binding, internalization and lysozomal association of ^{125}I-glucagon in isolated rat hepatocytes. *J. Clin. Invest.* **66**, 1081–1093.

Borle, A., and Briggs, F. (1968). Microdetermination of calcium in biological material by automatic fluorometric titration. *Anal. Chem.* **40**, 336–344.

Boynton, A. L., and Whitfield, J. F. (1976). Different calcium requirements for proliferation of conditionally and unconditionally tumorigenic mouse cells. *Proc. Natl. Acad. Sci. U.S.A.* **73**, 1651–1654.

Boynton, A. L., and Whitfield, J. F. (1979). The cyclic AMP-dependent initiation of DNA synthesis by T51B rat liver epithelioid cells. *J. Cell. Physiol.* **101**, 139–148.

Boynton, A. L., and Whitfield, J. F. (1980a). Possible involvement of arachidonic acid in the initiation of DNA synthesis by rat liver cells. *Exp. Cell Res.* **126**, 477–481.

Boynton, A. L., and Whitfield, J. F. (1980b). Stimulation of DNA synthesis in calcium-deprived T51B liver cells by the tumor promoters phenobarbital saccharin, and 12-O-tetradecanoylphorbol-13-acetate. *Cancer Res.* **40**, 4541–4545.

Boynton, A. L., and Whitfield, J. F. (1981). Cyclic AMP-dependent protein kinases and calmodulin stimulate the initiation of DNA synthesis by rat liver cells. *Adv. Cyclic Nucleotide Res.* **14**, 411–419.

Boynton, A. L., and Whitfield, J. F. (1983). The role of cyclic AMP in cell proliferation: A critical assessment of the evidence. *Adv. Cyclic Nucleotide Res.* **15**, 193–294.

Boynton, A. L., Whitfield, J. F., and Isaacs, R. J. (1976). Calcium-dependent stimulation of BALB/c 3T3 mouse cell DNA synthesis by a tumor-promoting phorbol ester (PMA). *J. Cell. Physiol.* **87**, 25–32.

Boynton, A. L., Whitfield, J. F., Isaacs, R. J., and Tremblay, R. G. (1977a). Different extracellular calcium requirements for proliferation of nonneoplastic, preneoplastic, and neoplastic mouse cells. *Cancer Res.* **37**, 2657–2661.

Boynton, A. L., Whitfield, J. F., Isaacs, R. J., and Tremblay, R. (1977b). The control of human WI-38 cell proliferation by extracellular calcium and its elimination by SV-40 virus-induced proliferative transformation. *J. Cell. Physiol.* **92**, 241–248.

Boynton, A. L., Whitfield, J. F., and MacManus, J. P. (1980). Calmodulin stimulates DNA synthesis by rat liver cells. *Biochem. Biophys. Res. Commun.* **95**, 745–749.

Boynton, A. L., Swierenga, S. H. H., and Whitfield, J. F. (1981a). The calcium independence of neoplastic cell proliferation: A promising *in vitro* measure of carcinogenic potential. *In* "Short-Term Tests for Chemical Carcinogens" (H. Stich and R. Sans, eds.), pp. 262–271. Springer-Verlag, Berlin and New York.

Boynton, A. L., Whitfield, J. F., MacManus, J. P., Armato, U., Tsang, B. K., and Jones, A. (1981b). Involvement of cyclic AMP and cyclic AMP-dependent protein kinases in the initiation of DNA synthesis by rat liver cells. *Exp. Cell Res.* **135**, 190–211.

Boynton, A. L., Kleine, L. P., Durkin, J. P., Whitfield, J. F., and Jones, A. (1982). Mediation by calcicalmodulin and cyclic AMP of tumor promoter-induced DNA synthesis in calcium-deprived rat liver cells. *In* "Ions, Cell Proliferation and Cancer" (A. L. Boynton, W. L. McKeehan, and J. F. Whitfield, eds.), pp. 417–431. Academic Press, New York.

Boynton, A. L., Kleine, L. P., and Whitfield, J. F. (1983). Ca/phospholipid-dependent protein kinase activity correlates to the ability of transformed liver cells to proliferate in Ca^{2+}-deficient medium. *Biochem. Biophys. Res. Commun.* **115**, 383–390.

Boynton, A. L., Kleine, L. P., and Whitfield, J. F. (1984). Relation between colony formation in calcium-deficient medium, colony formation in soft agar, and tumor formation by T51B rat liver cells. *Cancer Lett. (Shannon, Irel.)* **21**, 293–302.

Brostrom, M. A., Brotman, L. A., and Brostrom, G. O. (1982). Calcium-dependent adenylate cyclase of pituitary tumor cells. *Biochim. Biophys. Acta* **721**, 227–235.

Carpenter, G., King, L., Jr., and Cohen, S. (1978). Epidermal growth factor stimulates phosphorylation in membrane preparations *in vitro*. *Nature (London)* **276**, 409–410.

Castagna, M., Takai, Y., Kaibuchi, K., Sano, K., Kikkawa, U., and Nishizuka, Y. (1982).

Direct activation of calcium-activated, phospholipid-dependent protein kinase by tumor-promoting phorbol esters. *J. Biol. Chem.* **257**, 7847–7851.

Cheung, W. Y., ed. (1980). "Calcium and Cell Function: Calmodulin," Vol. 1. Academic Press, New York.

Christoffersen, T., and Berg, T. (1974). Glucagon control of cyclic AMP accumulation in isolated intact rat liver parenchymal cells *in vitro. Biochim. Biophys. Acta* **338**, 408–417.

Damluji, R., and Riley, P. A. (1979). Time dependency of a crucial effect of EGTA on DNA synthesis in cultures of Swiss 3T3 cells. *Exp. Cell Res.* **47**, 446–453.

Durkin, J. P., Boynton, A. L., and Whitfield, J. F. (1981). The *src* gene product (pp60[src]) of avian sarcoma virus rapidly induces DNA synthesis and proliferation of calcium-deprived rat cells. *Biochem. Biophys. Res. Commun.* **103**, 233–239.

Ebashi, S. (1980). Regulation of muscle contraction. *Proc. R. Soc. London, Ser B* **207**, 259–286.

Greengard, P. (1978). "Cyclic Nucleotides, Phosphorylated Proteins and Neuronal Function." Raven Press, New York.

Gwatkin, R. B. L. (1977). "Fertilization Mechanisms in Man and Mammals." Plenum, New York.

Hirata, F. (1981). The regulation of lipomodulin, a phospholipase inhibitory protein, in rabbit neutrophils by phosphorylation. *J. Biol. Chem.* **256**, 7730–7733.

Jones, A., Boynton, A. L., MacManus, J. P., and Whitfield, J. F. (1982). Ca-calmodulin mediates the DNA-synthetic response of calcium-deprived liver cells to the tumor promoter TPA. *Exp. Cell Res.* **138**, 87–93.

Kleine, L. P., Boynton, A. L., and Whitfield, J. F. (1985). A role for protein kinases in the initiation of DNA synthesis by rat liver cells. *In* "Ions, Membranes and the Electrochemical Control of Cell Functions" (A. Pilla and A. L. Boynton, eds.). Verlag Chemie, New York (in press).

Langer, G. A., and Frank, J. S. (1972). Lanthanum in heart cell culture. *J. Cell Biol.* **54**, 441–455.

Lapetina, E. G. (1982a). Regulation of arachidonic acid production: Role of phospholipase C and A₂. *Trends Pharmacol. Sci.* **3**, 115–118.

Lapetina, E. G. (1982b). Platelet-activating factor stimulates the phosphatidylinositol cycle. *J. Biol. Chem.* **257**, 7314–7317.

Martin, R. B., and Richardson, F. S. (1979). Lanthanides as probes for calcium in biological systems. *Q. Rev. Biophys.* **12**, 181–210.

Mufson, R. A., Laskin, J. D., Fisher, P. B., and Weinstein, I. B. (1979). Melittin shares certain effects with phorbol ester tumour promoters. *Nature (London)* **280**, 72–74.

Ng, K. W., Livesay, S. A., Larkins, R. G., and Martin, T. J. (1983). Calcitonin effects on growth and on selective activation of type II isoenzyme of cyclic adenosine 3′,5′-monophosphate-dependent protein kinase in T470 human breast cancer cells. *Cancer Res.* **43**, 794–800.

Niedel, J. E., Kuhn, L. J., and Vandenbark, G. R. (1983). Phorbol diester receptor copurifies with protein kinase C. *Proc. Natl. Acad. Sci. U.S.A.* **80**, 36–40.

Owen, N. E., and Villereal, M. L. (1982). Lys-bradykinin stimulates Na⁺ influx and DNA synthesis in cultured human fibroblasts. *Cell (Cambridge, Mass.)* **32**, 979–985.

Perris, A. D., and Whitfield, J. F. (1969). The mitogenic action of bradykinin on thymic lymphocytes and its dependence on calcium. *Proc. Soc. Exp. Biol. Med.* **130**, 1198–1201.

Rodbell, M. (1983). The actions of glucagon at its receptor: Regulation of adenylate cyclase. *In* "Glucagon I" (P. J. Lefèbvre, ed.), pp. 263–290. Springer-Verlag, Berlin and New York.

Rubin, R. P. (1982). "Calcium and Cellular Secretion." Plenum, New York.

Schatzmann, R. C., Wise, B. C., and Kuo, J. F. (1981). Phospholipid-sensitive calcium-dependent protein kinase: Inhibition by antipsychotic drugs. *Biochem. Biophys. Res. Commun.* **98,** 669–676.

Shier, W. T., and Durkin, J. P. (1982). Role of stimulation of arachidonic acid release in the proliferative response of 3T3 mouse fibroblasts to platelet derived growth factor. *J. Cell. Physiol.* **112,** 171–181.

Swierenga, S. H. H., Whitfield, J. F., and Boynton, A. L. (1978a). Age-related and carcinogen-induced alterations of the extracellular growth factor requirements for cell proliferation *in vitro. J. Cell. Physiol.* **94,** 171–180.

Swierenga, S. H. H., Whitfield, J. F., and Karasaki, S. (1978b). The loss of proliferative calcium dependence. A simple *in vitro* indicator of tumorigenicity. *Proc. Natl. Acad. Sci. U.S.A.* **75,** 6069–6072.

Swierenga, S. H. H., Whitfield, J. F., Boynton, A. L., MacManus, J. P., Rixon, R. H., Sikorska, M., Tsang, B. K., and Walker, P. R. (1980). Regulation of proliferation of normal and neoplastic rat liver cells by calcium and cyclic AMP. *Ann. N.Y. Acad. Sci.* **349,** 294–311.

Swierenga, S. H. H., Goyette, R., and Marceau, N. (1981). Change in cell surface morphology, fibronectin matrix formation and cycloskeletal organization of normal and tumorigenic liver cells. *Proc. Can. Fed. Biol. Sci.* p. 265.

Takai, Y., Kishimoto, A., Iwasa, Y., Kawahara, Y., Mori, T., and Nishizuka, Y. (1979). Calcium-dependent activation of a multifunctional protein kinase by membrane phospholipids. *J. Biol. Chem.* **254,** 3692–3695.

Tucker, R. W., Snowdowne, K. W., and Borle, A. (1983). Platelet derived growth factor produces transient increases in the intracellular concentration of free calcium in BALB/c 3T3 cells. *J. Cell Biol.* **97,** 3439.

Walenga, R. W., Opas, E. E., and Feinstein, M. B. (1981). Differential effects of calmodulin antagonists on phospholipases A_2 and C in thrombin-stimulated platelets. *J. Biol. Chem.* **256,** 12523–12528.

Whitfield, J. F., MacManus, J. P., and Gillan, D. J. (1970). Cyclic AMP mediation of bradykinin-induced stimulation of mitotic activity and DNA synthesis in thymocytes. *Proc. Soc. Exp. Biol. Med.* **133,** 1270–1274.

Whitfield, J. F., MacManus, J. P., Rixon, R. H., Boynton, A. L., Youdale, T., and Swierenga, S. H. H. (1976). The positive control of cell proliferation by the interplay of calcium ions and cyclic nucleotides: A review. *In Vitro* **12,** 1–18.

Whitfield, J. F., Boynton, A. L., MacManus, J. P., Rixon, R. H., Sikorska, M., Tsang, B. K., and Walker, P. R. (1980). The roles of calcium and cyclic AMP in cell proliferation. *Ann. N.Y. Acad. Sci.* **339,** 216–242.

6

Platelet Growth Factors: Presence and Biological Significance

ALLAN LIPTON

Division of Oncology and
The Specialized Cancer Research Center
The Milton S. Hershey Medical Center
The Pennsylvania State University
Hershey, Pennsylvania

I. INTRODUCTION

The concept that platelets contain mitogenic factors developed from the observation of Balk that chicken plasma was less effective than chicken serum in promoting the growth of chicken fibroblasts (Balk, 1971). He postulated either that the serum mitogenic factors were released from precursors in plasma or from thrombocytes when blood was clotted during the preparation of serum. Kohler and Lipton (1974) extended this observation and actually demonstrated that platelets contain growth factor for fibroblasts. Ross *et al.* (1974) presented similar results for smooth muscle cells.

CONTROL OF ANIMAL CELL PROLIFERATION,
VOLUME I

II. PLATELET-DERIVED GROWTH FACTOR FOR
MESENCHYMAL CELLS

A. Purification and Characterization

Extracts of frozen-thawed or thrombin treated platelets contain growth factor activity. This platelet-derived growth factor (PDGF) can stimulate the multiplication of fibroblasts (Kohler and Lipton, 1974), smooth muscle cells (Ross et al., 1974), and glial cells (Heldin et al., 1981). It does not promote the growth of arterial endothelial cells or cells of epithelial origin.

Purification of a PDGF (PDGF I) was facilitated by the observation that this molecule withstood heating for 10 min at 100°C and was a basic protein with an isoelectric point of about 9.8. This polypeptide growth factor that stimulates the proliferation of connective tissue cells has been purified by several groups. Unreduced PDGF has an apparent molecular weight on SDS–polyacrylamide gels of 27,500–31,000 (Antoniades et al., 1979; Deuel et al., 1981; Heldin et al., 1981; Raines and Ross, 1982). With the use of gel permeation chromatography, PDGF has been separated into two chemically defined components [PDGF I and PDGF II having molecular weights of 31,000 and 28,000, respectively (Antoniades, 1981; Deuel et al., 1981)] which have essentially identical amino acid compositions but differ in covalently bound carbohydrate (Deuel et al., 1981). Four components have been resolved on SDS–polyacrylamide gels of MWs 27,000, 28,500, 29,000, and 31,000 (Raines and Ross, 1982). Because PDGF is resistant to denaturation by SDS, it has been possible to show comigration of mitogenic activity with the 27,000–31,000 MW polypeptides eluted from gels run without reducing agents (Antoniades, 1981; Heldin et al., 1977; Raines and Ross, 1982).

PDGF I and II consist of two chains joined by disulfide bonds. Retention of the disulfide bonds in the native configuration is necessary for activity. Following cleavage of the disulfide bonds of PDGF I, mitogenic activity is lost and a number of polypeptides ranging in molecular weight from 11,000 to 17,500 are formed. These have been partially resolved on SDS–polyacrylamide gels (Antoniades, 1981; Deuel et al., 1981; Heldin et al., 1981; Raines and Ross, 1982), or by reverse-phase high-performance liquid chromatography (HPLC) in pyridine formate buffers using propanol as organic modifier (Johnsson et al., 1982).

B. Cellular Receptor for PDGF

Purified PDGF has been radioiodinated with full retention of biological activity using an iodine monochloride method (Bowen-Pope and Ross, 1982). A class of high-affinity receptors for PDGF has been demonstrated on responsive cells using ^{125}I-labeled ligand (Bowen-Pope and Ross, 1982; see Chapter 10; Heldin et al., 1981). The receptor is specific for PDGF in the sense that other growth factors such as epidermal growth factor (EGF), fibroblast growth factor (FGF), or insulin do not compete with ^{125}I-labeled PDGF for binding (Heldin et al., 1981). In addition, various unresponsive cells such as those of epithelial origin show no significant specific binding of ^{125}I-labeled PDGF.

Recently, the cellular receptor for PDGF has been purified (Heldin et al., 1981; J. S. Huang et al., 1983; see Chapter 10). The receptor has a molecular weight of ~180,000. A single class of specific high-affinity binding sites has been purified to near homogeneity for Swiss mouse 3T3 cell membranes. Both the solubilized and intact membrane receptors have an identical K_d for ^{125}I-labeled PDGF. Human foreskin fibroblasts have ~3 × 10^5 receptors/cell with a dissociation constant of about 1 nM (Heldin et al., 1981). The PDGF–receptor complex is internalized by the cell and subsequently degraded in the lysosomes (Heldin et al., 1982).

C. Intracellular Localization of PDGF

Subcellular localization of PDGF has been accomplished. It is located in the platelet α granules. It is released along with other α-granule constituents (platelet factor 4, β-thromboglobulin, and platelet fibrinogen) following stimulation with thrombin, arachidonic acid, or collagen (Kaplan et al., 1979; Witte et al., 1978). Megakaryocytes are platelet precursors in the bone marrow that are known to be active in protein synthesis. Growth factor activity as determined by the stimulation of [^{3}H]thymidine incorporation into DNA of quiescent 3T3 cells in culture has been found in guinea pig bone marrow. Quantitative dilution studies demonstrated that of the cells present in the guinea pig bone marrow, only the megakaryocyte possessed significant amounts of activity similar to PDGF (Chernoff et al., 1980). The amount of activity present in one megakaryocyte was equivalent to that present in 1000–5000 platelets, approximately the number of platelets shed from a single megakaryocyte (Chernoff et al., 1980). Thus, PDGF appears to have its origin in the bone marrow megakaryocyte.

D. Biological Roles of PDGF I and PDGF II

1. Cell Cycle

PDGF plays a critical role in the cell cycle of nontransformed fibroblasts and other mesenchymal cells. Cells exposed to PDGF are rendered "competent," i.e., they are potentially able to leave G_0 and enter the cell cycle (Pledger et al., 1978; Ross et al., 1974). Fibroblast growth factor from bovine brain or pituitary gland and a factor recently purified from bovine spinal cord can also recruit quiescent cells into the cell cycle (Jennings et al., 1979; see Chapter 3).

Progression of cells through G_1 and S requires the continued presence of platelet-poor plasma (Pledger et al., 1978; Vogel et al., 1978). Cells exposed to just PDGF will not synthesize new DNA. The concentration of plasma determines, in part, the rate of entry of cells into the S phase. Factors in plasma that can mediate the progression of cells through G_1 and S phase include insulin and somatomedin C (Clemmons and Van Wyk, 1981).

PDGF also stimulates many activities in cells that occur much earlier than increased thymidine incorporation. These include an increased rate of endocytosis of tracer molecules such as [^{14}C]sucrose within 1 hr of PDGF exposure (Davies and Ross, 1978), increased cholesterol synthesis (Antoniades et al., 1979), increased turnover of phospholipids (Habenicht et al., 1981), and increased binding of low-density lipoprotein (LDL) to specific high-affinity receptor for this lipoprotein (Antoniades et al., 1979). Finally, human PDGF stimulates release of arachidonic acid from cellular phospholipids (Shier and Durkin, 1982). Synthesis and release of prostaglandins is another early event beginning within minutes after exposure to PDGF (Shier, 1980; Shier and Durkin, 1982).

In addition, a number of cellular events which may or may not be related to cellular growth are stimulated by PDGF in vitro, such as protein synthesis (Burke and Ross, 1977), prostacyclin synthesis (Coughlin et al., 1980), endocytosis, and chemotaxis (Grotendorst et al., 1981). Also, the numbers of certain receptors on the cell surface are modulated by PDGF: the LDL, somatomedin C, and FSH receptors increase (Chait et al., 1980; (Clemmons et al., 1980; Mondschein and Schomberg, 1981; Witte and Cornicelli, 1980), whereas EGF receptors decrease after PDGF exposure (Heldin et al., 1982; Wrann et al., 1980).

2. Growth Control—Tyrosine Phosphorylation

Cells are known to contain enzymes, called kinases, that add phosphate groups to amino acids. Usually the phosphate group is added to serine and threonine, not to tyrosine. The first hint that an unusual chemical modification of proteins might be related to control of cell growth came from the initial observation that the transforming protein of Rous sarcoma virus, an RNA tumor virus, adds phosphate to tyrosines of certain proteins (Hunter and Sefton, 1980; see Chapter 11). Five different classes of RNA tumor viruses make transforming proteins that phosphorylate tyrosine. Recently, EGF was shown to stimulate the phosphorylation of 150,000- and 170,000-dalton membrane proteins in A-431 epidermal carcinoma cells and livers of normal mice. Phosphotyrosine was identified in hydrolysates of the protein phosphorylated by the membrane protein kinase activity expressed in target cells after exposure to EGF (Cohen et al., 1982; see Chapter 7). In similar fashion, PDGF stimulates the incorporation of ^{32}P from $[\gamma\text{-}^{32}P]ATP$ into a 170,000-dalton protein by an endogenous tyrosine-specific protein kinase in membrane preparations of Swiss mouse 3T3 cells. The similarity of PDGF and EGF in stimulating phosphotyrosine-specific protein kinase activity and the stimulation of a similar activity by viral transformation (src) genes suggest that a common mechanism may exist for the phenotypic expression of increased DNA synthesis and cell growth stimulated by these separate factors. One important difference between growth hormones and transforming viruses is that with growth hormones the signal that results in tyrosine phosphorylation is transient; it only persists as long as the hormones and their receptors are present. With transforming viruses, the signal is on all the time and the cells lose growth control.

3. Wound Healing

An essential step in wound healing is the multiplication of fibroblasts. These cells are responsible for the synthesis of extracellular material (Ross, 1975) and wound contraction (Gabbiani et al., 1972; Majno et al., 1971). The growth of fibroblasts is to a great extent under the control of hormones and specific growth factors such as PDGF and FGF. Many authors state that these factors have a role in the healing of wounds, but little experimental data are available.

In the wound experiments to be described, we employed female, random-bred Syrian hamsters (Charles River Breeding Laboratories). Animals were fed and watered *ad libitum* and housed singly in a temperature-controlled room. Hamsters ~6-months old and weighing 140–230 gm were caged singly prior to wounding.

The method of wounding was to induce general anesthesia with intraperitoneal Nembutal (85 mg/kg) and shave the back of the animal.

Two symmetrically placed full-thickness skin wounds were made with a 6-mm skin biopsy punch using a circular motion. Previous experiments in our laboratory have shown that the location of the wound affects the rate of wound contraction. Wounds placed cephalad on the back of the hamster had a significantly accelerated rate of wound contraction compared to caudal wounds. In these experiments two wounds were placed on the dorsum of each animal. Wounds were equidistant from the midline of the back and were at the level of the front limbs.

In each experiment 20 animals were used. Immediately after wounding, 20 µl of the mitogen was applied to one wound and 20 µl of the control solution was applied to the contralateral wound. Identical applications were made twice daily for the first 3 days after wounding for a total of six applications. To avoid potential locational differences in wound healing, half of the animals received the mitogen on the left side and the control solution on the right. This configuration was reversed for the other half.

The healing process was evaluated with two indices, wound size and time to complete healing. The product of the length and width of the wound yielded a measurement of wound size. This measurement was used to calculate the rate of wound contraction. Healing of each wound was judged complete when no scab was visible and complete epithelialization had occurred. Statistical analysis was performed using the Student's t test.

Several concentrations of dexamethasone (DEX) (Elkins-Sinn, Inc., Cherry Hill, New Jersey) in combination with insulin (INS) (purified crystalline bovine insulin—Eli Lilly, Indianapolis, Indiana) were tested before adding known growth factors. Initially, one wound was treated with 10 ng each of DEX and INS per application, and distilled H_2O was applied to the contralateral wound. There was a significant delay in total healing time of the wounds treated with DEX and INS (Table I). In the second experiment 1 pg of DEX and 10 ng of INS were applied, and no delay in wound healing was observed (Table I). In subsequent experiments with growth factors the latter concentrations of DEX and INS were used.

Wounds were next treated with PDGF (Collaborative Research Inc., Waltham, Massachusetts) (0.77 units/application) and DEX and INS. Daily observation revealed no visual differences in the rate of wound contraction between treated and control wounds. Wound measurements showed no significant difference in the rate of wound contraction between control wounds and wounds treated with PDGF and DEX and INS ($p > .1$) (Fig. 1). The animals were also observed daily to determine the point of complete healing of the wound. As seen in Fig.

TABLE I

Effect of Various Mitogens on the Time to Complete Wound Healing

Test material (amount per application)	Days ±SEM	Control	Days ±SEM	p Value
Insulin (10 ng) + dexamethasone (10 ng)	15.00 ± .99	H_2O	11.75 ± .62	<.01
Insulin (10 ng) + dexamethasone (1 pg)	11.06 ± .70	H_2O	11.94 ± .59	>.2
PDGF (0.77 units) + insulin (10 ng) + Dex (1 pg)	11.32 ± .56	H_2O	13.28 ± .93	>.05
FGF (154 ng) + insulin (10 ng) + Dex (1 pg)	12.60 ± .52	H_2O	12.10 ± .63	>.5
Purified thrombin (10 μg)	11.95 ± .55	15 M NaCl	10.70 ± .36	>.05
Defined medium F (modified)	11.65 ± .47	H_2O	12.15 ± .72	>.5
Liver cell supernatant (30.8 μg)	10.95 ± .65	Krebs Buffer	11.21 ± .66	>.5
EGF (770 ng) + insulin (10 ng) + Dex (1 pg)	10.95 ± .41	H_2O	12.05 ± .69	>.1
Human colostrum	11.95 ± .62	H_2O	11.05 ± .32	>.2
FGF (154 ng) + EGF (770 ng) + insulin (10 ng) + Dex (1 pg)	11.25 ± .38	H_2O	10.55 ± .41	>.2

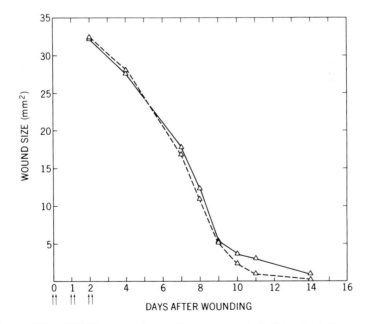

Fig. 1. Effect of PDGF on rate of wound contraction. Each data point is the mean of 20 animals. Wounds were treated with 0.77 units PDGF, INS (10 ng), and DEX (1 pg) in 20 μl of distilled H_2O (---), or with 20 μl of distilled H_2O alone (——). Arrows indicate the time of each 20-μl application to the wounds. No statistically significant differences were found in the rate of wound contraction between experimental and control wounds (p > .1).

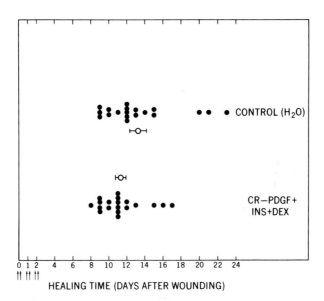

Fig. 2. Healing time of PDGF-treated wounds. Individual healing times and the mean ±SEM for 20 animals are presented. Wounds were treated with a solution of PDGF (0.77 units), INS (10 ng), and DEX (1 pg) in 20 µl of distilled H_2O, or with 20 µl of distilled H_2O alone. Arrows indicates the time of each 20-µl application to the wounds. Healing of each wound was judged complete when no scab was visible and complete epithelialization had occurred. No statistically significant difference was found in healing time between experimental and control wounds ($p > .05$). CR-PDGF, Collaborative Research PDGF.

2, there was no difference between control wounds and wounds treated with PDGF and DEX and INS ($p > .5$) (Table I).

We then added to wounds other preparations known to enhance the growth of fibroblasts. Purified human α-thrombin (John Fenton, Albany, New York) (10 µg/application), defined medium F for fibroblasts (Weinstein *et al.*, 1982) (20 µl/application), and liver cell supernatant (Witkowski *et al.*, 1979) (30.8 µg/application) also failed to accelerate wound healing (Table I).

One possible explanation for the negative results described above is that stimulation of epithelial cell proliferation is needed to accelerate wound healing. Epithelial cells from a variety of tissues have been induced to grow *in vitro* in the presence of EGF. EGF (Collaborative Research) (770 ng/application) with DEX and INS did not significantly accelerate wound contraction or healing time ($p > .1$) (Table I). This negative result is in agreement with the report of Greaves, who failed to accelerate the healing of epidermal wound in man using topically applied EGF (Greaves, 1980).

In the final set of experiments, we added both a fibroblast stimulant (FGF) plus an epithelial stimulant (EGF) to the same wound. FGF (154 ng/application) and EGF (770 ng/application) with DEX and INS were applied to the wounds. There was no significant difference in the rate of wound contraction ($p > .2$) (Fig. 3) and time to complete wound healing ($p > .2$) (Fig. 4) between treated and control wounds. Human colostrum (second day after birth), known to cause the growth of both fibroblasts and epithelial cells (Steimer et al., 1981), also did not alter these healing end points (Table I).

Thus, in the present experiments we were unable to accelerate healing of full-thickness skin wounds in the hamster using a variety of mitogens. One possible explanation for our negative results is that these wounds were already saturated with optimal amounts of mitogenic factors and therefore any additional growth-promoting preparation would have no effect. Another possible explanation is that growth-promoting macromolecules other than the ones tested are needed to accelerate wound healing. Recent data suggest that pros-

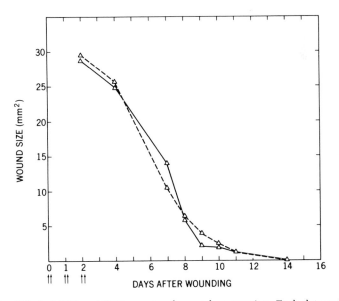

Fig. 3. Effect of EGF and FGF on rate of wound contraction. Each data point is the mean of 20 animals. Wounds were treated with 154 ng of FGF and 770 ng of EGF, INS (10 ng), and DEX (1 pg) in 20 µl of distilled H_2O (---), or with distilled H_2O alone (——). Arrows indicate the time of each 20-µl application to the wounds. No statistically significant differences were found in the rate of wound contraction between experimental and control wounds ($p > .2$).

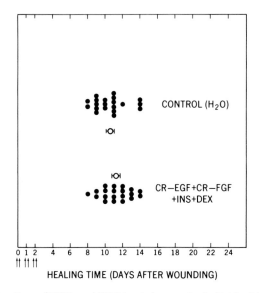

Fig. 4. Healing time of EGF- and FGF-treated wounds. Individual healing times and the mean ±SEM for 20 animals are presented. Wounds were treated with FGF (154 ng), EGF (770 ng), INS (10 ng), and DEX (1 pg) in 20 μl of distilled H_2O, or with distilled H_2O alone. Arrows indicate the time of each 20-μl application to the wounds. No statistically significant difference was found in healing time between experimental and control wounds ($p > .2$).

taglandins may be involved in the proliferation of fibroblasts (Otto *et al.*, 1982).

III. GROWTH FACTORS FOR TUMOR CELLS

A. Transforming Growth Factors from Tumor Cells

Certain transformed cells secrete mitogens such as multiplication-stimulating activity (insulin-like growth factors) (DeLarco and Todaro, 1978a; Dulak and Temin, 1973; see Chapter 4) and transforming growth factors, which do not appear to be structurally related to PDGF.

The transforming growth factors (TGFs) are a newly discovered class of cell growth agents. TGFs are low-molecular-weight polypeptides produced by rodent and human tumor cells, which are capable of reversibly stimulating nontransformed cells to grow as colonies in soft agar. Stimulation of anchorage-independent cell growth is the *in vitro* transformation parameter which most closely correlates with *in vivo* tumorigenicity (Kahn *et al.*, 1980). TGFs also have the capacity to stimulate the division of cell monolayers and to induce morphological cell changes.

TGF was first described in the culture medium of Moloney virus sarcoma transformed mouse cells and was called sarcoma growth factor (SGF) (DeLarco and Todaro, 1978b). After this initial observation, several mammalian cells have been shown to contain or release related growth factors displaying several common biological and physicochemical characteristics. *In vitro*, the spontaneous release of TGFs has only been observed with transformed cell lines from different mammalian species: human tumor cells (Todaro *et al.*, 1980), spontaneously transformed rat cells, rat cells transformed by Kirsten murine sarcoma virus (DeLarco *et al.*, 1981; Ozanne *et al.*, 1980) or simian virus 40 (SV40) (Kaplan *et al.*, 1982) or Rous sarcoma virus (Kryceve-Martinerie *et al.*, 1982), and chemically transformed mouse cells (Moses *et al.*, 1981). But *in vivo*, TGFs can be recovered by acid–ethanol extraction not only from grafted tumors in the rat and mouse (Kaplan *et al.*, 1982), but also from normal mouse organs. More recently, TGF activity has been identified in acid–ethanol extracts of human urine (Sherwin *et al.*, 1983). TGFs are acid- and heat-stable polypeptides which contain disulfide bonds (DeLarco and Todaro, 1978). TGFs isolated from human sarcoma and carcinoma cell lines have molecular weights in the size classes of 7000 and 20,000–23,000 (Todaro *et al.*, 1980). The larger form has not been detected in the conditioned medium of tumor cells growing in culture. In contrast, a low-molecular-weight TGF activity (6000–8000) is found in human urine, but a high-molecular-weight TGF activity (30,000–35,000) is found in the urine of many cancer patients (Sherwin *et al.*, 1983). TGFs can compete with EGF for binding to the EGF cell membrane receptor and thus may be functionally related to this molecule (Marquardt and Todaro, 1982; Todaro *et al.*, 1980). More recently, TGFs which do not compete with EGF for binding to its receptor have been identified in both normal and transformed rodent cells and have been shown to lack soft agar growth-promoting activity except in the presence of added EGF (Roberts *et al.*, 1982).

B. Oncogene Product—Relation to PDGF

A human osteosarcoma cell line, U-2 OS, when cultured under serum-free conditions, has been shown to produce a growth factor (osteosarcoma-derived growth factor, ODGF) for human-cultured glial cells (Westermark and Wasteson, 1975). The biological activity of ODGF resides in a cationic, relatively heat-resistant, reduction-susceptible protein with a molecular weight of 30,000. ODGF was further shown to be composed of two different polypeptide chains (about 13,000–14,000 and 16,000–17,000 daltons, respectively) linked by disulfide bonds (Heldin et al., 1980). ODGF is thus similar to PDGF. In addition, the growth factor released by SV40-transformed BHK cells also has PDGF-like properties (Bourne and Rozengurt, 1976; Burk, 1976).

As mentioned, PDGF has been purified to homogeneity. The PDGF molecule consists of two polypeptide chains that may be homologous in structure. The partial amino acid sequences of the two chains, now called PDGF I and PDGF II, are similar but differ in covalently bound carbohydrate (Deuel et al., 1981). Russell Doolittle of the University of California at San Diego has established a computer bank of protein sequences for his studies of protein evolution (Doolittle, 1981). A computer search revealed a region of virtual identity between a 90–104 amino acid stretch of PDGF and that of the predicted sequence of p28^{sis}, the protein product of the onc gene of simian sarcoma virus (sis gene) (Doolittle et al., 1983; Waterfield et al., 1983). The resemblance was particularly close for PDGF II. A total of 87% of the 70 amino acids in the known segments of this polypeptide matched those in the onc gene product sequence. This close match between the sequences of the sis gene product and PDGF leaves little doubt that the viral gene is derived either from a gene coding for PDGF or a closely related one. An extension of these observations is that a PDGF-like protein is present in lysates of simian sarcoma virus-transformed cells and two cell lines derived from human malignancies (S. S. Huang et al., 1983). Thus, we have the first known function of an oncogene product.

IV. CONCLUDING REMARKS

These exciting new biochemical discoveries have raised several possibilities as to the molecular basis for the development of malignancy. Prolonged abnormal expression of PDGF or a related protein in transformed cells could lead to uncontrolled growth through secretion of the growth factor. A second possible mechanism is direct intracellular release of PDGF. Such intracellular release of the mitogenic signal would bypass the need for a receptor interaction at the cell surface. By such mechanisms cells would be constitutively activated and result in the appearance of the malignant phenotype.

These observations would appear to offer several new strategies for cancer therapy, such as (1) a specific antibody to inhibit cancer cell-produced PDGF-like molecules or (2) a specific antibody to inhibit cancer cell receptors for PDGF.

ACKNOWLEDGMENTS

The author wishes to acknowledge the collaboration of Kim Leitzel, Cheryl Cano, and Dr. James Marks in the wound healing experiment reported in this review. He also wishes to thank Mrs. Judy Weigel for her help in preparing this manuscript.

REFERENCES

Antoniades, H. N. (1981). Human platelet derived growth factor purification of platelet derived growth factor-I and platelet derived growth factor-II and separation of their reduced subunits. *Proc. Natl. Acad. Sci. U.S.A.* **78**(12), 7314–7317.

Antoniades, H. N., Scher, C. D., and Stiles, C. D. (1979). Purification of human platelet derived growth factor. *Proc. Natl. Acad. Sci. U.S.A.* **76**(4), 1809–1813.

Balk, S. D. (1971). Calcium as a regulator of the proliferation of normal, but not of transformed, chicken fibroblasts in a plasma-containing medium. *Proc. Natl. Acad. Sci. U.S.A.* **68**, 271–276.

Bourne, H., and Rozengurt, E. (1976). An 18000 molecular weight polypeptide induces early events and stimulates DNA synthesis in cultured cells. *Proc. Natl. Acad. Sci. U.S.A.* **73**(12), 4555–4559.

Bowen-Pope, D. F., and Ross, R. (1982). Platelet derived growth factor 2: Specific binding to cultured cells. *J. Biol. Chem.* **257**(9), 5161–5171.

Burk, R. R. (1976). Induction of cell proliferation by a migration factor released from a transformed cell line. *Exp. Cell Res.* **101**, 293–298.

Burke, J. N., and Ross, R. (1977). Collagen synthesis by monkey arterial smooth muscle cells during proliferation and quiescence in culture. *Exp. Cell Res.* **107**(2), 387–396.

Chait, A., Ross., R., Albers, J. J., and Bierman, E. L. (1980). Platelet derived growth factor stimulates activity of low density lipoprotein receptors. *Proc. Natl. Acad. Sci. U.S.A.* **77**(7), 4084–4088.

Chernoff, A., Levine, R. F., and Goodman, D. S. (1980). Origin of platelet derived growth factor in megakaryocytes in guinea pigs. *J. Clin. Invest.* **65**(4), 926–930.

Clemmons, D. R., and Van Wyk, J. J. (1981). Somatomedin C and platelet derived growth factor stimulate human fibroblast replication. *J. Cell. Physiol.* **106**(3), 361–368.

Clemmons, D. R., Van Wyk, J. J., and Pledger, W. J. (1980). *Proc. Natl. Acad. Sci. U.S.A.* **77**(11), 6644–6648.

Cohen, S., Fava, R. A., and Sawyer, S. T. (1982). Purification and characterization of epidermal growth factor receptor protein kinase from normal mouse liver. *Proc. Natl. Acad. Sci. U.S.A.* **79**(20), 6237–6241.

Coughlin, S. R., Moskowitz, M. A., Zetter, B. R., Antoniades, H. N., and Levine L. (1980). Platelet dependent stimulation of prostacyclin synthesis by platelet growth factor. *Nature (London)* **288**, 600–602.

Davies, P. F., and Ross, R. (1978). Mediation of pinocytosis in cultured arterial smooth muscle and endothelial cells by platelet derived growth factor. *J. Cell Biol.* **79**(3), 663–671.

DeLarco, J. E., and Todaro, G. J. (1978a). A human fibrosarcoma cell line producing multiplication stimulating activity related peptides. *Nature (London)* **272**, 356–358.

DeLarco, J. E., and Todaro, G. J. (1978b). Growth factors from murine sarcoma virus transformed cells. *Proc. Natl. Acad. Sci. U.S.A.* **75**(8), 4001–4005.

DeLarco, J. E., Preston, Y. A., and Todaro, G. J. (1981). Properties of a sarcoma growth factor-like peptide from cells transformed by a temperature sensitive sarcoma virus. *J. Cell. Physiol.* **109**(1), 143–152.

Deuel, T. F., Huang, J. S., Proffitt, R. T., Baenziger, J. U., Chang, D., and Kennedy, B. B. (1981). Human platelet derived growth factor purification and resolution into 2 active protein fractions. *J. Biol. Chem.* **256**(17), 8896–8899.

Doolittle, R. F. (1981). Similar amino acid sequences chance on common ancestry. *Science* **214**, 149–159.

Doolittle, R. F., Hunkapiller, M. W., Hood, L. E., Devare, S. G., Robbins, K. C., Aaronson, S. A., and Antoniades, H. N. (1983). Simian sarcoma virus onc gene, V-sis, is derived from the gene (or genes) encoding a platelet-derived growth factor. *Science* **221**, 275–277.

Dulak, N. C., and Temin, H. M. (1973). Multiplication stimulating activity for chicken embryo fibroblasts from rat liver cell conditioned medium—a family of small polypeptides. *J. Cell. Physiol.* **81**(2), 161–170.

Gabbiani, G., Hirschel, B. J., Ryan, G. B., Statkou, P. R., and Majno, G. (1972). Granulation tissue as a contractile organ—A study of structure and function. *J. Exp. Med.* **135**(4), 719–734.

Greaves, M. W. (1980). Lack of effect of topically applied epidermal growth factor on epidermal growth in man in vivo. *Clin. Exp. Dermatol.* **5**(1), 101–104.

Grotendorst, G. R., Seppa, H. E. J., Kleinman, H. K., and Martin, G. R. (1981). Attachment of smooth muscle to collagen and their migration toward platelet derived growth factor. *Proc. Natl. Acad. Sci. U.S.A.* **78**(6), 3669–3672.

Habenicht, A. J. R., Glomset, J. A., King, W. C., Nist, C., Mitchell, C. D., and Ross, R. (1981). Early changes in phosphatidylinositol and arachidonic acid metabolism in quiescent Swiss 3T3 cells stimulated to divide by platelet derived growth factor. *J. Biol. Chem.* **256**(23), 12329–12335.

Heldin, C. H., Wasteson, A., and Westermark, B. (1977). Partial purification and characterization of platelet factors stimulating the multiplication of normal human glial cells. *Exp. Cell Res.* **109**(2), 429–438.

Heldin, C. H., Westermark, B., and Wasteson, A. (1980). Chemical and biological properties of a growth factor from human cultured osteosarcoma cells: Resemblance with platelet derived growth factor. *J. Cell. Physiol.* **105**(2), 235–246.

Heldin, C. H., Westermark, B., and Wasteson, A. (1981). Specific receptors for platelet derived growth factor on cells derived from connective tissue and glia. *Proc. Natl. Acad. Sci. U.S.A.* **78**(6), 3664–3668.

Heldin, C. H., Wasteson, A., and Westermark, B. (1982). Interaction of platelet derived growth factor with its fibroblast receptor. Demonstration of ligand degradation and receptor modulation. *J. Biol. Chem.* **257**(8), 4216–4221.

Huang, J. S., Huang, S. S., and Deuel, T. F. (1983). Platelet derived growth factor: Expression of related-growth factor activity and antigen in virus infected and in cultured malignant cells. *Abstr. Pop., 25th Annu. Meet., Am. Soc. Hematol.* Abstract No. 933.

Huang, S. S., Huang, J. S., and Deuel, T. F. (1983). Platelet derived growth factor: Receptor purification and initial characterization. *Abstr. Pop., 25th Annu. Meet., Am. Soc. Hematol.* Abstract No. 934.

Hunter, T., and Sefton, B. M. (1980). Transforming gene product of Rous sarcoma virus phosphorylates tyrosine. *Proc. Natl. Acad. Sci. U.S.A.* **77**(3), 1311–1315.

Jennings, T., Jones, R. D., and Lipton, A. (1979). A growth factor from spinal cord. *J. Cell. Physiol.* **100**(2), 273–278.

Johnsson, A., Heldin, C. H., Wasteson, A., and Westermark, B. (1982). Platelet derived growth factor identification of constituent Polypeptide chains. *Biochem. Biophys. Res. Commun.* **104**(1), 66–74.

Kahn, P., Simons, R. S., Klein, A. S., and Shin, S. (1980). Tumor formation by transformed cells in nude mice. *Cold Spring Harbor Symp. Quant. Biol.* **44**, 695–702.

Kaplan, K. L., Broekman, M. J., Chernoff, A., Lesznik, G. R., and Drillings, M. (1979). Platelet alpha granule protein studies on release and subcellular localization. *Blood* **53**(4), 604–618.

Kaplan, P. L., Anderson, M., and Ozanne, B. (1982). Transforming growth factor production enables cells to grow in the absence of serum: An autocrine system. *Proc. Natl. Acad. Sci. U.S.A.* **79**(2), 485–489.

Kohler, N., and Lipton, A. (1974). Platelets as a source of fibroblast growth promoting activity. *Exp. Cell Res.* **87**(2), 297–301.

Kryceve-Martinerie C., Lawrence, D. A., Crochet, J., Jullien P., and Vigier, P. (1982). Cells transformed by Rous sarcoma virus release transforming growth factors. *J. Cell. Physiol.* **113**(3), 365–372.

Majno, G., Gabbiani, G., Hirschel, B. J., Ryan, G. B., and Statkov, P. R. (1971). Contraction of granulation tissue *in vitro:* Similarity to smooth muscle. *Science* **173**, 548–550.

Marquardt, H., and Todaro, G. J. (1982). Human transforming growth factor production by a melanoma-cell line. Purification and initial characterization. *J. Biol. Chem.* **257**(9), 5220–5225.

Mondschein, J. S., and Schomberg, D. W. (1981). Growth factors modulate gonadotropin receptor induction in granulosa cell cultures. *Science* **211**, 1179–1180.

Moses, H. L., Branum, E. L., Proper, J. A., and Robinson, R. A. (1981). Transforming growth factor production by chemically transformed cells. *Cancer Res.* **41**(7), 2842–2848.

Otto, A. M., Nilsen-Hamilton, M., Boss, B. D., Ulrich, M. O., and De Asua, L. J. (1982). Prostaglandin E-1 and prostaglandin E-2 interact with prostaglandin F-2-alpha to regulate Initiation of DNA replication and cell division in Swiss 3T3 cells. *Proc. Natl. Acad. Sci. U.S.A.* **79**(16), 4992–4996.

Ozanne, B., Fulton, R. J., and Kaplan, P. L. (1980). Kirsten murine sarcoma virus transformed cell lines and a spontaneously transformed rat cell line produce transforming factors. *J. Cell. Physiol.* **105**(1), 163–180.

Pledger, W. J., Stiles, C. D., Antoniades, H. N., and Scher, C. D. (1978). An ordered sequence of events is required before BALB/c 3T3 cells become committed to DNA synthesis. *Proc. Natl. Acad. Sci. U.S.A.* **75**(6), 2839–2843.

Raines, E. W., and Ross, R. (1982). Platelet derived growth factor 1. High yield purification and evidence for multiple forms. *J. Biol. Chem.* **257**(9), 5154–5160.

Roberts, A. B., Anzano, M. A., Lamb, L. C., Smith, J. M., Frolik, C. A., Marquardt, H., Todaro, G. J., and Sporn, M. B. (1982). Isolation from murine sarcoma cells of novel transforming growth factors potentiated by epidermal growth factors. *Nature (London)* **295**, 417–419.

Ross, R. (1975). Connective tissue cell proliferation and synthesis of extracellular matrix—A review. *Philos. Trans. R. Soc. London, Ser. B* **271**, 247–259.

Ross, R., Glomset, J., Kariya, B., and Harker, L. (1974). A platelet dependent serum factor that stimulates the proliferation of arterial smooth muscle cells *in vitro. Proc. Natl. Acad. Sci. U.S.A.* **71**(4), 1207–1210.

Sherwin, S. A., Twardzik, D. R., Bohn, W. H., Cockley, K. D., and Todaro, G. J. (1983). High molecular weight transforming growth factor activity in the urine of patients with disseminated cancer. *Cancer Res.* **43**(1), 403–407.

Shier, W. T. (1980). Serum stimulation of phospholipase A-2 and prostaglandin release in 3T3 cells is associated with platelet derived growth promoting activity. *Proc. Natl. Acad. Sci. U.S.A.* **77**(1), 137–141.

Shier, W. T., and Durkin, J. P. (1982). Role of stimulation of arachidonic acid release in the proliferative response of 3T3 mouse fibroblasts to platelet derived growth factor. *J. Cell. Physiol.* **112**(2), 171–181.

Steimer, K. S., Packard, R., Holden, D., and Klagsbrun, M. (1981). The serum-free growth of cultured cells in bovine colostrum and milk obtained later in the lactation period. *J. Cell. Physiol.* **109**(2), 223–234.

Todaro, G. J., Fryling, C., and DeLarco, J. E. (1980). Transforming growth factors produced by certain human tumor cells: Polypeptides that interact with epidermal growth factor receptors. *Proc. Natl. Acad. Sci. U.S.A.* **77**(9), 5258–5262.

Vogel, A., Raines, E., Kariya, B., Rivest, M. J., and Ross, R. (1978). Coordinate control of 3T3 cell proliferation by platelet derived growth factor and plasma components. *Proc. Natl. Acad. Sci. U.S.A.* **75**(6), 2810–2814.

Waterfield, M. D., Scrace, G. T., Whittle, N., Stroobant, P., Johnsson, A. Wasteson, A., Westermark, B., Heldin, C. H., Huang, J. S., and Deuel, T. F. (1983). Platelet derived growth factor is structurally related to the putative transforming protein p28[sis] simian sarcoma virus. *Nature (London)* **304** 35–39.

Weinstein, R., Hoover, G. A., Majure, J., Van Der Spek, J. Stemerman, M. B., and Maciag, T. (1982). Growth of human foreskin fibroblasts in a serum-free defined medium without platelet derived growth factor. *J. Cell. Physiol.* **110**(1), 23–28.

Westermark, B., and Wasteson, A. (1975). The response of cultured human normal glial cells to growth factors. *Adv. Metab. Disorders* **8**, 85–100.

Witkowski, E., Schuler, M. F., Feldhoff, R. C., Jacob, S. I., Jefferson, L. S., and Lipton, A. (1979). Liver as a source of transformed cell growth factor. *Exp. Cell Res.* **124**(2), 261–268.

Witte, L. D., and Cornicelli, J. A. (1980). Platelet derived growth factor stimulates low density lipoprotein receptor activity in cultured human fibroblasts. *Proc. Natl. Acad. Sci. U.S.A.* **77**(10), 5962–5966.

Wrann, M., Fox, C. F., and Ross, R. (1980). Modulation of epidermal growth factor receptors on 3T3 cells by platelet derived growth factor. *Science* **210**, 1363–1365.

III

Receptors

7

The EGF Receptor

HARVEY R. HERSCHMAN

Department of Biological Chemistry and
Laboratory of Biomedical and Environmental Sciences
UCLA Center for Health Sciences
Los Angeles, California

I. IDENTIFICATION OF THE EGF RECEPTOR

The EGF ligand/EGF receptor system has been among the most valuable experimental models available for the study both of ligand–receptor interactions and the nature of the transmembrane signals generated in a ligand-induced biological response. The biology and biochemistry

169

of epidermal growth factor (EGF) have been described in detail in Chapter 3. In this chapter I will discuss our current understanding of the structure of the EGF receptor, its properties both as enzyme and substrate, its synthesis and degradation, current hypotheses on its mode of action in acting as a transducer of signal from hormone to nucleus, and as a causal component in certain cancers.

Although a wide range of cells respond to EGF, the biology of this growth factor has been elucidated primarily through the study of cultured human fibroblasts and murine 3T3 cells. The first demonstration of an EGF "receptor" was described by Hollenberg and Cuatrecasas (1973): When EGF labeled with ^{125}I was incubated with cultured human fibroblasts, these cells demonstrated saturable binding, competable by unlabeled EGF. The first physical description of the EGF receptor was presented in 1977. Das *et al.* (1977) cross-linked an heterobifunctional photoactivatable derivative of [^{125}I]EGF to the EGF receptor of intact 3T3 cells. When an SDS extract of the cells was subjected to electrophoresis on polyacrylamide gels, a radiolabeled band estimated at 190 kilodaltons was observed. The photoaffinity-labeled band was identified as the EGF receptor on the basis of (1) proportionality of labeling and binding, (2) blockage of labeling with native EGF, (3) absence of labeling in "down-regulated" cells (see below), and (4) absence of labeling on cells unable to bind to EGF.

A formal demonstration that the EGF receptor is necessary for the mediation of the biological response to EGF was provided by Pruss and Herschman (1977) when they isolated variant 3T3 cell lines unable to respond to EGF and demonstrated the variants were unable to bind the radioactive ligand. Bishayee *et al.* (1982) report that they can transfer EGF receptors to this cell line and restore the mitogenic response to EGF.

II. EGF-INDUCED RECEPTOR LOSS: DOWN-REGULATION OF THE EGF RECEPTOR

Carpenter and Cohen (1976) demonstrated that the binding of EGF to the receptor of human fibroblasts was a complex event, not a simple equilibrium binding reaction between ligand in solution and receptor at the cell surface. When human fibroblasts were presented with saturating levels of [^{125}I]EGF at 37°C, the amount of cell-associated EGF increased for approximately 30 min. Cell-associated EGF subsequently decreased over the next 3 hr to a level only 25% of the maximum, despite the continued presence of excess labeled EGF. This phenomenon did not occur when binding was carried out at 4°C. When labeled EGF was bound for a short time to the cells and excess hormone was then removed by washing, the fate of cell-bound EGF could be observed during subsequent incubation at 37°C. Radioactive EGF was released into the medium with a half-life of approximately 20 min. However, the labeled EGF could no longer bind to cells; it had been degraded to monoiodotyrosine. Carpenter and Cohen (1976) suggested that this [^{125}I]EGF degradation and loss of EGF binding activity occurred as a result of the EGF-induced internalization of hormone–receptor complexes and the subsequent lysosomal degradation of both EGF and its receptor. Aharonov et al. (1978) demonstrated the same phenomenon in 3T3 cells and, in support of this hypothesis, also showed that the EGF receptors remaining on the cell surface after "down-regulation" had the same affinity, but were reduced in number. They also showed that the initial internalization of EGF and its receptor was not sufficient to provoke the mitogenic response; removal of EGF after down-regulation abrogated the subsequent initiation of DNA synthesis. Despite the flurry of metabolic activity occurring in the initial down-regulation response, continued occupancy of the remaining EGF receptors was necessary for the mitogenic response.

Aharonov et al. (1978) suggested that the major role of ligand-induced receptor down-regulation may be to permit feedback regulation of the receptor level in the face of varying hormone concentrations. Knaur et al. (1984) have demonstrated that at steady state (i.e., after all down-regulation has occurred), there is a linear relationship between the mitogenic stimulation and the occupancy of EGF receptors. At steady state after down-regulation, they find no "spare" receptors—all receptors must be occupied for maximal mitogenic response.

III. THE PATHWAY OF EGF-INDUCED RECEPTOR INTERNALIZATION

Electron microscopic studies using EGF labeled with electron-dense materials (ferritin, colloidal gold) suggest the EGF receptor is initially diffusely distributed over the surface of the cell (Haigler et al., 1979; Hopkins and Boothroiyd, 1981). A similar conclusion can be drawn from the analysis of the distribution of fluorescent EGF derivatives on living cells (Schlessinger et al., 1978a; Haigler et al., 1978). When EGF binds to the cell the EGF receptors, which are able to move in the plane of the membrane (Schlessinger et al., 1978b), undergo a rapid, temperature-dependent, ligand-induced clustering (Schlessinger et al.,

1978a; Haigler et al., 1978, 1979). Hillman and Schlessinger (1982) demonstrated that EGF-induced diffusion of the EGF receptor in the plane of the membrane was not nearly as dependent on temperature as the endocytosis of the ligand–receptor complex. They suggest that lateral diffusion of the receptor is not the rate-limiting step in the endocytotic process.

After binding EGF, the clustered receptors apparently diffuse to coated pits on the cell surface (Gorden, 1978; Haigler et al., 1979). Morphological and biochemical studies demonstrated that the EGF–receptor complex at the cell surface is rapidly vesiculated in a coated vesicle and sequestered in the cytoplasm (Gorden, 1978; Fine et al., 1981). These EGF receptors are therefore no longer available to bind competing, unlabeled ligand, nor can they be stripped of their EGF by acid wash (Haigler et al., 1979); they are "sequestered" receptors. Endocytotic vesicles convey sequestered EGF and the EGF receptor to cellular lysosomes. EGF begins to appear in this subcellular compartment within 5–10 min after binding (Gorden, 1978; Fine et al., 1981). The EGF is subsequently degraded to low-molecular-weight components in this compartment.

The degradation process for labeled EGF has been quite well described. Matrisian et al. (1984) utilized a purified preparation of ^{125}I-labeled EGF to demonstrate that a series of intermediate cleavage products are generated en route to the lysosomes prior to degradation to low-molecular-weight compounds in this organelle. Peptide analysis of cleavage products (Planck et al., 1984) coupled with subcellular fractionation studies (Matrisian et al., 1984) suggest successive cleavages at the carboxy terminal end of the EGF molecule occur during receptor-mediated binding, internalization, and transport to the lysosome.

What about the loss of receptor? It is clear that receptor is sequestered by internalization and is no longer available for binding at the cell surface. Is EGF receptor also degraded as a consequence of ligand-induced internalization? Das and Fox (1978) followed the fate of cross-linked receptor–EGF complex and observed breakdown products of discrete sizes within the cells. The conclusions, however, had to be constrained by the considerations that (1) they were observing only a small percentage of the receptor, since the cross-linking reaction occurs at a relatively low efficiency, and (2) it was quite possible (and perhaps even likely) that the cross-linked complex might be processed differently than the native receptor.

Recently, Krupp et al. (1982) utilized a density shift procedure to measure the rates of synthesis and degradation of the EGF receptor in A-431 cells (see below) in the presence and absence of EGF. They report a decrease in receptor half-life from 16 to 4.5 hr as a consequence of exposure to EGF. Stoscheck and Carpenter (1984) have utilized an antibody to the EGF receptor to measure directly the disappearance of [^{35}S]methionine-labeled receptors from cells in the presence and absence of EGF. The normal half-life of EGF receptors in human fibroblasts was 10.2 hr. When these cells were exposed to EGF the half-life of prelabeled receptor was decreased to 1.2 hr. Thus, the receptor internalized as a consequence of EGF-induced endocytosis is, like EGF, rapidly degraded. The potential biological roles of EGF-induced receptor clustering, endocytosis of EGF or EGF receptor, and degradation of EGF or its receptor will be discussed subsequently.

IV. A CELL LINE ENRICHED IN EGF RECEPTORS

Fabricant *et al.* (1977) screened a number of human tumor cell lines for the presence of EGF receptors. One particular cell line known as A-431, a human epidermoid carcinoma cell line originally placed in culture in 1973 (Giard *et al.*, 1973), had a remarkably high level of EGF receptors. Binding studies demonstrated that A-431 cells have $2-3 \times 10^6$ EGF receptors per cell, a value $20-30$ times greater than that observed for the majority of cells that display EGF receptors. This cell line has been of great value in elucidating the biochemistry and enzymology of the EGF receptor. However, in contrast to many cell types, EGF inhibits the growth of A-431 cells (Barnes, 1982; Buss *et al.*, 1982). Consequently, if we assume the major causal biological role of EGF is to act as a mitogen (an assumption still unproved in a physiological context), the A-431 cell is not an appropriate model system. However, because (1) this cell line serves as a rich source for receptor purification and characterization and (2) one anticipates that many EGF-mediated biochemical events may be amplified in their response in these cells despite the inability of the cell to mount the appropriate physiological response, the A-431 cell line has been used extensively for the biochemical characterization of the EGF receptor and the study of EGF-mediated events.

V. TYROSINE KINASE AND THE EGF RECEPTOR

A. Enzymatic Activity of the EGF Receptor

Carpenter *et al.* (1978, 1979) demonstrated that EGF was able to stimulate the transfer of phosphate from [γ-^{32}P]ATP to an endogenous macromolecular acceptor(s) in membrane preparations of A-431 cells. The reaction, stimulated threefold in the presence of EGF, required either Mn^{2+} or Mg^{2+} ions, but was calcium and cyclic nucleotide independent. When the membrane proteins were analyzed by electrophoresis, several substrates were observed, including phosphorylated proteins of molecular weights estimated at 170,000, 150,000, 80,000 and 22,000 (King et al., 1980). These initial reports also demonstrated that the A-431 membrane preparation could phosphorylate, in an EGF-dependent manner, soluble exogenous substrates such as histone.

When EGF receptor from solubilized membrane preparations was purified by EGF-affinity chromatography (as discussed more fully below), the purified preparation contained both the EGF-dependent kinase activity and the EGF-binding activity, suggesting the ligand-binding and kinase activities might reside on a common molecule. Moreover, the data indicated that the 150-kilodalton protein capable of binding [^{125}I]EGF and demonstrating EGF-stimulated kinase activity was also a major substrate for the phosphorylation reaction. These observations suggested that binding of EGF might stimulate an autophosphorylation of the EGF receptor by a kinase activity intrinsic to the receptor itself.

The presumptive EGF receptor–kinase was labeled with either radioactive parafluorosulforylbenzoyl adenosine (Buhrow et al., 1982) or 8-azido-ATP (Navarro et al., 1982), and the covalent complex was then subjected to electrophoresis and autoradiography. The bands previously identified with EGF binding and kinase activity were radioactively labeled, suggesting the EGF receptor–kinase molecule has an ATP binding site. This modification inhibited both the autophosphorylation reaction (Buhrow et al., 1982) and phosphorylation of exogenous substrates (Buhrow et al., 1983).

When they more closely examined the products of EGF-stimulated phosphorylation of intact membranes from A-431 cells, histone phosphorylated by A-431 membranes in the presence of EGF, or the product of EGF-stimulated autophosphorylation of affinity-purified receptor, Ushiro and Cohen (1980) made the startling observation that EGF-mediated phosphorylation occurred almost exclusively on tyrosine residues. Dimethyl sulfoxide is also able to stimulate the tyrosyl phosphorylation of the EGF receptor in liver microsomal fractions (Rubin and Earp, 1983). Recent studies by H. S. Earp (unpublished) suggest that DMSO induces an alteration in EGF receptor conformation. Because the kinase activity of the receptor has been used as a marker in its purification, these two topics (kinase activity and purification) are closely interwoven. We shall, therefore, return to the topic of intrinsic tyrosine kinase activity in discussions of purification of the EGF receptor and in consideration of its mechanism of action.

In addition to autophosphorylation, the experiments of Carpenter *et al.* (1978, 1979) demonstrated that the EGF receptor–kinase could carry out the phosphorylation of exogenous substrates. The prototype tyrosyl kinase activities were found as properties of the products of the oncogenes of several oncornaviruses (see Chapters 1 and 11). Collett and Erikson (1978) demonstrated that the kinase associated with Rous sarcoma virus (pp60^{src}) could phosphorylate the heavy chain of an IgG antibody directed against this enzyme. When antibody to pp60^{src} was incubated with EGF receptor–kinase, the immunoglobulin was phosphorylated in an EGF-dependent manner on a tyrosine residue (Kudlow *et al.*, 1981; Chinkers and Cohen, 1981). These data suggest some structural relationship of pp60^{src} kinase and EGF receptor–kinase at the catalytic site. However, although anti-pp60^{src} could serve as a substrate for the EGF receptor–kinase, it was unable to precipitate either the EGF binding activity or the kinase activity.

A variety of molecules have now been shown to serve as substrates in cell-free systems for EGF-dependent tyrosyl phosphorylation by the EGF receptor–kinase. Cohen *et al.* (1982a) demonstrated that tubulin could be phosphorylated in an EGF-dependent fashion by affinity-purified EGF receptor–kinase. Baldwin *et al.* (1983a) demonstrated that human growth hormone could be phosphorylated on tyrosine by A-431 cell membranes. Gastrin and middle T antigen of polyoma virus share with human growth hormone a sequence in which a tyrosine residue is preceded by several adjacent glutamic acid residues; these molecules also will serve as substrates *in vitro* for EGF receptor–kinase (Baldwin *et al.*, 1983b; Segawa and Ito, 1983). Recently, Ghosh-Dastidar *et al.* (1984b) have purified a 94-kilodalton substrate for EGF receptor–kinase from human placental extracts. Interest in this substrate was sharpened by its low K_m, 10^{-7} M. Because this substrate had molecular properties similar to those of several steroid receptors, Ghosh-Dastidar *et al.* (1984a) used as substrate a purified preparation of progesterone receptor and demonstrated the EGF-dependent phosphorylation of this protein by EGF receptor–kinase. The progesterone receptor also had a K_m of 10^{-7} M in this reaction.

Fava and Cohen (1984) have isolated a 35-kilodalton protein from A-431 cells that, in the presence of calcium, serves as a substrate for the EGF receptor–kinase. Tyrosine phosphorylation is elevated by the presence of EGF. This protein demonstrates a reversible calcium-dependent binding to A-431 membrane vesicles. Phosphorylation of the 35-kilodalton protein requires calcium; autophosphorylation of the EGF receptor is calcium dependent. It is of interest to note that Bishayee *et al.* (1984) have recently reported that a 34-kilodalton peripheral membrane protein mitogenic for the EGF nonresponsive 3T3 derivative 3T3-NR6 (Pruss and Herschman, 1977) can be extracted from plasma membranes of A-431 cells.

Erikson and his colleagues have shown there is a 34-kilodalton protein that serves as a substrate for pp60^{src} kinase in vitro (Erikson and Erikson, 1980). Fava and Cohen (1984) conclude that their 35-kilodalton substrate is not the pp60^{src} substrate, since the anti-34-kilodalton antiserum (provided by Erikson) was unable to precipitate the 35-kilodalton protein from A-431 cells phosphorylated by the EGF receptor–kinase. (Erikson's laboratory has shown that their antiserum to the 34-kilodalton protein cross-reacts with the similar 34-kilodalton protein from A-431 cells.) Moreover, Fava and Cohen (1984) report that they were unable to phosphorylate the purified 34-kilodalton pp60^{src} kinase substrate from A-431 cells with EGF receptor–kinase. In this experiment, substantial phosphorylation of 35-kilodalton protein was observed. In contrast, Ghosh-Dastidar and Fox (1983) report that when the soluble protein fraction of A-431 cells is incubated with EGF receptor–kinase, they observe EGF-enhanced phosphorylation of a 34-kilodalton protein that can subsequently be precipitated with antibody to the pp60^{src} 34-kilodalton substrate. If both observations are correct, the data suggest there must be some additional cytoplasmic factor necessary for phosphorylation of the 34-kilodalton pp60^{src} substrate by EGF receptor–kinase. The observation that EGF treatment of intact A-431 cells with EGF stimulates phosphorylation of the pp60^{src} 34-kilodalton substrate (Erikson et al., 1981; see next section) makes these data all the more provocative. The enzymology of the kinase reaction has also been studied in some detail using synthetic peptides with sequences similar to that of the phosphorylated tyrosyl peptide of pp60^{src} (Pike et al., 1982; Cassel et al., 1983). It should be emphasized that despite the evidence that these various proteins and peptides serve as substrates for EGF receptor–kinase in vitro, there is no definitive evidence that their EGF-dependent phosphorylation is causal in any EGF-mediated biological response.

B. Cellular Phosphotyrosine in Response to EGF

Several groups have examined the phosphorylation of cellular substrates after exposure of intact cells to EGF. Hunter and Cooper (1981) reported a rapid (1 min) three- to fourfold increase in total phosphotyrosine content of A-431 cells in response to EGF administration. The EGF receptor–kinase was itself phosphorylated as a consequence of this exposure. When supernatants from cells exposed to EGF were subjected to two-dimensional gel electrophoresis two proteins, 81 kilodaltons and 39 kilodaltons, were extensively phosphorylated on tyrosine. The latter protein is apparently the same protein that serves as a major substrate for pp60^{src} kinase.

Erikson et al. (1981) demonstrated that A-431 cells treated with EGF had elevated levels of tyrosyl-phosphorylated 34-kilodalton protein. This protein is also the major substrate for pp60^{src} kinase. More recently, Cooper et al. (1982) compared the tyrosyl phosphorylation of proteins from 3T3 cells in response to EGF and platelet-derived growth factor (PDGF). Each growth factor induced tyrosyl phosphorylation of a constellation of proteins, some of which are unique to the individual growth factors, some of which are common in the response to the two mitogens. Nakamura et al. (1983) treated quiescent chick embryo fibroblasts with EGF, PDGF or serum and observed a common tyrosyl phosphorylation of a 42-kilodalton protein(s). In these cells phosphorylation of 42-kilodalton protein(s) accounted for nearly all the mitogen-induced increase in phosphotyrosine in the soluble proteins. They observed a similar increase in tyrosyl phosphorylation of a 42-kilodalton substrate in quiescent 3T3 cells exposed to EGF. Cooper et al. (1984) confirmed these observations.

It should be emphasized that there is no evidence that the tyrosyl phosphorylation of proteins (89 kilodaltons, 39 kilodaltons in A-431, 42 kilodaltons in chick embryo fibroblasts or 3T3 cells) in intact cells treated with EGF occurs as a result of the direct phosphorylation of these proteins by the EGF receptor–kinase. This is also true for the 34-kilodalton protein, despite the fact that one group has reported it to be a substrate for EGF receptor–kinase phosphorylation in vitro. Serum, PDGF, and/or EGF treatment of cells may stimulate, in some manner, a common tyrosyl protein kinase whose substrate is one of these proteins. As an example, both chick embryo fibroblasts and 3T3 cells mount a mitogenic response when exposed to the tumor promoter tetradecanoylphorbol acetate (TPA). Three different groups (Bishop et al., 1983; Gilmore and Martin, 1983; Cooper et al., 1984) have demonstrated that the mitogenic response to TPA is accompanied by tyrosyl phosphorylation of the 42-kilodalton protein. TPA appears to produce its major biological effects by activating C-kinase (discussed below), a calcium, phospholipid-dependent protein kinase with a rigid substrate specifically for serine and threonine residues. Thus, it is almost certainly the case that the TPA-induced tyrosyl phosphorylation of the 42-kilodalton protein(s) is due to C-kinase activation of another kinase with tyrosine specificity. Such a kinase cascade could also occur as a consequence of EGF or PDGF binding and result in phosphorylation of the 42-kilodalton protein rather than the direct phosphorylation of the

42-kilodalton protein (or any other cellular substrates) by EGF receptor–kinase or PDGF receptor–kinase.

VI. PURIFICATION AND CHARACTERIZATION OF THE EGF RECEPTOR

The initial estimates of the size of the EGF receptor were determined by electrophoretic analysis of affinity-labeled preparations and ranged between 140 and 190 kilodaltons (Das et al., 1977; Sahyoun et al., 1978). Hock et al. (1979) solubilized the affinity-labeled receptor in detergent and demonstrated the covalently labeled receptor bound to several lectin affinity columns, suggesting the EGF receptor might be a glycoprotein. They also demonstrated that the affinity-labeled receptor–EGF covalent complex could be bound to and eluted from an anti-EGF affinity resin. Unfortunately, this approach to receptor purification is limited because of the low efficiency of the affinity-labeling procedures. The glycoprotein nature of the receptor was subsequently confirmed by carbohydrate staining of receptor preparations subsequent to electrophoresis (King et al., 1980).

Nexo et al, (1979) demonstrated that although receptor solubilized in Nonident P-40 was unable to bind [^{125}I]EGF, when the solubilized receptor was adsorbed to Con A–Sepharose, EGF binding could be observed. Using this assay and more classical purification procedures (lectin-affinity chromatography, gel filtration chromatography, DEAE–cellulose chromatography, and Cibacron Blue–Sepharose chromatography), they obtained a 110-fold purification of EGF receptor from human placental membranes. In this partially purified preparation the major component, by analysis on SDS–PAGE, was a 160-kilodalton protein (Hock et al., 1979).

The unlabeled receptor from A-431 cells was solubilized in Triton X-100 with retention of both the EGF binding activity (assayed with a polyethylene glycol precipitation reaction) and EGF-dependent kinase activity (Cohen et al., 1980). When the Triton X-100 supernatant was passed over an EGF affinity column, both the EGF binding activity and the EGF-dependent kinase activity could be eluted with ammonia or ethanolamine, suggesting an association of these two activities. A major band at about 150 kilodaltons was seen with Coomasie Blue staining on SDS gels of this preparation, with faint bands (including one at 170 kilodaltons) also visible. When the affinity-purified receptor preparation was subjected to EGF-dependent phosphorylation with [γ-^{32}P]ATP, a phosphorylated doublet at 170 and 150 kilodaltons was observed, with the major phosphorylation substrate migrating at 150 kilodaltons.

Cohen et al. (1982b) subsequently demonstrated that EGF receptor could be isolated from A-431 cells in the predominantly 170-kilodalton form. With the use of a new procedure for preparing membrane vesicles from A-431 cells by osmotic shock, the affinity-purified receptor was almost entirely in the 170-kilodalton form. When the previous procedure was used, which involved scraping the cells from the culture dishes, the receptor was in the 150-kilodalton form. Both forms retained the autophosphorylation activity and the [^{125}I]EGF binding activity. When kinase activity was examined in more detail, the 170-kilodalton form of the EGF receptor–kinase isolated from vesicles had a much greater intrinsic kinase activity; i.e., it was much more active in the autophosphorylation reaction. In contrast, the 150-kilodalton form had a greater specific activity for tyrosine phosphorylation of exogenous substrate (anti-pp60^{src}, tubulin). An antibody prepared against the 170-kilodalton protein eluted from SDS–PAGE was able to precipitate kinase, EGF binding, and kinase substrate activities. Cohen et al. (1982b) postulated that the 150-kilodalton form of the receptor might be a proteolytic fragment of the 170-kilodalton "native" receptor, generated during the original membrane preparation. Two groups (Gates and King, 1982; Cassel and Glaser, 1982) subsequently demonstrated that an endogenous calcium-activated neutral protease is able to convert the EGF receptor–kinase from the larger to the smaller electrophoretic form. The enzyme is activated by scraping the cells in the presence of calcium. Yeaton et al. (1983) have recently partially purified and characterized the calcium-activated neutral protease in A-431 cells that cleaves the larger form of the EGF receptor. Their data suggest a protease-susceptible phosphorylated domain on the cytosolic portion of the receptor. The biological role, if any, of this proteolysis is unknown.

Nearly all the studies on highly purified receptor have been carried out with preparations from A-431 cells. A-431 is a tumor cell line, with an aberrant response to EGF. Cohen et al. (1982b) suggested that the receptor from A-431 cells might in some way be altered relative to the receptor of normal cells. They purified EGF receptor from rabbit liver, using EGF-affinity chromatography and lectin-affinity chromatography. Although the human and murine receptors could be distinguished antigenically, no other qualitative differences were observed in size, enzyme activity, or substrate activity.

Carpenter (1983) cites four lines of evidence suggesting the kinase and binding activities reside in a single protein molecule. These in-

clude (1) the copurification (through electrophoresis) of the two activities, (2) their coprecipitation by both conventional and monoclonal
antibodies, (3) the covalent labeling of the molecule with ATP analogues, and (4) the coordinate expression in cells with varying levels of
the receptor (Buss et al., 1982) of [^{125}I]EGF binding and EGF-dependent
protein kinase activity. While each of these observations could occur
with the individual activities segregated on two interacting proteins,
the most parsimonious explanation is that they are intrinsic activities
of a single molecule. They can, however, be functionally separated;
kinase activity can be inactivated by heating without affecting EGF
binding activity (Carpenter et al., 1979).

VII. THE EGF RECEPTOR AS SUBSTRATE

I have previously discussed the EGF-dependent tyrosine phosphorylation of the EGF receptor. The cleavage of the receptor by a
calcium-activated neutral protease, both in vitro and, perhaps, in intact
cells was also previously noted.

Recently, several groups (Cochet et al., 1984; Iwashita and Fox, 1984)
have suggested that the EGF receptor is a substrate for the calcium,
phospholipid-dependent protein kinase known as C-kinase, a protein
kinase with serine and threonine specificity. Iwashita and Fox (1984)
reported that intact A-431 cells treated with the tumor promoter TPA
had a two- to threefold increase in the level of phosphate labeling of the
EGF receptor (isolated with a monoclonal antibody). The relative abundance of phosphotyrosine residues was reduced following TPA treatment, in contrast to results observed with EGF.

TPA is thought to exert many, if not all of its biological effects by
activation of C-kinase. In the presence of calcium and phospholipid, C-
kinase is further activated by diacylglycerol. TPA can, in the presence
of calcium and phospholipid, replace diacylglycerol and form a complex with C-kinase, also further activating the enzyme. The experiments of Iwashita and Fox (1984) suggest that TPA may activate C-
kinase of A-431 cells and stimulate its phosphorylation of the EGF
receptor. In a complementary series of experiments, Cochet et al. (1984)
reported that partially purified C-kinase binds to A-431 membrane
preparations in the presence of calcium and phosphorylates the EGF
receptor, primarily at threonine residues. When purified EGF receptor
was used as substrate, a requirement for phospholipid and either TPA
or diacylglycerol was said to exist. These authors also reported that
TPA treatment of intact A-431 cells reduced the EGF-stimulated phos-

photyrosine level by approximately one-half both in total protein and on the EGF receptor. Phosphorylation of the EGF receptor by C-kinase also decreased the EGF-stimulated EGF receptor autophosphorylation *in vitro*. These results may explain the effects of TPA on EGF binding to cells, discussed subsequently (Section IX).

The pp60^{src} kinase is phosphorylated at a serine site by cAMP-dependent protein kinase (Gilmer *et al.*, 1982). Because of the similarities in enzyme activity and substrates between pp60^{src} kinase and the EGF receptor–kinase, Ghosh-Dastidar and Fox (1984) assayed purified EGF receptor as substrate for cAMP-dependent protein kinase. They report that when cAMP and cAMP-dependent kinase are added to a system containing EGF and its receptor, EGF receptor phosphorylation increased threefold. Surprisingly, enhanced phosphorylation of the EGF receptor in this system was absolutely dependent on the presence of EGF. In the absence of EGF, no phosphorylation was observed in a reaction mixture including cAMP, cAMP-dependent protein kinase, [γ-^{32}P]ATP, and EGF receptor. Phosphorylation at tyrosine residues was identical in the presence or absence of cAMP and the cAMP-dependent protein kinase; in their presence there was increased phosphorylation at both serine and threonine residues. Heat treatment that renders the receptor refractory to EGF-dependent autophosphorylation without destroying EGF binding activity also eliminated the cAMP kinase-dependent phosphorylation. The biological roles of the phosphorylation of the EGF receptor by C-kinase and cAMP-dependent protein kinase, while the subject of intense speculation, are presently unknown.

VIII. DNA NICKING ACTIVITY OF THE EGF RECEPTOR

Recently, Mroczkowski *et al.* (1984) have reported that their most purified preparations of EGF receptor, using EGF–Affi gel chromatography, wheat germ lectin–Sepharose chromatography, and DNA–cellulose chromatography, have the ability to nick supercoiled DNA. The purified receptor can convert pBR322 form I (supercoiled) DNA to form II DNA. The reaction, which occurs with either the 170 or the 150-kilodalton form of the EGF receptor, is ATP dependent. The activity is not, however, stimulated by addition of EGF, in contrast to the tyrosine kinase activity of the EGF receptor.

φX174 RF1 DNA, SV40 DNA, and PM2 DNA are also substrates for conversion of the supercoil to the nicked form. A variety of control experiments demonstrate that the receptor preparations have little or no nuclease activity in the presence or absence of either EGF or ATP. The nicking reaction is resistant to inactivation at temperatures that result in complete loss of tyrosine kinase activity. Mroczkowski et al. report that pp60^{src} (provided by Erikson and Brugge) had similar DNA nicking activity and had a "much higher" specific activity than that of the EGF receptor. These authors occasionally observe "the generation of high molecular weight DNA species (possibly representing concatinated DNA) exclusively in the presence of ATP and the EGF receptor" and suggest the activity may be both a nicking and a closing enzyme.

IX. POSTULATED CAUSAL EGF-MEDIATED RESPONSE(S) IN STIMULATION OF CELL DIVISION

A variety of receptor-mediated responses to EGF binding have been suggested to be causal in the mitogenic response (see Chapter 13). To date none has been conclusively proved to be both necessary and sufficient.

1. The possibility that internalized EGF or a fragment thereof is required for mitogenesis has been suggested. However, Schreiber et al. (1981a) reported that an IgM monoclonal antibody to the EGF receptor could induce both [^{3}H]thymidine incorporation and an increase in cell number in human fibroblasts. More recently, Hapgood et al. (1983) reported that this same antibody, like EGF, induced synthesis of prolactin in rat pituitary GH$_3$ cells. These data suggest that the information for eliciting the biological response lies in the receptor, that the EGF molecule plays no role beyond its effect as ligand for the receptor.

2. Many researchers consider EGF-stimulated tyrosine kinase activity to be among the most likely of candidates for a causal role in EGF mitogenesis. The circumstantial evidence is compelling: Tyrosine kinase activity is a still relatively infrequently encountered enzymatic activity. It is shared by receptors for molecules known to be involved in growth regulation (EGF, PDGF, insulin) and by the oncogene products of a number of transforming RNA viruses that cause derangement of normal cellular growth. As described previously, the EGF receptor–kinase and the pp60^{src} kinase presumably share some serologic cross-reactivity in that anti-pp60^{src} can serve as substrate for both enzymes. However, Schreiber et al. (1981b) report that cyanogen bromide-cleaved epidermal growth factor (CNBr-EGF) acts as an antagonist in that it can bind to the receptor and stimulate several EGF-mediated events, including phosphorylation of the EGF receptor, but is unable to stimulate DNA synthesis. This same group (Schreiber et al., 1983).

finds that a monovalent Fab fragment of their mitogenic IgM mono-clonal antibody also stimulates EGF receptor phosphorylation, but does not induce DNA synthesis. They conclude that the protein phos-phorylation induced by EGF is not a sufficient signal for mitogenesis.

3. EGF, like a number of other hormones, stimulates calcium influx and turnover of phosphatidylinositol (Sawyer and Cohen, 1981; Smith et al., 1983) with consequent generation of diacylglycerol. Many of these hormones are now thought to work, at least in part, by activation of C-kinase via increased intracellular calcium and diacylglycerol. However, Butler-Gralla and Herschman (1981) isolated a variant 3T3 cell line that retains TPA binding (Blumberg et al., 1981) and C-kinase activity (Bishop et al., 1985) but is unable to mount a mitogenic re-sponse to TPA. If EGF were to have its major pathway of mitogenesis mediated by diacylglycerol activation of C-kinase, one would expect this variant to also be unable to exhibit a mitogenic response to EGF. However, the variant is fully capable of dividing in response to EGF.

4. Schlessinger and his associates have performed a series of experi-ments from which they conclude receptor aggregation and/or inter-nalization may be the key event in the mitogenic response (reviewed by Schlessinger et al., 1982). CNBr-EGF bound to EGF receptors (albeit with reduced affinity) but could not induce mitogenesis. In contrast to native EGF, CNBr-EGF did not induce receptor clustering. When bivalent antibody (but not monovalent Fab fragments) to EGF was add-ed to cells exposed to CNBr-EGF, both microaggregation of the EGF-receptor complex and mitogenic activity were restored (Shechter et al., 1979).

Neither clustering of receptors (analyzed by fluorescence micros-copy) nor mitogenesis was observed when cells were exposed to a univalent Fab fragment of the monoclonal antibody that acts as an EGF agonist. If these Fab fragments were cross-linked with anti-murine Ig antibodies, receptor aggregation and mitogenic responsiveness were again restored (Schreiber et al., 1983). These investigators regard "re-ceptor clustering . . . as a necessary and sufficient signal for the induc-tion of DNA synthesis." In the same study, Schreiber et al. (1983) describe a second monoclonal to the EGF receptor. This IgG molecule does not prevent EGF binding to the receptor (and thus is binding to a site distinct from the EGF binding site). This monoclonal antibody is unable to stimulate either receptor clustering or mitogenesis. When this antibody is cross-linked with anti-murine Ig antibodies, it too can elicit stimulation of DNA synthesis. The authors suggest that occupancy of the hormone binding site is not essential for the mitogenic response, only appropriate receptor clustering. Unfortunately, they do not report whether this latter antibody in either the disperse or aggregated condition on target cells enhances the tyrosine phosphorylation of the EGF receptor.

In a similar experiment, Das et al. (1984) found that a polyclonal antibody to the EGF receptor, able to stimulate phosphorylation of the solubilized receptor, could not stimulate DNA synthesis in human fibroblasts either by itself or following sequential addition of rabbit anti-mouse IgG to ensure clustering. They conclude that "receptor clustering and kinase activation are not sufficient for progression through the mitogenic pathway." Das et al. also do not report on the ability of this antibody to stimulate phosphorylation of the EGF receptor in intact cells. The differing specificities of the antibodies of Das et al. (1984) and Schreiber et al. (1983), monoclonal versus polyclonal, may account for the differences observed in mitogenic capacity of cross-linked antibody to the EGF receptor.

The experiments of Schlessinger and his colleagues leave open the possibility that internalization of the receptor or a portion of the receptor is necessary to act at an internal site in the cell in order to elicit a mitogenic response (see Chapter 13). It is quite difficult to rule out in an intact cell system a small amount of receptor endocytosis. It is thus consistent with these observations to suggest that some portion of the receptor, internalized as a consequence of either EGF or antibody-induced aggregation, may have an intracellular role in the mitogenic response. It is, of course, quite interesting to speculate that the recently described ATP-dependent interaction of the EGF receptor with super-coiled DNA (Mroczkowski et al., 1984) might, after EGF-induced internalization, have a role in subsequent stimulation of DNA synthesis and cell division.

X. MODULATION OF THE EGF RECEPTOR BY ENDOGENOUS AND EXOGENOUS AGENTS

A variety of agents are able to modulate the expression of the EGF receptor on target cells. The biological roles of these regulatory events are not understood, nor are their mechanisms. The following discussion is not meant to be exhaustive; it does, however, cover the major types of modulations of which we are currently aware.

1. TPA was one of the earliest agents for which modulation of the EGF receptor was observed. Lee and Weinstein (1978) demonstrated that the binding of $[^{125}I]EGF$ to HeLa cells was inhibited by TPA. Although the original experiments were interpreted to be the result of a direct competition for EGF receptors by TPA, a number of subsequent

studies demonstrated that in the presence of TPA, the affinity of the EGF receptor for EGF was substantially reduced by interaction between TPA and some binding site other than the EGF receptor. Magun and his colleagues (1980; Matrisan et al., 1981) have suggested that the ability of TPA to synergistically stimulate DNA synthesis with EGF (at low concentrations) may be a consequence of its ability to temporarily decrease the binding of EGF, reducing EGF degradation and thus maintaining a stimulatory level of EGF for a longer period of time.

When the interaction of TPA with C-kinase was discovered (Castagna et al., 1982), it was thought that the modulation of EGF binding by TPA would likely be a consequence of C-kinase activation. Subsequently (as described previously), Cochet et al. (1984) reported that the EGF receptor is, indeed, a substrate for C-kinase.

2. Transformation of murine cells by murine or feline RNA tumor viruses drastically reduces the level of EGF receptors on the transformed cells (Todaro et al., 1976). Although cells transformed by RNA tumor viruses subsequently produce a new product that interacts with the EGF receptor (described below), down-regulation of the EGF receptor by this agent does not appear to be the sole explanation of the phenomenon (Pruss et al., 1978). It is interesting to note in this regard that Cherington et al. (1979), in evaluating macromolecular requirements for growth of normal versus transformed cells, concluded that one of the initial steps in establishment of the reduced serum requirement of chemically transformed cells is relaxation of the EGF requirement.

3. Ivanovic and Weinstein (1982) reported that treatment of murine fibroblasts with polyaromatic hydrocarbons that bind to the aryl-hydrocarbon hydroxylase receptor also leads to the loss of EGF binding activity. In a subsequent study (Kärenlampi et al., 1983) on a murine hepatoma cell line, the time course was more extensively analyzed. Both groups report that, in contrast to TPA where the inhibition is very rapid and involves a reduction in affinity, the polyaromatic hydrocarbon-induced decrease occurs over a substantial time period, perhaps up to 24 hr. Ivanovic and Weinstein (1982) also demonstrated that the polyaromatic hydrocarbon-induced loss of EGF binding was due to a diminished number of EGF receptors rather than a reduced affinity.

4. Differentiation of some cell populations leads to reductions in EGF receptor levels. The rat PC12 pheochromocytoma cell line has receptors for both nerve growth factor (NGF) and EGF. Exposure to NGF promotes morphological and biochemical differentiation of the cells into a more mature, neuronal form (Greene and Shooter, 1980). During the NGF-induced maturation procedure (over 3 days or so), the level of EGF receptors on the cells drops dramatically (Huff and Guroff, 1979).

Recently, Lim and Hauschka (1984) used a system in which proliferating myoblast cells could be committed to terminal differentiation by removal of fibroblast growth factor from the medium. They found that the level of EGF receptors declined by 95% within 24 hr. Several lines of investigation suggested the decline in EGF receptors is related to differentiation rather than to cell cycle arrest. [In another context Moses and his co-workers have demonstrated that cell cycle position does influence the level of EGF binding (Robinson et al., 1981, 1982).] The observations with pheochromocytoma and myoblast cells suggest that reduction of EGF receptors may be a way in which terminally differentiating cells establish their refractory nature with regard to mitogenic responsiveness.

5. In contrast to the long-term requirement for NGF modulation of EGF receptors, Wrann et al. (1980) report that PDGF application to Swiss 3T3 cells causes a rapid (2 hr) transient decrease in the ability of these cells to bind EGF. In subsequent studies, Wharton et al. (1982) suggested that the number of EGF receptors is reduced with no loss in affinity. Although Wrann et al. postulated that this might occur as the result of a common mechanism for cellular internalization of both growth factors, subsequent experiments with purified PDGF showed that the reciprocal reaction (reduction of PDGF receptor levels as a consequence of exposure to EGF) did not occur.

Cholera toxin, which by itself has no effect on EGF binding, greatly enhanced the ability of PDGF to decrease EGF binding on BALB/c 3T3 cells and also enhanced the mitogenic activity of PDGF (Wharton et al., 1982). The mitogenic activity of EGF is also greatly enhanced by cholera toxin (Pruss and Herschman, 1978). In the BALB/c 3T3 system PDGF-stimulated DNA synthesis is dependent on EGF. Cholera toxin was able to alleviate the EGF requirement. These authors (Wharton et al., 1982; Leof et al., 1982) suggest that the PDGF-induced down-regulation of EGF receptors, along with agents elevating cyclic AMP, can mimic the action of EGF. These data lend support to the argument that occupation of the EGF receptor is not necessary for mitogenesis if the EGF receptor can be aggregated and internalized by an alternative mechanism.

More recently, Collins et al. (1983) and Bowen-Pope et al. (1983) have reinvestigated the modulation of EGF receptors on Swiss 3T3 cells by PDGF. Both groups report that the decrease in EGF binding following administration of PDGF does not cause down-regulation of EGF receptors. They found that reduced binding of EGF after cells are exposed to PDGF is the result of a change in the affinity of EGF receptors, similar to that seen following TPA administration. In the three publications in which Scatchard analysis of the data was performed three qualitatively different results were obtained: (1) linear Scatchard plots suggesting a loss of EGF receptors as a consequence of PDGF binding (Wharton et al., 1982), (2) linear Scatchard plots indicating that PDGF reduces the affinity of the EGF receptor without decreasing their number (Bowen-Pope et al., 1983), and (3) curvilinear Scatchard plots that suggest the PDGF-induced reduction in EGF binding is due to the loss of a "high-affinity" binding site. The former study was carried out with BALB/c 3T3 cells, while the latter two reports were for Swiss 3T3 cells. While differences in cell strains may account for some of the discrepancies in these observations, additional experiments utilizing other methods to analyze receptor levels (e.g., with antireceptor antibodies) are required to clarify the important question of how one mitogen modifies expression of the receptor for another mitogen.

6. Murine cells transformed by RNA tumor viruses and many human tumor cells produce an activity, known as transforming growth factor (DeLarco and Todaro, 1978), capable of causing normal cells to transiently express aspects of the transformed phenotype. The activity is, in many cases, separable into two polypeptides known as α-TGF and β-TGF (Anzano et al., 1983). The α-TGF molecule, which bears sequence homology to the EGF molecule primarily in its alignment of its cystinyl residues (Marquart et al., 1983), binds to the EGF receptor as effectively as EGF and stimulates the EGF pathway (receptor phosphorylation, ion transport, mitogenesis). When the EGF receptor is occupied either by EGF or by α-TGF, β-TGF is able to elicit growth in soft agar (Roberts et al., 1983). Recently, Assoian et al. (1984) demonstrated that treatment of NRK cells with β-TGF induced an increase in the number of EGF receptors. The elevation of EGF receptor levels was time, temperature, and concentration dependent and could be blocked by inhibitors of either protein synthesis or carbohydrate processing.

Since both α-TGF and β-TGF activity have now been found in normal tissues, their specific relevance to cancer is questionable. While the biological role of these agents (like EGF itself) is unclear, their requirement for cooperative interaction at two distinct sites, one of which is the EGF receptor, to elicit biological activity is quite provocative.

XI. SEQUENCE HOMOLOGY BETWEEN THE EGF RECEPTOR AND THE AVIAN ERYTHROBLASTOSIS VIRUS v-erb-B ONCOGENE PRODUCT

Downward et al. (1984) have recently reported partial sequence data for the human EGF receptor. Receptor was purified either by immunoaffinity procedures (using monoclonal antisera) or by EGF-affinity chromatography from both A-431 cells and human placental membranes. Fourteen peptides obtained by preparative reverse-phase HPLC were sequenced with a gas-phase sequencing system. The amino acid sequences of these EGF receptor peptides were then compared with an oncogene sequence data base. Six of the fourteen peptides demonstrated remarkable homology to amino acid sequences predicted for the protein product of the transforming protein coded by the v-erb-B gene of avian erythroblastosis virus (AEV). The six peptides derived from the EGF receptor have a total of 83 amino acids. Of these, 74 residues are identical to those of the v-erb-B gene product; four were conservative substitutions. No insertions or deletions were required to obtain this match. Since the v-erb-B gene is found in an avian retrovirus, this degree of homology with the human EGF receptor is, as the authors comment, "remarkable." The authors suggest that "it is likely that the v-erb-B sequences were mainly acquired by AEV from those cellular sequences which encode the avian EGF receptor." They consequently suggest that the cellular homologue of the v-erb-B gene, c-erb-B, codes for the human EGF receptor.

Eight of the fourteen sequenced peptides of the EGF receptor did not show any homology with the v-erb-B gene product. This latter protein is a 68-kilodalton phosphorylated membrane glycoprotein (Privalsky et al., 1983; Hayman et al., 1983). It is likely that these eight peptides are derived from the N-terminal portion of the c-erb-B gene, a portion, not present in the v-erb-B gene found in AEV, that codes for the EGF/α-TGF binding site of the receptor.

It is of interest that the portion of the EGF receptor gene present in v-erb-B is thought to possess both the site of receptor tyrosine phosphorylation and the presumptive catalytic site for tyrosine kinase activity. While the EGF receptor had previously been postulated by many to be "a good candidate for a viral oncogene" product (Newmark, 1984), the expectation was that it would be likely that one of the oncogenes encoding a tyrosine kinase activity would be derived from the EGF receptor gene. However, although there has been some intensive searching, tyrosine kinase activity has not yet been found associated with the v-erb-B oncogene product, despite some sequence similarity with the src oncogene and other oncogenes encoding tyrosine kinases.

XII. THE EGF RECEPTOR GENE

A. Mapping the Gene for the Human EGF Receptor

Shimizu and his associates have used somatic cell hybridization procedures to demonstrate dominance of the EGF receptor gene. Hybrid cells formed by fusion of human fibroblast cell lines positive for EGF receptor (EGFR) with the murine L cell derivative A9, deficient in EGF binding activity, were able to bind and degrade [^{125}I]EGF (Shimizu et al., 1980). Somatic cell hybridization studies by this group suggested that the gene for the EGF receptor is located on human chromosome 7. In subsequent studies utilizing human fibroblasts with translocations of portions of chromosome 7 hybridized with murine A9 cells, the region containing the EGF receptor has been narrowed to the p13→q22 region of this chromosome (Kondo and Shimizu, 1983). Carlin and Knowles (1982) have shown that a surface glycoprotein known as SA7 is identical to the EGFR. Because SA7 has been mapped to the 7pter→p12 region of chromosome 7 (Knowles et al., 1977; Kondo and Shimizu, 1983), Kondo and Shimizu suggest it is possible that the EGF receptor gene may be found in the small region p13→p12 of chromosome 7.

More recently, Shimizu et al. (1984) have used somatic cell hybridization techniques to analyze the basis for the overproduction of EGF receptors in A-431 cells. The cells have two copies of intact chromosome 7 and two distinct translocation chromosomes involving chromosome 7. After fusing A-431 cells to A9 cells, they isolated 10 independent clones and scored for EGF binding, retention of chromosome 7, and retention of the translocation chromosome M4. They concluded that the presence of the translocation chromosome M4 is correlated with the elevated expression of EGF receptors. However, the difference in EGF binding observed with cells retaining only chromosome 7 and those retaining only chromosome M4 was from three- to fivefold; the difference in EGF binding between A-431 cells and most normal cells is usually greater than 20-fold. The data are provocative, however, since overproduction of a cellular oncogene product as a consequence of chromosomal translocation is thought to be the causal (or at least contributory) event in several types of cancer.

B. Cloning of the c-erb-B/EGF Receptor Gene

The report of Downward et al. (1984) provides compelling evidence for the concept that the v-erb-B oncogene is derived from a portion of the gene encoding the EGF receptor or a closely related gene. Using a recombinant chick DNA library, Vennström and Bishop (1982) have cloned cellular DNA sequences containing the v-erb-B coding sequence. Heteroduplex mapping of the λ-phage clones containing c-erb-B (EGF receptor) coding sequences revealed a very complex pattern; the data suggest that the avian genomic sequences homologous to v-erb-B contain 11 coding regions (covering 2.1 kb), separated by 10 intervening sequences collectively totaling 21 kb in length. Since the v-erb-B oncogene is thought to represent only approximately one-half to two-thirds of the coding sequences of the EGF receptor, these data suggest that the intact EGF receptor gene is a large, complex structure. Two large RNA molecules are transcribed from c-erb-B in normal chick embryos, one approximately 9 kb in length and one 12 kb.

Jansson et al. (1983) have recently described the isolation from a recombinant DNA library of human fetal liver DNA of a cloned human genomic sequence homologous to v-erb-B. However, because of the presence of repetitive sequences, little analysis of this clone was presented. Using this human c-erb-B clone as probe, Spurr et al. (1984) employed somatic cell hybrids to map the human c-erb-B gene. They found that the human c-erb-B gene maps to the identical region of chromosome 7 previously reported for the EGF receptor gene. These data strongly support the hypothesis that the c-erb-B gene, or a closely linked, highly homologous gene, codes for the EGF receptor.

Lin et al. (1984) have recently used polyclonal antibodies to clone fragments of the cDNA for the EGF receptor of A-431 cells. The resulting probes were then used to isolate additional cDNA sequences using a "messenger sandwich" technique. Sequence studies of these cDNA clones have confirmed and extended the homology data reported by Downward et al. (1984); in a region encoding 254 residues, only six amino acid differences were observed between the avian v-erb-B sequence and the human EGF receptor. The EGF receptor cDNA clone and the avian v-erb-B gene are about 85% homologous in nucleotide sequence over a 762 base pair region encoding a presumptive portion of the kinase domain. When the EGF receptor cDNA was used to analyze messenger RNA species in A-431 cells, 12 kb and 9 kb sequences were present in large quantity. Genomic blots suggested A-431 cells have amplified copies of a single gene for the EGF receptor. Human fibroblasts had reduced levels of both EGF receptor genes and the 12 kb and 9 kb messages. A-431 cells also produce a related 4.5 kb mRNA containing only 5′ coding sequences, postulated to code for a secreted EGF receptor-related peptide produced by A-431 cells (Weber et al., 1984). Ullrich et al. (1984) used synthetic polynucleotides to clone a cDNA containing the complete coding region for one EGF receptor.

C. Insertional Activation of the c-erb-B Oncogene by Avian Leukosis Virus

Avian leukosis virus (ALV) is a naturally occurring chicken cancer virus that can induce a variety of tumors in chickens. Unlike AEV, ALV requires a long latency period and does not contain an oncogene. A variety of experimental observations suggest that ALV induces cancers by insertion of the viral long-terminal repeat (LTR) near a cellular on-cogene, leading to aberrant expression of the product of the cellular oncogene.

Erythroblastosis, the same disease caused by the acutely transform-ing AEV virus carrying the v-erb-B oncogene, is among the various types of cancer that ALV can induce in chickens. Fung et al. (1983) have now shown that the molecular basis for the induction of erythroblastosis by ALV is the insertion of the viral LTR in the vicinity of the c-erb-B gene, leading to the activation of c-erb-B. They isolated a molecular clone from the genome of an ALV-induced erythroblastosis tumor and demonstrated that the viral LTR was inserted 5′ to the most 5′ exon of the c-erb-B genomic segment described by Vennström and Bishop (1982). Restriction mapping demonstrated that the LTR was inserted only 1.5–2 kb from this 5′ exon. It will be of great interest to determine where this LTR insertion occurs with respect to the start of transcription of the EGF receptor (assuming c-erb-B and the EGF recep-tor gene are, indeed, the same). If the 5′ end of the c-erb-B/EGF receptor gene is similar in structure to the already cloned portion, it is likely that this insertion occurs downstream from the start of normal transcription and, like the v-erb-B oncogene present in AEV, will result in the pro-duction of a truncated EGF receptor protein.

XIII. THE FUTURE

It is likely that the cloning of the structural gene for the EGF receptor will be published before this review. Analysis of the structure of this gene will clarify the sequence relationship between v-erb-B, c-erb-B, the EGF receptor gene, and the transformants elicited by the insertion of the longer terminal repeat of ALV in the c-erb-B gene. Manipulation of the structure of the EGF receptor gene by deletion analysis and in vitro mutagenesis procedures should allow, by transfection analysis, the identification of the portions of the receptor molecule required for (1) tyrosine kinase activity, (2) EGF binding, (3) substrate sites for auto-phosphorylation, C-kinase, and cAMP-dependent protein kinase, (4) receptor clustering, (5) EGF-mediated internalization, and (6) DNA nicking activity. Transfection of appropriately modified EGF receptor gene constructs into variant cell lines that exhibit growth control but do not have a functional EGF receptor (Pruss and Herschman, 1977; Butler-Gralla and Herschman, 1981; Terwilliger and Herschman, 1984) should define which of these (or other) properties of the EGF receptor are required for EGF-mediated initiation of cellular proliferation in a controlled fashion, and which regions of the truncated EGF receptor gene cause cellular transformation. In the space of slightly over a decade, we will have seen the transition from the EGF receptor as a concept to its molecular description, a determination of its essential elements in eliciting EGF-mediated biological responses, and a molecular characterization of how alteration in its structure can result in cancer.

ACKNOWLEDGMENTS

The preparation of this chapter was supported by award GM24797 from the National Institutes of Health and Contract DE AM03 76 SF00012 between the Regents of the University of California and the Department of Energy. I thank Drs. Carpenter, Cohen, Cunningham, Das, Earp, Fox, Gill, Hauschka, Hunter, Magun, Rosenfeld, and Weber for providing manuscripts prior to publication.

REFERENCES

Aharonov, A., Pruss, R. M., and Herschman, H. R. (1978). Epidermal growth factor: Relationship between receptor regulation and mitogenesis in 3T3 cells. *J. Biol. Chem.* **253**, 3970–3977.

Anzano, M. A., Roberts, A. B., Smith, J. M., Sporn, M. B., and DeLarco, J. E. (1983). Sarcoma growth factor from conditioned medium of virally transformed cells is composed of both type α and type β transforming growth factors. *Proc. Natl. Acad. Sci. U.S.A.* **80**, 6264–6268.

Assoian, R. K., Frolik, C. A., Roberts, A. B., Miller, D. M., and Sporn, M. B. (1984). Transforming growth factor-β controls receptor levels for epidermal growth factor in NRK fibroblasts. *Cell (Cambridge, Mass.)* **36**, 35–41.

Baldwin, G. S., Grego, B., Hearn, M. T. W., Knesel, J. A., Morgan, F. J., and Simpson, R. J. (1983a). Phosphorylation of human growth hormone by the epidermal growth factor-stimulated tyrosine kinase. *Proc. Natl. Acad. Sci. U.S.A.* **80**, 5276–5280.

Baldwin, G. S., Knesel, J., and Monckton, J. M, (1983b). Phosphorylation of gastrin-17 by epidermal growth factor-stimulated tyrosine kinase. *Nature (London)* **301**, 435–437.

Barnes, D. W. (1982). Epidermal growth factor inhibits growth of A-431 human epidermoid carcinoma in serum-free cell culture. *J. Cell Biol.* **93**, 1–4.

Bishayee, S., Feinman, J., Pittenger, M., Michael, H., and Das, M. (1982). Cell surface insertion of exogenous epidermal growth factor receptors into receptor-mutant cells: Demonstration of insertion in the absence of added fusogenic agents. *Proc. Natl. Acad. Sci. U.S.A.* **79**, 1893–1897.

Bishayee, S., Matesic, D., and Das, M. (1984). Identification of a 34,000 dalton mitogenic protein associated with plasma membranes from human A-431 epidermoid carcinoma cells. *Proc. Natl. Acad. Sci. U.S.A.* **81**, 3399–3403.

Bishop, R., Martinez, R., Nakamura, K. D., and Weber, M. J. (1983). A tumor promoter stimulates phosphorylation on tyrosine. *Biochem. Biophys. Res. Commun.* **115**, 536–543.

Bishop, R., Martinez, R., Weber, M. J., Blackshear, P. J., Beatty, S., Lim, R., and Herschman, H. R. (1985). Protein phosphorylation in a TPA nonproliferative variant of 3T3 cells. *Mol. Cell. Biol.* (submitted for publication).

Blumberg, P. M., Butler-Gralla, E., and Herschman, H. R. (1981). Analysis of phorbol ester receptors in phorbol ester unresponsive 3T3 cell variants. *Biochem. Biophys. Res. Commun.* **102**, 818–823.

Bowen-Pope, D. F., DiCorleto, P. E., and Ross, R. (1983). Interactions between the receptors for platelet-derived growth factor and epidermal growth factor. *J. Cell Biol.* **96**, 679–683.

Buhrow, S. A., Cohen, S., and Staros, J. V. (1982). Affinity labeling of the protein kinase associated with the epidermal growth factor receptor in membrane vesicles from A-431 cells. *J. Biol. Chem.* **257**, 4019–4022.

Buhrow, S. A., Cohen, S., Garbers, D. L., and Staros, J. V. (1983). Characterization of the interaction of 5'-p-fluorosulfonylbenzoyl adenosine with the epidermal growth factor receptor/protein kinase in A-431 cell membranes. *J. Biol. Chem.* **258**, 7824–7827.

Buss, J. E., Kudlow, J. E., Lazar, C. S., and Gill, G. N. (1982). Altered epidermal growth factor (EGF)-stimulated protein kinase activity in variant A-431 cells with altered growth responses to EGF. *Proc. Natl. Acad. Sci. U.S.A.* **79**, 2574–2578.

Butler-Gralla, E., and Herschman, H. R. (1981). Variants of 3T3 cells lacking mitogenic response to the tumor promoter tetradecanoylphorbol acetate. *J. Cell. Physiol.* **107**, 59–68.

Carlin, C. R., and Knowles, B. B. (1982). Identity of human epidermal growth factor (EGF) receptor with glycoprotein SA-7: Evidence for differential phosphorylation of the two components of the EGF receptor from A-431 cells. *Proc. Natl. Acad. Sci. U.S.A.* **79**, 5026–5030.

Carpenter, G. (1983). The biochemistry and physiology of the receptor-kinase for epidermal growth factor. *Mol. Cell. Endocrinol.* **31**, 1–19.

Carpenter, G., and Cohen, S. (1976). ^{125}I-labeled human epidermal growth factor binding, internalization, and degradation in human fibroblasts. *J. Cell Biol.* **71**, 159–171.

Carpenter, G., King, L., Jr., and Cohen, S. (1978). Epidermal growth factor stimulates phosphorylation in membrane preparations *in vitro*. *Nature (London)* **276**, 409–410.

Carpenter, G., King, L., Jr., and Cohen, S. (1979). Rapid enhancement of protein phosphorylation in A-431 cell membrane preparations by epidermal growth factor. *J. Biol. Chem.* **254**, 4884–4891.

Cassel, D., and Glaser, L. (1982). Proteolytic cleavage of epidermal growth factor receptor. *J. Biol. Chem.* **257**, 9845–9848.

Cassel, D., Pike, L. J., Grant, G. A., Krebs, E. G., and Glaser, L. (1983). Interaction of epidermal growth factor-dependent protein kinase with endogenous membrane proteins and soluble peptide substrate. *J. Biol. Chem.* **258**, 2945–2950.

Castagna, M., Takai, Y., Kaibuchi, K., Sano, S., Kikkawa, U., and Nishizuka, Y. (1982). Direct activation of calcium-activated, phospholipid-dependent protein kinase by tumor-producing phorbol esters. *J. Biol. Chem.* **257**, 7847–7851.

Cherington, P. V., Smith, B. L., and Pardee, A. B. (1979). Loss of epidermal growth factor requirement and malignant transformation. *Proc. Natl. Acad. Sci. U.S.A.* **76**, 3937–3941.

Chinkers, M., and Cohen, S. (1981). Purified EGF receptor-kinase interacts specifically with antibodies to Rous sarcoma virus transforming protein. *Nature (London)* **290**, 516–519.

Cochet, C., Gill, G. N., Meisenhelder, J., Cooper, J. A., and Hunter, T. (1984). C-Kinase phosphorylates the epidermal growth factor receptor and reduces its epidermal growth factor-stimulated tyrosine protein kinase activity. *J. Biol. Chem.* **259**, 2553–2558.

Cohen, S., Carpenter, G., and King, L., Jr. (1980). Epidermal growth factor receptor–protein kinase interactions. *J. Biol. Chem.* **255**, 4834–4842.

Cohen, S., Fava, R. A., and Sawyer, S. T. (1982a). Purification and characterization of epidermal growth factor receptor/protein kinase from normal mouse liver. *Proc. Natl. Acad. Sci. U.S.A.* **79**, 6237–6241.

Cohen, S., Ushiro, H., Stoscheck, C., and Chinkers, M. (1982b). A native 170,000 epidermal growth factor receptor–kinase complex from shed plasma membrane vesicles. *J. Biol. Chem.* **257**, 1523–1531.

Collett, M. S., and Erikson, R. L. (1978). Protein kinase activity associated with the avian sarcoma virus src gene product. *Proc. Natl. Acad. Sci. U.S.A.* **75**, 2021–2024.

Collins, M. K. L., Sinnett-Smith, J. W., and Rozengurt, E. (1983). Platelet-derived growth factor treatment decreases the affinity of the epidermal growth factor receptors of Swiss 3T3 cells. *J. Biol. Chem.* **258**, 11689–11693.

Cooper, J. A., Bowen-Pope, D. F., Raines, E., Ross, R., and Hunter, T. (1982). Similar effects of platelet-derived growth factor and epidermal growth factor on the phosphorylation of tyrosine in cellular proteins. *Cell (Cambridge, Mass.)* **31**, 263–273.

Cooper, J. A., Sefton, B. M., and Hunter, T. (1984). Diverse mitogenic agents induce the phosphorylation of two related 42,000 dalton proteins on tyrosine in quiescent chick cells. *Mol. Cell. Biol.* **4**, 30–37.

Das, M., and Fox, C. F. (1978). Molecular mechanisms of mitogen action: Processing of receptor induced by epidermal growth factor. *Proc. Natl. Acad. Sci. U.S.A.* **75**, 2644–2648.

Das, M., Miyakawa, T., Fox, C. F., Pruss, R. M., Aharonov, A., and Herschman, H. R. (1977). Specific radiolabeling of a cell surface receptor for epidermal growth factor. *Proc. Natl. Acad. Sci. U.S.A.* **74**, 2790–2794.

Das, M., Knowles, B., Biswas, R., and Bishayee, S. (1984). Receptor modulating properties of an antibody directed against the epidermal growth factor receptor. *Eur. J. Biochem.* **141**, 429–434.

DeLarco, J. E., and Todaro, G. J. (1978). Growth factors from murine sarcoma virus-transformed cells. *Proc. Natl. Acad. Sci. U.S.A.* **75**, 4001.

Downward, J., Yarden, Y., Mayes, E., Scrace, G., Totty, N., Stockwell, P., Ullrich, A., Schlessinger, J., and Waterfield, M. D. (1984). Close similarity of epidermal growth factor receptor and v-erb-B oncogene protein sequences. *Nature (London)* **307**, 521–527.

Erikson, E., and Erikson, R. L. (1980). Identification of a cellular protein substrate phosphorylated by the avian sarcoma virus-transforming gene product. *Cell (Cambridge, Mass.)* **21**, 829–836.

Erikson, E., Shealy, D. J., and Erikson, R. L. (1981). Evidence that viral transforming gene products and epidermal growth factor stimulate phosphorylation of the same cellular protein with similar specificity. *J. Biol. Chem.* **256**, 11381–11384.

Fabricant, R. N., DeLarco, J. E., and Todaro, G. J. (1977). Nerve growth factor receptors on human melanoma cells in culture. *Proc. Natl. Acad. Sci. U.S.A.* **74**, 565–569.

Fava, R. A., and Cohen, S. (1984). Isolation of a calcium-dependent 35-kilodalton substrate for the epidermal growth factor receptor/kinase from A-431 cells. *J. Biol. Chem.* **259**, 2636–2645.

Fine, R. E., Goldenberg, R., Sorrentino, J., and Herschman, H. R. (1981). Subcellular structures involved in internalization and degradation of epidermal growth factor. *J. Supramol. Struct.* **15**, 235–251.

Fung, Y.-K. T., Lewis, W. G., Crittenden, L. B., and Kung, H.-J. (1983). Activation of the cellular oncogene c-erb-B by LTR insertion: Molecular basis for induction of erythroblastosis by avian leukosis virus. *Cell (Cambridge, Mass.)* **33**, 357–368.

Gates, R. E., and King, L. E., Jr. (1982) Calcium facilitates endogenous proteolysis of the EGF receptor–kinase. *Mol. Cell. Endocrinol.* **27**, 263–276.

Ghosh-Dastidar, P., and Fox, C. F. (1983). Epidermal growth factor and epidermal growth factor receptor-dependent phosphorylation of a $M_r=34,000$ protein substrate for pp60^{src}. *J. Biol. Chem.* **258**, 2041–2044.

Ghosh-Dastidar, P., and Fox, C. F. (1984). cAMP-dependent protein kinase stimulates EGF-dependent phosphorylation of EGF receptors. *J. Biol. Chem.* **259**, 3864–3869.

Ghosh-Dastidar, P., Coty, W. A., Griest, R. E., Woo, D. D. L., and Fox, C. F. (1984a). Progesterone receptor subunits are high-affinity substrates for phosphorylation by epidermal growth factor receptor. *Proc. Natl. Acad. Sci. U.S.A.* **81**, 1654–1658.

Ghosh-Dastidar, P., Woo, D. D. L., and Fox, C. F. (1984b). Purification and properties of a high-affinity 94 kDa substrate for epidermal growth factor receptor kinase. *J. Cell Biochem.* **8**, 301–310.

Giard, D. J., Aaronson, S. A., Todaro, G. J., Arnstein, P., Kersey, J. H., Dosik, H., and Parks, W. P. (1973). *In vitro* cultivation of human tumors: Establishment of cell lines derived from a series of solid tumors. *JNCI, J. Natl. Cancer Inst.* **51**, 1417–1423.

Gilmer, T. M., Parsons, J. T., and Erikson, R. L. (1982). Construction of plasmids for expression of rous sarcoma virus transforming protein pp60^{src} in *Escherichia coli*. *Proc. Natl. Acad. Sci. U.S.A.* **79**, 2152–2156.

Gilmore, T., and Martin, G. S. (1983). Phorbol ester and diacylglycerol induce protein phosphorylation at tyrosine. *Nature (London)* **306**, 487–490.

Gorden, P., Carpenter, J. L., Cohen, S., and Orci, L. (1978). Epidermal growth factor: Morphological demonstration of binding, internalization, and lysosomal association in human fibroblasts. *Proc. Natl. Acad. Sci. U.S.A.* **75**, 5025–5029.

Greene, L. A., and Shooter, E. M. (1980). The nerve growth factor: Biochemistry, synthesis, and mechanism of action. *Annu. Rev. Neurosci.* **3**, 353–402.

Haigler, H. T., Ash, J. F., Singer, S. J., and Cohen, S. (1978). Visualization by fluorescence of the binding and internalization of epidermal growth factor in human carcinoma cells A-431. *Proc. Natl. Acad. Sci. U.S.A.* **75**, 3317–3321.

Haigler, H. T., McKanna, J. A., and Cohen, S. (1979). Direct visualization of the binding and internalization of a ferritin conjugate of epidermal growth factor in human carcinoma cells A-431. *J. Cell Biol.* **81**, 382–395.

Haigler, H. T., Maxfield, F. R., Willingham, M. C., and Pastan, I. (1980). Dansylcadaverine inhibits internalization of ^{125}I-epidermal growth factor in BALB 3T3 cells. *J. Biol. Chem.* **255**, 1239–1241.

Hapgood, J., Libermann, T. A., Lax, I., Yarden, Y., Schreiber, A. B., Naor, Z., and Schlessinger, J. (1983). Monoclonal antibodies against epidermal growth factor receptor induce prolactin synthesis in cultured rat pituitary cells (GH$_3$). *Proc. Natl. Acad. Sci. U.S.A.* **80**, 6451–6455.

Hayman, M. J., Ramsay, G. M., Savin, K., and Kitchener, G. (1983). Identification and characterization of the avian erythroblastosis virus erb-B gene product as a membrane glycoprotein. *Cell (Cambridge, Mass.)* **32**, 579–588.

Hillman, G. M., and Schlessinger, J. (1982). Lateral diffusion of epidermal growth factor complexed to its surface receptor does not account for the thermal sensitivity of patch formation and endocytosis. *Biochemistry* **21**, 1667–1672.

Hock, R. A., Nexo, E., and Hollenberg, M. D. (1979). Isolation of the human placenta receptor for epidermal growth factor—urogastrone. *Nature (London)* **277**, 403–405.

Hollenberg, M. D., and Cuatrecasas, P. (1973). Epidermal growth factor: Receptors in human fibroblasts and modulation of action by cholera toxin. *Proc. Natl. Acad. Sci. U.S.A.* **70**, 2964–2968.

Hopkins, C. R., and Boothroyd, B. (1981). Early events following the binding of epidermal growth factor to surface receptors on ovarian granulosa cells. *Eur. J. Cell Biol.* **24**, 259–265.

Huff, K. R., and Guroff, G. (1979). Nerve growth factor-induced reduction in epidermal growth factor responsiveness and epidermal growth factor receptors in PC12 cells: An aspect of cell differentiation. *Biochem. Biophys. Res. Commun.* **89**, 175–180.

Hunter, T., and Cooper, J. A. (1981). Epidermal growth factor induces rapid tyrosine phosphorylation of proteins in A-431 human tumor cells. *Cell* **24**, 741–752.

Ivanovic, V., and Weinstein, I. B. (1982). Benzo[a]pyrene and other inducers of cytochrome P_1-450 inhibit binding of epidermal growth factor to cell surface receptors. *Carcinogenesis (N.Y.)* **3**, 505–510.

Iwashita, S., and Fox, C. F. (1984). Epidermal growth factor and potent phorbol tumor promoters induce epidermal growth factor receptor phosphorylation in a similar but distinctively different manner in human epidermoid carcinoma A-431 cells. *J. Biol. Chem.* **259**, 2559–2567.

Jansson, M., Phillipson, L., and Vennström, B. (1983). Isolation and characterization of multiple human genes homologous to the oncogenes of avian erythroblastosis virus. *EMBO J.* **2**, 561–565.

Kärenlampi, S. O., Eisen, H. J., Hankinson, O., and Nebert, D. W. (1983). Effects of cytochrome P_1-450 inducers on the cell-surface receptors for epidermal growth factor, phorbol 12,13-dibutyrate, or insulin of cultured mouse hepatoma cells. *J. Biol. Chem.* **258**, 10378–10383.

King, L. E., Jr., Carpenter, G., and Cohen, S. (1980). characterization by electrophoresis of epidermal growth factor stimulated phosphorylation using A-431 membranes. *Biochemistry* **19**, 1524–1528.

Knauer, D. J., Wiley, H. S., and Cunningham, D. D. (1984). Relationship between epidermal growth factor receptor occupancy and mitogenic response: Quantitative analysis using a steady-state model system. *J. Biol. Chem.* **259**, 5623–5631.

Knowles, B. B., Solter, D., Trinchieri, G., Maloney, K. M., Ford, S. R., and Aden, D. P. (1977). Complement-mediated antiserum cytotoxic reactions to human chromosome 7 coded antigen(s): Immunoselection of rearranged human chromosome 7 in human–mouse somatic cell hybrids. *J. Exp. Med.* **145**, 314–326.

Kondo, I., and Shimizu, N. (1983). Mapping of the human gene for epidermal growth factor receptor (EGFR) on the P13 q22 region of chromosome 7. *Cytogenet. Cell Genet.* **35**, 9–14.

Krupp, M. N., Connolly, D. T., and Lane, M. D. (1982). Synthesis, turnover, and down-regulation of epidermal growth factor receptors in human A-431 epidermoid carcinoma cells and skin fibroblasts. *J. Biol. Chem.* **257**, 11489–11496.

Kudlow, J. E., Buss, J. E., and Gill, G. N. (1981). Anti-pp60[src] antibodies are substrates for EGF-stimulated protein kinase. *Nature (London)* **290**, 519–521.

Lee, L.-S., and Weinstein, I. B. (1978). Tumor-promoting phorbol esters inhibit binding of epidermal growth factor to cellular receptors. *Science* **202**, 313–315.

Leof, E. B., Olashaw, N. E., Pledger, W. J., and O'Keefe, E. J. (1982). Cyclic AMP potentiates down regulation of epidermal growth factor receptors by platelet-derived growth factor. *Biochem. Biophys. Res. Commun.* **109**, 83–91.

Lim, R. W., and Hauschka, S. D. (1984). A rapid decrease in EGF binding capacity accompanies the terminal differentiation of mouse myoblasts *in vivo. J. Cell Biol.* **98**, 739–747.

Lin, C. R., Chen, W., Kruijer, W., Stolarsky, L., Jonas, V., Weber, W., Wedner, J., Lazar, C. S., Evans, R. M., Verma, I., Gill, G. N., and Rosenfeld, M. G. (1984). Expression cloning of human EGF receptor cDNA reveals gene amplification and three related mRNA products in A-431 cells. *Science* **224**, 843–848.

Magun, B. E., Matrisian, L. M., and Bowden, G. T. (1980). Epidermal growth factor: Ability of tumor promoter to alter its degradation, receptor affinity and receptor number. *J. Biol. Chem.* **255**, 6373–6381.

Marquardt, H., Hunkapiller, M. W., Hood, L. E., Twardzik, D. R., DeLarco, J. E., Stephenson, J. R., and Todaro, G. J. (1983). Transforming growth factors produced by retrovirus-transformed rodent fibroblasts and human melanoma cells: Amino acid sequence homology with epidermal growth factor. *Proc. Natl. Acad. Sci. U.S.A.* **80**, 4684–4688.

Matrisian, L. M., Bowden, G. T., and Magun, B. E. (1981). Mechanism of synergistic induction of DNA synthesis by epidermal growth factor and tumor promoters. *J. Cell. Physiol.* **108**, 417–425.

Matrisian, L. M., Planck, S. R., and Magun, B. E. (1984). Intracellular processing of epidermal growth factor. I. Acidification of ^{125}I-epidermal growth factor in intracellular organelles. *J. Biol. Chem.* **259**, 3047–3052.

Mroczkowski, B., Mosig, G., and Cohen, S. (1984). An ATP-dependent interaction of the EGF receptor and supercoiled DNA. *Nature (London)* **309**, 270–272.

Nakamura, K. O., Martinez, R., and Weber, M. J. (1983). Tyrosine phosphorylation of specific proteins after mitogen stimulation of chicken embryo fibroblasts. *Mol. Cell. Biol.* **3**, 389–390.

Navarro, J., Abdel, A., Oharry, M., and Racker, E. (1982). Inhibition of tyrosine protein kinase by halomethyl ketones. *Biochemistry* **21**, 6138–6144.

Newmark, P. (1984). Cell and cancer biology meld. *Nature (London)* **307**, 499.

Nexo, E., Hock, R. A., and Hollenberg, M. D. (1979). Lectin-agarose immobilization, a new method for detecting soluble membrane receptors. *J. Biol. Chem.* **254**, 8740–8743.

Pike, L. J., Marquardt, H., Todaro, G. J., Gallis, B., Casnellie, J. E., Bornstein, P., and Krebs, E. G. (1982). Transforming growth factor and epidermal growth factor stimulate the phosphorylation of a synthetic, tyrosine-containing peptide in a similar manner. *J. Biol. Chem.* **257**, 14628–14631.

Planck, S. R., Finch, J. S., and Magun, B. E. (1984). Intracellular processing of epidermal growth factor. II. Intracellular cleavage of the COOH-terminal region of ^{125}I-epidermal growth factor. *J. Biol. Chem.* **259**, 3053–3057.

Privalsky, M. L., Sealy, L., Bishop, J. M., McGrath, J. P., and Levinson, A. D. (1983). The product of the avian erythroblastosis virus erb-B locus is a glycoprotein. *Cell (Cambridge, Mass.)* **32**, 1257–1267.

Pruss, R. M., and Herschman, H. R. (1977). Variants of 3T3 cells lacking mitogenic response to epidermal growth factor. *Proc. Natl. Acad. Sci. U.S.A.* **74**, 3918–3921.

Pruss, R. M., and Herschman, H. R. (1979). Cholera toxin stimulates division of 3T3 cells. *J. Cell. Physiol.* **98**, 469–474.

Pruss, R. M., Herschman, H. R., and Klement, V. (1978). 3T3 variants lacking receptors for epidermal growth factor are susceptible to transformation by Kirsten sarcoma virus. *Nature (London)* **274**, 272–274.

Roberts, A. B., Frolik, C. A., Anzano, M. A., and Sporn, M. B. (1983). Transforming growth factors from neoplastic and nonneoplastic tissues. *Fed. Proc., Fed. Am. Soc. Exp. Biol.* **42**, 2621–2626.

Robinson, R. A., Volkenant, M. E., Ryan, R. J., and Moses, H. L. (1981). Decreased epidermal growth factor binding in cells growth arrested in G_1 by nutrient deficiency. *J. Cell. Physiol.* **109**, 517–524.

Robinson, R. A., Branum, E. L., Volkenant, M. E., and Moses, H. L. (1982). Cell cycle variation in ^{125}I-labeled epidermal growth factor binding in chemically transformed cells. *Cancer Res.* **42**, 2633–2638.

Rubin, R. A., and Earp, H. S. (1983). Dimethyl sulfoxide stimulates tyrosine residue phosphorylation of rat liver epidermal growth factor receptor. *Science* **219**, 60–63.

Sahyoun, N., Hock, R. A., and Hollenberg, M. D. (1978). Insulin and epidermal growth factor-urogastrone: Affinity cross-linking to specific binding sites in rat liver membranes. *Proc. Natl. Acad. Sci. U.S.A.* **75**, 1675–1679.

Sawyer, S. T., and Cohen, S. (1981). Enhancement of calcium uptake and phosphatidylinositol turnover by epidermal growth factor in A-431 cells. *Biochemistry* **20**, 6280–6286.

Schlessinger, J., Schechter, Y., Willingham, M. C., and Pastan, I. (1978a). Direct visualization of binding, aggregation and internalization of insulin and epidermal growth factor on living fibroblastic cells. *Proc. Natl. Acad. Sci. U.S.A.* **75**, 2659–2663.

Schlessinger, J., Schechter, Y., Cuatrecasas, P., Willingham, M., and Pastan, I. (1978b). Quantitative determination of the lateral diffusion coefficients of the hormone–receptor complexes of insulin and epidermal growth factor on the plasma membrane of cultured fibroblasts. *Proc. Natl. Acad. Sci. U.S.A.* **75**, 5353–5357.

Schlessinger, J., Schreiber, A. B., Levi, A., Lax, I., Libermann, T., and Yarden, Y. (1982). Regulation of cell proliferation by epidermal growth factor. *CRC Crit. Rev. Biochem.* **14**, 93–111.

Schreiber, A. B., Lax, I., Yarden, Y., Eshhar, Z., and Schlessinger, J. (1981a). Monoclonal antibodies against receptor for epidermal growth factor-induced early and delayed effects of epidermal growth factor. *Proc. Natl. Acad. Sci. U.S.A.* **78**, 7535–7539.

Schreiber, A. B., Yarden, Y., and Schlessinger, J. (1981b). A nonmitogenic analogue of epidermal growth factor enhances the phosphorylation of membrane proteins. *Biochem. Biophys. Res. Commun.* **101**, 517–523.

Schreiber, A. B., Libermann, T. A., Lax, I., Yarden, Y., and Schlessinger, J. (1983). Biological role of epidermal growth factor-receptor clustering. *J. Biol. Chem.* **258**, 846–853.

Segawa, K., and Ito, Y. (1983). Enhancement of polyoma virus middle T antigen tyrosine phosphorylation by epidermal growth factor. *Nature (London)* **304**, 742–744.

Shechter, Y., Hernaez, L., Schlessinger, Y., and Cuatrecasas, P. (1979). Local aggregation of hormone–receptor complexes is required for activation by epidermal growth factor. *Nature (London)* **278**, 835–838.

Shimizu, N., Behzadian, M. A., and Shimizu, Y. (1980). Genetics of cell surface receptors for bioactive peptides: Binding of epidermal growth factor is associated with the presence of human chromosome 7 in human–mouse cell hybrids. *Proc. Natl. Acad. Sci. U.S.A.* **77**, 3600–3604.

Shimizu, N., Kondo, I., Gamou, S., Behzadian, M. A., and Shimizu, Y. (1984). Genetic analysis of hyperproduction of epidermal growth factor receptors in human epidermoid carcinoma A-431 cells. *Somat. Cell Mol. Genet.* **10**, 45–53.

Smith, K. B., Losonczy, I., Pannerselvam, M., Fehnel, P., and Salomon, D. (1983). Effect of 12-0-tetradecanoylphorbol-13-acetate (TPA) on the growth inhibitory and increased phosphatidylinositol (PI) responses induced by epidermal growth factor (EGF) in A-431 cells. *J. Cell. Physiol.* **117**, 91–100.

Spurr, N. K., Solomon, E., Jansson, M., Sheer, O., Goodfellow, P. N., Bodmer, W. P., and Vennström, B. (1984). Chromosomal localisation of the human homologues to the oncogenes erb-A and B. *EMBO J.* **3**, 159–163.

Stoscheck, C. M., and Carpenter, G. (1984). "Down regulation" of EGF receptors: Direct demonstration of receptor degradation in human fibroblasts. *J. Cell Biol.* **98**, 1048–1053.

Terwilliger, E., and Herschman, H. R. (1984). 3T3 variants unable to bind epidermal growth factor cannot complement in co-culture. *Biochem. Biophys. Res. Commun.* **118**, 60–64.

Todaro, G., DeLarco, G., and Cohen, S. (1976). Transformation by murine and feline sarcoma viruses specifically blocks binding of epidermal growth factor to cells. *Nature (London)* **264**, 26–31.

Ullrich, A., Coussens, L., Hayflick, J. S., Dull, T. J., Gray, A., Tam, A. W., Lee, J., Yarden, Y., Liberman, T. A., Schlessinger, J., Mays, E. L. V., Whittle, N., Waterfield, M. D., and Seeburg, P. H. (1984). Human epidermal growth factor cDNA sequence and abberrant expression of the amplified gene in A431 epidermoid carcinoma cells. *Nature (London)* **309**, 418–425.

Ushiro, H., and Cohen, S. (1980). Identification of phosphotyrosine as a product of epidermal growth factor-activated protein kinase in A-431 cell membranes. *J. Biol. Chem.* **255**, 8363–8365.

Vennström, B., and Bishop, J. M. (1982). Isolation and characterization of chicken DNA homologous to the two putative oncogenes of avian erythroblastosis virus. *Cell (Cambridge, Mass.)* **28**, 135–143.

Weber, W., Gill, G. N., and Spiess, J. (1984). Production of an epidermal growth factor receptor-related protein. *Science* **224**, 294–297.

Wharton, W., Leof, E., Pledger, W. J., and O'Keefe, E. J. (1982). Modulation of the epidermal growth factor receptor by platelet-derived growth factor and choleragen: Effects on mitogenesis. *Proc. Natl. Acad. Sci. U.S.A.* **79**, 5567–5571.

Wrann, M., Fox, C. F., and Ross, R. (1980). Modulation of epidermal growth factor receptors on 3T3 cells by platelet-derived growth factor. *Science* **210**, 1363–1365.

Yeaton, R. W., Lipari, M. T., and Fox, C. F. (1983). Calcium-mediated degradation of epidermal growth factor receptor in dislodged A-431 cells and membrane preparations. *J. Biol. Chem.* **258**, 9254–9261.

8

Effect of Insulin on Growth in Vivo and Cells in Culture

GEORGE L. KING AND C. RONALD KAHN

Research Division
Joslin Diabetes Center
Brigham and Women's Hospital
Harvard Medical School
Boston, Massachusetts

I. INTRODUCTION

I. Insulin is a polypeptide hormone of 51 amino acids secreted from the β cells present in the islets of Langerhans in the pancreas. Insulin

201

TABLE I
Partial List of the Actions of Insulin

Metabolic	Growth-promoting
Hexose transport and glycogen synthesis	RNA and DNA synthesis
Anti-lipolysis	Cellular proliferation
Amino acid transport	
Protein synthesis	
Ion transport	

has been shown to possess a variety of biological activities that are essential for the maintenance of life in higher animals (Cahill, 1971; Czech, 1977; Kahn, 1979). Most of the biological actions of insulin identified in vivo can be reproduced with isolated cell systems, indicating the direct involvement of insulin in initiating each of these effects (Cahill, 1971; Czech, 1977; Fajans, 1979; Kahn, 1979; Steiner et al., 1972). The biological activities of insulin may be separated into acute metabolic and chronic growth-promoting effects (Table I). In general, the metabolic effects of insulin, such as its ability to stimulate glucose transport and metabolism, inhibition of lipolysis, and enzyme activation, are observed at low insulin concentrations (0.1–1 nM), and the time of onset for these actions is very rapid, usually within minutes after exposure of cells to insulin (Blundell et al., 1972; King and Kahn, 1981; King et al., 1980; Steiner et al., 1972). In contrast, the growth-promoting effects of insulin, such as the stimulation of DNA synthesis and cell proliferation, are usually observed at high concentrations (0.1–1 μM), and the incubation time required for insulin to act is measured in hours to days (Blundell, et al., 1972; Cahill, 1971; Czech, 1977; Fajans, 1979; Kahn, 1979; King and Kahn, 1981; King et al., 1980; Petrides and Bohler 1980; Rechler et al., 1974; Steiner et al., 1972).

In the past, insulin's growth-promoting effect has been treated primarily as a laboratory and pharmacological phenomenon, and most of the research and reviews on the mechanism of insulin action have concentrated on insulin's metabolic effects. However, recent studies from our laboratory and others have begun to investigate the growth-promoting aspect of insulin actions in detail using cells in tissue culture (King et al., 1980, 1982a,b, 1983a,b; King and Kahn, 1981; Danho et al., 1981; Foley et al., 1982; Pedersen, 1977; Rechler et al., 1974, 1981). These studies have increased our understanding of the mechanisms by which insulin is influencing the growth of cells. This chapter will review the recent progress on the mechanisms of insulin's growth-promoting effects as studied on cells in culture. The data supporting a physiological role for its growth effects will be elaborated. Last, we will consider the exciting recent findings that the insulin receptor itself contains tyrosine kinase activity similar to that observed in receptors for some other growth factors (see Chapters 7 and 10) and the role of this activity in insulin's growth-promoting effect.

II. INSULIN GROWTH EFFECTS *IN VIVO*

Evidence for the growth-promoting effects of insulin *in vivo* has been primarily surmised from studies of various diseases associated with hyperinsulinemia. The most well-known example of a possible growth effect of insulin *in vivo* is the finding of macrosomia and organomegaly in the infants of poorly controlled diabetic mothers (Pedersen, 1977) and infants suffering from Beckwith-Wiedeman syndrome, β-cell hyperplasia, and nesidioblastosis (Hill, 1978). The hyperinsulinemia found in these syndromes has been postulated to be responsible for the macrosomia and organomegaly. Insulin's effect on somatic growth appears to start after the twenty-ninth to thirtieth week of gestation, since fetuses with pancreatic agenesis have both normal weight and length up to the thirtieth week of pregnancy, whereas after that time they gradually fall from curves of normal embryonic development (Hill, 1978). The mechanism by which insulin acts on growth in the third trimester is not clear.

The "hyperglycemia–hyperinsulinemia" hypothesis was proposed by Pedersen (1954) to explain the macrosomia of offspring from diabetic mothers. This hypothesis still appears to be a plausible explanation with some modifications. This hypothesis states that hyperglycemia in the diabetic mother leads to fetal hyperglycemia, since glucose is permeable across the placental membrane. Although insulin does not cross the placenta, the resulting fetal hyperglycemia stimulates the fetal islets which results in hypertrophy and hyperinsulinemia. Hyperinsulinemia could increase the uptake and utilization of glucose and amino acids which would ultimately enhance fetal growth (Hill, 1978). Alternatively, insulin can also stimulate growth directly by increasing DNA synthesis and cellular proliferation (Hill, 1978; Susa et al., 1979). Both mechanisms probably contribute to insulin's effect on fetal macrosomia.

Animal models of diabetes, including rhesus monkeys treated with streptozotocin, also exhibit macrosomia and selective organmegaly similar to human infants (Susa et al., 1979). Susa et al. (1979), using this model, asked whether the macrosomia was due to excessive substrates or a direct effect of insulin in growth. To study this question, rhesus monkeys were implanted with minipumps which infused insulin into the fetus and thus produced a hyperinsulinemic, but euglycemic state. The fetuses with hyperinsulinemia had a 34% increase in body weight and enlargement of the liver, placenta, heart, and spleen (Susa, et al., 1979). The enlargement of the liver was due primarily to hyperplasia and not hypertrophy of cells, although liver glycogen content was mildly elevated. These findings closely resemble what is found in the offspring of diabetic mothers. The authors concluded from these data that insulin can directly induce cellular growth in vivo and is responsible for the findings of macrosomia and organmegaly in diabetes. These pathological findings can still be due to a relative increase in the utilization of substrate that is facilitated by insulin rather than a direct effect on growth, since these studies only measured glucose levels and did not measure the absolute amount of glucose used.

The finding of hepatomegaly in the offspring of diabetic mothers has led to the postulate that insulin is an important growth factor in vivo for the liver. In hepatocyte cultures and hepatoma cell lines, insulin has been shown to stimulate DNA synthesis and cellular proliferation at both pharmacological and physiological ranges (Susa et al., 1979). Several authors have tested the growth effect of insulin in vivo by measuring its ability to stimulate DNA and cell growth in fetal rat hepatocytes (Leffert, 1974). Starzl et al. (1976) showed that high levels of insulin help to preserve hepatic ultrastructure and cellular proliferation when pancreatic blood flow has been directed from the liver. Other studies have suggested that a combination of glucagon and insulin is required (Starzl et al., 1976). Nevertheless, these studies suggest strongly that insulin has an important role in the maintenance of hepatic integrity and regeneration.

Insulin's growth effects in vivo have also been suggested to be pathophysiologically important in the proliferation of aortic smooth cells observed in atherosclerosis. Stout (1979) has postulated the moderately elevated insulin levels observed in the treatment of diabetes or naturally occurring in types I and II diabetics might be responsible for the accelerated rate of atherosclerosis as observed in diabetes. Recently, our laboratory has shown that insulin can affect the metabolism and growth of retinal capillary endothelial cells and pericytes (King et al., 1983b), which will be described later in the review. It is quite likely that insulin plays a role in vascular maintenance in a variety of ways. Even the islet itself may require insulin for growth (McEvoy and Hegre, 1978). Transplanted pancreatic cells exhibit improved survival with

insulin treatment and control of glucose (McEvoy and Hegre, 1978). Whether this is due to the metabolic or growth effect of insulin is not clear.

Insulin may be important in the growth of children as well. The growth rate of children who had craniophargomia resection returned to normal in spite of their persistent deficiency in growth hormone (Holmes et al., 1968). Bucher et al. (1983) recently studied the levels of insulin-like growth factors (IGFs) and insulin in these children. They reported that in a subject of these children both growth hormones and IGF I were deficient. Since these children had mildly elevated levels of insulin levels after oral glucose levels, the authors postulated that the mild hyperinsulinema may be responsible for the maintenance of a normal growth rate.

Therefore, insulin's growth effect is probably important physiologically in vivo both during fetal and normal development. Insulin in vivo appears to affect growth by providing substrates, stimulating cellular replication and growth, and influencing the levels of somatomedins or other growth factors.

III. INSULIN GROWTH EFFECTS ON CELLS IN CULTURE

The growth-promoting effect of insulin on cells grown in tissue culture was first reported by Gey and Thalhimer (1924), who observed that insulin increased the growth of fibroblasts harvested from chick embryos. Fifteen years later, Latta and Bucholz (1939) confirmed this finding and showed that this effect of insulin was observed at high concentrations $(1-2$ IU/ml $\sim 10^{-6}$ M). Shortly thereafter, von Haam and Cappel (1940), using fibroblasts derived from mouse embryonic heart, showed the first dose–response curve of insulin for stimulation of cell growth. A clear effect of insulin was observed at 1 mIU/ml $(1 \times 10^{-8}$ M). However, this still had to be regarded as a pharmacologic effect, since plasma insulin concentrations are usually between 10 and 100 μIU/ml $(1 \times 10^{-9}$ M). Since that time, the requirement for pharmacological doses of insulin for demonstration of a growth effect has been confirmed repeatedly. A probable explanation for the gross disparity of the concentration requirement between the metabolic and growth effects of insulin will be presented later in the chapter.

Von Haam and Cappel (1940) also observed that the addition of insulin to the mouse fibroblast for 10 days caused "differentiation" of the cells (i.e., enlargement and accumulation of fat granules). This is perhaps the earliest findings of insulin as a maturation or differentiation factor and was rediscovered in recent studies of 3T3-L1 cells which differentiate into adipocyte-like cells in the presence of insulin (Karlson et al., 1979; Van Obberghen et al., 1979). These investigators further observed that insulin's effect varied in different cell types and in different media. Thus, insulin could stimulate DNA synthesis in human fibroblasts grown in minimal media, whereas no effect was seen if insulin was added to media fortified by embryonic extracts and 20% serum (von Haam and Cappel, 1940). However, insulin did not have any effect on growth of HeLa cells either alone or in medium with 20% serum. These data pointed out that the responsiveness of cells to insulin's growth effect varies greatly, and transformed cells behave differently from nontransformed cells in their responses to hormones or growth factors.

The inconsistent effect of insulin to stimulate DNA synthesis and cell replication was clarified by a series of subsequent studies using both HeLa cells and mouse L cells (Leslie and Davidson, 1952; Leslie and Paul, 1954; Leslie et al., 1957; Paul and Pearson, 1960). These studies revealed that the growth effect of insulin was best observed in the cells that had been incubated in either serum-free medium or a medium containing low concentrations of serum. The serum-sparing nature of insulin was further studied by Temin (1967) and Griffiths (1971; 1972), who showed that insulin alone in serum-free media had a rather weak growth-promoting effect when compared to serum. When a small amount of serum was added with insulin, the effectiveness of insulin was enhanced, suggesting that serum contains other growth factors which are more potent than insulin and may enhance its action. In addition, Temin (1967) showed that some transformed cells will respond to insulin in serum-free media, whereas the same cells in a nontransformed state require some serum component in addition to insulin before significant cellular replication will occur. Thus, the process of transformation appears to reduce the cellular requirement for growth factors as well as the growth-promoting effect of insulin itself.

Since the early, classical studies of insulin's growth-promoting effects, there have been a large number of publications which have centered on this topic. There are many types of cells in which insulin can either stimulate DNA synthesis or cellular replication. Rechler et al. (1974) showed that insulin can stimulate DNA synthesis, i.e., [3H]thymidine incorporation in human fibroblasts. The characteristics of the insulin effect are similar to those seen in most other diploid cells. The effect of insulin is much weaker than that of serum, and the concentrations of insulin required (10^{-8}–10^{-6}) are high as compared to the concentrations required for metabolic effects (Figs. 1 and 2). The dose–response curves of insulin for the growth and metabolic effects

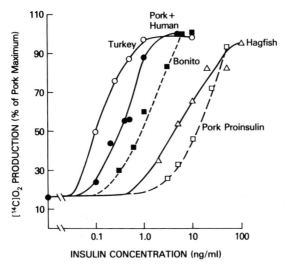

Fig. 1. Relative potencies of various insulin analogues on the stimulation of glucose oxidation in rat adipocytes, a metabolic effect of insulin. From King and Kahn (1981). Reprinted by permission from *Nature*, Vol. 292, pp. 644–646. Copyright © 1981 Macmillan Journals Limited.

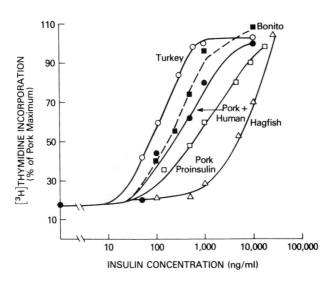

Fig. 2. Relative potencies of various insulin analogues on the stimulation of [³H]thymidine incorporation into the DNA of human fibroblast, a growth-promoting effect of insulin. From King and Kahn (1981). Reprinted by permission from *Nature*, Vol. 292, pp. 644–646. Copyright © 1981 Macmillan Journals Limited.

also exhibit very different slopes. This is apparent in Figs. 1 and 2 in which we have the effect of insulin on glucose oxidation in adipocytes and its effect on [³H]thymidine incorporation into the DNA of human fibroblasts (King and Kahn, 1981). Clearly 100 to 1000 times more insulin is required for the growth effect than the metabolic effect, and the shape of the dose–response curve for metabolic effects is rather sharp, whereas that for the growth response is broad and extends over several logs of concentration. The time course for the onset of insulin effect on DNA synthesis in the human fibroblast also demonstrates the characteristic 8- to 12-hr lag before the onset of effect (Fig. 3) (Rechler *et al.*, 1974). This delay of onset for the DNA synthesis appears to hold true for all cells and other growth factors as well (Gospodarowicz and

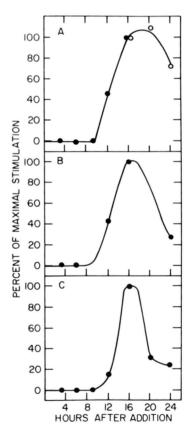

Fig. 3. Time of onset for the stimulation of DNA synthesis in human fibroblasts for insulin (A), NSILA (IGFs) (B), and serum (C). From Rechler *et al.* (1974). Copyright © 1974 The Williams and Wilkins Company, Baltimore.

Moran, 1976; Pledger et al., 1979; Zapf et al., 1978a). The effect of insulin on DNA synthesis can be shown by autoradiography to be due to its ability to increase the fraction of cells which progress into the S phase of the cell cycle (Rechler et al., 1974).

Insulin alone does not produce a stimulation of DNA synthesis in all cell types. In some cells, such as mouse 3T3-L1 cells or mammary epithelial cells, insulin is unable to stimulate DNA synthesis (Rozengurt and Mendoza, 1980). However, when insulin is added along with other hormones such as dexamethasone, a synergistic effect on DNA synthesis or cell replication is observed. In Cloudman S91 melanoma cells, insulin is a potent inhibitor of growth, and this effect occurs at rather low concentrations (Kahn et al., 1980). The exact reason for these differences in insulin effect remains unknown, but will be considered later.

In a few cell types, insulin has been shown to stimulate cellular growth at low concentrations similar to those required for its metabolic effects $(10^{-10}–10^{-9}$ M). This type of response has been observed in a rat hepatoma cell line (Koontz, 1980; Koontz and Iwahashi, 1981) and in normal pericytes isolated from bovine retinal capillaries and placed into culture (King et al., 1983b). Thus, insulin's effect on cell growth in tissue culture varies a good deal, depending on the cell type. Although very high concentrations of insulin are required in most cells, there are a few interesting exceptions (Leffert, 1974).

The requirement of pharmacological concentrations of insulin for its growth effects has puzzled many investigators. One possible reason which has been frequently considered was that a significant amount of insulin could have been degraded over 12–20 hr of preincubation required for the growth activity bioassays (Hayashi et al., 1978). Such degradation could be due to the high cysteine content of the media, which may disrupt the disulfide bonds in the insulin molecule, and/or degradative proteases present in the cultured cells which could modify insulin structure and decrease its biological potency. However, assays of culture medium indicate that with high insulin concentrations, (i.e., 10^{-8} M), only about 5% is degraded after incubation with human fibroblasts for 24 hr at 37°C (King and Kahn, 1981; Mott et al., 1979). Thus, degradation does not explain the need for high insulin concentrations.

Another plausible explanation for the need for pharmacological levels of insulin was that insulin itself is not an actual growth factor, but that a minor contaminant of the insulin preparation was responsible. Since very large amounts of insulin are used for some experiments $(1 \times 10^{-6}$ M), even impurity of 0.1% would result in a concentration of 1×10^{-9} M, a concentration at which many other growth factors have been shown to be active for the stimulation of DNA synthesis (Gospodaro-wicz and Moran, 1976; Pledger et al., 1979; Zapf et al., 1978a). This seems unlikely for several reasons. Various studies have shown that all of the known minor impurities in the insulin preparations, such as proinsulin, glucagon, or C-peptide, have either no or very little effect on growth. Also, increasingly purified insulin preparations exhibit increased potency for growth. Proof that the growth-promoting effect of insulin was due to the hormone itself, however, finally came from studies using insulin preparations which had been chemically synthesized directly from amino acids (Petrides and Bohler, 1980; King et al., 1983a). Thus, we and others showed that synthetic human insulin was equally potent with monocomponent insulin purified from the pancreas in both metabolic effects, assayed by the stimulation of glucose oxidation in rat adipocytes, and growth effects, indicated by its ability to stimulate [^{3}H]thymidine incorporation into the DNA of human fibroblasts (Figs. 1 and 2). These data conclusively demonstrated that the insulin molecule itself can stimulate DNA synthesis, since any contaminants occurring during the chemical synthesis would not be the same as those found in the insulin purified from the pancreas.

IV. MECHANISM OF INSULIN'S GROWTH-PROMOTING ACTION

Although the exact mechanisms involved in insulin action remain unknown, certain aspects of insulin's growth effect have been elucidated over the past several years. An understanding of this problem, however, is not possible without comparing insulin's activities to a family of related polypeptide hormones, the IGFs.

A. Insulin-Like Growth Factors

The IGFs (see Chapter 4) are a family of polypeptide hormones which are defined by their ability to mimic insulin's biological effects, but they are sufficiently different in structure such that they are not recognized by specific anti-insulin antibodies (Rinderknecht and Humbel, 1976; Van Wyk and Underwood, 1978; Zapf et al., 1978a). Since the first description by Salmon and Daughaday (1956) the IGFs have been structurally characterized and their biological activities studied both in vivo and in vitro (Phillips and Vassilopoulous-Sellin, 1980; Rinderknecht and Humbel, 1976; Van Wyk and Underwood, 1978; Zapf et al., 1978a).

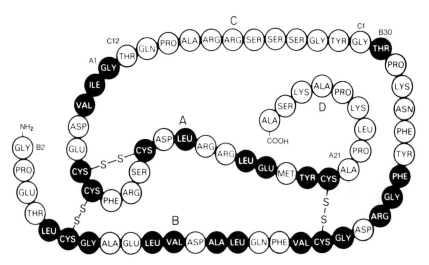

Fig. 4. Amino acid sequence of IGF I (Rinderknecht and Humbel, 1978b). From King and Kahn (1984). Copyright © 1984 John Wiley and Sons, Inc.

Structurally, two IGFs have been defined, IGF I and IGF II. IGF I and II are polypeptides of 70 and 67 amino acid residues, respectively, and are identical in 62% of their primary amino acid sequence (Rinderknecht and Humbel, 1978a,b). The structure of IGF I is shown in Fig. 4. Compared to insulin, it is identical at 49% of the amino acid residues. Physiologically, the IGFs, especially IGF I, appear to be controlled by the level of growth hormone. This finding coupled with infusion studies in animals suggest that IGFs may be one of the mediators of the effect of growth hormone on bone growth. In plasma, over 90% of the IGFs circulate bound to specific binding proteins (Phillips and Vassilopoulous-Sellin, 1980; Rinderknecht and Humbel, 1976; Van Wyk and Underwood, 1978; White *et al.*, 1981; Zapf *et al.*, 1982a) and in this respect are quite different from insulin and most other polypeptide hormones.

The IGFs appear to be identical to two other classes of polypeptide hormones. The somatomedins, isolated from human and rodent plasma, have the same spectrum of biological and chemical properties as the IGFs (Phillips and Vassilopoulous-Sellin, 1980; Rinderknecht and Humbel, 1976; Van Wyk and Underwood, 1978; Zapf *et al.*, 1978a). Klapper *et al.* (1983) suggest that somatomedin C is indistinguishable immunologically and structurally from IGF I. A similar group of insulin-like growth factors has also been isolated from a line of cultured liver cells derived from the Buffalo rat. These cells were suspected of making a growth factor, since they were capable of growing in serum-free medium (Nissley and Rechler, 1978). Media from these cells contain a family of insulin-like growth factors referred to as multiplication-stimulating activity or MSA (Fajans, 1979; Nissley and Rechler, 1978; Rechler et al., 1981). One component of MSA has been purified and sequenced and differs from human IGF II by only five amino acids (Marquardt et al., 1981).

Foresch, Van Wyk and their colleagues have shown that the IGFs (and somatomedians) are capable of mimicking all of insulin's biological activities on metabolism and growth in a variety of cells (Zapf et al., 1978a). The major difference between the IGFs and insulin is the dose–response curves for the various activities. Thus, the IGFs stimulate DNA synthesis at very low concentrations of 1×10^{-9} M, while the concentrations required for metabolic responses are 400- to 1000-fold higher (King et al., 1980; King and Kahn, 1981; Zapf et al., 1978a), whereas with insulin the converse is true.

B. Receptors for Insulin and the Insulin-Like Growth Factors

Polypeptide hormones initiate their biological effects by binding to specific receptors on the cellular membrane (Kahn, 1975; Roth, 1979). This interaction can be studied directly using ^{125}I-labeled hormones and a variety of cell membrane preparations (Kahn, 1975; Roth, 1979). Insulin binds to its receptor with high affinity (Fig. 5) (K_d about 1×10^{-9} M) and with biological specificity. Using analogues of insulin, it was shown that the relative affinity of each of these insulins for binding to the receptor correlated very highly with the potency of that analogue for stimulating the metabolic effects of insulin, such as glucose oxidation (DeMeyts et al., 1978; Kahn, 1975; Roth, 1979).

Megyesi et al. (1974) using ^{125}I-labeled preparations of insulin and a partially purified IGF, showed that specific and different receptors existed on a variety of cells to which each hormone can bind with high affinity (Megyesi et al., 1974, 1978). A similar finding was made by Van Wyk and his colleagues for somatomedin C. Since these peptides are similar in structure, insulin and the IGFs (somatomedins) are also able to bind to each other's receptors, although with a much lower affinity.

It is now clear that at least two types of IGF receptors can be distinguished based on their relative reactivity with IGF I and IGF II and their ability to interact with insulin (Foley et al., 1982; Nissley and Rechler, 1978; Rechler et al., 1980; 1981; Rosenfeld et al., 1980; Van Wyk and Underwood, 1978; Zapf et al., 1978a,b). The type I IGF recep-

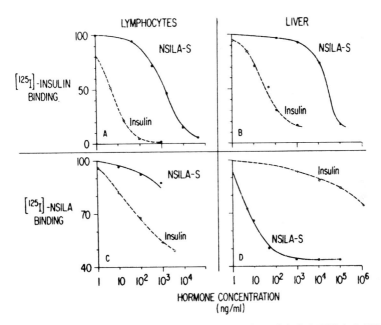

Fig. 5. Competition of ^{125}I-labeled insulin (top) and ^{125}I-labeled NSILA-S (IGFs) in lymphocytes and liver membranes by unlabeled insulin and NSILA-S (IGFs). From Kahn (1975).

tor reacts preferentially with IGF I (somatomedin C), as compared to IGF II (MSA), and also binds insulin with low affinity. The type II IGF receptor reacts preferentially with IGF II, as compared to IGF I, and does not recognize insulin even at very high concentrations. Some cells (such as cultured human lymphocytes) possess only IGF I and insulin receptors (Rosenfeld et al., 1980), while others (liver cells) possess only IGF II and insulin receptors (Megyesi et al., 1978), and some cells (human fibroblasts) possess all three types of receptors: IGF I, IGF II, and insulin receptors (Nissley and Rechler, 1978).

Using affinity-labeled techniques, Massague et al. (1981), Kasuga et al., (1981) and Bharmick et al. (1981) have begun to elucidate the structure of these three different receptor types (Fig. 6). The insulin receptor consists of two subunits, α and β, linked by disulfide bonds. The α-subunit contains the insulin binding site and has a molecular weight of about 135,000 Pilch and Czech, 1980. The β-subunit has a molecular weight about 95,000 and appears to possess a tyrosine-specific protein kinase activity (Kasuga et al., 1982a; Petruzzelli et al., 1982) (see Section V,B). In the native receptor, these subunits are linked via disulfide bonds to form oligomers, the most prominent of which is

TYPE I: NOT INHIBITED BY INSULIN

TYPE II: INHIBITED BY INSULIN

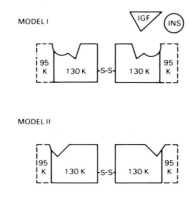

INSULIN RECEPTORS

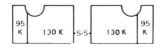

Fig. 6. Postulated subunits for the structures of receptors for insulin, IGF I, and IGF II. From King and Kahn (1984). Copyright 1984 John Wiley and Sons, Inc.

thought to be an α_2-β_2 structure of molecular weight 350,000. This structure has been confirmed by gel electrophoresis of immunoprecipitates of the labeled receptor (Van Obberghen et al., 1981) and by purification of the receptor to near homogeneity (Petruzzelli et al., 1982). The IGF I receptor appears to be very similar to the insulin receptor in structure, with an M_r of 130,000 binding subunit and an M_r of 90,000 tyrosine kinase subunit (Marquardt et al., 1981). Again, gel electrophoresis under nonreducing conditions suggests that these are jointed in the native receptor to form a complex of $M_r > 300,000$. In addition to their similar subunit structure, both the insulin and IGF I receptor are glycoproteins, as suggested by lectin chromatography, and both possess similar lectin specificities. Both are also derived from

single-chain precursors or proreceptors (Hedo *et al.*, 1983). Despite this similarity of structure, the IGF I and insulin receptors can be distinguished by the relative abilities of insulin and IGFs to inhibit the binding of the labeled ligands (Kasuga *et al.*, 1981) and by some anti-receptor antibodies (Kasuga *et al.*, 1983b).

The IGF II receptor is very different from receptors of insulin and IGF I. It appears on both reduced and nonreduced gel electrophoresis to be a single polypeptide. Under nonreducing conditions, the M_r is 220,000, and this increases to an apparent molecular weight of 260,000 in the presence of thiotreitol, suggesting the presence of intrachain disulfide bonds (Kasuga *et al.*, 1981; Massague *et al.*, 1981). Interestingly, with the use of the technique of radiation inactivation (Harmon *et al.*, 1980), IGF II receptor and the insulin receptor appear to have similar functional sizes (90,000–110,000), despite this apparent difference in chemical structure.

C. Anti-Receptor Antibodies as Probes of Receptor Function

It is clear from the foregoing discussion that insulin and the IGFs could produce their metabolic and growth effects through the insulin receptor, or the IGF receptors, or both. To distinguish these possible pathways, another reagent is needed that will interact specifically with insulin or IGF receptors and will either mimic or block the effect of the hormones initiated via that receptor (Fig. 7). We have studied this

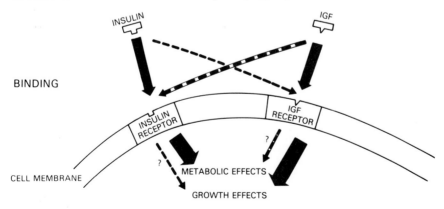

Fig. 7.   Drawing of the possible mechanism of action for the metabolic and growth effects of insulin and IGFs. From King and Kahn (1984). Copyright 1984 John Wiley and Sons, Inc.

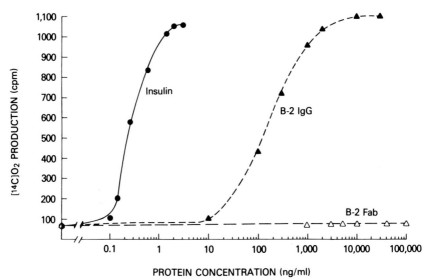

Fig. 8. Effect of insulin, bivalent anti-insulin receptor antibodies (B-2, IgG), and monovalent B-2 Fab on the stimulation of glucose oxidation in rat adipocytes. From King *et al.* (1980). Reproduced from *The Journal of Clinical Investigation*, 1980, Vol. 66, pp. 130–140, by copyright permission of The American Society for Clinical Investigation.

question by using antibodies to the insulin receptor. These antibodies are from the sera of some patients with a severe form of insulin resistance associated with acanthosis nigricans first described by Kahn *et al.* (1977, 1981). These autoantibodies, mostly of the IgG class, are capable of binding to the insulin receptor with a very high affinity (Fig. 8) (Kahn *et al.*, 1977, 1978). In addition, these antibodies can immunoprecipitate the solubilized insulin receptor (Van Obberghen *et al.*, 1981) and mimic insulin's biological actions (Table II) (Kahn *et al.*, 1977, 1981). This biological activity, however, requires bivalence. The monovalent Fab fragments generated by treating IgGs with papain bind to the insulin receptor and inhibit insulin binding, but are incapable of stimulating biological effects (Fig. 8) (King *et al.*, 1980; Kahn *et al.*, 1978). The intact IgG and the monovalent Fab fragment bind relatively specifically to the insulin receptor (Foley *et al.*, 1982; King *et al.*, 1980) and do not inhibit MSA (IGF II) binding except at very high concentrations. The autoantibody to the insulin receptor provides a specific probe for insulin receptor function with agonist activity when intact, and with antagonist activity when monovalent Fab fragments are prepared from the IgG. The intact IgG mimics insulin action in glucose transport in adipocytes, but has no effect on DNA synthesis in fibroblasts (Figs. 8 and 9). Likewise, when Fab anti-receptor antibody is added to isolated rat adipocytes, insulin binding is inhibited and the

TABLE II

Insulin-Like Effects of Anti-Insulin Receptor Antibody

Biological effects	Cell types studied
Stimulation of transport processes	
2-Deoxyglucose	Adipocytes, 3T3-L1 cells, muscles, human fibroblasts
Amino acid (A1B)	Adipocytes
Stimulation of enzymatic processes	
Insulin-receptor phosphorylation	3T3-L1 cells
Cytoplasmic protein phosphorylation	Adipocytes, 3T3-L1 cells
Glycogen synthase	Adipocytes, hepatocytes, muscle
Tyrosine aminotransferase	Hepatoma cells
Pyruvate dehydrogenase	Adipocytes
Acetyl-CoA carboxylase	Adipocytes
Lipoprotein lipase	3T3-L1 cells
Stimulation of glucose metabolism	
Glucose incorporation into glycogen	Adipocytes, 3T3-L1 cells, muscle
Glucose incorporation into lipids	Adipocytes
Glucose oxidation to CO_2	Adipocytes
Stimulation of macromolecular synthesis	
Leucine incorporation into protein	Adipocytes, human fibroblasts
Uridine incorporation into RNA	Hepatoma cells
Thymidine incorporation into DNA	Hepatoma cells, melanoma cells, retinal pericytes
Miscellaneous effects	
Inhibition of lipolysis	Adipocytes
Stimulation of glucosaminoglycan secretion	Chondrosarcoma cells

dose–response curves for both insulin and MSA are shifted to the right, whereas in the monovalent Fab fragments did not have any effect on the dose–response curves of either insulin or MSA (rat IGF II) for stimulation of DNA synthesis in fibroblasts (Fig. 10) (King et al., 1980). This suggests that both insulin and the IGFs are mediating their metabolic effects through the insulin receptor (Fig. 10). On the other hand, the growth effects of insulin or IGFs in fibroblasts appear not to be mediated by the insulin receptor (Fig. 11). These data support the hypothesis that the difference of insulin requirement for metabolic and growth effects is because the insulin is mediating these effects through different receptors. The metabolic effects are mediated through the insulin receptor where the growth effects are mediated through one of the growth factor receptors which binds insulin with low affinity (King et al., 1980).

A few caveats and exceptions to the above concept deserve comment. First, recent studies suggest that many anti-insulin receptor antibodies, including the antibodies used in the above studies, cross-react with IGF

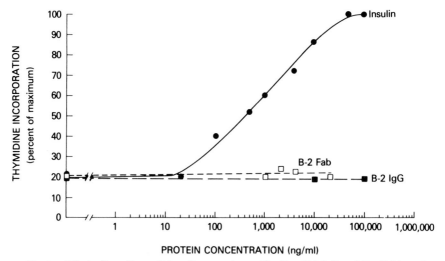

Fig. 9.   Effect of insulin, anti-insulin receptor antibodies (B-2 IgG and B-2 Fab) on the stimulation of DNA synthesis in human fibroblasts. From King *et al.* (1980). Reproduced from *The Journal of Clinical Investigation*, 1980, Vol. 66, pp. 130–140, by copyright permission of The American Society for Clinical Investigation.

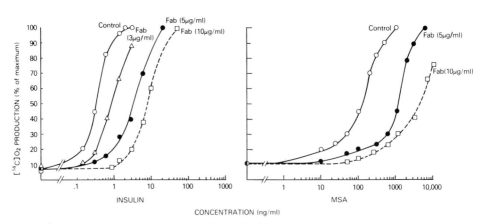

Fig. 10.   Effect of monovalent Fab anti-receptor antibody on the dose–response curves for insulin (left) and MSA (IGF II, right) on the stimulation of glucose oxidation in rat adipocytes. From King *et al.* (1980). Reproduced from *The Journal of Clinical Investigation*, 1980, Vol. 66, pp. 130–140, by copyright permission of The American Society for Clinical Investigation.

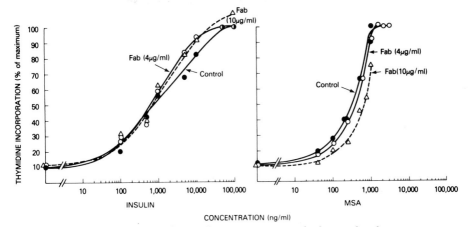

Fig. 11. Effect of monovalent Fab anti-receptor antibody on the dose–response curves for insulin (left) and MSA (IGF II, right) on the stimulation of DNA synthesis in human fibroblasts. From King *et al.* (1980). Reproduced from *The Journal of Clinical Investigation*, 1980, Vol. 66, pp. 130–140, by copyright permission of The American Society for Clinical Investigation.

I receptors, even though they are without effect on IGF II receptors (Kasuga *et al.*, 1981). Thus, the failure of the monovalent Fab to block-the growth effect of insulin and MSA in human fibroblasts suggests that, in this cell, DNA synthesis requires neither the insulin nor the IGF I receptors. Second, in other cells, insulin may initiate its growth effect via the insulin receptor. For example, we have recently studied insulin effects on pericytes isolated from bovine retinal capillaries. These cells are postulated to provide structural support to the capillary endothelial cells, and loss of these cells is one of the earliest lesions present in diabetic retinopathy (Kuwabara and Cogan, 1963). Unlike other diploid cells, the pericyte has the unusual property of being able to respond to insulin for growth at a very low concentration of insulin $(1 \times 10^{-9}$ M) (Fig. 12). The dose–response curve, however, is still very broad, extending from 10^{-9} to 10^{-7} M. Direct binding studies showed that these cells have insulin receptors as well as receptors for both IGF I and II. Interestingly, autoantibodies to the insulin receptor, previously unable to stimulate DNA synthesis in fibroblasts, are capable of stimulating DNA synthesis in the pericyte, supporting the idea that in this cell insulin can mediate its growth effects through its own receptor. Others have presented similar data suggesting a direct effect of the insulin receptor on DNA synthesis in a cultured hepatoma cell line (Koontz, 1980; Massague *et al.*, 1982).

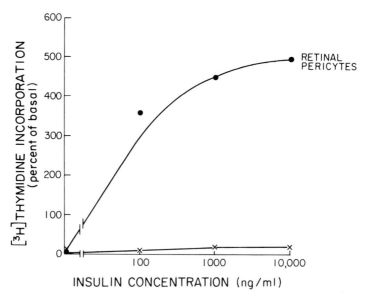

Fig. 12. Effect of insulin on the stimulation of DNA synthesis in bovine retinal capillary pericytes. From King and Kahn (1984). Copyright 1984 John Wiley and Sons, Inc.

D. Mapping the Bioactive Sites of Insulin

Although the insulin molecule is usually depicted as a simple linear two-chain structure, elegant X-ray crystallographic studies by Hodgkins and Mercola (1973) have shown that its three-dimensional structure is quite complex (Fig. 13). Using both naturally occurring and chemically derived analogues of insulin, Blundell *et al.* (1972) have defined the specific regions of the insulin molecule on this structure which appear to be involved in binding to the insulin receptor and producing metabolic effects, the major areas of immunogenicity, and the areas involved in dimer and hexamer formation. The insulin receptor binding site is a small portion of the molecule which includes the N- and C-terminal portions of the A chain (A1 and A21) and most of the C-terminal part of the B chain (B22 and B30). These regions coincide on one surface of the molecule in its three-dimensional structure (Fig. 13) and account for the metabolic activity of the molecule. Thus, the relative affinity of an insulin analogue for binding to the insulin receptor is closely correlated wth its biological potency in stimulating metabolic activities. Our study cited above had suggested that the growth effect of insulin may be mediated through another receptor. It was important to determine if the site responsible for the growth-promoting activity

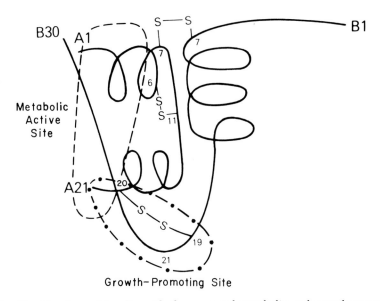

Fig. 13. Structure of insulin with the proposed metabolic and growth-promoting sites. From King and Kahn (1984). Copyright 1984 John Wiley and Sons, Inc.

might be different from that responsible for the metabolic activity of the molecule.

We have studied and compared the relative potency for metabolic and growth-promoting effects of more than 20 different analogues of insulin and three different IGFs (Table III) (King and Kahn, 1981). An example of the dose–response curves for metabolic and growth activities of several analogues is shown. The relative potency of these insulin analogues is different for these two different effects. For example, the order of relative potency in the stimulation of glucose oxidation for these analogues was turkey insulin > pork insulin > bonito insulin > proinsulin > DOP insulin, whereas for stimulation of DNA the order was turkey > bonito > pork > proinsulin > DOP (King and Kahn, 1981). These insulins differ from each other only in terms of relative affinity for binding to their respective receptors, since the maximal level of stimulation of each of these insulin analogues was the same.

These studies revealed several interesting trends (Table III). First, metabolic and growth effects of various insulins diverged in their relative potencies. Relatively speaking, all insulins with the exception of hagfish insulin were more potent for growth than for metabolic activity. This was true for both naturally occurring and chemically synthesized analogues. The most extreme divergence occurs with the IGFs (IGF I,

TABLE III

Comparison of Insulins and IGFs in Metabolic and Growth Assays[a]

| | ED$_{50}$ Pork/ED$_{50}$ analogue $\times$ 100 | | |
	Stimulation of glucose oxidation A	Stimulation of thymidine incorporation B	B/A
Pork (monocomponent)	100	100	1.0
Beef	100	98	1.0
Human (synthetic)	94	120	1.3
Mouse	82	130	1.6
Sheep	75	100	1.3
Horse	87	112	1.3
Proinsulin (pork)	2	51	26
Proinsulin (beef)	3	63	21
Turkey	240	350	1.5
Bonito (fish)	42	160	3.8
Hagfish	7	8	1.1
Guinea pig[b]	3	12	4
Porcupine[b]	3	110	37
Coypus[b]	1	190	190
Casiragua[b]	2	43	22
Desoctapeptide (DOP) insulin	0.4	30	75
Des-Ala des-Asp insulin	5	10	2
Des-Gly A1 des-Phe B1 insulin	0.8	14	18
Butoxycarbonyl A1 insulin	50	80	1.6
A1-B29 adipoyl insulin	6	98	16
Insulin-like growth factor I (IGF I)	0.2	860	4300
Insulin-like growth factor II (IGF II)	1.0	430	430
Multiplication-stimulating activity (MSA)	2.0	120	600

[a]From King and Kahn (1981). Reprinted by permission from *Nature*, Vol. 292, pp. 644–646. Copyright © 1981 Macmillan Journals Limited.

[b]ED$_{50}$ values for the hystricomorphs were derived by taking the concentration of insulin needed to stimulate thymidine incorporation that corresponds to the half maximum of pork insulin's dose–response curve.

IGF II, and MSA), since they have less than 1% of pork insulin's metabolic effects, but are four to eightfold more potent than pork insulin for growth effect (Figs. 14 and 15). Another way of measuring the divergence between these two effects is to compare the ratios of growth to metabolic potencies of the insulin (Table III, column B/A), using pork insulin as a standard. If the metabolic and growth effects of an insulin analogue are equally altered as compared to pork insulin, then the B/A ratio would be 1. However, if the growth effect of an analogue is less

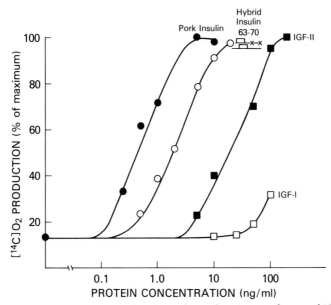

Fig. 14. Dose–response curves of pork insulin and various analogues of IGFs on the stimulation of glucose oxidation in rat adipocytes. From King *et al.* (1982b).

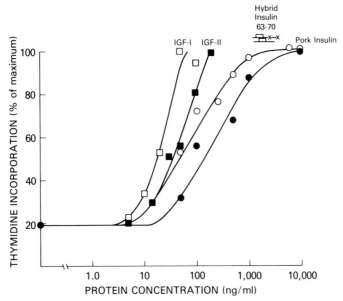

Fig. 15. Dose–response curves of pork insulin and various analogues of IGFs on the stimulation of DNA synthesis in human fibroblasts. From King *et al.* (1982b).

affected by structural changes than its metabolic effect, the B/A ratio would be greater than 1. Note that this is particularly true for the IGFs, proinsulin, desalanine-desasparagine (DAA) insulin, and desoctapeptide (DOP) insulin.

The second interesting finding is that the insulin from the hystricomorphs (guinea pig, porcupine, coypu, and casiragua) is very different from all the other insulin or IGFs tested (Fig. 16). All of these insulins have less than 5% of the metabolic activity of pork insulin. Dose–response curves for their growth-promoting response, however, are interesting in two ways. First, the hystricomorph insulins have retained more growth-promoting activity than metabolic activity at low concentrations. Second, at higher concentrations, these insulins actually stimulate DNA synthesis to a greater maximum than all the other insulins or IGFs.

From these studies, several conclusions can be drawn. First, the growth-promoting actions of insulin can tolerate a greater change in the structure than the metabolic effect. Second, there is nonparallel evolution between these two functions of the insulin molecule. Third, the insulins of the hystricomorphs are uniquely active with regard to growth-promoting effects as compared to other insulins and IGFs. Fourth, the sites for metabolic and growth effects occupy different regions in the insulin molecule, and thus the two activities behave differently in response to structural changes.

Since the primary structure of most of the insulins and IGFs is known, it is possible to correlate them with the bioactivity data in order to determine if some features of the insulin molecule important for growth-promoting effects of the insulin molecule are overlapping, since all of the structural changes that decrease insulin's metabolic effects also lessen its growth-promoting properties. Even some changes outside the major area of bioactivity influence both effects in parallel. For example, in turkey insulin a substitution of the histidine residue at A8 (A chain, residue 8) enhances both its metabolic and growth effect.

However, there are some striking differences between the two bioactivity sites. The B22–B30 region of insulin has been shown to be crucial for metabolic activity (Blundell et al., 1972). Thus, in DOP insulin in which the portion of the B chain is removed, metabolic effects are reduced by 200-fold. The growth effect of DOP insulin, on the other hand, is not greatly altered. Other areas on the insulin molecule which might be important for growth are those which share similarities between the IGFs and the insulin analogues with highest growth-promoting effects. These areas are: B10 residue (asparagine, glutamine, and glutamic acid in IGFs and hystricomorphs; histidine in other insulins), B13 residue (aspartic acid in IGF and hystricomorphs, glutamic acid in

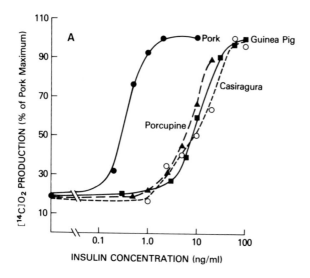

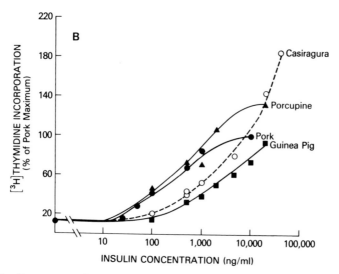

Fig. 16. Comparative dose–response curves of insulins from pork and the hystricomorphs on the stimulation of glucose oxidation in rat adipocytes (A) and DNA synthesis in human fibroblasts (B). From King *et al.* (1983a).

others), B25 residue (tyrosine in coypu insulin, casiragua insulin, and IGFs, phenylanine in others), and A4 residue (aspartic acid in hystricomorph insulins, glutamic acid in others). Another area that the IGFs and hystricomorph insulins differed from other insulins is the region of B20–21 (asparatate in hystricomorph insulins and IGF, glutamate in other insulins).

E. Bioactivity of Insulin–IGF Hybrid Molecules

To further define the structural features that are important for the growth-promoting effect of insulin and IGFs, we have studied a molecular hybrid molecule which contains part of the IGF molecule linked to an insulin backbone. The IGFS are single-chain polypeptide hormones which may be divided into four regions, termed A, B, C, and D (Fig. 4) (Rinderknecht and Humbel, 1978a,b). Regions A and B of IGF are 50% homologous in amino acid sequence to the A and B chains of insulin. The C region of the IGF was named to correspond to the C-peptide of proinsulin, but differs in size (8 amino acids in IGF vs. 33 in proinsulin) and does not contain basic amino acids which allow cleavage to a two-chain molecule. The D region consists of eight and five amino acids in IGF I and IGF II, respectively, and are unique to IGFs (Rinderknecht and Humbel, 1978a,b). It is located at the carboxy terminal of the A chain and is thought to lie in close proximity to the B21–B22 region in the three-dimensional structure (Blundell et al., 1978).

The insulin–IGF I hybrid molecule was synthesized by fusing the octapeptide sequence of IGF I D region to the carboxy terminal of A chain of a human insulin molecule (Danho et al., 1981). The resultant hybrid molecule is human insulin with IGF I D region connected to the A 21 residue. Compared to insulin, the insulin–IGF I hybrid has only 28% and 20% of the potency in insulin receptor assay and metabolic effects, respectively. In radioimmunoassay with anti-pork insulin antibodies, the hybrid had only 11% the potency of pork insulin. By contrast, the insulin–IGF I hybrid was actually more potent than pork insulin by two-to fivefold in a growth-promoting assay measuring [3H]thymidine incorporation into DNA of human fibroblasts (Figs. 14 and 15). These data suggest that the D region octapeptide is important for the IGF I in multiple ways. By blocking the A21 residue of insulin, IGF I lowers its metabolic activity and increases growth-promoting activity. Because the D region of IGF I and IGF II have little similarity, this may also play a role in determining relative affinity for the IGF I and IGF II receptors.

From the studies of the various analogues of insulin and IGFs and the

insulin–IGF I hybrid, we can postulate a location for the growth-promoting site on the insulin molecule. We believe the site is probably a
region centered on the B20–B22 area of the molecule because in this
region the D peptide as well as primary amino acid differences are
maximally concentrated. Both the D-peptide and C-peptide also block
regions of the insulin molecule important for its metabolic activity.
This region is schematically illustrated in Fig. 13.

F. A Relationship between Hystricomorph Insulins and Platelet-Derived Growth Factors

Among analogues studied, the hystricormorph insulins differed from
other insulins and IGFs in terms of their effect on DNA synthesis in
human fibroblasts. As noted, the hystricomorph insulins were able to
stimulate DNA synthesis in human fibroblasts to a greater maximum
than other insulins and IGFs (Fig. 16) (King and Kahn, 1981; King et al.,
1983a). In fact, at a concentration of 50 µg/ml, guinea pig insulin stimulated DNA synthesis two-to threefold greater than pork insulin. These
data were confirmed using autoradiography to quantitate the number of
cells which incorporate [^{3}H]thymidine into nuclear DNA as the cells
progress into the S phase of the cell cycle (Fig. 17). Again, guinea pig
insulin was more potent than pork insulin or the IGFs. In most cells at
maximal concentrations, pork insulin and IGFs are not additive. Interestingly, the growth effect of guinea pig and pork insulins were
additive in fibroblasts. Similar increases in DNA synthesis were observed when guinea pig insulin and MSA are added together. These
data suggest that hystricomorph insulins are mediating their effects
through a different receptor or pathway from insulin or the IGFs (King
and Kahn, 1981; King et al., 1983a).

When pork insulin and MSA are added to fibroblasts in the presence
of 5% platelet-poor plasma, no synergistic effects are observed. This is
similar to previous studies of the IGFs and indicates their independent
action from other plasma components on stimulation of DNA synthesis. When guinea pig insulin is added to platelet-poor plasma, a
large synergistic effect is seen. This effect of plasma to enhance the
action of guinea pig insulin is very similar to the effect of plasma on a
growth factor of different origin, namely, platelet-derived growth factor
(PDGF) (Gospodarowicz and Moran, 1976, King and Kahn, 1981; King
et al., 1983a).

To determine whether the unusual growth effects of the guinea pig
insulin are due to the presence of a different receptor on the surface of
the cell, we studied the binding of guinea pig insulin to human fibro-

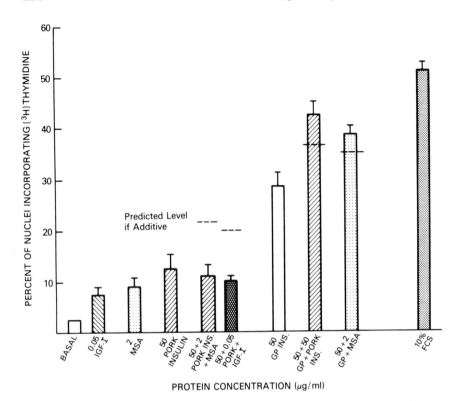

Fig. 17. Effect of various insulins and IGFs on the stimulation of human fibroblasts' progression into S phase of the cell cycle. From King *et al.* (1983a).

blasts (Fig. 18). With the use of ^{125}I-labeled guinea pig insulin, 3% of the tracer is bound per 10^6 cells. This binding was inhibited by un-labeled guinea pig insulin, with 50% displacement occurring at a con-centration of 0.8 μg/ml (140 nM). Casiragua insulin, another hys-tricomorph insulin, was two to threefold more potent than guinea pig insulin in inhibiting labeled guinea pig insulin binding, whereas pork insulin and MSA (rat IGF II) did not show any competition, even at very high concentration.

Since the binding studies suggested that the receptor under study has high specificity but low affinity for guinea pig insulin, it seemed likely that the receptor involved could be a receptor for another growth factor. A variety of known growth factors were tested and only PDGF was found to be active against this receptor. Four preparations of PDGF that varied in purity and potency for stimulation of growth competed with

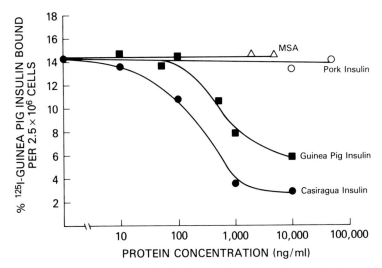

Fig. 18.   Competition of ^{125}I-labeled guinea pig binding to human fibroblasts by un-labeled insulins and IGF II (MSA). From King *et al.* (1983a).

guinea pig insulin in accordance with their biological effects (Fig. 19). PDGF was able to displace 50% of the ^{125}I-labeled guinea pig insulin at 8 ng/ml, indicating that PDGF is binding to this receptor at high af-finity. Likewise, serum competed for guinea pig insulin binding in concentrations similar to those which produce growth effects, whereas platelet-poor plasma was very weak in this regard. Recently, in collab-oration with E. H. Heldin, we showed that guinea pig insulin competed with ^{125}I-labeled PDGF for binding with albeit low affinity.

Taken together, these data suggest that the unusual growth effect of hystricomorph insulins is due to their ability to bind to PDGF receptor. Although PDGF is known to alter the binding of receptors for other growth factors by noncompetitive mechanisms (Clemmons *et al.*, 1980; Heldin *et al.*, 1982; Wrann *et al.*, 1980), PDGF was able to compete for guinea pig insulin binding even at 4°C, suggesting that PDGF and guinea pig are interacting at the receptor level in a process which does not require metabolic activity of the cell. The exact mechanism by which two apparently different polypeptides of 32,000 and 6,000 mo-lecular weight could share a receptor is unknown (Antoniades, 1981; Deuel *et al.*, 1981; Heldin *et al.*, 1979; Raines and Ross, 1982; Water-field *et al.*, 1983). However, it is possible that when the three-dimen-sional structure of PDGF is known, some homology between these two

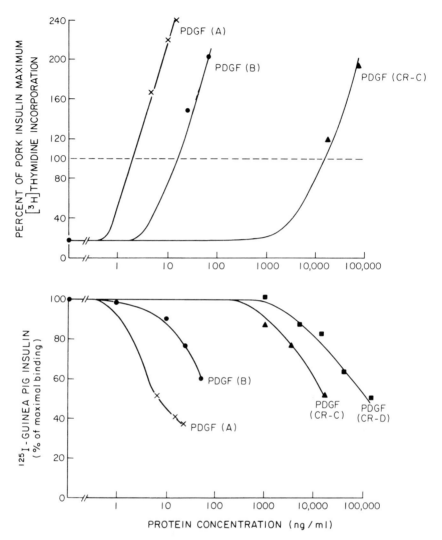

Fig. 19. Effect of various preparations of PDGF on the stimulation of DNA synthesis (top) and the competition with ^{125}I-labeled guinea pig insulin for binding (bottom) in human fibroblasts. From King *et al.* (1983a).

polypeptides will be found. This explanation is plausible, since recently tight homology has been found between the primary sequences of PDGF and a viral oncogene, *v-sis*. Certainly the bioactivity data suggest a common evolutionary source for hystricomorph insulins and PDGF.

G. Interaction between IGF and Insulin Receptors

In addition to stimulating DNA synthesis in cells through the IGF receptors, insulin has been found to influence the binding of the IGFs to their receptor. This was first observed in studies using rat adipocytes where competition experiments revealed that insulin did not inhibit the binding of ^{125}I-labeled IGF I, IGF II, and MSA to their receptors, but actually produced a modest increase in tracer binding (King et al., 1981a; Schoenle et al., 1977; Zapf et al., 1978b). This effect of insulin occurred at physiologic concentrations, i.e., between 10^{-10} and 10^{-9} M (Fig. 20). Competition curves and Scatchard analysis of the binding data indicate the effect of insulin is to increase the affinity of IGFs for the receptors. Recently, however, immunoprecipitation experiments

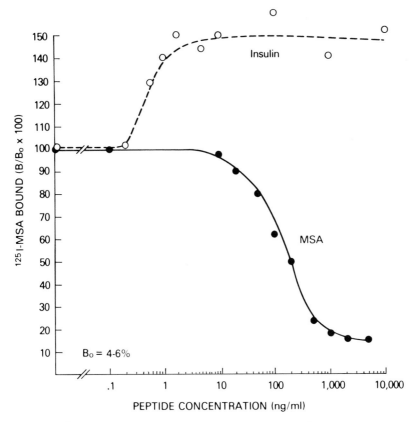

Fig. 20. Effect of unlabeled insulin and MSA (IGF II) on the binding of ^{125}I-labeled MSA to rat adipocytes. From King et al. (1982a).

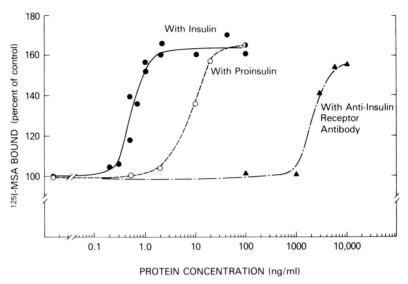

Fig. 21. Effect of insulin, proinsulin, and anti-insulin receptor antibody on the binding of ^{125}I-labeled MSA (IGF II) on rat adipocytes. From King *et al.* (1982a).

using specific antibodies to IGF I receptors have suggested that the effect of insulin is to induce a transloaction of IGF II receptors from intracellular membranes to the plasma membranes in a manner analogous to the effect of insulin on glucose transport (Oppenheimer *et al.*, 1983). Although the exact mechanism of this effect is unknown, some metabolic events appear to be involved, since the effect is not observed in broken cell preparations.

The effect of insulin to increase IGF binding occurs at low hormone concentrations, suggesting that this effect of insulin is mediated through the insulin receptor (King *et al.*, 1982a). This observation is supported by studies using insulin analogues and specific antibodies to the insulin receptors. Proinsulin, which is only 5% as potent as pork insulin in binding to the insulin receptor and in metabolic effects, can also increase ^{125}I-labeled MSA binding to adipocytes, but at concentrations 20-fold higher than those required for pork insulin. Autoantibodies to the insulin receptor which can bind only to insulin receptors and mimic insulin's metabolic effect can also mimic this insulin effect (Fig. 21). In addition, the monovalent Fab fragment made from the anti-insulin receptor antibody that acts as a competitive inhibitor of insulin action inhibits insulin's effect in increasing MSA binding. These studies strongly suggest that insulin is increasing MSA binding by acting through the insulin receptor (King *et al.*, 1982a).

V. POST-RECEPTOR STEPS OF INSULIN'S GROWTH EFFECTS

The various steps that occur following the binding of insulin to its receptor or to the IGF receptor until the beginning of the S phase of a cell cycle are not yet understood. The only clear and consistent finding in the action of insulin and most growth factors is a delay of 12–18 hr (Rechler *et al.*, 1974). Further, insulin must be present for most of this 12–18 hr for the maximal effect to be observed. This suggests an on-going signal event rather than a "triggering event" upon initial hormone binding. Several recent observations must also be considered when discussing the growth-promoting actions of insulin.

A. Hormone Internalization

Recent studies on the fate of hormones and growth factors have revealed that most polypeptide hormones, including growth hormone (Gorden *et al.*, 1980), epidermal growth factor (Gorden *et al.*, 1978), and insulin (Schlessinger *et al.*, 1978), initiate the aggregation of their respective receptors and are internalized by specific receptor-mediated endocytosis after binding. This process is schematically illustrated in Fig. 22. The process of internalization is rapid and can take place in a few minutes at 37°C. Once inside the cells, the hormone–receptor com-

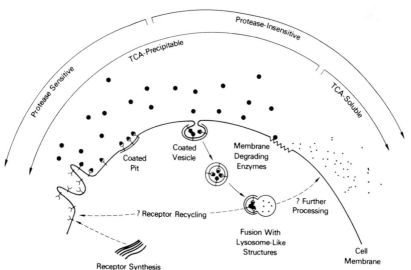

Fig. 22. Schematic drawing of the internalization and the processing of peptide hormone receptor complex. From King and Kahn (1984). Copyright 1984 John Wiley and Sons, Inc.

plexes fuse with lysosome-like structures and are either degraded or recycled (Gorden et al., 1980; Schlessinger et al., 1978; Haigler et al., 1978; Posner et al., 1981). Some authors have postulated that this involved process is linked to insulin's chronic or growth effects (Haigler et al., 1978).

In our opinion, several lines of reasoning seem to be against this hypothesis. First is the data on the time of onset of insulin's growth effect. As noted, this effect of insulin requires the continued presence of the hormone for at least 6 hr (Rechler et al., 1974). The process of internalization, on the other hand, occurs over several minutes and is largely complete by 1 hr (Gorden et al., 1978, 1980; Schlessinger et al., 1978; Haigler et al., 1978; Posner et al., 1981). Second, Carpentier et al. (1979) have shown that anti-insulin receptor antibodies and their monovalent Fab fragments are also internalized by this receptor-mediated mechanism and are processes in a manner very similar to insulin itself. However, neither the bivalent IgG anti-insulin receptor antibody nor its monovalent Fab fragments can mimic insulin's growth-promoting effects in most cells (King et al., 1980; King and Kahn, 1981).

Last, in an experiment in collaboration with Dr. Haden Coon (National Institutes of Health), porcine insulin or diluent was microinjected directly into the perinuclear area of 400–500 human fibroblasts in monolayer culture (Fig. 23). The amount of insulin injected into the cell was equal to the amount internalized when the cells are exposed to 10^{-6} M insulin (i.e., the concentration which can produce a maximal effect on stimulation of DNA synthesis in human fibroblasts). After the cells were microinjected with insulin or diluent, the cells were incubated for 24 hr in a medium supplemented with [^{3}H]thymidine. The culture dishes were then covered with emulsion and developed several days later. The number of cells that incorporated [^{3}H]thymidine are an indication of cells that have moved from the G_0 phase into the S phase. No difference in the percentage of cells in S phase was observed between cells injected with insulin or with diluent. In addition, no significant difference in cell viability in injected and noninjected cells was found. Although these studies provide only negative data, they suggest that internalization of insulin is not a major factor in production of its growth effect. Further studies are obviously needed to define this question more clearly.

B. Receptor Phosphorylation

Over the past few years, a possible clue to the mechanism of action of insulin (Kasuga et al., 1982a; Haring et al., 1984a), EGF (Haring et al., 1984a), and PDGF has been uncovered (Clemmons et al., 1980; Heldin

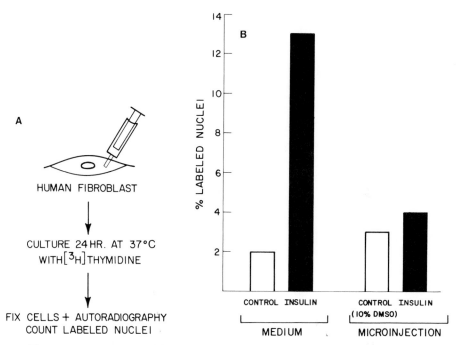

Fig. 23. Microinjection of insulin on medium. (A) Schematic drawing of microinjection of insulin into human fibroblasts. (B) Effect of insulin on the growth of human fibroblasts. From King and Kahn (1984). Copyright 1984 John Wiley and Sons, Inc.

et al., 1979). Each of these receptors possess a tyrosine-specific protein kinase activity and undergoes autophosphorylation upon ligand binding (see Chapters 7 and 10). Insulin-receptor phosphorylation has been demonstrated in both intact and broken cell systems (Kasuga *et al.*, 1982a,b).

In the intact cell experiments, the ATP pool is first labeled by preincubation of cells with [^{32}P]orthophosphate for 2 hr. Insulin receptors can then be isolated before or after insulin treatment using lectin chromatography and immunoprecipitation, and the receptor subunits identified by SDS–polyacrylamide gel electrophoresis (Kasuga *et al.*, 1982a,c). Using this technique, Kasuga *et al.* (1982a,c) observed that the β-subunit of the insulin receptor appeared under posttranslational covalent modification by phosphorylation (Fig. 24). Evidence that this was indeed the β-subunit of the insulin receptor included the findings that (1) its migration in gel electrophoresis under reducing conditions was identical to that of the β-subunit of the receptor, (2) on nonreducing gels, the radioactivity was associated with proteins of M_r 350,000 and 520,000, (3) the protein was precipitated by four different sera

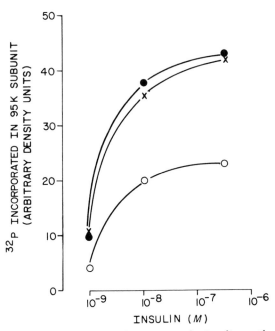

Fig. 24. Phosphorylation of the insulin receptors by insulin on the three types of Cloudman S91 melanoma cells (○) III, (X) 46, and (●) IA. From Haring *et al.* (1984a). Reproduced from *The Journal of Cell Biology*, 1984, Vol. 99, pp. 900–908, by copyright permission of The Rockefeller University Press.

containing antibodies to the insulin receptor, and (4) insulin stimulated the phosphorylation three- to fivefold. Phosphoamino acid analysis in the basal state revealed predominately phosphoserine and a small amount of phosphothreonine (Kasuga *et al.*, 1982c). After insulin stimulation, there appeared to be an increase in the level of phosphoserine and *de novo* appearance of phosphotyrosine (Kasuga *et al.*, 1982c).

The tyrosine phosphorylation of the insulin receptors appears to be catalyzed by a protein kinase activity present in the receptor itself (Kasuga *et al.*, 1983a; Roth and Cassell, 1983). This autophosphorylation can be observed *in vitro* when [γ-^{32}P]ATP and insulin are incubated with solubilized crude receptor preparations from hepatoma cells (Kasuga *et al.*, 1982c, 1983a), cultured lymphocytes (Kasuga *et al.*, 1982a), 3T3-L1 cells (Van Obberghen and Kowalski, 1982), freshly isolated hepatocytes (Haring *et al.*, 1982), adipocytes (Zick *et al.*, 1983), and normal rat liver membranes (Pawelek *et al.*, 1982), as well as partially purified and highly purified preparations of insulin receptors from human placenta (Kasuga *et al.*, 1983a; Van Obberghen and

Kowalski, 1982). This phosphorylation reaction is sensitive to insulin when added *in vitro*, with a three- to tenfold increase in the amount of ^{32}P incorporated into the receptor. In all cases, the phosphorylation occurs exclusively in the tyrosine residues and predominately on the β-subunit of the receptor. This reaction is quite rapid, reaching a maximum within a few minutes in the presence of insulin. $[\gamma\text{-}^{32}P]ATP$, but not $[\gamma\text{-}^{32}P]GTP$, will act as a phosphate donor for the reaction, and Mn^{2+} is required as a cofactor (Kasuga *et al.*, 1982c; Pawelek *et al.*, 1982).

Evidence that the receptor itself is a tyrosine kinase is threefold. First, receptor kinase activity and binding activity copurify and both are retained in a preparation of receptor purified to near homogeneity (Kasuga *et al.*, 1983a; Roth and Cassell, 1983; Van Obberghen and Kowalski, 1982). Second, if the partially purified receptor is immunoprecipitated using the anti-receptor antibody, the kinase activity is in the immunoprecipitate rather than in the supernatant (Haring *et al.*, 1984b). Finally, several groups have succeeded in labeling the β-subunit of the receptor with affinity analogues of ATP, demonstrating an ATP binding site on this subunit (Roth and Cassell, 1983).

Although receptor phosphorylation has been recognized for only a relatively short period of time, at least seven different receptor proteins have been shown to undergo this type of covalent modification. The most interesting of the receptor-mediated phosphorylations are those which, like insulin receptors, occur at tyrosine residues and are stimulated upon ligand binding. Such reactions have also been observed to occur for the receptors for epidermal growth factor (EGF) (Cohen *et al.*, 1981; see Chapter 7), a protein presumed to be the receptor for PDGF (Ek *et al.*, 1982; Nishimura *et al.*, 1982; see Chapter 10) and the receptor for IGF I (Jacob *et al.*, 1983). In contrast to the insulin receptor–kinase, the EGF receptor–kinase and the PDGF receptor–kinase are single peptide chains of M_r 160,000 and 180,000 respectively, which possess both the hormone binding site and the kinase active site (Cohen *et al.*, 1981; Pike *et al.*, 1983). The gene products of several avian and mammalian retroviruses have also been shown to be associated with tyrosine protein kinase activity (Hunter and Sefton, 1980; Bishop and Varnus, 1982; see Chapter 11).

C. Receptor Phosphorylation and the Growth-Promoting Effect of Insulin

In an effort to determine the functional significance of receptor phosphorylation insulin action, Haring *et al.* (1984a) have performed studies of insulin binding and receptor phosphorylation in cell lines in which insulin action has been altered by mutagenesis. The lines are the Cloudman S91 melanoma and two of its variants initially isolated by Pawelek et al. (Kahn et al., 1980a). In contrast to most other cells, insulin inhibits cell replication in the wild-type Cloudman melanoma (1A). Thus, it has been possible to select two types of variants with respect to insulin response by treatment of the cells with a mutagen (EMS) followed by culture in medium containing insulin. One type is a variant in which insulin stimulates growth (type 46) and the other variant (type III) is termed insulin resistant, since insulin neither stimulates nor inhibits growth (Kahn et al., 1980a).

^{125}I-labeled insulin binding is similar for the wild-type (1A) and the insulin-stimulated variant (Type 46). The insulin-resistant variant (lll), on the other hand, exhibited approximately a 30% decrease in insulin binding. This was due to a decrease in receptor affinity with no change in receptor concentration (Fig. 25) (Haring et al., 1984a).

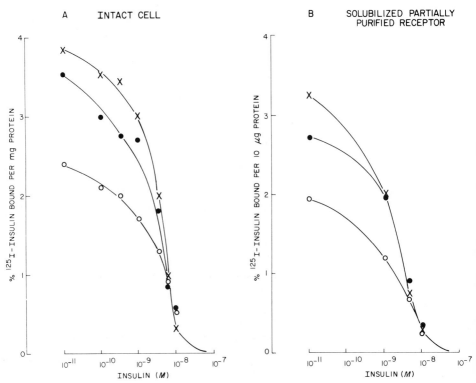

Fig. 25. ^{125}I-labeled insulin binding to Cloudman S91 melanoma cells (○) III, (X) 46, and (●) IA. From Haring et al. (1984a). Reproduced from *The Journal of Cell Biology*, 1984, Vol. 99, pp. 900–908, by copyright permission of the Rockefeller University Press.

Insulin stimulated phosphorylation of the 95K subunit of its receptor in all three cell types with similar kinetics. The amount of ^{32}P incorporated into the 95K band of the insulin-resistant cell line (lll), however is ~50% of the amount present in the receptors of the two other cell lines (Fig. 24). The difference in the (lll) line was reflected throughout the entire dose–response curve and throughout the entire time course of phosphorylation (Haring et al., 1984a). J. Pawelek (personal communication) have also identified a 90K nonreceptor protein which is phosphorylated in the wild-type cell and insulin-dependent variant, but is markedly decreased in the insulin-resistant variant.

These data suggest that the insulin-resistant melanoma (type lll) possesses a defect in the insulin receptor which alters both its binding and autophosphorylation properties and also insulin's effect on growth on phosphorylation of other cellular proteins. This suggests a possible role of receptor autophosphorylation in both the binding and signaling function of the insulin receptor.

D. Relationship between the Early and the Growth Effects of Insulin

Although stimulation of DNA synthesis and cellular replication by serum or insulin usually requires 8–12 hr for onset of effect, many studies have reported that within minutes after the addition of these growth factors to quiescent cells a variety of activities can be observed. These include an increase uptake of hexose (Rozengurt and Stein, 1977), uridine (Rosengurt and Stein, 1977), Mg^{2+} (Sanui and Rubin, 1977), Ca^{2+} (Clausen, 1977), K^+ (Rozengurt and Heppel, 1975), $^{86}Rb^+$ (Rozengurt and Heppel, 1975), Na^+ (Fehlmann and Freychet, 1981), and amino acids (Fehlmann et al., 1979). Similarly, protein synthesis is stimulated before cellular replication occurs (Nishimura et al., 1982). Studies by many authors have shown that these early biological effects of growth factors and serum appear to be very important for the ultimate stimulation of DNA synthesis and cellular replication. However, it is also clear that the stimulation of these cellular activities does not automatically start a chain of events that leads to DNA synthesis and cellular growth. In fact, the effect of insulin on the metabolism and growth of cells provides an excellent example supporting the conclusion that these early events are not the key steps in signaling cells onto an irreversible path of replication.

Insulin is similar to other growth factors with respect to these biological effects when studied in cultured cells (Cahill, 1971; Czech, 1977; Fajans, 1979; Kahn, 1979; Steiner et al., 1972). At concentrations less than 10^{-9} M, a concentration at which insulin is binding to its own receptor on the cells, insulin has been shown to be able to stimulate glucose transport in many types of cells including adipocytes, muscle cells, and fibroblasts (Cahill, 1971; Czech, 1977; Fajans, 1979; Kahn, 1979; Steiner et al., 1972). In isolated hepatocytes and other cells, increases Na^+ influx by stimulating Na^+,K^+-ATPase (Fehlmann and Freychet, 1981; see Chapter 13). Thus, the effect of insulin is rapid and reaches a maximum at concentrations less than 5 nM. Fehlmann and Freychet (1981) found a close concordance between this biological effect of insulin and the occupancy of insulin receptors. Since an increase in Na^+ influx appears to be important in the initiation of cellular proliferation in many types of cells such as rat hepatocytes and cultured mouse fibroblasts, this effect of insulin has been suggested to be a crucial early step of insulin's effect on growth.

However, the exact relationship between insulin's effect on intracellular levels of Na^+ and K^+ and other biological actions is unknown. Some studies have reported that the inhibition of Na^+,K^+-ATPase will result in an inhibition of hexose transport, suggesting that insulin's effect on Na^+ and K^+ levels is related to its metabolic activity (Czech, 1977; Blatt et al., 1972). Arguments also have been made for a role for ion flux in terms of insulin's effect on uridine uptake and phosphorylation, amino acid transport, protein synthesis, and RNA synthesis. On the other hand, Fehlmann and Freychet (1981) have shown that the inhibition of Na^+,K^+-ATPase by insulin does not inhibit insulin's effect on amino acid transport as indicated by measuring of aminoisobutyric acid transport in hepatocytes.

The association between these early biological effects and insulin's stimulation of cell growth remains unclear for several reasons. In most cells, including fibroblasts, muscle cells, chrondrosarcoma cells, and hepatocytes, insulin can stimulate both fast metabolic events, such as Na^+ influx, and chronic growth effects, such as DNA synthesis. However, unlike other growth factors, two distinct ranges of insulin concentrations are required, as discussed above. We have also provided some evidence that the growth effects and metabolic events are mediated through different receptors (King et al., 1980). Therefore, the observations that there are earlier effects of insulin which occur with low concentrations of the hormone may only be relevant to its metabolic effects. This hypothesis would explain the findings that insulin can stimulate many of these rapid events at low concentrations, yet it cannot stimulate DNA synthesis in most cell types except at high concentrations. This postulation does not rule out the possibility that early effects of insulin on Na^+ influx may play a crucial role leading to cellular replication because it is possible that insulin could stimulate Na^+ influx and then mediate its growth effects by binding at low affinity to a receptor of an IGF. Obviously, further data are needed in regard to the role of Na^+ influx on cellular proliferation.

VI. SUMMARY

It is clear that the growth-promoting action of insulin is very complex and is similar to the actions of other growth factors. In tissue culture, insulin can induce many biological effects, but its growth effects require long incubation time and relatively high hormone concentrations. At least three receptors (insulin, IGF I, and IGF II) may be involved in these actions, and in addition there are interactions among these receptors. The biochemical studies on the insulin receptor and its tyrosine subunit have provided strong evolutionary links between insulin and other growth factors. However, among all the growth factors that have been studied, insulin is unusual in many ways. It is the only peptide hormone for which detailed information on its production, secretion, and physiological metabolism is known. Physiologically, insulin is the only one among the known growth factors which is known to be crucial for survival. Not surprisingly, the plasma concentration of insulin is normally controlled within rather narrow limits. Furthermore, virtually every cell type that can be grown in a defined medium in culture requires insulin. However, it is likely that the in vivo growth effects of insulin are even more important than previously believed. The complications of diabetes have provided a glimpse as to the potential importance of insulin's growth-promoting actions.

From these data, it is tempting to speculate that insulin during its early evolutionary development may have been used as a growth promoter. Since organism growth is usually closely linked to the availability of fuel substrates, the concentration of growth-promoter insulin may have become regulated by the environmental concentrations of glucose or amino acids. As complex organisms evolved, multiple and different genes for insulin evolved to form insulin and IGFs. Insulin retained its sensitivity to glucose and became prominent in its metabolic activities. IGFs became specialized for growth. Due to their common evolutionary origin, insulin and IGFs have retained similar machinery for biological actions. This would explain the finding of tyrosine kinase activity in receptors for insulin and IGF, similarly to all other growth factors.

The study of insulin action is entering a very exciting era. For the first time, detailed biochemical and biological analysis on hormone receptors and post-receptor events can be studied using very sophisticated techniques. These studies should provide information both on insulin's mechanism of action and its role in physiological states.

ACKNOWLEDGMENTS

The authors wish to acknowledge grant support from the National Institutes of Health (R01 EY05110-01 and AM 31036), the American Diabetes Association, and the Juvenile Diabetes Foundation. We also wish to thank Patricia Morrison for excellent secretarial assistance.

REFERENCES

Antoniades, H. A. (1981). Human platelet-derived growth factor: Purification of PDGF-I and PDGF-II and separation of their reduced subunits. *Proc. Natl. Acad. Sci. U.S.A.* **78**, 7314–7317..

Bharmick, B., Bala, R. M., and Hollenberg, M. D. (1981). Somatomedin receptor of human placenta: Solublization photo-labeling, partial purification and comparison with insulin receptor. *Proc. Natl. Acad. Sci. U.S.A.* **78**, 4279–4283.

Bishop, M. J., and Varus, H. (1982). Functions and origins of retroviral transforming genes. *In* "RNA Tumor Viruses" (R. Weiss, N. Teich, H. Varmus, and J. Coffin, eds.), pp. 999–1008. Cold Spring Harbor Lab. Cold Spring Harbor, New York.

Blatt, L. M., McVerry, P. N., and Kim, K. H. (1972). Regulation of hepatic glycogen synthase of rama catesbeinana: Inhibition of the action of insulin by ouabain. *J. Biol. Chem.* **247**, 6551–6554.

Blundell, T. L., Dodson, G. C., Hodgkins, D. C., and Marcola, R. A. (1972). Insulin: The structure in the crystal and its reflection in chemistry and biology. *Adv. Protein Chem.* **26**, 229–402.

Blundell, T. L., Bedarker, S., Rinderknedt, E., and Humbel, R. E. (1978). Insulin-like growth factor: A model for tertiary structure accounting for immunoreactivity and receptor binding. *Proc Natl. Acad. Sci. U.S.A.* **75**, 180–184.

Bucher H., Zapf J., Torreani, T., Prader, A., Froesch, E. R., and Illig, R. (1983). Insulin-like growth factors I and II, prolactin and insulin or decreased longitudinal growth after operation for craniopharyngioma. *N. Engl. J. Med.* **309**, 1142–1146.

Cahill G. F., Jr. (1971). Physiology of insulin in man. *Diabetes* **20**, 785–799.

Carpentier, J. L., Van Obberghen, E., Gorden, P., and Orci, L. (1979). ^{125}I-Anti-insulin receptor antibody binding to cultured human lymphocytes: Morphological events are similar to the binding of ^{125}I-insulin. *Diabetes* **28**, 354 (Abstr.).

Clausen, T. (1977). Calcium, glucose transport and insulin action. *FEBS-Symp.*, **42**, 481–486.

Clemmons, D. R., Van Wyke, J. J., and Pledger, W. J. (1980). Sequential addition of platelet factor and plasma to BALB/c 3T3 fibroblast cultures stimulates somatomedin-C binding early in cell cycle. *Proc. Natl. Acad. Sci. U.S.A.* **77**, 6644–6648.

Cohen, S., Chinker, M., and Ushiro (1981). EGF-receptor protein kinase phosphorylates tyrosine and may be related to the transforming kinase of Rous sarcoma virus. *Cold Spring Harbor, Conf. Cell Proliferation* **8**, 801–808.

Czech, M. P. (1977). Molecular basis of insulin action. *Annu. Rev. Biochem.* **46,** 359–384.

Danho, W., Bullesbach, E. E., Gattner, H. G., King, G. L., and Kahn, C. R. (1981). Synthesis and biological properties of "hybrids" between human insulin and insulin-like growth factor. I. *Pept. Synth, Struct., Funct., Proc. Am. Pept. Symp., 7th, 1981,* pp. 113–122.

DeMeyts, P., Van Obberghen, E., Roth, J., Wollmer, A., and Brandenburg, D., (1978). Mapping of the residues responsible for the negative cooperativity of the receptor-binding region of insulin. *Nature (London)* **273,** 504–509.

Deuel, T. F., Huang, J. S., Proffitt, R. T., Baenziger, J. U., Chang, D., and Kennedy, B. B. (1981). Human Platelet-derived growth factor, purification and resolution in two active protein fractions. *J. Biol. Chem.* **256,** 8896–8899.

Ek, B., Westermark, B., Wasteson, A., and Heldin, C. H. (1982). Stimulation of tyrosine-specific phosphorylation by platelet-derived growth factor. *Nature (London)* **295,** 419–420.

Fajans, S. S. (1979). Diabetes Mellitus: Description, etiology and Pathogenesis, natural history and testing procedures. *In* "Endocrinology" (L. J. DeGroot, ed), Vol 2, pp. 1007–1024. Grune & Stratton, New York.

Fehlmann, M., and Freychet, P. (1981). Insulin and glucagon stimulation of $(Na^+ - K^+)$-ATPase transport activity in isolated rat hepatocytes. *J. Biol. Chem.* **256,** 7449–7453.

Fehlman, M., LeCam, A., and Freychet, P. (1979). Insulin and glucagon stimulation of amino acid transport in isolated rat hepatocytes: Synthesis of a high affinity component of transport. *J. Biol. Chem.* **254,** 10431–10437.

Foley, T. P., Nissley, S. P., Stevens, R. L., King, G. L., Hascall, V. C., Humbel, R. E., Short, P. A., and Rechler, M. M., (1982). Demonstration of receptors for insulin and insulin-like growth factors on swarm rat chondrocytes: Evidence that insulin stimulates proteoglycan synthesis through the insulin receptor. *J. Biol. Chem.* **257,** 663–669.

Gey, G. O., and Thalhimer, W. (1924). Observations on the effects of insulin introduced into the medium of tissue cultures. *JAMA J. Am. Med. Assoc.* **82,** 1609.

Gorden, P., Carpentier, J. L., Cohen, S., and Orci, L. (1978). Epidermal growth factor: Morphological demonstration of binding; internalization and lysosomal association in human fibroblasts. *Proc. Natl. Acad. Sci. U.S.A.* **75,** 5025–5029.

Gorden, P., Carpentier, J. L., Freychet, P., and Orci, L. (1980). Internalization of polypeptide hormones: Mechanism, intracellular localization and significance. *Diabetologia* **18,** 263–274.

Gospodarowicz, D., and Moran, J. S. (1976). Growth factor in mammalian cell culture. *Annu. Rev. Biochem.* **45,** 531–538.

Griffiths, J. B. (1971). The effect of medium changes on the growth and metabolism of the human diploid cells, w 1-38. *J. Cell Sci.* **8,** 43–52.

Griffiths, J. B. (1972). Role of serum, insulin and amino acid concentration in contact inhibition of growth of human cells in culture. *Exp. Cell Res.* **75,** 47–56.

Haigler, H. T., Ash, J. F., Singer, S. J., and Cohen, S. (1978). Visualization of fluorescence of the binding and internalization of epidermal growth factor in human carcinoma cells A-431. *Proc. Natl. Acad. Sci. U.S.A.* **75,** 3317–3321.

Haring, H. U., Kasuga, M., and Kahn C. R. (1982). Insulin receptor phosphorylation in intact adipocytes and in a cell-free system. *Biochem. Biophys. Res. Commun.* **108,** 1538–1545.

Haring, H. U. Kasuga, M., Pawelek, J., and Kahn, C. R. (1984a). Abnormality of insulin binding and receptor phosphorylation in insulin resistant melanoma cells. *J. Cell Biol.* **99**, 900–908.

Haring, H. U., Kasuga, M., White, M. F., Crettaz, M., and Kahn, C. R. (1984b). Phosphorylation and dephosphorylation of the insulin receptor: Evidence against an intrinsic phosphatase activity. *Biochemistry* **23**, 3298–3306.

Harmon, J. T., Kahn, C. R., Kempner, E. S., and Schlegel, W. (1980). Characterization of the insulin receptor in its membrane environment by radiation inactivation. *J. Biol. Chem.* **255**, 3412–3419.

Hayashi I., Larner, J., and Sato, G. (1978). Hormonal growth control of cells in culture. *In Vitro* **14**, 23–30.

Hedo, J. A., Kahn, C. R., Hayashi, M., Yamada, K. M., and Kasuga, M., (1983). Biosynthesis and glycosylation of the insulin receptor: Evidence for a single polypeptide precursor of the two major subunits. *J. Biol. Chem.* **258**, 10020–10023.

Heldin C. H., Westermark, B., and Wasteson, A. (1979). Platelet-derived growth factor: Purification and partial characterization. *Proc. Natl. Acad. Sci. U.S.A.* **76**, 3722–3726.

Heldin C. H., Wasteson, A., and Westermark, B. (1982). Interaction of platelet-derived growth factor with its fibroblast receptor. *J. Biol. Chem.* **257**, 4216–4221.

Hill, P. E. (1978). Effect of insulin on fetal growth. *Semin. Perinat.* **2**, 319–328.

Hodgkins, D. C., and Mercola, D. (1973). *In* "Handbook of Physiology" (R. O. Greep and E. B. Astwood, eds.), Washington, D.C. American Physiol. Soc., Sect. 7, Vol. II, pp. 139–159.

Holmes, L. B., Frantz, A. G., Rabkin, M. T., Soeldner, J. S., and Crawford, J. D. (1968). Normal growth with subnormal growth hormone levels. *N. Engl. J. Med.* **279**, 559–566.

Hunter, T., and Sefton, B. M. (1980). Transforming gene product of Rous sarcoma virus phosphorylates tyrosine. *Proc. Natl. Acad. Sci. U.S.A.* **77**, 1311–1315.

Jacob, J., Cull, F. C., Earp, H. S., Svobada, M. E., Van Wyk, J. J., and Cuatrecases, P. (1983). Somatomedin-c stimulates the phosphorylation of the β-subunit of its own receptor membranes. *J. Biol. Chem.* **258**, 9581–9584.

Kahn, C. R. (1975). Membrane receptors for polypeptide hormones. *Methods Membr. Biol.* **3**, 81–146.

Kahn, C. R. (1979). What is the molecular basis for insulin action? *Trends Biochem. Sci.* **263**, 4–7.

Kahn, C. R., Baird, K., Flier, J. S., and Jarrett, D. B. (1977). Effect of autoantibodies to the insulin receptor on isolated adipocytes: Studies of insulin binding and insulin action. *J. Clin. Invest.* **60**, 1094–1106.

Kahn, C. R., Baird, K. L., Jarrett, D. B., and Flier, J. S. (1978). Direct demonstration that receptor crosslinking or aggregation is important in insulin action. *Proc. Natl. Acad. Sci. U.S.A.* **75**, 4209–4213.

Kahn, C. R., Murray, M., and Pawelek, J. (1980). Inhibition of proliferation of Cloudman S91 melanoma cells by insulin and characterization of some insulin-resistant varieties. *J. Cell. Phys.* **103**, 109–119.

Kahn, C. R., Baird, K., and Flier, J. S. (1981). Insulin receptors, receptor antibodies and the mechanism of insulin action. *Recent Prog. Horm. Res.* **37**, 477–538.

Karlson, F. A., Van Obberghen, E., Grunfeld, C., and Kahn, C. R. (1979). Desensitization of the insulin receptor at an early post-receptor step by prolonged exposure to anti-receptor antibody. *Proc. Natl. Acad. Sci. U.S.A.* **76**, 809–813.

Kasuga, M., Van Obberghen, E., Nissley, S. P., and Rechler, M. M. (1981). Demonstration of two subtypes of insulin-like growth factor receptors by affinity cross-linking. *J. Biol. Chem.* **256,** 5305–5308.

Kasuga, M., Karlsson, F. A., and Kahn, C. R. (1982a). Insulin stimulates the phosphorylation of the 95,000 dalton subunit of its own receptor. *Science* **215,** 185–187.

Kasuga, M., Zick, Y., Blithe, D. L., Crettaz, M., and Kahn, C. R. (1982b). Insulin stimulates tyrosine phosphorylation of the insulin receptor in a "cell-free" system. *Nature (London)* **298,** 667–669.

Kasuga, M., Zick, Y., Blithe, D. L., Karlsson, F. A., Haring, H. U., and Kahn, C. R. (1982c). Insulin stimulation of phosphorylation of the β-subunit of the insulin receptor: Formation of both phosphoserine and phosphotyrosine. *J. Biol. Chem.* **257,** 9891–9894.

Kasuga, M., Fujita-Yamaguchi, Y., Blithe, D. L., and Kahn, C. R. (1983a). Tyrosine-specific protein kinase activity is associated with the purified insulin receptor. *Proc. Natl. Acad. Sci. U.S.A.* **80,** 2137–2141.

Kasuga, M., Sasaki, N., Kahn, C. R., Nissley, S. P., and Rechler, M. M. (1983b). Antireceptor antibodies as probes of insulin-like growth factor receptor structure. *J. Clin. Invest.* **72,** 1459–1469.

King, G. L., and Kahn, C. R. (1981). Non-parallel evolution of metabolic and growth-promoting functions of insulin. *Nature (London)* **292** 644–646.

King, G. L., and Kahn, C. R. (1984). The growth-promoting effects of insulin. *In* "Growth and Maturation Factors" (G. Guroff, ed.), Vol. 2, pp. 223–266. Wiley, New York.

King, G. L., Kahn, C. R., Rechler, M. M., and Nissley, S. P. (1980). Direct demonstration of separate receptors for growth and metabolic activities of insulin and multiplication-stimulating activity (an insulin-like growth factor using antibodies to the insulin receptor). *J. Clin. Invest.* **66,** 130–140.

King, G. L., Kahn, C. R., and Rechler, M. M. (1982a). Interactions of insulin receptors with receptors for insulin-like growth factors in rat adipocytes. *J. Biol. Chem.* **257,** 10001–10006.

King, G. L., Kahn, C. R., and Samuels, B. (1982b). Synthesis and characterization of molecular hybrids of insulin and insulin-like growth factor I. *J. Biol. Chem.* **257,** 10869–10873.

King, G. L., Kahn, C. R., and Heldin, C. H. (1983a). Sharing of biological effect and receptors between guinea pig insulin and platelet-derived growth factor. *Proc. Natl. Acad. Sci. U.S.A.* **80,** 1308–1312.

King, G. L., Buzney, S. M., Kahn, C. R., Hetos, N., Buchwald, S., and Rand, L. (1983b). Differential responsiveness to insulin of endothelial and support cells from micro- and macrovessels. *J. Clin. Invest.* **71,** 974–979.

Klapper, D. G., Svoboda, M. E., and Van Wyk, J. J. (1983). The sequence analysis of somatomedin C: Conformation of identity with insulin-like growth factor I. *Endocrinology (Baltimore)* **112,** 2215–2217.

Koontz, J. W. (1980). Insulin as a potent specific growth factor in a rat hepatoma cell line. *J. Supramol. Struct.* **4,** 171.

Koontz, J. W., and Iwahashi, M. (1981). Insulin as a potent, specific growth factor in a rat hepatoma cell line. *Science* **211,** 947–947.

Kuwabara, F., and Cogan, D. G. (1963). Retinal vascular patterns. *Arch. Ophthalmol. (Chicago)*.

Latta, J. S., and Bucholz, D. J. (1939). *Arch. Exp. Zellforsch. Besonders Gewebezwecht.* **23,** 146–151.

Leffert, H. L. (1974). Growth control of differential fetal rat hepatocytes in primary monolayer culture. VIII. Hormonal control of DNA synthesis and its possible significance
to the problem of liver regulation. *J. Cell Biol.* **62,** 792–881.

Leslie, I., and Davidson, J. N. (1952). The effect of insulin on cellular composition and
growth of chick-heart explants. *Biochem. J.* **49,** XLI–XLII (abstr.).

Leslie, I., and Paul, J. (1954). The action of insulin on the composition of cells and
medium during culture of chick-heart explants. *J. Endocrinol.* **11,** 110–124.

Leslie, I., Fulton, W. C., and Sinclair, R. (1957). The metabolism of human embryonic and
malignant cells—Their response to insulin. *Biochim. Biophys. Acta* **24,** 365–380.

Leslie, I., Fulton, W. C., and Sinclair, R. (1957). The metabolism of human embryonic and
malignant cells—Their response to insulin. *Biochim. Biophys. Acta* **24,** 365–
380.

Liberman, I., Abrams, R., and Ove, P. (1963). Changes in the metabolism of ribonucleic
acid preceding the synthesis of deoxyribonucleic acid in mammalian cells cultured
from the animal. *J. Biol. Chem.* **238,** 2141–2149.

McEvoy, R. C., and Hegre, O. D. (1978). Syngeneic transplantation of fetal rat pancreas II.
Effect of insulin treatment on the growth and differentiation of pancreatic implants
fifteen days after transplantation. *Diabetes* **27,** 988–995.

Marquardt, H., Todaro, G. J., Henderson, L. E., and Oroszlan, S. (1981). Purification and
primary structure of a polypeptide with multiplication stimulating activity from rat
liver cell cultures. *J. Biol. Chem.* **256,** 6859–6865.

Massague, J., Guillette, B. J., and Czech, M. P. (1981). Affinity labeling of multiplication
stimulating activity receptors in membranes from rat and human tissues. *J. Biol.
Chem.* **250,** 2122–2125.

Massague, J., Blinderman, L. A., and Czech, M. P. (1982). The high affinity insulin
receptor mediates growth stimulation in rat hepatoma cells. *J. Biol. Chem.* **257,**
13958–13963.

Megyesi, K., Kahn, C. R., Roth, J., Froesch, E. R., Humbel, R. E., Zapf, J., and Neville, D.
M., Jr. (1974). Insulin and nonsuppressible insulin-like activity (NSILA-s): Evidence for separate plasma membrane receptor sites. *Biochem. Biophys. Res. Commun.* **57,** 307–315.

Megyesi, K., Kahn, C. R., Roth, J., Neville, D. M., Jr., Nissley, S. P., Humbel, R. E., and
Froesch, E. R. (1978). The NSILA receptor in liver membranes: Characterization and
comparison with the insulin receptor. *J. Biol. Chem.* **250,** 8990–8996.

Mott, D.M., Howard, B. V., and Bennett, P. H. (1979). Stoichiometric binding and regulation of insulin receptors on human diploid fibroblasts using physiologic insulin
levels. *J. Biol. Chem.* **254,** 8762–8767.

Nishimura, J., Huang, J. S., and Deuel, T. F. (1982). Platelet-derived growth factor stimulates tyrosine-specific protein kinase activity in Swiss mouse 3T3 cell membranes.
Proc. Natl. Acad. Sci. U.S.A. **79,** 4303–4307.

Nissley, S. P., and Rechler, M. M. (1978). Multiplication stimulting activity (MSA): A
somatomedin-like polypeptide from cultured rat line cells. *Natl. Cancer Inst.
Monogr.* **48,** 167–177.

Oppenheimer, C. L., Pessio, J. E., Massague, J., Gitomer, W., and Czech, M. P. (1983).
Insulin action rapidly modulates the apparent affinity of the insulin-like growth
factor II receptor. *J. Biol. Chem.* **258,** 4824–4830.

Paul, J., and Pearson, E. S. (1960). The action of insulin on the metabolism of cell culture.
J. Endocrinol. **21,** 287–294.

Pawelek, J., Murray, M., and Fleischmann, R. (1982). Growth of cells in hormonally
defined media. *Cold Spring Harbor Conf. Cell Proliferation* **9,** 911–916.

Pedersen, J. (1977). "The Pregnant Diabetic and Her Newborn." Williams & Wilkins, Baltimore, Maryland.

Petrides, P. E., and Bohler, P., (1980). The mitogenic activity of insulin: An intrinsic property of the molecule. Biochem. Biophys. Res. Commun. 95, 1138–1144.

Petruzzelli, L. M., Ganguly, S., Smith, C. J., Cobb, M. H., Rubin, C. S., and Rosen, O. M. (1982). Insulin activates a tyrosine-specific protein kinase in extracts of 3T3-L1 adipocytes and human placenta. Proc. Natl. Acad. Sci. U.S.A. 79, 6792–6796.

Phillips, L., and Vassilopoulous-Sellin, R. (1980). Somatomedians. Part I. N. Engl. J. Med. 302, 371–438.

Pike, L. J., Bowen-Pope, D. F., Ross, R., and Krebs, E. G. (1983). Characterization of platelet-derived growth factor stimulated phosphorylation in cell membranes. J. Biol. Chem. 258, 9383–9390.

Pilch, P. F., and Czech, M. P. (1980). The subunit structure of the high affinity insulin receptor. J. Biol. Chem. 255, 1722–1731.

Pledger, W. J., Stiles, C. D., Antoniades, H. N., and Scher, C. D. (1979). Introduction of DNA synthesis in BALB/c 3T3 cells by serum components: Re-evaluation of the commitment process. Proc. Natl. Acad. Sci. U.S.A. 74, 4481–4485.

Posner, B. I., Bergeron, J. J. M., Josefsbery, Z., Khan, M. N., Khan, R. J., Patel, B., Sikstroth, R. A., and Verma, A. K. (1981). Polypeptide hormones: Intracellular receptors and internalization. Recent Prog. Horm. Res. 37, 539–582.

Raines, E. W., and Ross, R. (1982). Platelet-derived growth factor I. High yield purification and evidence for multiple forms. J. Biol. Chem. 257, 5154–5171.

Rechler, M. M., Goldfine, I. D., Podskalny, J., and Wells, C. A. (1974). DNA synthesis in human fibroblasts: Stimulation by insulin and by non-suppressible insulin-like activity. J. Clin. Endocrinol. Metab. 39, 512–521.

Rechler, M. M., Zapf, J., Nissley, S. P., Froesch, E. R., Moses, A. C., Podskalny, J. M., Schillings, E. E., and Humbel, R. E. (1980). Interactions of insulin-like growth factors I and II and multiplication-stimulating activity with receptors and serum carrier proteins. Endocrinology (Baltimore) 107, 1451–1459.

Rechler, M. M., Nissley, S. P., King, G. L., Moses, A., Schilling, E. E., Romonus, J., Short, P. A., and White, R. M. (1981). Multiplication stimulating activity (MSA) from the BRL3A rat liver cell line: Relation to human somatomedins and insulin. J. Supramol. Struct. 292, 644–646.

Rinderknecht, E., and Humbel, R. (1976). Polypeptides with nonsuppressible insulin-like cell growth-promoting activities in human serum: Isolation, chemical characterization and some biological properties of forms I and II. Proc. Natl. Acad. Sci. U.S.A. 73, 2365–2369.

Rinderknecht, E., and Humbel, R. E. (1978a). Primary structure of human insulin-like growth factor II. FEBS Lett. 89, 283–286.

Rinderknecht, E., and Humbel, R. E. (1978b). The amino acid sequence of human insulin-like growth factor I and its structural homology with proinsulin. J. Biol. Chem. 253, 2769–2776.

Rosenfeld, R. G., Thorsson, A. V., and Hintz, R. L. (1980). Characteristics of a specific receptor for somatomedin-C (SM-C) on cultured human lymphocytes: Evidence that SM-C modulates homologous receptor concentration. Endocrinology (Baltimore) 107, 1841–1848.

Roth, J. (1979). Receptors for peptide hormones. In "Endocrinology" (L. DeGroot, ed.), Vol. 3, pp. 2037–2054. Grune & Stratton, New York.

Roth, R. A., and Cassell, D. J. (1983). Insulin receptor: Evidence that it is a protein kinase. Science 219, 299–301.

Rozengurt, E., and Heppel, L. A. (1975). Serum rapidly stimulates Ouabain-sensitive $^{86}Rb^+$ influx in quiescent 3T3 cells. *Proc. Natl. Acad. Sci. U.S.A.* **72**, 4492–4495.

Rozengurt, E., and Mendoza I. S. (1980). Monovalention fluxes and the control of cell proliferation in cultire fibrobeasts. *Ann. N. Y. Acad. Sci.* **339**, 175–190.

Rozengurt, E., and Stein, W. D. (1977). Regulation of uridine uptake by serum and insulin in density-inhibited 3T3 cells. *Biochim. Biophys. Acta* **464**, 417–432.

Salmon, W. D., and Daughaday, W. H. (1956). A hormonally controlled serum factor which stimulates sulfate incorporation by cartilage *in vitro*. *J. Lab. Clin. Med.* **49**, 825–836.

Sanui, H., and Rubin, H. (1977). Correlated effects of external magnesium on cation content and DNA synthesis in cultured chicken embryo fibroblasts. *J. Cell. Physiol.* **92**, 23–30.

Schlessinger, J., Schecter, Y., Willingham, M. C., and Pastan, I. (1978). Direct visualization of binding, aggregation and internalization of insulin and epidermal growth factor on living fibroblastic cells. *Proc. Natl. Acad. Sci. U.S.A.* **75**, 2659–2663.

Schoenle, E., Zapf, J., and Froesch, E. R. (1977). Effects of insulin and NSILA on adipocytes of normal and diabetic rats: Receptor binding, glucose transport and glucose metabolism. *Diabetabologia* **13**, 243–249.

Schuldiner, S., and Rozengurt, E. (1982). Na^+/H^+ antiport in Swiss 3T3 cells: Mitogenic stimulation leads to cytoplasmic alkanization. *Proc. Natl. Acad. Sci. U.S.A.* **79**, 7778–7782.

Starzl, T. E., Porter, K. A., Watanabe, K., and Putnam, C. N. (1976). Effects of insulin, glucagon and insulin/glucagon infusions on liver morphology and cell division after complete portacaval shunt in dogs. *Lancet* **1**, 821–825.

Stiner, D. F., Rubenstein, A. H., and Melani, F. (1972). Circulating proinsulin: Immunology, measurement and biological activity. *In* "Handbook of Physiology" (R. V. Greep and E. B. Astwood, eds.), Sect. 7, Vol. I, pp. 515–528. Am. Physiol. Soc., Washington, D.C.

Stout, R. W. (1979). Diabetes and atherosclerosis—the role of insulin. *Diabetologia* **16**, 141–150.

Susa, J. B., McCormick, K. L., Widness, J. A., Singer, D. B., Oh, W., Admsons, K., and Schwartz, R. (1979). Chronic hyperinsulinemia in the fetal rhesus monkey: Effects on fetal growth and composition. *Diabetes* **28**, 1058–1063.

Temin, H. M. (1967). Studies on carcinogenesis by avian sarcoma viruses. VI: Differential multiplication of infected and of converted cells in response to insulin. *J. Cell. Physiol.* **69**, 377–384.

Van Obberghen, E., and Kowalski, A. (1982). Phosphorylation of the hepatic insulin receptor: Stimulating effect of insulin on intact cells and in a cell-free system. *FEBS Lett.* **143**, 179–182.

Van Obberghen, E., Spooner, P. M., Kahn, C. R., Chernick, S. S., Garrison, M. M., Karlsson, F. A., and Grunfeld, C. (1979). Insulin-receptor antibodies mimic a late insulin effect. *Nature (London)* **280**, 500–504.

Van Obberghen, E., Kasuga, M., Le Cam, A., Hedo, J. A., Itin, A., and Harrison, L. C., (1981) Biosynthetic labeling of the insulin receptor: Studies of subunits in cultured human IM-9 lymphocytes. *Proc. Natl. Acad. Sci. U.S.A.* **78**, 1052–1056.

Van Wyk, J. J., and Underwood, L. E. (1978). The somatomedins and their actions. *In* "Biochemical Actions of Hormones" (G. Litwãck, ed.), Vol. 5, pp. 101–148. Academic Press, New York.

von Haam, E., and Cappel, L. (1940). Effect of hormones upon cells grown *in vitro*. I. The effect of sex hormones upon fibroblasts. *Am. J. Cancer* **39**, 350–358.

Waterfield, M. D., Scrace, G. T., Whittle, N., Stroobast, P., Johnson, A., Wasteen, A., Heldin, C. H., Huang, J. S., and Deuel, T. F. (1983). Platelet-derived growth factor is structurally related to the putative transforming protein p28[sis] of simian sarcoma virus. *Nature (London)* **304,** 35–39.

White, R. M., Nissley, S. P., Moses, A. C., Rechler, M. M., and Johnsonbaugh, R. E. (1981). The growth hormone dependence of a somatomedin-binding protein in human serum. *J. Clin. Endocrinol. Metab.* **53,** 49–57.

Wrann, M., Fox, C. F., and Ross, R. (1980). Motivation of epidermal growth factor receptors on 3T3 cells by platelet-derived growth factor. *Science* **210,** 1363–1365.

Zapf, J. E., Rinderknecht, R. E., and Froesch, E. R. (1978a). Nonsuppressible insulin-like activity (NSILA) from human serum: Recent accomplishments and their physiologic implications. *Metab., Clin. Exp.* **27,** 1803–1928.

Zapf, J. E., Schoenle, E., and Froesch, E. R. (1978b). Insulin-like growth factors I and II: Some biological actions and receptor binding characteristics of two purified constituents of nonsuppressible insulin-like activity of human serum. *Eur. J. Biochem.* **87,** 285–296.

Zick, Y., Kasuga, M., Kahn, C. R., and Roth, J. (1983). Characterization of insulin-mediated phosphorylation of the insulin receptor in a cell-free system. *J. Biol. Chem.* **258,** 75–80.

9

Glucagon Receptors and Their Functions

SUZANNE K. BECKNER, RICHARD HORUK,[1]
FREDERICK J. DARFLER, AND MICHAEL C. LIN

Laboratory of Cellular and Developmental Biology
National Institute of Arthritis, Diabetes,
  and Digestive and Kidney Diseases
National Institutes of Health
Bethesda, Maryland

[1]Present address: Department of Medicine, University of California, San Diego, La Jolla, California 92093.

251

CONTROL OF ANIMAL CELL PROLIFERATION,
VOLUME I

I. INTRODUCTION

Glucagon receptors were among the first hormone receptors to be extensively studied, primarily in liver. Although glucagon does stimulate the proliferation of some cells (Murakami *et al.*, 1981; Miyazaki *et al.*, 1982) and is required for liver regeneration (Leffert *et al.*, 1983), the hormone is not generally associated with cell growth. However, unlike the growth-promoting hormones, the mechanism of action of glucagon is well established. Like a number of polypeptide hormones and neurotransmitters, the cellular effects of glucagon are mediated by adenylate cyclase. This complex is composed of at least three well-defined components: the hormone receptor (which is located on the external surface of the plasma membrane), guanine nucleotide regulatory proteins N_s and N_i (which stimulate and inhibit activity, respectively), and the catalytic unit, which converts ATP to cAMP (Rodbell, 1980). This second messenger activates a specific protein kinase, which then initiates a cascade of protein phosphorylation that regulates the activity of a number of key regulatory enzymes (Sutherland *et al.*, 1965). Such regulation, particularly of those enzymes involved in glucose metabolism, is necessary for normal cell growth and function.

The major targets of glucagon are the liver and kidney. Glucagon, via the cAMP cascade, stimulates gluconeogenesis, glycogenolysis, and lipolysis and inhibits glycogen synthesis and lipogenesis in liver (Exton, 1981) and adipocytes (Rodbell, 1972); stimulates liver mitochondrial function (Sutton and Pollak, 1980), and activity of the enzymes of the urea cycle in kidney (Katz and Lindheimer, 1977) and liver (R. C. Lin *et al.*, 1982), the activity of 3-hydroxy-3-methylglutaryl-coenzyme A (HMG-CoA) reductase (the rate-limiting step of cholesterol biosynthesis) in liver (Edwards *et al.*, 1979); and regulates solute transport in kidney tubule epithelium (Kolanowski, 1983).

The glucagon receptor and the components of adenylate cyclase all exist in close association with the plasma membrane. It has therefore been possible to examine the regulation of this system in purified membrane preparations. Such studies have elucidated the importance of structural, ionic, and hydrophobic interactions and the contribution of membrane components to the interaction of glucagon with the adenylate cyclase complex. However, the membrane association of the individual components has hampered efforts to isolate, purify, and reconstitute resolved components of the complex. The catalytic unit of adenylate cyclase has been fairly well studied in the soluble form. More recently, the guanine nucleotide regulatory proteins have been purified and characterized. Although some adenylate cyclase-linked hormone receptors have been recently purified, to date glucagon receptors have not yet been isolated. It is likely that a fair degree of homology exists between receptors which regulate adenylate cyclase, so information from one system may be applicable to others. This review will summarize what is currently known about the glucagon-responsive adenylate cyclase system and the cellular effects of the hormone.

II. GENERAL CHARACTERISTICS OF GLUCAGON INTERACTION WITH ITS RECEPTORS

The first step in any hormone action is binding to specific cell surface receptors. The interaction between a hormone and its receptor is normally studied by radiolabeled hormone binding. Obviously, the homogeneity and purity of the radiolabeled hormone must be considered in the interpretation of binding data. Both [125]I- and ^{3}H-labeled glucagon have been used as tracers to characterize glucagon binding sites on intact hepatocytes, established cell lines, and purified membrane preparations. The low specific activity of ^{3}H-labeled glucagon (Lin et al., 1977) makes it less practical for routine use, so preparations of ^{125}I-labeled glucagon (of variable purity and specific activity) have generally been utilized. More recently, ^{125}I-labeled monoiodoglucagon has been prepared by high-performance liquid chromatography (HPLC) (Rojas et al., 1983). This procedure effectively separates ^{125}I-labeled Tyr[10]-monoiodoglucagon from oxidized ^{125}I-labeled glucagon and native glucagon.

A. Affinity and Potency of Radiolabeled Glucagon

Glucagon can be iodinated at tyrosine residues 10 and 13. Several iodotyrosyl glucagons have been prepared which exhibit different affinities and potencies compared to the native hormone (Bromer et al., 1973; Desbuquois, 1975). Since iodination of glucagon increases the hydrophobicity of the molecule (due to the incorporation of the bulky iodine atom), the affinity of iodoglucagon for the receptor is greater than that of the native hormone. However, the potency of iodoglucagon for adenylate cyclase stimulation is pH dependent (Lin et al., 1976), since the electrophilic substitution of iodine lowers the pK of the tyrosine hydroxyl from 10 to 8.2. Since preparations of ^{125}I-labeled glucagon are not always homogeneous and estimates of concentration are not always exact, it is probably fortuitous that the initial study showed an identical concentration dependency for ^{125}I-labeled gluca-

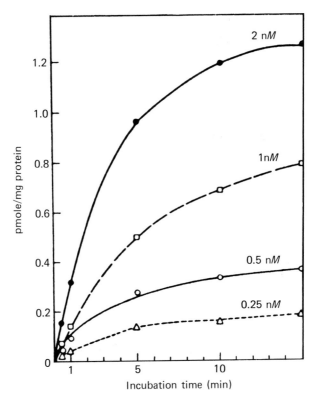

Fig. 1. Binding of ^{125}I-labeled glucagon to hepatic membranes. Rat liver membranes were incubated with different concentrations of ^{125}I-labeled glucagon at 30°C for the indicated times. The nonspecific binding was determined in the presence of 1 μM native glucagon as described. From Rodbell *et al.* (1974).

gon binding and adenylate cyclase activation by native glucagon (Rodbell *et al.*, 1971a).

The activation of adenylate cyclase in isolated membranes by glucagon is rapid under optimal conditions; a steady state is reached within 30 sec after the addition of hormone (Rodbell *et al.*, 1974). However, as seen in Fig. 1, the binding of ^{125}I-labeled glucagon appeared to be much slower, requiring a minimum of 5–10 min under identical conditions (Rodbell *et al.*, 1974). This discrepancy disappeared when HPLC-purified monoiodoglucagon was used (Rojas *et al.*, 1983), confirming the necessity of homogeneous preparations of radiolabeled hormones for clear interpretation of binding studies.

Binding studies using ^{3}H-labeled glucagon revealed two distinct classes of glucagon binding sites on rat liver membranes (Lin *et al.*,

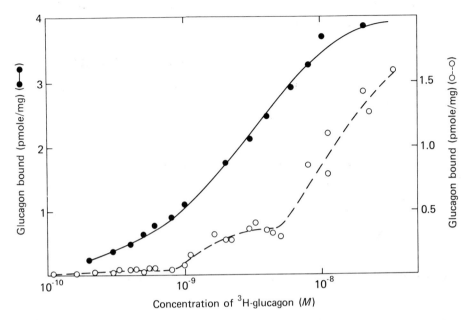

Fig. 2. Effect of ³H-labeled glucagon binding. Membranes were incubated for 4 and 15 min with (○) or without (●) 100 μM GTP, respectively, with the indicated concentration of ³H-labeled glucagon. Nonspecific binding was determined as in Fig. 1. From Lin et al. (1977).

1977). The concentration dependency for binding to the high-affinity sites corresponded to the range over which glucagon activated adenylate cyclase (Fig. 2). This suggested that only a fraction of the total glucagon binding sites are involved in adenylate cyclase stimulation under physiological conditions. It is possible that the lower affinity binding sites function as spare receptors or are desensitized forms of the receptor (see Section VI).

Based on binding studies performed by various groups, the dissociation constant for the interaction of glucagon with its receptor is about 1–5 nM in the absence of nucleotides (Lin et al., 1977). In contrast to the binding of β-adrenergic hormones (Williams et al., 1978; Bird and Maguire, 1978), glucagon binding is unaffected by Mg^{2+} (Rodbell et al., 1971b). The number of glucagon receptors on isolated hepatocytes is about 200,000 sites/cell (Sonne et al., 1978). The binding capacity of rat liver membranes was estimated to be 1–5 pmol/mg membrane protein (Rodbell et al., 1971a). The kidney is another major target for glucagon; however, the affinity is lower by a factor of 10 and the number of ɜlucagon binding sites on Madin Darby canine kidney cells is ~5000–10,000 sites/cell (M. C. Lin, unpublished observation).

B. Effect of Guanine Nucleotides on Glucagon Receptors

One of the most important developments in the study of adenylate cyclase-coupled hormones was the discovery of the guanine nucleotide requirement for hormone activation (Rodbell, 1980). Additionally, guanine nucleotides also affect the kinetics of glucagon binding in several ways.

Guanine nucleotides at micromolar concentrations increased the rate of binding of ^{125}I-labeled glucagon, but decreased the binding affinity without altering the number of binding sites (Rodbell et al., 1971b). This initial observation was confirmed by using ^{3}H-labeled glucagon. In fact, as seen in Fig. 3, in the presence of GTP the binding kinetics of ^{3}H-labeled glucagon paralleled those of adenylate cyclase activation (Lin et al., 1977). Under these conditions, the binding of ^{3}H-labeled glucagon reached equilibrium in less than 30 sec, which is nearly iden-

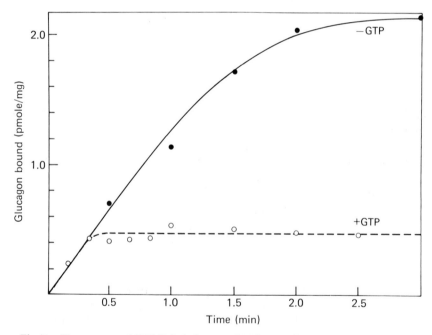

Fig. 3. Time course of ^{3}H-labeled glucagon binding in the presence and absence of GTP. Membranes were incubated with 3 nM ^{3}H-labeled glucagon for the indicated time and nonspecific binding determined as in Fig. 1. From Lin et al. (1977).

tical to the time required to observe adenylate cyclase activation by native glucagon.

III. REGULATION OF ADENYLATE CYCLASE ACTIVATION BY GLUCAGON

Because of its regulation by a wide variety of hormones and neurotransmitters, the hormone-sensitive adenylate cyclase system and its components have been extensively studied and reviewed (Ross et al., 1983; Cooper, 1983; Rodbell, 1981). It is generally accepted that the components (hormone receptor, guanine nucleotide regulatory proteins, and the catalytic unit of the enzyme) are not prelinked within the membrane. Data from target size analysis suggest that the separated units may exist as large multimeric complexes in the membrane (Schlegel et al., 1979) which break down and interact only after exposure to hormone (Rodbell et al., 1975; Houslay et al., 1977).

The stimulation of adenylate cyclase by hormones can be divided into several distinct steps. The binding of the hormone to the receptor occurs on the outer surface of the membrane. The interaction of this complex with the regulatory proteins is a transmembrane event, while the stimulation of adenylate cyclase by the regulatory protein occurs on the inner surface of the bilayer. Activity can therefore be regulated by factors affecting any of these components, including the membrane itself. Numerous models have been proposed to describe such interactions within the membrane (Swillens and Dumont, 1980). Alteration of binding and adenylate cyclase activation by modification of the glucagon molecule is discussed below (see Section IV).

A. The Catalytic Unit (C)

Adenylate cyclase has been solubilized from a number of sources with nonionic detergents. Unfortunately, hormone responsiveness is generally lost and catalytic activity is not as great as in particulate preparations (Pohl et al., 1971a; Neer, 1977; Neer et al., 1980). Aggregated catalytic units have been separated from other membrane components of both rabbit liver (Ross, 1981) and bovine caudate nucleus (Bender and Neer, 1983), but not yet in the monomeric form. Unlike the particulate enzyme, the activity of isolated C is stimulated more by Mn^{2+} than Mg^{2+}. Mn^{2+} activates C through a divalent cation site distinct from the active site, which requires Mg^{2+}-ATP (Mg^{2+} and Mn^{2+} are equally effective) as the active substrate (Bender and Neer,

1983). The diterpene forskolin (Seamon *et al.*, 1981), which activates virtually all mammalian adenylate cyclases, activates resolved C, but also potentiates stimulation of C by regulatory proteins (see Section III, B), suggesting a complex mechanism of action.

B. Guanine Nucleotide Regulatory Proteins, N_s and N_i

Guanine nucleotides are absolutely required for adenylate cyclase activity. These effects are mediated by the guanine nucleotide regulatory proteins N_s and N_i, which stimulate and inhibit activity, respectively (Gilman, 1984). Both N_s and N_i are dimers composed of α (45,000 and 41,000 daltons, respectively) and β (35,000 daltons) subunits (Northup *et al.*, 1980; Katada *et al.*, 1984a). The two α subunits are clearly distinct, but both bind guanine nucleotides and divalent cations, which are required for activation (Northup *et al.*, 1982; Bokoch *et al.*, 1984). The α subunit of N_s can be ADP ribosylated by cholera toxin, which persistently stimulates adenylate cyclase (Vaughan and Moss, 1978); fluoride also acts on this protein to stimulate adenylate cyclase (Sternweis and Gilman, 1982). The α subunit of N_i can be ADP ribosylated by pertussis toxin, which abolishes GTP-dependent inhibition of adenylate cyclase (Bokoch *et al.*, 1984).

The 35,000-dalton β subunits of N_s and N_i are identical (Katada *et al.*, 1984b). Studies with isolated N_s and N_i subunits suggest that the β subunit ultimately regulates adenylate cyclase by regulating the activity of N_s and N_i (Northup *et al.*, 1983b). A current model (Gilman, 1984) proposes that hormones which stimulate adenylate cyclase activity cause the dissociation of the 35,000-dalton protein from N_s, thereby stimulating catalysis (Northup *et al.*, 1983a). On the other hand, hormones which inhibit adenylate cyclase activity stimulate the dissociation of the 35,000-dalton protein from N_i, which binds to N_s and prevents its interaction with C (Katada *et al.*, 1984b).

Current data support such a model. That the α subunit of N_s (in the presence of GTP) is necessary and sufficient to activate C (Northup *et al.*, 1983a) demonstrates that it is the α subunit of N_s which interacts with C to catalyze the conversion of ATP to cAMP (Northup *et al.*, 1983b). The hydrolysis of GTP bound to N_s occurs during this interaction and GTPase activity has been demonstrated with purified N_s reconstituted into lipid vesicles (Brandt *et al.*, 1983). Furthermore, the addition of purified β subunit to N_s decreases the activity of adenylate cyclase (Northup *et al.*, 1983b).

A third subunit of 5000 daltons (γ) has also been found in association with both N_s and N_i (Hildebrandt *et al.*, 1984; Condina *et al.*, 1984).

However, the function of this protein has not been determined, suggesting that the regulation of adenylate cyclase may be even more complex.

As noted, guanine nucleotides decrease the affinity of receptors for glucagon (Rodbell et al., 1971b), and at first glance, the dual effects of GTP on binding and adenylate cyclase activation would appear contradictory. However, kinetic analysis of the β-adrenergic responsive adenylate cyclase has led Swillens and Dumont (1980) to propose that the hydrolysis of GTP releases the hormone–receptor complex from N_s, which decreases the affinity of the hormone for the receptor. In other words, a hormone–receptor–N_s complex (high affinity) is required for activation of adenylate cyclase. Following the binding and hydrolysis of GTP (with concomitant catalysis), the hormone signal is terminated by lowered receptor affinity. Studies with intact cells (see Section VI) support these ideas, as in all cases the occupation of a very small number of receptors (high affinity) seems sufficient to produce maximal stimulation of cAMP production. In the intact cell, GTP concentrations would not be limiting. Evidence suggests that two separate GTP sites regulate the effects of guanine nucleotides on binding and adenylate cyclase activation (Lad et al., 1977; Iyengar et al., 1979).

C. Modulation of Activity by Membrane Lipids

From the previous discussion, it is clear that the interaction of the components of the hormone-responsive adenylate cyclase depends on the membrane environment. The role of membrane lipids in the regulation of adenylate cyclase has recently been reviewed (Houslay and Gordon, 1983). Not surprisingly, benzyl alcohol, which fluidizes membranes, enhanced glucagon activation of adenylate cyclase (Dipple and Houslay, 1978). In contrast, cholesterol, which increases lipid order and decreases membrane fluidity, inhibited the effects of glucagon on adenylate cyclase (Whetton et al., 1983). Studies comparing the effects of several of these agents on the various components of the glucagon-responsive adenylate cyclase have futher demonstrated the regulatory role of surface charge and hydrophobic interactions (Rubalcava et al., 1983).

It has been suggested that phospholipids play a specific role in the activation of adenylate cyclase by glucagon (Pohl et al., 1971b). Hydrolysis of acidic phospholipids markedly decreased glucagon stimulation of adenylate cyclase, while hydrolysis of neutral phospholipids was without effect (Rubalcava and Rodbell, 1973). However, secondary effects of such membrane perturbations cannot be ruled out.

More recently, β-adrenergic receptors and purified N_s have been functionally reconstituted in lipid vesicles (Pedersen and Ross, 1982). Further studies such as these with purified components should define more clearly the specific role of lipids in the regulation of adenylate cyclase activity besides their obvious effects on membrane fluidity.

D. Glucagon Receptor Fusion Studies

Since functional glucagon receptors have not yet been purified (see Section V), it is not possible to study their regulation of adenylate cyclase in an artificial membrane. However, it is possible to examine glucagon receptors in partially purified membrane preparations by utilization of the cell fusion technique first described by Citri and Schramm (1980). Glucagon-responsive adenylate cyclase has been demonstrated in polyethylene glycol-fused hybrids of liver membranes with Friend erythroleukemia cells (Schramm, 1979) and membranes from bovine caudate nucleus (Tolkovsky and Martin, 1982). The same technique has been utilized to demonstrate that differences between normal and spontaneously transformed RL-PR-C cell glucagon responsiveness resulted from a loss of high-affinity glucagon receptors (Reilly and Blecher, 1981).

IV. GLUCAGON STRUCTURE–FUNCTION RELATIONSHIPS

A. Glucagon Molecule

Glucagon is a single polypeptide of 29 amino acid residues (Fig. 4) produced in the α cells of the pancreas. The amino acid sequences of glucagon from a number of mammals, birds, and fish have been determined (Sundby, 1976; Markusson *et al.*, 1972; Pollock and Kimmel, 1975; Trakatellis *et al.*, 1975) and, with the exceptions of guinea pig, angler fish, and shark (Sundby, 1976), are highly conserved. As noted, structural, ionic, and hydrophobic factors influence the interaction of glucagon with its receptor.

Glucagon has a mainly flexible conformation with little or no secondary structure in dilute aqueous solution (Wagman *et al.*, 1980), but becomes more ordered upon self-association (Gratzer *et al.*, 1972) or under certain experimental conditions in the presence of phospholipids or detergents (Bornet and Edelhoch, 1971; Schneider and Edelhoch, 1972). In concentrated solutions, an equilibrium exists between random conformers and an α-helical trimer (Blundell, 1983). The interaction of glucagon with its receptor is thought to involve

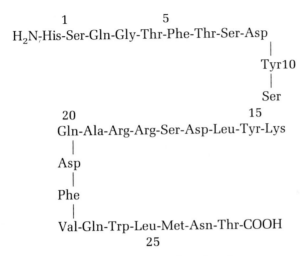

1   5
H₂N-His-Ser-Gln-Gly-Thr-Phe-Thr-Ser-Asp
                                        |
                                     Tyr10
                                        |
                                      Ser
         20   15
    Gln-Ala-Arg-Arg-Ser-Asp-Leu-Tyr-Lys
      |
    Asp
      |
    Phe
      |
    Val-Gln-Trp-Leu-Met-Asn-Thr-COOH
                      25

Fig. 4. Primary structure of porcine glucagon.

hydrophobic contacts between hormone and receptor which induce an α-helical structure in the hormone as does its interaction with phospholipids (Blundell, 1983).

B. Structure–Function Studies

Considerable research has been directed toward determining which residues of glucagon are essential for receptor binding and which are involved in the activation of adenylate cyclase. Studies with chemically modified, enzymatically modified, and synthetic glucagon analogues have helped to partially resolve this question. However, there are several limitations to such an approach. Any observed alteration in biological activity could easily be explained by conformational changes in the glucagon molecule itself. Even glucagon, which has little secondary structure in dilute aqueous solution, may require a certain flexibility in regions of the hormone which interact and change conformation upon binding to the receptor. Blocking of certain functional groups could seriously impair this flexibility and lead to decreased ability of the analogue to interact with the receptor. Additionally, strenuous efforts must be made to ensure that there is no contamination of the modified product by unreacted native hormone.

Despite these problems, numerous glucagon derivatives have been synthesized. Only a few of the more well-studied derivatives will be discussed here. For more extensive coverage, the review by Bromer (1983) is recommended.

In general, modification of the amino terminal, Asp[9,15,21], or Thr[29] results in a decrease in biological activity. Similarly, neutralization of the positive charge of Lys[12] (Carrey and Epand, 1982) or Arg[17,18] (Bromer, 1983) causes a large decrease in both binding and biological activity. Blundell (1983) has speculated that such modification increases the helical content of the molecule by interaction with adjacent tyrosine residues, thus stabilizing the α helix at the expense of receptor binding. It is likely that the flexible conformation of glucagon in solution is required for receptor binding and that the hormone–receptor complex is stabilized by the formation of an α helix between the two. However, retention of the positive charge on Lys[12] preserves almost full binding and bioactivity (Wright and Rodbell, 1980a). The positive charge at this residue would destabilize α-helix formation (Blundell, 1983) and may additionally form an ionic interaction with the receptor. Similarly, retention of a positive charge on Arg[17,18] has little effect on binding and actually enhances adenylate cyclase stimulation (Epand and Cote, 1976).

Chemical modification of Tyr[10,13], as previously discussed, probably enhances potency due to the increased hydrophobicity of the residues caused by the iodine atom (Lin *et al.*, 1976). However, this can only partially explain the change, since nitration of Tyr[10,13] also increases the potency of the derivative (Patterson and Bromer, 1973), yet the nitro group is more hydrophilic than iodine.

In contrast, attachment of bulky residues to Trp[25] or Met[27] (Epand and Cote, 1976; Wright and Rodbell, 1980b; Demoliou-Mason and Epand, 1980; Felts *et al.*, 1970) has little effect on binding or adenylate cyclase activation, suggesting that the structure at this end of the molecule does not contribute to the hormone's actions.

C. Glucagon Fragments and Antagonists

Substitutions and modifications of the glucagon molecule reveal the importance of secondary and tertiary structure in glucagon action. It is also possible to determine which portions of the glucagon molecule are important for receptor interactions and which are involved in the transmission of the hormone signal by utilizing glucagon fragments.

Numerous studies suggest that the carboxyl terminal region of glucagon is important for binding, but not for adenylate cyclase stimulation (Epand *et al.*, 1981). Cleavage of Asn[28] and Thr[29] (glucagon$_{1-27}$) results in a parallel loss of binding and adenylate cyclase stimulation (Lin *et al.*, 1975). The same is true for glucagon$_{1-21}$ (Wright *et al.*, 1978), where the eight carboxyl terminal residues were removed. On the other hand, removal of the amino terminal histidine residue (des-His[1] glucagon) results in a fragment with a slight reduction in binding activity, but a markedly reduced biological activity (Lin et al., 1975).

Taken together, these studies suggest that the glucagon receptor contains domains that distinguish between the amino terminal region of the hormone, which is responsible for activating adenylate cyclase, and the rest of the glucagon molecule, which is responsible for binding activity. Epand (1983) has proposed a general model for the stabilization of the hormone–receptor complex by a conformational change in the receptor. Based on structure–function studies, such a conformational change could be induced by small segments of several polypeptide hormones, including glucagon.

Unfortunately, none of the modifications of glucagon has resulted in a molecule which acts as a specific high-affinity antagonist. However, several modifications of glucagon fragments have resulted in compounds with reduced intrinsic biological activity, which effectively block glucagon activation of adenylate cyclase (Hruby, 1982). Two of these compounds, (des-His[1]) (2-N-trinitrophenylserine, homo-Arg[12])-glucagon and (1-N-trinitrophenyl-His[1], homo-Arg[12])-glucagon, were relatively potent glucagon antagonists with K_i values of 38 and 6.9 nM, respectively.

V. ISOLATION AND PURIFICATION OF GLUCAGON RECEPTORS

Attempts to isolate and purify glucagon receptors have been hampered by the lack of high-affinity glucagon analogues and the lability of receptor activity following detergent treatment.

Early attempts to purify glucagon receptors relied on detergent solubilization of membrane binding sites prelabeled with saturating amounts of ^{125}I-labeled glucagon (Giorgio et al., 1974; Welton et al., 1977). However, detergent solubilization disrupts hormone–receptor complexes and greatly reduces receptor function. More recently, it has been possible to covalently label the receptor with photoaffinity analogues of glucagon. Using the biologically inactive analogue N^E-4-azido-2-nitrophenyl glucagon, Bregman and Levy (1977) reported the labeling of a 28,000-dalton protein from rat liver plasma membranes. Unfortunately, no attempt was made to determine the specificity of this labeling by preincubation of the membranes with native hormone, making it difficult to assess its biological significance. Recently, Johnson et al. (1981) have cross-linked ^{125}I-labeled glucagon to a 53,000-dalton protein with hydroxysuccinimidyl 4-azidobenzoate. Demoliou-

Mason and Epand (1982) have synthesized a photoaffinity analogue of glucagon by attaching a nitroazidophenyl group to Trp[25]. With the use of this photoanalogue, proteins in the range of 50,000–70,000 daltons were labeled (Demoliou-Mason and Epand, 1982).

Wright *et al.* (1984) have recently described a new photoaffinity analogue of glucagon, N^E-4-azidophenylamidinoglucagon (APA-glucagon), in which the photosensitive group is attached to Lys[12]. This analogue binds to rat liver plasma membranes and activates adenylate cyclase with an affinity and potency similar to that of native glucagon. Irradiation of labeled APA-glucagon bound to membranes results in the incorporation of label into a 55,000-dalton protein, as determined by SDS–PAGE (Fig. 5). Incorporation could be blocked by the addition of excess unlabeled glucagon or GTP (see Section II,B), suggesting that this 55,000-dalton protein is a component of the glucagon receptor. This

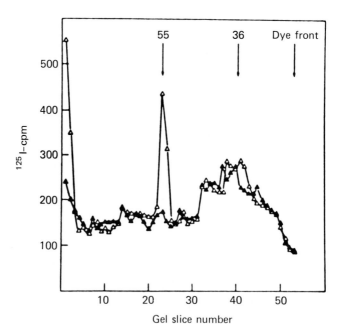

Fig. 5. SDS–PAGE of isoelectric focused purified glucagon receptor. The major peak of [125I]-labeled glucagon protein which migrated with a p*I* of 5.75–5.95 was eluted from isoelectric focusing gel and run on SDS–PAGE. Material was derived from labeled membranes incubated in the presence (▲) or absence (△) of 1 μM glucagon as described. From Horuk and Wright (1983).

protein was further demonstrated to be an acidic protein (Horuk and Wright, 1983) with an isoelectric point of 5.85. That the glucagon receptor is a glycoprotein is suggested by the studies of Herberg et al. (1984) in which CHAPS-solubilized liver membrane extracts incubated with ^{125}I-labeled glucagon were selectively absorbed onto wheat germ lectin Sepharose. The binding of ^{125}I-labeled glucagon extracts was sensitive to N-acetylglucosamine and markedly reduced by excess unlabeled glucagon.

The 55,000-dalton component just described (Horuk and Wright, 1983) represents the SDS denatured component of the glucagon receptor. The estimated size of the glucagon receptor isolated from liver membranes with Lubrol was 100,000 daltons (Rodbell, 1980), suggesting that the receptor is a multimeric protein composed of at least one 55,000-dalton subunit. Further evidence for this comes from target size analysis and radiation inactivation studies of the rat hepatic adenylate cyclase complex, which suggest that the glucagon receptor is a 120,000-dalton protein (Schlegel et al., 1979). It is open to speculation whether two 55,000-dalton proteins are needed to form an active receptor or whether another protein, which is not labeled by photoaffinity derivatives, is necessary.

Definitive proof that this 55,000-dalton protein can act as a physiological glucagon receptor must await its reconstitution into a cell line lacking glucagon receptors with concomitant restoration of biological activity (Schramm, 1979). The large-scale purification of glucagon receptors in an active form should be facilitated by the photoaffinity probes previously described.

VI. GLUCAGON INTERACTIONS
WITH HORMONE-SENSITIVE CELLS

In order to examine physiological consequences of the interaction of glucagon with its receptor, a good intact cell system is required. Freshly isolated hepatocytes or primary cultures of hepatocytes would provide the best systems to examine glucagon receptors in intact cells. Although conditions have been developed to maintain viability and cell function of isolated hepatocytes (Williams et al., 1977; Dickson and Pogson, 1977; Leffert and Paul, 1972; Armato et al., 1978; Freeman et al., 1981), in general such preparations are less than optimal due to heterogenous cell populations, cell damage during isolation, and loss of viability. Although differentiated functions can be maintained in primary cultures (Tanaka *et al.*, 1978), such cells eventually deteriorate and lose most of their differentiated functions. Consequently, several glucagon-responsive cell lines have been established.

A. Hepatocytes

The ability of glucagon and various analogues to stimulate glucose production in isolated hepatocytes was shown to parallel their effects on the stimulation of adenylate cyclase in hepatocyte membrane preparations (Hruby *et al.*, 1981).

Studies with isolated rat hepatocytes demonstrated that glucagon and cAMP stimulate phosphorylation of pyruvate kinase (type L), a key regulatory enzyme in glucose metabolism. This was not the case with chicken hepatocytes in which the type K isozyme of pyruvate kinase predominates, suggesting that only the type L isozyme is regulated by cAMP-dependent phosphorylation and dephosphorylation (Ochs and Harris, 1978). The phosphorylation of rat hepatocyte pyruvate kinase by glucagon was stimulated by glucocorticoids (Postle and Bloxham, 1982), while stimulation of cAMP accumulation by glucagon was inhibited by angiotensin II (Cardenas-Tanus *et al.*, 1982). The stimulation of the urea cycle (R. C. Lin *et al.*, 1982; Rabier *et al.*, 1982) and HMG-CoA reductase (Edwards *et al.*, 1979) by glucagon were also demonstrated in primary cultures of rat hepatocytes. Additionally, glucagon was shown to stimulate hepatocyte K^+ transport (Ihlenfeldt, 1981), Ruthenium Red-insensitive Ca^{2+} transport (Reinhart and Bygrave, 1981), and citrulline synthesis (Verhoeven *et al.*, 1982) in hepatocytes.

Isolated hepatocytes have also proved useful for developmental studies. Fetal hepatocytes are much less sensitive to glucagon compared to adult hepatocytes (Blazquez *et al.*, 1976). That the guanine nucleotide regulatory component (N_s) is present early in liver development (Sicard and Aprille, 1977) suggests that glucagon receptors appear later in fetal development. The effects of obesity and dietary state on insulin and glucagon receptors have also been investigated in hepatocytes (Broer *et al.*, 1977).

Glucagon binding studies with isolated hepatocytes have confirmed results obtained with purified rat liver membranes. Again, at least two classes of glucagon binding sites exist (Bonnevie-Nielsen and Tager, 1983); a small number (1%) with high affinity and a large number with low affinity for the hormone (Fig. 6). Furthermore, the relationship between binding and stimulation of cAMP accumulation was nonlinear; maximal activity occurred (Fig. 7) with occupancy of only a limited number of glucagon binding sites (Sonne *et al.*, 1978). Glucagon

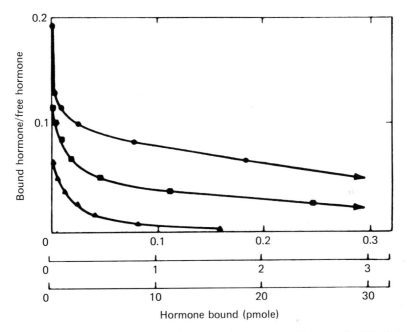

Fig. 6. Scatchard plot of glucagon binding to hepatocytes. The specific ^{125}I-labeled glucagon bound to hepatocytes was determined as described (Bonnevie-Nielsen and Tager, 1983). Data were plotted on the upper (●), middle (■) and lower (▲) abscissa scale. From Bonnievie-Nielsen and Tager (1983).

inhibition of glucose conversion to glycogen could be attributed to occupancy of the high-affinity receptor sites (Bonnevie-Nielsen and Tager, 1983).

The phenomenon of desensitization of hormone responsiveness has been described for several hormones, including glucagon. Infusion of glucagon into rats or incubation of liver slices with glucagon for 1–2 hr caused a 50% decrease in cAMP responsiveness to glucagon (DeRubertis and Craven, 1976). Further experiments utilizing liver membranes demonstrated a 100% desensitization to glucagon, accompanied by a 50% down-regulation of glucagon receptors, with no change in affinity for glucagon (Iyengar et al., 1980). Similar results were obtained with cultured rat hepatocytes (Santos and Blazquez, 1982). The desensitization of hepatocytes to glucagon was mimicked by cAMP and reversed by incubation with hormone-free media. Neither desensitization nor resensitization was dependent on protein or RNA synthesis (Gurr and Ruh, 1980).

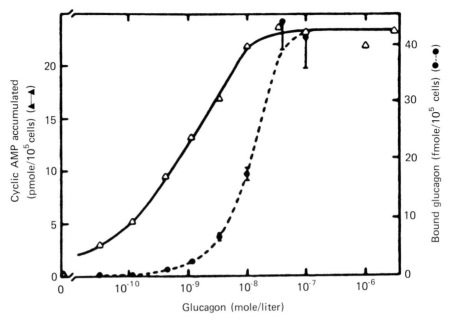

Fig. 7. Concentration dependence of glucagon binding and stimulation of adenylate cyclase. Isolated hepatocytes were incubated as described (Sonne *et al.*, 1978). At zero times a mixture of ^{125}I-labeled glucagon and unlabeled glucagon was added to give the indicated concentration. After 3 min of incubation at 37°C, 0.25 ml aliquots in triplicate were removed for determination of binding. Trichloroacetic acid was added to the rest of the sample for determination of accumulated cAMP. The cAMP data are expressed as cAMP accumulated due to the addition of glucagon (basal value subtracted). The binding data are expressed as amount of specifically bound glucagon. From Sonne *et al.* (1978).

B. Established Liver Cell Lines

Although primary cultures and freshly isolated hepatocytes provide the most physiological system for study of glucagon regulation of metabolic function, their use is limited to short-term studies due to the problems previously discussed. Bausher and Schaffer (1974) have established a cloned line of stable diploid rat hepatocytes (RL-PR-C). Like primary hepatocytes, these cells have two classes (Fig. 8) of glucagon binding sites (Reilly and Blecher, 1981) with high (K_d = 10 nM, 10,000 sites/cell) and low (K_d = 500 nM, 100,000 sites/cell) affinity. Again, the relationship between binding and stimulation of cAMP production was not linear (Reilly *et al.*, 1980). Maximal stimulation occurred with occupancy of less than 3% of the available glucagon binding sites. These cells have also proved useful for studying the phenomena of down-regulation and desensitization to insulin (Petersen *et al.*, 1978) and glucagon (Reilly *et al.*, 1980). In contrast to hepatocytes, desensitization of RL-PR-C cells was not accompanied by receptor down-regulation. However, like hepatocytes, desensitization was reversible and not affected by cycloheximide (Reilly *et al.*, 1980). Following ~300 population doublings, RL-PR-C cells spontaneously transform (Schaeffer and Polifka, 1975). These transformed cells exhibit a large increase (300,000 sites/cell) in the number of low-affinity binding sites (Fig. 8) and a total loss of detectable high-affinity binding and activation of adenylate cyclase by glucagon (Reilly and Blecher, 1981). Transformation of RL-

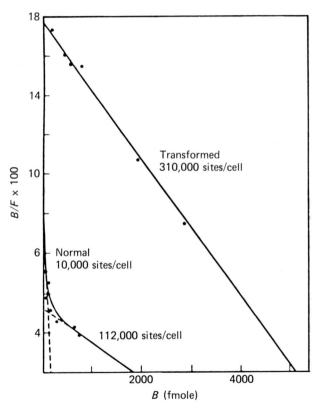

Fig. 8. Scatchard plots of glucagon binding to normal and transformed RL-PR-C hepatocytes. Cells were incubated with 1.5 nM ^{125}I-labeled glucagon and 0–10 μM unlabeled glucagon for 1 hr at 22°C. The plots were generated from specific binding data by the least-squares linear regression method. K_a, association constant, is the slope of the curve. B and F, bound and free glucagon, respectively. From Reilly and Blecher (1981).

PR-C cells is also accompanied by alterations in cAMP metabolism (Beckner *et al.*, 1980). Thus, this cell line represents a good model system to study the physiological functions of glucagon receptors as well as their relationship to differentiation and transformation.

Several stable cell clones (RLA) have been established by transformation of fetal liver cells with a temperature-sensitive mutant of simian virus 40 (Schlegel-Haueter *et al.*, 1980; Chou and Schlegel-Haueter, 1981). RLA255-4 and RLA209-15 have glucagon receptors coupled to adenylate cyclase. Although these lines selectively express some differentiated functions under temperature-controlled conditions, glucagon responsiveness was evident at both the permissive and restrictive temperatures (Schlegel-Haueter *et al.*, 1980).

Several lines of hepatoma cells have been established for use in carcinogenesis studies. Although some of these lines have functional adenylate cyclase, generally the coupling of glucagon to physiological function has been lost in such cell lines (Wimalasena *et al.*, 1980).

C. Kidney Cell Lines

The distal tubule of the kidney represents yet another physiological target of glucagon (Bailly *et al.*, 1980). A well-characterized kidney cell line is the Madin Darby canine kidney (MDCK) line (Gaush *et al.*, 1966). These cells retain normal kidney solute transport functions in culture (Cereijido *et al.*, 1978; McRoberts *et al.*, 1982), which are regulated by cAMP (Lever, 1979). MDCK cells respond to a number of polypeptide hormones which elevate cAMP, including glucagon (Rindler *et al.*, 1979). A line of MDCK cells transformed with Harvey murine sarcoma virus has been established (Scolnick *et al.*, 1976). In this transformed line, p21, a GTP-binding protein (Shih *et al.*, 1980), which is the oncogene product of Harvey sarcoma virus, can be detected associated with the plasma membrane (Willingham *et al.*, 1980). Transformation also results in the selective loss of glucagon receptors and responsiveness (M. C. Lin *et al.*, 1982) while all other hormone responses are still evident. Because glucagon responsiveness can be induced in this line (M. C. Lin *et al.*, 1982; Beckner *et al.*, 1985a), this system offers a convenient model to examine biochemical factors which regulate the development of glucagon receptors as well as receptor biogenesis.

The stimulation of cAMP formation by glucagon and a glucagon fragment in normal MDCK cells closely resembled that of liver and kidney membranes (Beckner *et al.*, 1985b). Responsiveness to native glucagon occurred over a concentration range of 0.1 nM–1 μM, and stimulation was half maximal at 10 nM. The glucagon fragment, $glucagon_{1-27}$, has a reduced potency, but full agonist activity. Interestingly, the glucagon responsiveness of the induced transformed cells was noticeably different from that of normal cells. The potency of native glucagon was reduced 10-fold (half-maximal activation at 100 nM) and responsiveness to glucagon occurred over a concentration range of 10 nM–1 μM. Furthermore, $glucagon_{1-27}$ had a reduced potency relative to the native hormone, as observed in normal cells, but this analogue was only a partial agonist in the induced system (Beckner et al., 1985b). These data suggest that, like hepatocytes, normal MDCK cells may have heterogenous populations of glucagon binding sites, with differing affinity for the hormone, while only the low-affinity site may be induced in transformed MDCK cells.

While the binding of glucagon to its receptor in liver membranes has been well characterized (see Section II), glucagon binding to cells in culture has been less well studied. However, conditions have been established to measure glucagon binding in MDCK cells (M. C. Lin et al., 1982, 1985) and RL-PR-C hepatocytes (Reilly and Blecher, 1981). Unfortunately, the low number of binding sites compared to liver and the high nonspecific background make rigorous Scatchard analysis difficult.

VII. CONCLUDING REMARKS

Clearly, our knowledge of glucagon receptors is incomplete. A complete purification of glucagon receptors is essential for understanding the structure as well as the organization of this receptor and its relationship to the other components of the adenylate cyclase system. It is possible to monitor glucagon receptors by binding of radioligands and adenylate cyclase activation. However, these approaches are not feasible for examining nonfunctional or denatured receptors. The development of antibodies against different functional domains of the receptor would provide a powerful tool for examining the glucagon-responsive adenylate cyclase system under more physiological conditions. Hopefully, receptor isolation will proceed at a faster pace with the use of recently developed affinity probes.

The regulation of cellular function by glucagon represents a very fertile area of research. Although not directly involved in proliferation, glucagon regulation of protein phosphorylation ultimately regulates many key enzymes of metabolism.

Glucagon receptors and metabolic regulation represent a very specialized differentiated function often lost during transformation. An understanding of the biochemical mechanisms underlying receptor loss as well as receptor biogenesis and processing during normal development and differentiation is crucial to our understanding of developmental biology. Continued examination of model systems such as MDCK and RL-PR-C cell lines should provide insight into the complex regulation of cellular function by glucagon.

ACKNOWLEDGMENTS

The authors are grateful to Dr. Martin Rodbell for helpful discussions during the preparation of this manuscript.

REFERENCES

Armato, U., Andreis, P. G., Draghi, E., and Mengato, L. (1978). Neonatal rat hepatocytes in primary culture free from density dependent regulation of growth. *In Vitro* **14,** 479–484.

Bailly, C., Imbert-Teboul, M., Chabardes, D., Hus-Chitharel, Montegut, M., Clique, A., and Morel, F. (1980). The distal nephron of rat kidney; a target site for glucagon. *Proc. Natl. Acad. Sci. U.S.A.* **77,** 3422–3424.

Bauscher, J., and Schaeffer, W. (1974). A diploid rat liver cell culture. *In Vitro* **9,** 286–293.

Beckner, S., Reilly, R., Martinez, A., and Blecher, M. (1980). Alterations of cAMP metabolism and hormone responsiveness of cloned differentiated rat liver cells (RL-PR-C) upon spontaneous transformation. *Exp. Cell Res.* **128,** 151–158.

Beckner, S., Darfler, F., and Lin, M. C. (1985a). Induction of glucagon responsiveness in transformed MDCK cells unresponsive to glucagon. *In* "Methods in Enzymology" **109,** 356–360.

Beckner, S., Wright, D., and Lin, M. C. (1985b). Glucagon responsiveness induced in transformed MDCK cells differs from that of parental. Submitted for publication.

Bender, J. L., and Neer, E. J. (1983). Properties of the adenylate cyclase catalytic unit from caudate nucleus. *J. Biol. Chem.* **258,** 2432–2439.

Bird, S., and Maguire, M. (1978). The agonist-specific effect of magnesium ion on binding by beta-adrenergic receptors in S49 lymphoma cells. *J. Biol. Chem.* **253,** 8826–8878.

Blazquez, E., Rubalcava, B., Montesano, R., Orci, L., and Unger, R. (1976). Development of insulin and glucagon binding and the adenylate cyclase response in liver membranes of the prenatal, postnatal, and adult rat; evidence of glucagon "resistance." *Endocrinology (Baltimore)* **98,** 1014–1023.

Blundell, T. L. (1983). The conformation of glucagon. *Handb. Exp. Pharmacol.* **66,** 37–56.

Bokoch, G. M., Katada, T., Northup, J. K., Ui, M., and Gilman, A. G. (1984). Purification and properties of the inhibitory guanine nucleotide binding regulatory component of adenylate cyclase. *J. Biol. Chem.* **259,** 3560–3567.

Bonnevie-Nielsen, V., and Tager, H. (1983). Glucagon receptors on isolated hepatocytes and hepatocyte membrane vesicles. *J. Biol. Chem.* **258,** 11313–11320.

Bornet, H., and Edelhoch, H. (1971). Polypeptide hormone interaction. *J. Biol. Chem.* **246,** 1785–1792.

Brandt, D. R., Asano, T., Pedersen, S. E., and Ross, E. M. (1983). Reconstitution of catecholamine-stimulated guanosinetriphosphatase activity. *Biochemistry* **22**, 4357–4362.

Bregman, M. D., and Levy, D. (1977). Labelling of glucagon binding components in hepatocyte plasma membrane. *Biochem. Biophys. Res. Commun.* **78**, 584–590.

Broer, Y., Freychet, P., and Rosselin, G. (1977). Insulin and glucagon–receptor interactions in the genetically obese Zucker rat: Studies of hormone binding and glucagon-stimulated cyclic AMP levels in isolated hepatocytes. *Endocrinology (Baltimore)* **101**, 236–249.

Bromer, W. W. (1983). Chemical characteristics of glucagon. *Handb. Exp. Pharmacol.* **66**, 1–22.

Bromer, W. W., Boucher, M. E., and Patterson, J. M. (1973). Glucagon structure and function: II. Increased activity of iodoglucagon. *Biochem. Biophys. Res. Commun.* **53**, 134–139.

Cardenas-Tanus, R. J., Huerta-Bahena, J., and Garcia-Sainz, J. A. (1982). Angiotensin II inhibits the accumulation of cyclic AMP produced by glucagon but not its metabolic effects. *FEBS Lett.* **143**, 1–4.

Carrey, D., and Epand, R. M. (1982). The role of nonspecific hydrophobic interactions in the biological activity of N-acyl derivatives of glucagon. *J. Biol. Chem.* **257**, 10624–10630.

Cereijido, M., Robbins, E. S., Dolan, W. J., Rotunno, C. A., and Sabatini, D. D. (1978). Polarized monolayers formed by epithelial cells on a permeable and translucent support. *J. Cell Biol.* **77**, 853–860.

Chou, J. Y., and Schlegel-Haueter, S. E. (1981). Study of liver differentiation *in vitro*. *J. Cell Biol.* **89**, 216–222.

Citri, Y., and Schramm, M. (1980). Resolution, reconstitution and kinetics of the primary action of a hormone receptor. *Nature (London)* **287**, 297–300.

Condina, J., Hildebrandt, J. D., Sekura, R. D., Birnbaumer, M., Bryan, J., Manclark, C. R., Iyengar, R., and Birnbaumer, L. (1984). N_s and N_i, the stimulatory and inhibitory regulatory components of adenylyl cyclases. Purification of the human erythrocyte proteins without the use of activating regulatory ligands. *J. Biol. Chem.* **259**, 5871–5886.

Cooper, D. M. F. (1983). Receptor-mediated stimulation and inhibition of adenylate cyclase. *Curr. Top. Membr. Transp.* **18**, 67–84.

Demoliou-Mason, C., and Epand, R. M. (1980). Binding of a glucagon photo affinity label to rat liver plasma membranes and its effect on adenylate cyclase activity. *Biochemistry* **19**, 4539–4546.

Demoliou-Mason, C., and Epand, R. M. (1982). Identification of the glucagon receptor by covalent labeling with a radiolabeled photoreactive glucagon analogue. *Biochemistry* **21**, 1996–2004.

DeRubertis, F. R., and Craven, P. (1976). Reduced sensitivity of the hepatic adenylate cyclase-cAMP system to glucagon during sustained hormonal stimulation. *J. Clin. Invest.* **57**, 435–443.

Desbuquois, B. (1975). Iodoglucagon, preparation and characteristics. *Eur. J. Biochem.* **53**, 569–580.

Dickson, A. J., and Pogson, C. I. (1977). The metabolic integrity of hepatocytes in sustained incubations. *FEBS Lett.* **83**, 27–32.

Dipple, I., and Houslay, M. D. (1978). The activity of glucagon-stimulated adenylate cyclase from rat liver plasma membranes is modulated by the fluidity of its lipid environment. *Biochem. J.* **174**, 179–190.

Edwards, P. A., Lemongello, D., and Fogelman, A. M. (1979). The effect of glucagon norepinephrine and dibutyryl cyclic AMP on cholesterol efflux and on the activity of 3-hydroxy-3-methylglutaryl-CoA reductase in rat hepatocytes. *J. Lipid Res.* **20**, 2–7.

Epand, R. M. (1983). Relationships among several different nonhomologous polypeptide hormones. *Mol. Cell Biochem.* **57**, 41–47.

Epand, R. M., and Cote,T. E. (1976). Conformational and biological properties of a covalently linked dimer of glucagon. *Biochim. Biophys. Acta* **453**, 365–373.

Epand, R. M., Rosselin, G., Hui Bon Hoa, D., Cote, T. E., and Laburthe, M. (1981). Structural requirements for glucagon receptor binding and activation of adenylate cyclase in liver. *J. Biol. Chem.* **256**, 1128–1132.

Exton, J. H. (1981). The effects of glucagon on hepatic glycogen metabolism and gluconeogenesis. *In* "Current Endocrinology, Glucagon" (R. H. Unger and L. Orci, eds.), pp. 196–219. Am. Elsevier, New York.

Felts, P. W., Ferguson, M. E. C., Hagey, K. A., Stitt, E. S., and Mitchell, W. M. (1970). Studies on the structure–function relationships of glucagon. *Diabetologia* **6**, 44–45.

Freeman, A. E., Engvall, E., Hirata, K., Yoshida, Y., Kottel, R. H., Hilborn, V., and Ruoslahti, E. (1981). Differentiation of fetal liver cells *in vitro. Proc. Natl. Acad. Sci. U.S.A.* **78**, 3659–3663.

Gaush, C. R., Hard, W. L., and Smith, T. F. (1966). Characterization of an established line of canine kidney cells. *Proc. Soc. Exp. Biol. Med.* **122**, 931–935.

Gilman, A. G. (1984). G proteins and dual control of adenylate cyclase. *Cell (Cambridge, Mass.)* **36**, 577–579.

Giorgio, N., Johnson, C. B., and Blecher, M. (1974). Hormone receptors: Properties of glucagon-binding proteins isolated from liver plasma membranes. *J. Biol. Chem.* **249**, 428–437.

Gratzer, W. B., Creeth, J. M., and Beaven, G. H. (1972). Presence of trimers in glucagon solution. *Eur. J. Biochem.* **31**, 505–509.

Gurr, J. A., and Ruh, T. A. (1980). Desensitization of primary cultures of adult rat liver parenchymal cells to stimulation of cAMP production by glucagon and epinephrine. *Endocrinology (Baltimore)* **107**, 1309–1319.

Herberg, J. T., Condina, J., Rich, K. A., Rojas, F. J. and Iyengar, R. (1984). The hepatic glucagon receptor: Solubilization, characterization, and development of an affinity adsorption assay for the soluble receptor. *J. Biol. Chem.* **259**, 9285–9294.

Hildebrandt, J. D., Condina, J., Risinger, R., and Birnbaumer, L. (1984). Identification of the gamma subunit associated with the adenylyl cyclase regulatory protein N_s and N_i. *J. Biol. Chem.* **259**, 2039–2042.

Horuk, R., and Wright, D. E. (1983). Partial purification and characterization of the glucagon receptor. *FEBS Lett.* **155**, 213–217.

Houslay, M. D., and Gordon, L. M. (1983). The activity of adenylate cyclase is regulated by the nature of its lipid environment. *Curr. Top. Membr. Transp.* **8**, 179–231.

Houslay, M. D., Ellory, J. C., Smith, G. A., Hesketh, T. R., Stein, J. M., Warren, G. B., and Metcalfe, J. C. (1977). Exchange of partners in glucagon receptor–adenylate cyclase complexes: Physical evidence for the independent mobile receptor model. *Biochim. Biophys. Acta* **467**, 208–219.

Hruby, V. J. (1982). Structure–conformation–activity studies of glucagon and semi-synthetic glucagon analogs. *Mol. Cell. Biochem.* **44**, 49–64.

Hruby, V. J., Agarwal, N. S., Griffen, A., Bregman, M. D., Nugent, C. A., and Brendel, K. (1981). Glucagon structure–function relationships using isolated rat hepatocytes. *Biochim. Biophys. Acta* **674**, 383–390.

Ihlenfeldt, M. J. A. (1981). Stimulation of Rb^+ transport by glucagon in isolated rat hepatocytes. *J. Biol. Chem.* **256**, 2213–2218.

Iyengar, R., Swartz, T. L., and Birnbaumer, L. (1979). Coupling of glucagon receptor to adenylyl cyclase: Requirement for receptor-related guanyl nucleotide binding site for coupling of receptor to the enzyme. *J. Biol. Chem.* **254**, 1119–1123.

Iyengar, R., Mintz, P. W., Swartz, T. L., and Birnbaumer, L. (1980). Divalent cation-induced desensitization of glucagon-stimulable adenylate cyclase in rat liver plasma membranes. *J. Biol. Chem.* **255**, 11875–11882.

Johnson, C. L., MacAndrew, V. I. and Pilch, P. F. (1981). Identification of the glucagon receptor in rat liver plasma membrane by photoaffinity cross linking. *Proc. Natl. Sci. U.S.A.* **78**, 875–878.

Katada, T., Bokoch, G., Northup, J. K., Ui, M., and Gilman, A. G. (1984a). The inhibitory guanine nucleotide binding regulatory component of adenylate cyclase. Properties and function of the purified protein. *J. Biol. Chem.* **259**, 3568–3577.

Katada, T., Northup, J. K., Bokoch, G. M., Ui, M., and Gilman, A. G. (1984b). The inhibitory guanine nucleotide binding regulatory component of adenylate cyclase. Subunit dissociation and guanine nucleotide dependent hormonal inhibition. *J. Biol. Chem.* **259**, 3578–3585.

Katz, A. L., and Lindheimer, M. D. (1977). Actions of hormones on the kidney. *Annu. Rev. Physiol.* **39**, 97–134.

Kolanowski, J. (1983). Influence of glucagon on water and electrolyte metabolism. *Handb. Exp. Pharmacol.* **66**, 525–536.

Lad, P. M., Welton, A. F., and Rodbell, M. (1977). Evidence for distinct guanine nucleotide sites in the regulation of glucagon receptor and of adenylate cyclase activity. *J. Biol. Chem.* **252**, 5942–5946.

Leffert, H. L., and Paul, D. (1972). Studies on primary cultures of differentiated fetal liver cells. *J. Cell Biol.* **52**, 559–568.

Leffert, H. L., Koch, K. S., Lad, P. J., de Hemptinne, B., and Skelly, H. (1983). Glucagon and liver regeneration. *Handb. Exp. Pharmacol.* **66**, 453–484.

Lever, J. (1979). Inducers of mammalian cell differentiation stimulate dome formation in a differentiated kidney epithelial cell line (MDCK). *Proc. Natl. Acad. Sci. U.S.A.* **76**, 1323–1327.

Lin, M. C., Wright, D. E., Hruby, V. J., and Rodbell, M. (1975). Structure–function relationships in glucagon: Properties of highly purified des-His-, monoiodo- and [des-Asn[28], Thr[29]] homoserine lactone [27]-glucagon. *Biochemistry* **14**, 1559–1563.

Lin, M. C., Nicosia, S., and Rodbell, M. (1976). Effects of iodination of tyrosyl residues on the binding and action of glucagon at its receptor. *Biochemistry* **15**, 4537–4546.

Lin, M. C., Nicosia, S., Lad, P. M., and Rodbell, M. (1977). Effects of GTP on binding of [³H]glucagon to receptors in rat hepatic plasma membranes. *J. Biol. Chem.* **252**, 2790–2792.

Lin, M. C., Koh, S. M., Dykman, D. D., Beckner, S. K., and Shih, T. Y. (1982). Loss and restoration of glucagon receptors and responsiveness in a transformed kidney cell line. *Exp. Cell Res.* **142**, 181–189.

Lin, M. C., Beckner, S. K., and Darfler, F. J. (1985). Characterization of hormone-sensitive Madin-Darby canine kidney cells. *In* "Methods in Enzymology" **109**, 360–365.

Lin, R. C., Snodgrass, P. J., and Rabier, D. (1982). Induction of urea cycle enzymes by glucagon and dexamethasone in monolayer cultures of adult rat hepatocytes. *J. Biol. Chem.* **257**, 5061–5067.

McRoberts, J. A., Erlinger, S., Rindler, M. J., and Saier, M. H. (1982). Furosemide-sensitive salt transport in the Madin-Darby canine kidney cell line. *J. Biol. Chem.* **257**, 2260–2266.

Markusson, J., Frandson, E., Harding, L. G., and Sundby, F. (1972). Crystallization, amino acid composition and immunology. *Horm. Metab. Res.* **4**, 360–363.

Miyazaki, K., Masui, H., and Sato, G. H. (1982). Control factors for keratinization of human bronchogenic epidermoid carcinoma cells. *In* "Growth of Cells in Hormonally Defined Media" (G. H. Sato, A. B. Pardee, and D. A. Sirbasku, eds.) Vol. 9, pp. 657–664. Cold Spring Harbor, New York.

Murakami, H., Masui, H., Sato, G., and Raschke, W. C. (1981). Growth of mouse plasmacytoma cells in serum-free, hormone-supplemented medium: Procedure for the determination of hormone and growth factor requirements for cell growth. *Anal. Biochem.* **114,** 422–428.

Neer, E. J. (1977). Solubilization and characterization of adenylyl cyclase: *In* "Receptors and Hormone Action" (B. W. O'Malley and L. Birnbaumer, eds.). Vol. 1, pp. 463–484. Academic Press, New York.

Neer, E. J., Echeverria, D., and Knox, S. (1980). Increase in the size of soluble brain adenylate cyclase with activation by Gpp(NH)p. *J. Biol. Chem.* **255,** 9782–9789.

Northup, J. K., Sternweis, P. C., Smigel, M. D., Schleifer, L. S., Ross, E. M., and Gilman, A. G. (1980). Purification of the regulatory component of adenylate cyclase. *Proc. Natl. Acad. Sci. U.S.A.* **77,** 6516–6520.

Northup, J. K., Smigel, M. D., and Gilman, A. G. (1982). The guanine nucleotide activating site of the regulatory component of adenylate cyclase. Identification by ligand binding. *J. Biol. Chem.* **257,** 11416–11423.

Northup, J. K., Sternweis, P. C., and Gilman, A. G. (1983a). The subunits of the stimulatory regulatory component of adenylate cyclase. Resolution, activity and properties of the 35,000-dalton (β) subunit. *J. Biol. Chem.* **258,** 11361–11368.

Northup, J. K., Smigel, M. D., Sternweis, P. C., and Gilman, A. G. (1983b). The subunits of the stimulatory regulatory component of adenylate cyclase. Resolution of the activated 45,000-dalton (α) subunit. *J. Biol. Chem.* **258,** 11369–11376.

Ochs, R. S., and Harris, R. A. (1978). Studies on the relationship between glycolysis, lipogenesis, gluconeogenesis and pyruvate kinase activity of rat and chicken hepatocytes. *Arch. Biochem. Biophys.* **190,** 193–201.

Patterson, J. M., and Bromer, W. W. (1973). Glucagon structure and function. Preparation and characterization of nitroglucagon and aminoglucagon. *J. Biol. Chem.* **248,** 8337–8342.

Pedersen, S. E., and Ross, E. M. (1982). Functional reconstitution of β-adrenergic receptors and the stimulatory GTP-binding protein of adenylate cyclase. *Proc. Natl. Acad. Sci. U.S.A.* **79,** 7228–7232.

Petersen, B., Beckner, S. K., and Blecher, M. (1978). Hormone receptors. Characterization of insulin receptors in a new line of cloned neonatal rat hepatocytes. *Biochim. Biophys. Acta* **542,** 470–485.

Pohl, S. L., Birnbaumer, L., and Rodbell, M. (1971a). The glucagon-sensitive adenyl cyclase system in plasma membranes of rat liver. Properties. *J. Biol. Chem.* **246,** 1849–1856.

Pohl, S. L., Krans, H. M. J., Kozyreff, V., Birnbaumer, L., and Rodbell, M. (1971b). The glucagon-sensitive adenyl cyclase system in plasma membranes of rat liver: Evidence for a role of membrane lipids. *J. Biol. Chem.* **246,** 4447–4454.

Pollock, H. G., and Kimmel, J. (1975). Chicken glucagon–isolation and amino acid sequence. *J. Biol. Chem.* **250,** 9377–9380.

Postle, A. D., and Bloxham, D. P. (1982). Glucocorticoid hormones have a permissive role in the phosphorylation of L-type pyruvate kinase by glucagon. *Eur. J. Biochem.* **124,** 103–108.

Rabier, D., Briand, P., Petit, F., Parvy, P., Kamoun, P., and Cathelineau, L. (1982). Acute effects of glucagon on citrulline biosynthesis. *Biochem. J.* **206,** 627–631.

Reilly, T. M., and Blecher, M. (1981). Restoration of glucagon responsiveness in spontaneously transformed rat hepatocytes (RL-PR-C) by fusion with normal progenitor cells and rat liver plasma membranes. *Proc. Natl. Acad. Sci. U.S.A.* **78**, 182–186.

Reilly, T. M., Beckner, S. K., and Blecher, M. (1980). Uncoupling of the glucagon receptor–adenylate cyclase system by glucagon in cloned differentiated rat hepatocytes. *J. Recept. Res.* **1**, 277–311.

Reinhart, P. H., and Bygrave, F. L. (1981). Glucagon stimulation of Ruthenium Red-insensitive calcium ion transport in developing rat liver. *Biochem. J.* **194**, 541–549.

Rindler, M. J., Chuman, L. M., Shaffer, L., and Saier, M. H. (1979). Retention of differentiated properties in an established dog kidney epithelial cell line (MDCK). *J. Cell Biol.* **81**, 635–648.

Rodbell, M. (1972). The glucagon sensitive adenylate cyclase system. *In* "Glucagon: Molecular Physiology, Clinical and Therapeutic Implications" (P. J. Lefebvre and R. H. Unger, eds.), pp. 61–76. Pergamon, Oxford.

Rodbell, M. (1980). The role of hormone receptors and GTP-regulatory proteins in membrane transduction. *Nature (London)* **284**, 17–22.

Rodbell, M. (1981). The actions of glucagon on the adenylate cyclase system. *In* "Current Endocrinology, Glucagon" (R. H. Unger and L. Orci, eds.), pp. 178–193. Am. Elsevier, New York.

Rodbell, M., Krans, H. M., Pohl, S. L., and Birnbaumer, L. (1971a). The glucagon sensitive adenyl cyclase system in plasma membranes of rat liver. Binding of glucagon: Method of assay and specificity. *J. Biol. Chem.* **246**, 1861–1871.

Rodbell, M., Krans, H. M. J., Pohl, S. L., and Birnbaumer, L. (1971b). The glucagon-sensitive adenyl cyclase system in plasma membranes of rat liver: Effects of guanyl nucleotides on binding of [^{125}I]glucagon. *J. Biol. Chem.* **246**, 1872–1876.

Rodbell, M., Lin, M. C., and Salomon, Y. (1974). Evidence for interdependent action of glucagon and nucleotides on the hepatic adenylate cyclase system. *J. Biol. Chem.* **249**, 59–65.

Rodbell, M., Lin, M. C., Salomon, Y., Londos, C., Harwood, J. P., Martin, B. R., Rendell, M., and Berman, M. (1975). Role of adenine and guanine nucleotides in the activity and response of adenylate cyclase systems to hormones: Evidence for multisite transition states. *Adv. Cyclic Nucleotide Res.* **5**, 3–29.

Rojas, F. J., Swartz, T. L., Iyengar, R., Garber, A. J., and Birnbaumer, L. (1983). Monoiodoglucagon: Synthesis, purification by high pressure liquid chromatography, and characteristics as a receptor probe. *Endocrinology (Baltimore)* **133**, 711–719.

Ross, E. M. (1981). Physical separation of the catalytic and regulatory proteins of hepatic adenylate cyclase. *J. Biol. Chem.* **256**, 1949–1953.

Ross, E. M., Pedersen, S. E., and Floria, V. A. (1983). Hormone-sensitive adenylate cyclase: Identity, function, and regulation of the protein components. *Curr. Top. Membr. Transp.* **18**, 109–142.

Rubalcava, B., and Rodbell, M. (1973). The role of acidic phospholipids in glucagon action on rat liver adenylate cyclase. *J. Biol. Chem.* **248**, 3831–3837.

Rubalcava, B., Grajales, M. O., Cerbon, J., and Pliego, J. A. (1983). The role of surface charge and hydrophobic interaction in the activation of rat liver adenylate cyclase. *Biochim. Biophys. Acta* **759**, 243–249.

Santos, A., and Blazquez, E. (1982). Direct evidence of a glucagon-dependent regulation of the concentration of glucagon receptors in the liver. *Eur. J. Biochem.* **121**, 671–677.

Schaeffer, W. I., and Polifka, M. D. (1975). A diploid rat liver cell culture. *Exp. Cell Res.* **95**, 167–175.

Schlegel, W., Kempner, E. S., and Rodbell, M. (1979). Activation of adenylate cyclase in hepatic membranes involves interactions of the catalytic unit with multimeric complexes of regulatory proteins. *J. Biol. Chem.* **254**, 5168–5176.

Schlegel-Haueter, S. E., Schlegel, W., and Chou, J. Y. (1980). Establishment of a fetal rat liver cell line that retains differentiated liver functions. *Proc. Natl. Acad. Sci. U.S.A.* **77**, 2731–2734.

Schneider, A. B., and Edelhoch, H. (1972). Polypeptide hormone interaction. II. Glucagon binding to lysolecithin. *J. Biol. Chem.* **247**, 4986–4991.

Schramm, M. (1979). Transfer of glucagon receptor from liver membranes to a foreign adenylate cyclase by a membrane fusion procedure. *Proc. Natl. Acad. Sci. U.S.A.* **76**, 1174–1178.

Scolnick, E. M., Williams, D., Maryak, J., Vass, W., Goldberg, R. J., and Parks, W. P. (1976). Type C particle-positive and type C particle-negative rat cell lines: Characterization of the coding capacity of endogenous sarcoma virus-specific RNA. *J. Virol.* **20**, 570–582.

Seamon, K. B., Padgett, W., and Daly, J. W. (1981). Forskolin: Unique diterpene activator of adenylate cyclase in membranes and intact cells. *Proc. Natl. Acad. Sci. U.S.A.* **78**, 3363–3367.

Shih, T. Y., Papageorge, A. G., Stokes, P. E., Weeks, M. O., and Scolnick, E. M. (1980). Guanine nucleotide-binding and autophosphorylating activities associated with the p21 protein of Harvey murine sarcoma virus. *Nature (London)* **287**, 686–691.

Sicard, R. E., and Aprille, J. R. (1977). Adenylate cyclase and cyclic nucleotide phosphodiesterases in the developing rat liver. *Biochim. Biophys. Acta* **500**, 235–245.

Sonne, E., Berg, T., and Christoffersen, T. (1978). Binding of ^{125}I-labeled glucagon and glucagon-stimulated accumulation of adenosine 3':5'-monophosphate in isolated intact rat hepatocytes. *J. Biol. Chem.* **253**, 3203–3210.

Sternweis, P. C., and Gilman, A. G. G. (1982). Aluminum: A requirement for activation of the regulatory component by fluoride. *Proc. Natl. Acad. Sci. U.S.A.* **79**, 4888–4891.

Sundby, F. (1976). Species variation in the primary structure of glucagon. *Metab. Clin. Exp.* **25**, Suppl. 1, 1319–1321.

Sutherland, E. W., Oye, I., and Butcher, R. W. (1965). The action of epinephrine and the role of the adenyl cyclase system in hormone action. *Recent Prog. Horm. Res.* **21**, 623–646.

Sutton, R., and Pollak, J. K. (1980). Hormone-initiated maturation of rat liver mitochondria after birth. *Biochem. J.* **186**, 361–367.

Swillens, S., and Dumont, J. E. (1980). A unifying model of current concepts and data on adenylate cyclase activation by β-adrenergic agonists. *Life Sci.* **27**, 1013–1028.

Tanaka, K., Sato, M., Tomita, Y., and Ichihara, A. (1978). Biochemical studies on liver functions in primary cultured hepatocytes of adult rats. *J. Biochem. (Tokyo)* **84**, 937–946.

Tolkovsky, A. M., and Martin, B. R. (1982). The glucagon receptor from liver can be functionally fused to caudate nucleus adenylate cyclase. *FEBS Lett.* **150**, 337–342.

Trakatellis, A. C., Tada, K., Yamaji, K., and Gardiki-Kouidai (1975). Isolation and partial characterization of angler fish proglucagon. *Biochemistry* **14**, 1508–1512.

Vaughan, M., and Moss, J. (1978). Mechanism of action of choleragen. *J. Supramol. Struct.* **8**, 473–488.

Verhoeven, A. J., Hensgens, H. E. S. J., Meijer, A. J., and Tager, J. M. (1982). On the nature of the stimulation by glucagon of citrulline synthesis in rat liver mitochondria. *FEBS Lett.* **140**, 270–275.

Wagman, M. E., Dobson, C. M., and Karplus, M. (1980). Proton NMR studies of the association and folding of glucagon in solution. *FEBS Lett.* **119**, 265–270.

Welton, A. F., Lad, P. M., Newby, A. C., Yamamura, H., Nicosia, S., and Rodbell, M. (1977). *J. Biol. Chem.* **252,** 5947–5950.

Whetton, A. D., Gordon, L. M., and Houslay, M. D. (1983). Elevated membrane cholesterol concentrations inhibit glucagon-stimulated adenylate cyclase. *Biochem. J.* **210,** 437–449.

Williams, G. M., Bermudez, E., and Scaramuzzino, D. (1977). Rat hepatocyte primary cell cultures. *In Vitro* **13,** 809–817.

Williams, L. T., Mullikin, D., and Lefkowitz, R. J. (1978). Magnesium dependence of agonist binding to adenylate cyclase-coupled hormone receptors. *J. Biol. Chem.* **253,** 2984–2989.

Willingham, M. C., Pastan, I., Shih, T. Y., and Scolnick, E. M. (1980). Localization of the *src* gene product of the Harvey strain of MSV to plasma membrane of transformed cells by electron microscopic immunocytochemistry. *Cell (Cambridge, Mass.)* **19,** 1005–1014.

Wimalasena, J., Leichtling, B. H., Lewis, E. J., Langan, T. A., and Wicks, W. D. (1980). Coordinate regulation of adenylate cyclase, protein kinase, and specific enzyme synthesis by cholera toxin in hormonally unresponsive hepatoma cells. *Arch. Biochem. Biophys.* **205,** 595–605.

Wright, D. E., and Rodbell, M. (1980a). Properties of amidinated glucagons. *Eur. J. Biochem.* **111,** 11–16.

Wright, D. E., and Rodbell, M. (1980b). Preparation of 2-thioltryptophan-glucagon and (tryptophan-S-glucagon). *J. Biol. Chem.* **255,** 10884–10887.

Wright, D. E., Hruby, V. J., and Rodbell, M. (1978). A reassessment of structure–function relationships in glucagon. *J. Biol. Chem.* **253,** 6338–6341.

Wright, D. E., Horuk, R., and Rodbell, M. (1984). Photoaffinity labeling of the glucagon receptor with a new glucagon analog. *Eur. J. Biochem.* **141,** 63–67.

10

The Platelet-Derived Growth Factor Receptor

DANIEL F. BOWEN-POPE, RONALD A. SEIFERT,
AND RUSSELL ROSS
Department of Pathology
School of Medicine
University of Washington
Seattle, Washington

I. INTRODUCTION

Platelet-derived growth factor (PDGF) is a mitogenic polypeptide found within the α granules of blood platelets and released from the α granules when the platelets are activated (see Chapter 6). PDGF has been purified from washed human platelets (Heldin *et al.*, 1979a; An-

281

toniades *et al.*, 1979) and from platelet-rich plasma (Deuel *et al.*, 1981; Raines and Ross, 1982). Material from these sources is a basic (p$I \simeq 10$) 30,000-dalton protein composed of two subunits of about 14,000 and 17,000 daltons joined by disulfide bonds. *In vitro*, PDGF is released from platelets during the preparation of the whole blood serum which is used as a standard supplement to defined nutrient media for culturing animal cells. The magnitude of the contribution of PDGF in serum to the total mitogenic potency of the serum depends on the test cells and on the source of the serum—as will be discussed, not all cell types express PDGF receptors. *In vivo*, it is likely that PDGF is released from platelets at sites of vascular damage and that it contributes toward the cell proliferation and connective tissue formation seen in healing wounds and in arteriosclerotic lesions (Ross and Glomset, 1976).

In addition to its release from blood platelets, PDGF or related molecules able to use the PDGF receptor may be secreted by specific cell types as a normal aspect of their differentiated phenotype (DiCorleto and Bowen-Pope, 1983; Gudas *et al.*, 1983; Seifert *et al.*, 1985; Rizzino and Bowen-Pope, 1985) or as a consequence of oncogenic transformation (Heldin *et al.*, 1980; Dicker *et al.*, 1981; Nister *et al.*, 1982; Deuel *et al.*, 1983; Bowen-Pope *et al.*, 1984a). The role that PDGF plays in various physiological processes will thus depend on the distribution of PDGF-responsive cell types relative to sites of platelet activation and to sites of cellular synthesis and secretion of PDGF. The common denominator for these processes, in the context of this chapter, will be the expression by cells of functional PDGF receptors.

II. SPECIFIC AND NONSPECIFIC BINDING

Binding interactions between receptors and ligands involve the same physical forces involved in the interactions between any two molecular entities: forces of ionic attraction, hydrophobic forces, van der Waals forces, etc. (Grossberg, 1977). In order to promote interactions between ligand and receptor at the very low ligand concentrations usually achieved in hormone systems, the binding forces are usually organized such that the overall affinity of the interaction between receptor and ligand is much greater than that of the interaction between ligand and other components of the system. On the other hand, the number of nonreceptor binding sites in a system can vastly exceed the number of receptors, favoring significant ligand binding to these sites merely through mass action. Distinguishing between these binding interactions can be difficult.

PDGF has molecular properties which give it a tendency to adhere "nonspecifically" to many surfaces. It is highly positively charged (p$I \simeq$ 10) (Heldin et al., 1977; Ross et al., 1979; Antoniades et al., 1979; Deuel et al., 1981) at physiological pH as well as possessing at least some hydrophobic regions (Heldin et al., 1979b; Ross et al., 1979). Binding of ^{125}I-PDGF to laboratory equipment can be greatly reduced by coating the surfaces with Surfasil (Pierce). Tissue culture surfaces cannot be treated in this fashion, however, since the treatment interferes with the ability of cells to attach and spread. Nonspecific binding can be greatly reduced by the presence of other soluble proteins. If low concentrations of ^{125}I-PDGF in medium containing 0.25% bovine serum albumin (BSA) are incubated in plastic cell culture dishes (without cells), about 11% of the total input ^{125}I-PDGF will bind to the plastic culture surface. About 50% of this nonreceptor-mediated binding has one of the properties of "specific" binding, since it competes with high concentrations of unlabeled PDGF. The amount of dish-bound ^{125}I-PDGF is reduced about fivefold if the plastic culture dish has been preincubated for several days in culture medium containing calf serum. During this period, most of the dish sites capable of binding ^{125}I-PDGF seem to become blocked by serum proteins. Since most cultures to which ^{125}I-PDGF binding is to be measured have included a period of cell proliferation in medium containing serum, the contribution of plastic-bound ^{125}I-PDGF to total bound ^{125}I-PDGF is not unmanageable. The contribution of plastic-bound ^{125}I-PDGF can be minimized further by selectively solubilizing cell-bound ^{125}I-PDGF using a solution of 1% nonionic detergent (Triton X-100) rather than NaOH. The nonionic detergent solubilizes the cell-bound ^{125}I-PDGF without solubilizing the plastic-bound ^{125}I-PDGF (Heldin et al., 1981; Bowen-Pope and Ross, 1982, 1985). With the use of this extraction procedure, nonspecific binding as a percentage of total input ^{125}I-PDGF is relatively constant at about 0.7% (Bowen-Pope and Ross, 1985). Therefore, depending on the number of specific PDGF receptors present, the nonspecific binding as a percentage of total binding varies from about 2% for confluent cultures of 3T3 cells to 15–20% for subconfluent cultures of diploid human fibroblasts.

The nonspecific binding of PDGF to surfaces and/or molecules other than the PDGF receptor is of considerable interest apart from its effects in complicating analysis of PDGF receptors. The PDGF that binds to the tissue culture plastic and/or to extracellular components could constitute a substantial reserve of PDGF if it is, or could be, biologically active. PDGF-responsive cells which are plated onto culture dishes that have been preincubated with pure PDGF grow considerably more rapidly in medium containing low mitogen levels than do cells plated onto dishes that have not been pretreated with PDGF.

This effect is observed both on the rate of cell proliferation (our unpublished observations) and on the percentage of labeled nuclei (Smith *et al.*, 1982). The PDGF which is adsorbed to the culture dish is resistant to removal by brief incubation with 6 M urea, methanol and Triton X-100 (Smith *et al.*, 1982). It has been proposed that the adsorbed PDGF is thus so tightly bound that it remains bound to the plastic during incubation with cells and is biologically active in this state. Although this is possible, it is also possible that the PDGF may desorb from the plastic before interacting productively with cell surface PDGF receptors. This pathway does occur to at least some extent, since when ^{125}I-PDGF has been adsorbed to the plastic dish, both degraded and nondegraded ^{125}I-PDGF accumulates in the culture medium during subsequent growth of the cells (our unpublished observations). In any case, in considering the biological significance of adsorbed PDGF, it is probably not important to know whether that PDGF is active *in situ* or only after desorption, or both. The biological significance *in vivo* may be to prolong the mitogenic signal resulting from the release of PDGF from platelets by localizing PDGF which is not immediately bound to PDGF mitogen receptors. Without such localization, the PDGF will be cleared from the circulation with a half-life of less than 2 min (Bowen-Pope *et al.*, 1984b). The biological significance of such binding in culture relates to the interpretation of experiments designed to investigate the mechanism of action of PDGF and the synergism between different classes of growth factors. It had been proposed that PDGF has "competence" activity; i.e., relatively brief exposure of cells to PDGF is sufficient to induce intracellular changes which render them capable of entering into the S phase of the cell cycle if a member of the "progression" class of growth factors is also present (Pledger *et al.*, 1977, 1978). Given the tenacity with which PDGF and other substances can bind to various components of the culture system, an alternative explanation for the apparent induction of a stable competent state was that the signal itself persisted through the rinsing and medium change intended to remove it (Dicker and Rozengurt, 1981; Van Obberghen-Schilling *et al.*, 1982). Singh *et al.* (1983) have used antibodies against PDGF to neutralize residual attached PDGF and confirmed the original hypothesis of induction of a persistent state independent of the persistence of adsorbed PDGF. Be this as it may, the physical persistence of PDGF adsorbed to components other than the PDGF receptors is a factor to be considered when evaluating experiments involving addition and removal of PDGF.

III. BASIC BINDING PROPERTIES OF THE PDGF RECEPTOR

A. Affinity

Studies in which increasing concentrations of ^{125}I-PDGF are incubated with PDGF-responsive cells have provided clear evidence for high-affinity PDGF binding sites with an apparent K_D of 10^{-11}–10^{-9} M (Heldin et al., 1981; Bowen-Pope and Ross, 1982; Huang et al., 1982; Williams et al., 1982). The very high affinity of binding is reflected in the very slow rate of dissociation of bound PDGF—virtually no ^{125}I-PDGF dissociates over a period of 4–8 hr (Heldin et al., 1982; Bowen-Pope and Ross, 1985). The bound ^{125}I-PDGF does not become irreversibly bound, since it can be dissociated at any time by rinsing the cultures with a solution of acidified saline (pH 3.7) which dissociates receptor-bound ^{125}I-PDGF without damaging the PDGF receptors (Bowen-Pope et al., 1983; Bowen-Pope and Ross, 1985).

Although the reported K_D values for the PDGF receptor are all quite low, the actual values have varied considerably, from 10^{-11} M (Bowen-Pope and Ross, 1982) to 10^{-10} M (Williams et al., 1982) and 10^{-9} M (Heldin et al., 1981; Huang et al., 1982). One possible explanation for these differences would be different estimates of the actual concentration of PDGF used in these experiments. Although it is certainly difficult to unambiguously determine true protein concentrations when only relatively small amounts of purified growth factor are available, it is unlikely that this is the only explanation for the discrepancy. The most likely explanation for the differences in the estimated K_D is deviation from true equilibrium binding conditions. The problems are caused by the very low concentrations of PDGF which need to be employed compared with the relatively large "concentration" of PDGF receptors expressed by many cell types. These problems are considerably greater in studying PDGF binding than when studying, for example, EGF binding, because the much higher affinity of PDGF binding necessitates working at lower concentrations of ligand. For example, a 2.4 cm^2 cell culture well containing 4×10^4 Swiss 3T3 cells (confluent but not dense), each of which expresses ~120,000 PDGF receptors per cell, will contain 4.8×10^9 PDGF receptors. By comparison, 1.0 ml of binding medium containing ^{125}I-PDGF at 10^{-11} M, i.e., near the K_D, will contain 6×10^9 molecules of PDGF. Clearly, there will be severe problems due to depletion of ligand during the measurement. Unless corrected for, this depletion will cause an overestimation of the true K_D (an underestimation of true affinity). Even under conditions in which the total number of receptors is considerably less than the amount of PDGF present, there can be problems which result from local depletion of ligand in the immediate vicinity of the cell monolayer (Maroudas, 1974). Many of these problems, and our attempts to minimize them, have been reported in considerable detail elsewhere (Bowen-Pope and Ross, 1985) and will only be summarized here.

The first consideration is to keep the number of receptors per assay as low as possible. Cell types, like the mouse 3T3 cell line, which express relatively large numbers of PDGF receptors per cell should be assayed at a sparse cell density to determine an accurate K_D.

The second consideration is to keep the volume of labeling solution as large as possible, e.g., 1.0 ml per 2.4 cm^2 culture well. This minimizes bulk-phase depletion of the ligand.

The third consideration is to allow the binding interaction to approach equilibrium. Given the likelihood that depletion of ligand will occur first in the layer of fluid immediately in contact with the cell monolayer, we have performed all incubations with continuous oscillatory mixing to attempt to minimize "local depletion." This increases the rate of binding of low concentrations of ^{125}I-PDGF by several fold. We have found that even with continuous mixing, equilibrium binding of high, but not of low concentrations of ^{125}I-PDGF is achieved by 4 hr (Bowen-Pope and Ross, 1985). This situation results in a relative underestimation of ^{125}I-PDGF binding at low concentrations and hence to an underestimation of binding affinity. The problem seems to have been generally overlooked because single high concentrations of ^{125}I-PDGF have been used to determine the kinetics of ^{125}I-PDGF binding. Binding incubation times chosen on the basis of those measurements underestimate the time needed for equilibrium binding at low concentrations.

Given the above problems in determining the "true" affinity of PDGF binding, it is possible to state with confidence only that the K_D for binding to the PDGF receptor is very low, in the range of 10^{-11} M (0.33 ng/ml). Figure 1 shows that the binding properties of the PDGF receptor seem to have been very highly conserved during evolution, since human ^{125}I-PDGF is bound with comparable affinity by cells from a

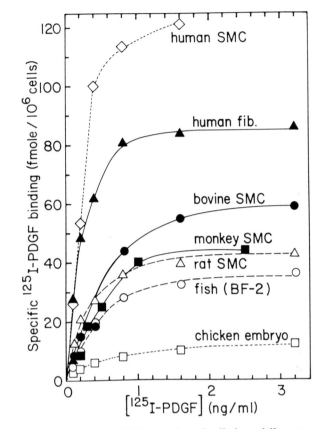

Fig. 1. Specific binding of ^{125}I-PDGF to cultured cells from different species. Mono-layer cultures of each cell type were prepared in 24-well culture trays. When the cultures were confluent, the medium was replaced with culture medium containing 2% plasma-derived serum or 2% calf whole blood serum from which the PDGF had been removed by cation exchange chromatography. After 2 days, the cultures were rinsed once with ice-cold saline, then incubated for 4 hr at 4°C with gentle oscillation in 1 ml of medium containing 0.25% BSA and the concentration of ^{125}I-PDGF shown on the abscissa, or with ^{125}I-PDGF plus a 100-fold excess of unlabeled PDGF to determine nonspecific binding (always less than 15% of total binding), or with medium alone to determine cell number by electronic particle counting. The cell types examined were (open diamond), adult human aortic smooth muscle; (closed triangle), adult human foreskin fibroblasts; (closed circle), bovine aortic smooth muscle; (closed square), adult monkey (*Macaca nemestrina*) aortic smooth muscle; (open triangle), adult rat aortic smooth muscle; (open circle), bluegill fry fish cell line BF-2 (American Type Culture Collection); (open square), second-ary cultures of chicken embryo cells.

wide range of species, including nonmammalian species (fish and chickens).

B. Number of PDGF Receptors per Cell and Cell-Type Distribution of Receptors

There is better agreement among different investigators as to the number of PDGF receptors per cell than as to the affinity of binding. Studies to date suggest that PDGF receptors are expressed only on "connective tissue" cell types, including dermal and tendon fibroblasts, vascular smooth muscle cells, glial cells, and chondrocytes (Table I). Most such cells express about 25,000 to 150,000 receptors per cell (Table I). To date, no cell type has been found which overproduces PDGF receptors to the extent that A-431 cells overproduce EGF receptors (Fabricant *et al.*, 1977). The highest producer of PDGF receptors ($\sim$6 $\times$ 10^5/cell, designated 3T3 HR, in Bowen-Pope and Ross, 1982) discovered to date is the NR-6 3T3 cell variant, selected by Pruss and Herschman (1977) on the basis of its inability to respond mitogenically to EGF and which was found to express no functional EGF receptors. Whether there is any connection between lack of expression of EGF receptors and increased expression of PDGF receptors is not known, although at least one other variant 3T3 cell line (TNR-2; Butler-Gralla and Herschman, 1981) which does not bind EGF does not show elevated numbers of PDGF receptors (our unpublished observations).

There are several cell types whose expression, or lack of expression, of PDGF receptors requires special comment. The ability of different blood leukocytes to express PDGF receptors is probably the most problematic. Deuel *et al.* (1982) reported that PDGF is chemotactic for monocytes and neutrophils. However, we have been unable to obtain a convincing chemotactic response to PDGF by human peripheral blood monocytes prepared by counterflow centrifugation (elutriation) (K. Shimokado, unpublished observations), nor have we been able to demonstrate specific ^{125}I-PDGF binding to such monocytes. One explanation for the disparity would be that the PDGF receptors on monocytes become blocked by PDGF released from platelets during our isolation of the monocytes. Such preexposure of test cells to ligand is known to interfere with measurement of a chemotactic response. In any case, there are some data which suggest that chemotactic response of leukocytes, where observed, may be mediated through cellular interactions different from that of PDGF with its mitogen receptor. Williams *et al.* (1983) reported that PDGF which has been reduced to its component chains was active in inducing leukocyte chemotaxis, even though it

TABLE I
Expression of PDGF Receptors by Different Cell Types

Cell type[a]	Method of detection[b]	Number of receptors/cell[c] ($\times 10^{-4}$)	Purity of PDGF[d]	Reference
Human cells				
Foreskin fibroblasts	a	30	p	Heldin et al., 1981
	a,b	4.5	p	Bowen-Pope and Ross, 1982
	d	+	p	Ek et al., 1982
	d	+	p	Ek and Heldin, 1982
	e	+	p	Nilsson et al., 1983a
	a	+	p	Bowen-Pope et al., 1983
	d	+	p	Pike et al., 1983
	c,d	+	p	Heldin et al., 1983
Dermal fibroblasts	b	+	s	Rutherford and Ross, 1976
	b	+	pp	Slayback et al., 1977
	a	+	p	Heldin et al., 1981
	e	+	p	Rosenfeld et al., 1984
Brain glial cells				
4-787CG	b	+	pp	Westermark and Wasteson, 1976
4-787CG	b	+	p	Heldin et al., 1979b
U-1508CG	a	+	p	Heldin et al., 1981
U-1508CG	d	+	p	Ek et al., 1982
Fetal lung fibroblasts (WI-38)	c	+	p	Glenn et al., 1982
WI-38 transformed by SV40	a	4.8	p	Bowen-Pope et al., 1984
WI-38 transformed by SV40	a	$\simeq 0^*$	p	Bowen-Pope et al., 1984
Bone marrow fibroblasts	a	5.0	p	Bowen-Pope et al., 1984
Bone marrow fibroblasts transformed by SV40	a	0.4^*	p	Bowen-Pope et al., 1984

(continued)

TABLE I (Continued)

Cell type[a]	Method of detection[b]	Number of receptors/cell[c] ($\times 10^{-4}$)	Purity of PDGF[d]	Reference
Bone marrow stromal cells able to support hematopoiesis	a,b	+	p	Rosenfeld et al., 1985
Glomerular capillary endothelial cells	b	+	p	Striker et al., 1984
Aortic smooth muscle	b	+	s	Ross et al., 1974
	b	+	s	Witte et al., 1978
	a,b	+	p	Bowen-Pope and Ross, 1982
Vas deferens smooth muscle	a,b	3	p	D. F. Bowen-Pope and N. S. Laden, unpublished observations
Osteosarcoma (4-393 OS)	a	+	p	Heldin et al., 1981
Glioblastoma (4-251 MG)	a	+	p	Heldin et al., 1981
Peripheral blood monocytes	b	+	p	Deuel et al., 1982
	b	+	p	Williams et al., 1983
Peripheral blood neutrophils	b	+	p	Deuel et al., 1982
	b	+	p	Williams et al., 1983
Peripheral blood lymphocytes	a	0	p	Heldin et al., 1981
	a	0	p	Our unpublished observations
Platelets	a	0	p	Bowen-Pope et al., 1985
Erythrocytes	a	0	p	Bowen-Pope et al., 1985
Umbilical vein endothelial cells	b	0*	s	Haudenschild et al., 1976
	b	0*	s	Wall et al., 1978
	a	0*	p	Heldin et al., 1981
Kidney glomerular epithelial cells	a,b	0	p	Our unpublished observations
Kidney mesangial cells	a,b	0	p	Our unpublished observations
	a,b	0	p	Our unpublished observations
Retinal glial cells	b	0	p	Burke, 1982
Peripheral Schwann cells	b	0	p	Lemke and Brokes, 1983

Fetal myoblasts and myotubes (H275F6)	a	0	p	D. F. Bowen-Pope, S. D. Hauschka, and R. Ross, unpublished
Epidermal carcinoma A-431	a	0*	p	Heldin et al., 1981
	a	0*	p	Bowen-Pope and Ross, 1982
	a	0*	p	Huang et al., 1982
	e	0*	p	Rosenfeld et al., 1984
Thyroid carcinoma (SW 1736)	a	0	p	Heldin et al., 1981
Squamous lung carcinoma (U-1752)	a	0	p	Heldin et al., 1981
Neuroblastoma (SH)	a	0	p	Heldin et al., 1981
Hepatoma (Hep G2)	a	0*	p	Bowen-Pope et al., 1984a
Bladder carcinoma (T24)	a	0*	p	Bowen-Pope et al., 1984a
Rodent cells				
3T3 (mouse)	b	+	s	Kohler and Lipton, 1974
	a	+	p	Heldin et al., 1981
	a,b	6.3	p	DiCorleto and Bowen-Pope, 1983
Swiss 3T3 (mouse)	b	+	pp	Vogel et al., 1980
	b	+	p	Deuel et al., 1981
	b	+	p	Raines and Ross, 1982
	a,b	13	p	Bowen-Pope and Ross, 1982
	d	+	p	Nishimura et al., 1982
	a,b	40	p	Huang et al., 1982
	c	+	p	Glenn et al., 1982
	a	+	p	Bowen-Pope et al., 1983
	d	+	p	Pike et al., 1983
	e	+	p	Rosenfeld et al., 1984
	a	7.2	p	Bowen-Pope et al., 1984a
	a,b,d	6.1	p	Bremer et al., 1984
Swiss 3T3 transformed by SV40	a	0.6*	p	Bowen-Pope et al., 1984a
Swiss 3T3 selected against response to PDGF	a,b	0.25*	p	Bowen-Pope and Ross, 1982
Swiss 3T3 clone D1	a	> 12	p	Bowen-Pope et al., 1984a
Spontaneously transformed 3T3 D1	a	≈ 0*	p	Bowen-Pope et al., 1984a

(continued)

TABLE I (*Continued*)

Cell type[a]	Method of detection[b]	Number of receptors/cell[c] ($\times 10^{-4}$)	Purity of PDGF[d]	Reference
NIH 3T3 (mouse)	a	6	p	Bowen-Pope et al., 1984a
NIH 3T3 transformed by Kirsten murine sarcoma virus	a	3*	p	Bowen-Pope et al., 1984a
NIH 3T3 transformed by simian sarcoma virus	a	0.9*	p	Bowen-Pope et al., 1984a
BALB/c 3T3 (mouse)	b		pp	Antoniades et al., 1975
	b		p	Antoniades et al., 1979
	a		p	Singh et al., 1982
	b		p	Smith et al., 1982
	a	11.4	p	Bowen-Pope et al., 1984a
BALB/c 3T3 transformed by SV40	a	1.8*	p	Bowen-Pope et al., 1984a
BALB/c 3T3 transformed by Moloney murine sarcoma virus	a	≈ 0*	p	Bowen-Pope et al., 1984a
C3H 10 $t_{1/2}$ (mouse)	a	12	p	Bowen-Pope et al., 1984a
C3H 10 $t_{1/2}$ transformed by methylcholanthrene	a	1.6	p	Bowen-Pope et al., 1984a
NR-6 (mouse)	a,b	39	0	Bowen-Pope and Ross, 1982
	d	+	p	Pike et al., 1983
NRK (rat)	a	3.6	p	Bowen-Pope et al., 1984a
	a	≈ 0	p	Bowen-Pope et al., 1984a
BHK (hamster)	a	0.4	p	Our unpublished observations
CHO (hamster)	a	0.6	p	Our unpublished observations
Adult rat aortic smooth muscle	a	2.1	p	Seifert et al., 1985
	b	+	p	Nilsson et al., 1983b
	b	+	pp	Weinstein et al., 1981
1- to 2-week rat aortic smooth muscle	a	+/0*	p	Seifert et al., 1985
	b	+	p	Nilsson et al., 1983b

Cell type				Reference
Calvaria cells (mouse)	b	+	p	Tashjian et al., 1982
Calvaria cells (rat)	b	+	pp	Canalis, 1981
Cloned myoblasts (mouse) (MM14DZ)	a	0	p	D. F. Bowen-Pope, S. D. Hauschka, R. Ross, unpublished
Hepatocytes (mouse)	a	0	p	Our unpublished observations
Blastocyst endoderm-like cells (mouse)	b	+	p	Rizzino and Bowen-Pope, 1985
Differentiated endoderm-like teratocarcinoma (mouse line PC13)	a,b	+	a	Rizzino and Bowen-Pope, 1985
Undifferentiated PC13	a,b	0*	a	Rizzino and Bowen-Pope, 1985
Differentiated endoderm-like teratocarcinoma (mouse line F9)	a,b	0.8	a	Rizzino and Bowen-Pope, 1985
Undifferentiated F9	a,b	0*	a	Rizzino and Bowen-Pope, 1985
Differentiated endoderm-like teratocarcinoma (mouse line PSA 5E)	a,b	0.5		Rizzino and Bowen-Pope, 1985
(mouse line PYS 2)	a,b	2.3		Rizzino and Bowen-Pope, 1985
Other mammalian cells				
Aortic smooth muscle (monkey)	b	+	s	Rutherford and Ross, 1976
	a,b	+	p	Bowen-Pope and Ross, 1982
	c	+	p	Glenn et al., 1982
Aortic smooth muscle (bovine)	a,b	5	p	Williams et al., 1982
	a	0.7–1.5	p	DiCorleto and Bowen-Pope, 1983
	a	3.5	p	This chapter, Fig. 1
Aortic smooth muscle (porcine)	a	+	p	Heldin et al., 1981
Tendon fibroblasts (fetal monkey)	a	2.4	p	Our unpublished observations
Dermal fibroblasts (monkey)	b	+	s	Rutherford and Ross, 1976
Articular chondrocytes (porcine)	a,b	+	p	Our unpublished observations
Aortic endothelial cells (bovine)	a,b	0*	p	Bowen-Pope and Ross, 1982
Kidney epithelioid (MDCK) (canine)	a	0	p	Our unpublished observations
Adrenal microvascular pericytes (bovine)	b	0	p	Bernstein et al., 1982

(continued)

TABLE I (Continued)

Cell type[a]	Method of detection[b]	Number of receptors/cell[c] ($\times 1^{-4}$)	Purity of PDGF[d]	Reference
Aortic endothelial cells (monkey)	a	0	p	Our unpublished observations
Nonmammalian cells				
1° chick embryo fibroblasts	b	+	s	Balk, 1971
	a	0.7	p	This chapter, Fig. 1
Bluegill fish (BF-2)	a	2.1	p	This chapter, Fig. 1

[a]The cell types analyzed have been grouped by species (or group of species) and, within a species, by cell type. Cell types expressing PDGF receptors are listed first within each category. To facilitate comparisons, transformed derivatives of a cell type expressing PDGF receptors are listed immediately after the parental cell type, even though they may not express receptors.

[b]The methods used to detect PDGF receptors have been divided into five groups (second column): (a) specific ^{125}I-PDGF binding; (b) biological response to PDGF; (c) affinity labeling of a ≈170,000-dalton membrane protein using ^{125}I-PDGF; (d) stimulation of phosphorylation of a 170,000-dalton membrane by PDGF; and (e) visualization of PDGF binding by light or electron microscopy.

[c]Number of receptors per cell. Detection method (a) yields estimates of the number of receptors per cell (third column). The other methods distinguish only between expression (+) or lack of expression (0) of receptors. In cases indicated by *, the test cells were found to secrete a substance able to bind to PDGF receptors. In these cases, the number of detected receptors will probably underestimate the level of synthesis of receptors (see text, Sections V, B and C).

[d]The purity of the PDGF used to detect PDGF receptors is described in the fourth column as (p) pure, (pp) partially purified, or (s) serum or platelet extract. All experiments using detection methods a, c, d, and e use pure PDGF. Measurement of biological response to serum or platelet extract, and later to partially purified PDGF, provided the basis for the discovery and characterization of PDGF. Some of the most important early observations, but by no means all of them, are listed in this table.

does not stimulate a mitogenic response by 3T3 cells and does not compete for ^{125}I-PDGF binding to PDGF mitogen receptors (Williams et al., 1982). This suggests that leukocytes may not be responding to PDGF via a receptor identical to that expressed by mitogenically responsive cell types.

A second mesenchymal cell type which seems not to express PDGF receptors is the arterial endothelial cell (Bowen-Pope and Ross, 1982). Arterial endothelial cells from monkey, pig, and cow aortas and from human umbilical vein do not bind significant amounts of ^{125}I-PDGF. It is impossible to be certain, however, that these cells do not synthesize PDGF receptors. The problem is that the endothelial cells synthesize and secrete levels of PDGF-like molecules which would be sufficient to block and down regulate any PDGF receptors which might be synthesized (DiCorleto and Bowen-Pope, 1983; Bowen-Pope et al., 1983). To date, we have not found conditions which specifically prevent synthesis of PDGF by arterial endothelial cells so that we cannot measure receptor levels under conditions in which there is no endogenous production of PDGF. Synthesis of PDGF-like molecules and lack of expression of available PDGF receptors may not, however, be a universal characteristic of vascular endothelial cells. We have recently found that capillary endothelial cells isolated from human glomeruli respond to PDGF (Striker et al., 1984).

A third mesenchymal cell type which does not express PDGF receptors is skeletal muscle cells. We have found that mouse and human muscle cells do not respond to PDGF or bind ^{125}I-PDGF, either before or after differentiation into multinucleate muscle fibers in culture (D. F. Bowen-Pope, S. D. Hauschka, and R. Ross, unpublished observations). This contrasts with the results with EGF in which myoblasts express EGF receptors that are lost during commitment to form muscle fibers (Lim and Hauschka, 1984). It also contrasts with the result with FGF, which is a potent mitogen for myocytes until they commit to form myotubes (Lim and Hauschka, 1984).

C. Binding Specificity

Competitive binding studies suggest that the PDGF receptor does not recognize other growth factors (Table II) or other platelet α-granule proteins. The one possible breach in hormone specificity seems to be insulin from the hystricomorphs (guinea pig, porcupine, coypu), which has a mitogenic potency more similar to PDGF than to insulin from other species (King et al., 1983). PDGF competes with ^{125}I-histricomorph insulin for binding to human fibroblasts, suggesting that the two molecules use the same receptor.

TABLE II
Specificity of ^{125}I-PDGF Binding

Substance tested[a]	$I_{0.5}$, or maximum concentration tested	Reference
Not inhibitory		
Epidermal growth	10^{-9} M	Williams et al., 1982
factor (EGF)	5 μg/ml	Heldin et al., 1981
	8 μg/ml	Huang et al., 1982
	5 μg/ml	Singh et al., 1983
	2.6 μg/ml	Bowen-Pope and Ross, 1982
Nerve growth factor (NGF)	10^{-9} M	Williams et al., 1982
Partially purified	10^{-9} M	Williams et al., 1982
fibroblast growth	5 μg/ml	Heldin et al., 1981
factor (FGF)	0.8 μg/ml gave slight inhibition	Huang et al., 1982
	5 μg/ml	Singh et al., 1983
	10 μg/ml	Bowen-Pope and Ross, 1982
Insulin (bovine and	5 μg/ml	Heldin et al., 1981
porcine)	4 μg/ml	Huang et al., 1982
	1000 μg/ml	Bowen-Pope and Ross, 1982
Bovine serum albumin (BSA)	1.6×10^{-2} g/ml	Bowen-Pope and Ross, 1982
Human thrombospondin	5×10^{-5} g/ml	Bowen-Pope and Ross, 1982
Human β-thromboglobulin	5 μg/ml	Singh et al., 1983
Human platelet	6.4×10^{-5} g/ml	Bowen-Pope and Ross, 1982
factor 4	100 μg/ml	Huang et al., 1982
	5 μg/ml	Singh et al., 1983
Bacitracin	5 μg/ml	Heldin et al., 1981
	5 μg/ml	Heldin et al., 1981
Cytochrome c	100 μg/ml	Huang et al., 1982
Lysozyme	5 μg/ml	Heldin et al., 1981
Chymotrypsinogen A	5 μg/ml	Heldin et al., 1981
Calcium pyrophosphate precipitates	1 mM	Bowen-Pope and Rubin, 1983
Ribonuclease	5 μg/ml	Heldin et al., 1981
Histone B	10^{-8} M	Williams et al., 1982
Thrombin	8 μg/ml	Huang et al., 1982
Inhibitory		
Histone II B	36% inhibition at 100 μg/ml	Huang et al., 1982
Polylysine	$I_{0.5} = 30$ μg/ml	Huang et al., 1982
Protamine	$I_{0.5} = 2$ μg/ml	Huang et al., 1982
Insulin (guinea pig)	$I_{0.5} = 2$–5 ng/ml	King and Kahn; see Chapter 8

[a]The substances in this table have been checked for inhibition of ^{125}I-PDGF binding to the PDGF receptor. Substances in the upper group had no effect on ^{125}I-PDGF binding. The concentration listed is the highest concentration tested. Substances in the second group did reduce ^{125}I-PDGF binding. Concentrations ($I_{0.5}$) giving 50% reduction in binding are recorded.

Many basic proteins have also been tested on the speculation that part of the interaction between the highly cationic PDGF molecule and its receptor will involve charge–charge interactions. None of the moderately cationic molecules tested, including bacitracin, cytochrome c, lysozyme, chymotrypsinogen A, or ribonuclease, inhibited binding (Table II). However, the very cationic molecules histone IIB, polylysine, and protamine did reduce binding (Table II). Our experience with polylysine has been that cells begin to exhibit morphological signs of damage at concentrations of polylysine which affect ^{125}I-PDGF binding. However, protamine shows no evidence of toxicity. The ability of histone, polylysine, and protamine to inhibit ^{125}I-PDGF binding supports the possibility that part of the binding interaction is mediated by an attraction between a negatively charged domain on the receptor and a positively charged domain in PDGF. However, this mode of interaction is only one of the functions contributing to the high-affinity binding of PDGF since, on a weight basis, these cationic peptides are 10^4 to 10^6-fold less effective than PDGF in binding to the PDGF receptor.

Although the receptors for PDGF and for EGF are distinct entities, there is evidence for close interaction between these two receptors. In cell types (e.g., fibroblasts) which express both PDGF and EGF receptors, the affinity of the EGF receptor for ^{125}I-EGF is reduced up to fivefold by preincubation of the cells with PDGF (Bowen-Pope et al., 1983; Collins et al., 1983). The effect is produced at 37°C and, albeit more slowly, at 4°C and persists when the PDGF is dissociated from PDGF receptors by rinsing with acidified saline (Bowen-Pope et al., 1983). The effect does not result from the direct action of PDGF on EGF receptors; since it is not observed with cell types (e.g., A-431) which express EGF but not PDGF receptors. This form of relatively direct interaction between receptors for two different growth factors may provide one mechanism through which cells can integrate the consequences of simultaneous exposure to more than one stimulus.

D. Conditions Affecting Binding

For initial characterization of ^{125}I-PDGF binding, we measured binding in medium which was able to support a mitogenic response to PDGF (Bowen-Pope and Ross, 1982). This medium consisted of Ham's F-12 medium buffered at pH 7.4 with 25 mM HEPES and containing 2% calf serum chromatographed to remove PDGF. As will be discussed in the next section, these studies demonstrated that the ^{125}I-PDGF binding site observed is the PDGF receptor. In order to simplify preparation of binding medium, we have now replaced the chromatographed serum with 0.25% BSA with no change in results (Bowen-Pope and Ross,

1985). High concentrations of plasma or serum in fact can interfere with binding to the PDGF receptor. Plasma contains components which themselves bind PDGF and which prevent this bound PDGF from interacting with the PDGF receptor (Huang *et al.*, 1983, 1984; Raines *et al.*, 1984; Bowen-Pope *et al.*, 1984b).

In order to determine the dependence of ^{125}I-PDGF binding on specific cations, while minimizing indirect effects on cells of exposure to different ionic conditions, we reduced the binding period to 1 hr at 4°C. Omission of Ca^{2+} and Mg^{2+} did not affect ^{125}I-PDGF binding to 3T3 cells under these conditions even in the presence of up to 0.1 mM EDTA (Bowen-Pope and Ross, 1982). ^{125}I-PDGF binding was reduced to less than 20% of control if both monovalent and divalent cations were replaced with isotonic sucrose. Binding was only partially restored by addition of the divalent salts but was brought to 90% of control by adding 50% of physiological levels of monovalent salts.

^{125}I-PDGF binding is not affected by changes in pH within a broad range between pH 6.0 and 8.0 (Bowen-Pope and Ross, 1982, 1985). Below pH 6.0, binding is greatly reduced. Prebound ^{125}I-PDGF can be efficiently dissociated from cell surface receptors without permanently reducing the binding capacity of the receptors (Bowen-Pope *et al.*, 1983; Bowen-Pope and Ross, 1985). This latter procedure is useful in distinguishing between ^{125}I-PDGF bound at the cell surface (can be dissociated by acidified saline) and ^{125}I-PDGF located within the cell (cannot be dissociated).

IV. THE PDGF BINDING SITE IS A FUNCTIONAL RECEPTOR

There are several examples in the literature of hormone binding sites which satisfy the requirements of saturability, high affinity, and, to a certain extent, specificity, and yet which are not hormone receptors (e.g., see Cuatrecasas and Hollenberg, 1976). Evidence that the ^{125}I-PDGF binding site previously described is the receptor which mediates mitogenic stimulation by PDGF includes the following: (1) The concentration dependence of ^{125}I-PDGF binding and of mitogenic stimulation by PDGF are very similar (Bowen-Pope and Ross, 1982; Huang *et al.*, 1982; Williams *et al.*, 1982). There is no evidence for the existence of "spare receptors" such as have been proposed for insulin and EGF to account for saturation of ligand binding at concentrations 10- to 100-fold higher than needed for maximal biological response (Cuatrecasas and Hollenberg, 1976; Adamson and Rees, 1981). (2) Only cell types which express high-affinity PDGF binding sites are mitogenically responsive to PDGF (see Table I). Epithelial cells neither bind nor respond to PDGF. This argues that the binding site is not merely a common cell surface component able to bind cationic molecules. (3) The strict ligand specificity of the binding site argues against a fortuitous binding interaction. The most potent binding competitor other than PDGF, protamine, is an antagonist of PDGF's chemotactic activity (Deuel et al., 1982), consistent with the involvement of the PDGF binding site in mediating this biological response. (4) As will be discussed, the ^{125}I-PDGF binding site is "down regulated" during exposure of cells to PDGF at 37°C. This response is characteristic of mitogen receptors (Kaplan, 1981).

V. VARIABLES WHICH AFFECT EXPRESSION OR PROPERTIES OF PDGF RECEPTORS

A. Culture Variables

The extent to which the common tissue culture variables (cell density, growth rate, time after plating) affect the numbers of PDGF receptors expressed per cell has not been systematically investigated. For other mitogens, e.g., EGF, such variables have been reported to have significant effects, but not always the same effect with each cell type. For example, BALB/c 3T3 cells were reported to express 10-fold more EGF receptors at high than at low cell density (Pratt and Pastan, 1978), while Holley et al. (1977) found that the number of EGF receptors per BSC-1 cell decreased with increasing cell density. This suggests that caution should be exercised in interpreting the significance of observed differences between different cell types in number of PDGF receptors per cell except where the differences are very large.

We have investigated the possibility that the senescence of cells passaged extensively in culture results from a defect in the ability of the senescent cells to bind and respond to PDGF, one of the major mitogens in the usual serum growth supplement. We have not found a significant reduction in the number of PDGF receptors expressed by young and senescent cells of many cell types (monkey and human vascular smooth muscle, adult human foreskin, and dermal fibroblasts, WI-38 fetal lung fibroblasts), even though the senescent cells no longer respond mitogenically to PDGF (D. F. Bowen-Pope and Ross, unpublished observations).

B. Oncogenic Transformation

One variable which has a dramatic effect on expression of available PDGF receptors is oncogenic transformation (see Chapter 1). Figure 2

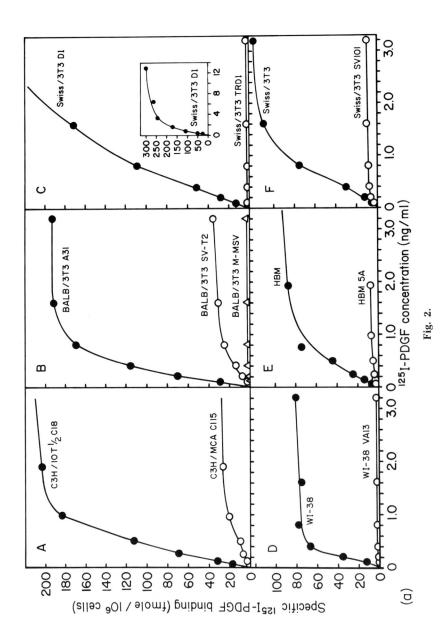

Fig. 2.

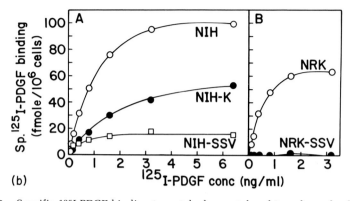

(b)

Fig. 2. Specific ^{125}I-PDGF binding to matched parental and transformed cell lines. Cultures were plated in 2.4 cm² wells in DMEM containing 5% calf serum. When slightly subconfluent, the plating medium was replaced with DMEM containing 2% PDGF-deficient calf serum and incubated for an additional 2 days. (a) Specific binding was measured by incubation for 3 hr at 4°C with gentle shaking in 1.0 ml/well binding medium containing the indicated concentration of ^{125}I-PDGF plus 20 μg/ml partially purified PDGF (the equivalent of ~400 ng/ml pure PDGF) to determine nonspecific binding, or in binding medium without ^{125}I-PDGF to determine the number of cells per well at the conclusion of the assay. Binding was terminated by rinsing four times with binding rinse and cell-bound ^{125}I-PDGF determined by extracting each culture with 1 ml 0.1% Triton X-100 with 0.1% BSA and gamma counting the entire extract. All measurements were made using triplicate wells and are plotted as the mean. The cell types used are identified in the figure. (b) Binding and analysis was performed as in Fig. 1 except that binding was performed at 37°C for 45 min without shaking and all rinse solutions were used at 37°C to minimize cell detachment. From Bowen-Pope et al. (1984).

shows many examples of [125]-PDGF binding to PDGF-responsive parental cell types compared with binding to oncogenically transformed derivative cell lines. In each case, the number of receptors available to bind exogenous ^{125}I-PDGF is greatly reduced. The mechanism of this phenotypic change may be the synthesis and secretion of PDGF-like molecules by the transformed cells. We have found that in each case (except transformation by methylcholanthrene), culture medium conditioned by the transformed cells, but not by their normal counterparts, contains significant levels of material which is able to compete with ^{125}I-PDGF for binding to PDGF receptors on test cells (Bowen-Pope et al., 1984a). At least a portion of this material is antigenically related to PDGF and is active as a mitogen. The simplest explanation for the reduced binding of exogenous ^{125}I-PDGF to these transformed cells is that they synthesize a PDGF-like molecule(s) which occupies and down regulates the cell's own PDGF receptors, possibly as part of an autocrine stimulation system (Sporn and Todaro, 1980). At the moment, however, we cannot rule out the possibility that oncogenic transformation independently regulates expression of PDGF receptors.

C. Differentiation

The same type of changes in PDGF phenotype seen as a consequence of oncogenic transformation have also been observed in some examples of normal development. Embryonal carcinoma cells (teratocarcinoma stem cells) have been used as model systems for studying early embryonic development (Martin, 1980). Embryonal carcinoma cell lines F9, PSA-1-G, and PC-13, which share many of the properties of cells of the mouse blastocyst inner cell mass, do not bind ^{125}I-PDGF and do secrete PDGF-like molecules into their culture medium (Rizzino and Bowen-Pope, 1985; Gudas et al., 1983). When the embryonal cells are induced by addition of retinoic acid to differentiate into endoderm-like cells, they cease secreting PDGF and become able to bind and respond to exogenously added PDGF.

At a later stage in development we have found (Seifert et al., 1985) a second instance of developmental regulation of PDGF phenotype. Smooth muscle cells isolated from the aortas of 2-week-old rats show very low levels of high-affinity PDGF binding, and secrete substantial amounts of PDGF-like material into the culture medium. By contrast, smooth muscle cells isolated from the aortas of 3-month-old rats express significant levels of PDGF receptors and secrete low to undetectable levels of PDGF. Each of these phenotypes is stable in culture over a period of at least 2 months and 20 doublings. The influences experienced by the smooth muscle cells *in vivo* which induce the conversion from one phenotype to the other are thus not reproduced in culture. If aortic smooth muscle cells in 2-week-old rats do synthesize PDGF *in vivo*, it is possible that this PDGF plays a role in the growth of the aorta which is still occurring at this time (Berry et al., 1972; Looker and Berry, 1972; Gerrity and Cliff, 1975). This would then be one of the first examples of autocrine growth stimulation occurring in a normal tissue.

D. Membrane Composition

Gangliosides are carbohydrate-rich complex lipids found in mammalian cell plasma membranes. The possibility that gangliosides may play some role in the regulation of cell growth was suggested by changes in ganglioside concentration and metabolism which occur with oncogenic transformation and density-dependent inhibition of growth (e.g., see Hakamori, 1975). We have found that exogenously added gangliosides can specifically affect the behavior of the PDGF receptor system (Bremer et al., 1984). Ganglioside G_{M1} but not Neu-AcnLc$_4$, inhibits the proliferative response of 3T3 cells to PDGF but not to EGF. The locus of the inhibitory effect seems to be the PDGF receptor itself. G_{M1} decreases the K_D for ^{125}I-PDGF binding to whole cells and, when added directly to in vitro phosphorylation assays (see later), greatly decreases the ability of PDGF to stimulate autophosphorylation of the PDGF receptor. The ability of G_{M1} to affect the properties of the PDGF in vitro argues that the effects of G_{M1} seen in intact cells are not indirect consequences of effects of G_{M1} on cell metabolism which secondarily affect the PDGF receptor. Such an influence of specific membrane components on growth factor receptors could play a role in integrating the effects on cell growth of changes in membrane structure and composition and of exposure to growth factors.

VI. PHYSICAL AND BIOCHEMICAL PROPERTIES OF THE PDGF RECEPTOR

A. Size and Structure

The size of the PDGF receptor has been determined by affinity cross-linking studies (Glenn et al., 1982). In these experiments, ^{125}I-PDGF was incubated with test cells at 4°C and nonbound ^{125}I-PDGF rinsed away such that greater than 90% of the cell-associated ^{125}I-PDGF was specifically bound to PDGF receptors. A bifunctional cross-linking reagent (EGS or DSS) was then added to covalently stabilize interactions between adjacent proteins (with reactive groups separated by less than 11 Å for DSS or 15 Å for EGS). Analysis of such cross-linked complexes by SDS–polyacrylamide gel electrophoresis indicated that the monomeric size of the PDGF receptor was about 164,000 daltons. The size of the receptor is not altered by reduction, suggesting that it does not consist of disulfide-bonded subunits. This finding has been confirmed (Heldin et al., 1983) using a photoactivatable cross-linking reagent which is first reacted with ^{125}I-PDGF, incubated with test cells, and the cross-linking reaction triggered by exposure to light. This method gives less efficient labeling of PDGF receptors, but is not susceptible to the formation of multimolecular aggregates. The receptor size determined by this method was 175,000–195,000 (Heldin et al., 1983). Analysis of the PDGF receptor (equated with the 170,000-dalton membrane protein phosphorylated in response to PDGF; see below) by two-dimensional gel electrophoresis suggests that the receptor has a net negative charge at physiological pH (pI = 5.3–5.5; Pike *et al.*, 1983). This is consistent with the existence of a negatively charged domain involved in binding PDGF (see above).

Since the cross-linking studies described above analyzed receptor size by SDS–polyacrylamide gel electrophoresis, i.e., under conditions which disrupt noncovalent interactions between proteins, this method would not determine whether the 164,000–195,000 receptor protein is normally found in close, but noncovalent association with other membrane protein(s). To look for such an association, Heldin *et al.* (1983) incubated fibroblasts with ^{125}I-PDGF, then solubilized the cells with the nonionic detergent Triton X-100 under conditions which do not disrupt strong, noncovalent interactions between proteins. When analyzed by molecular seiving chromatography on Sepharose 6B, the ^{125}I-PDGF migrated with a size which suggested that it was associated with a 200,000-dalton cell component. This is close enough to the size of the monomeric ^{125}I-PDGF binding site to suggest that the PDGF receptor exists in the cell membrane as monomer and not as a multisubunit complex.

As are many membrane proteins, the PDGF receptor appears to be a glycoprotein. Heldin *et al.* (1983) have tested a broad spectrum of carbohydrate-binding lectins for ability to inhibit binding of ^{125}I-PDGF. The results suggest the presence of galactose and N-acetylglucosamine near to the ^{125}I-PDGF binding site.

B. Tyrosine Protein Kinase Activity

The ability to catalyze transfer of phosphate groups from ATP to the tyrosine moieties of proteins is rare among all protein kinases (less than 0.03% of cellular protein-bound phosphate is on tyrosine; e.g., see Cooper *et al.*, 1982), but this activity is widely distributed among oncogene proteins and growth factor receptors. In those cases, the receptor/transforming gene products seem to be substrates for their own phosphotyrosine kinase activity (reviewed by Sefton and Hunter, 1984; see Chapter 11). There is mounting evidence that the PDGF receptor possesses PDGF-activatable tyrosine kinase activity. Incubation of membranes from PDGF-responsive cells with [γ-^{32}P]ATP, with or without added PDGF, followed by analysis of the phosphorylated proteins by SDS–polyacrylamide gel electrophoresis and autoradiography, demonstrates specific increases in the extent of phosphorylation of certain proteins in the presence of PDGF. The size of the major protein substrate(s), variously estimated as 185,000 (Ek and Heldin, 1982; Heldin *et al.*, 1983), 175,000 (Ek *et al.*, 1982), and 170,000 (Nishimura *et al.*, 1982; Pike *et al.*, 1983) is comparable to the size of the PDGF receptor estimated from affinity cross-linking studies (see above). This phosphorylated protein contains phosphoserine as well as phosphotyrosine (Ek and Heldin, 1982; Nishimura et al., 1982; Pike et al., 1983), possibly as a result of phosphorylation by a serine protein kinase which is preactivated or is activated as a secondary consequence of activation of the PDGF receptor. We have calculated that in the presence of PDGF, the 170,000-dalton protein incorporates 1–2 mol of phosphate per mol of protein, suggesting that autophosphorylation is not restricted to a minor subpopulation of PDGF receptors. Phosphorylation of the 170,000-dalton protein does not depend on cAMP-dependent kinase(s), since it is not affected by either a purified catalytic subunit of cAMP-dependent protein kinase or by a specific heat-stable inhibitor of cAMP-dependent protein kinase (Pike et al., 1983).

It has not yet been conclusively demonstrated that the PDGF receptor is itself a tyrosine protein kinase with autophosphorylation activity. However, considerable evidence favors this possibility: (1) the similarity in size of the PDGF receptor and the substrate for PDGF-stimulated phosphorylation; (2) the stimulation of phosphorylation of the protein by PDGF; (3) the phosphorylation of this protein only in cells which express PDGF receptors, and the proportionality between amount of phosphoprotein and number of PDGF receptors expressed (Pike et al., 1983); (4) the loss of this phosphoprotein in parallel with loss of the PDGF receptor during PDGF-induced down-regulation (Pike et al., 1983); (5) the ability of the same lectins to bind the phosphoprotein and PDGF receptors (Heldin et al., 1983); (6) the ability of PDGF, or PDGF covalently attached to Sepharose, to bind both the PDGF receptor and the phosphoprotein (Heldin et al., 1983); and (7) the ability of specific gangliosides to affect both ^{125}I-PDGF binding and PDGF-stimulated phosphorylation (Bremer et al., 1984, see below).

The relationship between the PDGF-stimulated tyrosine kinase activity of the receptor and the role of the receptor in initiating the biological response to PDGF is only conjectural. On the one hand, the autophosphorylation may be involved in signaling the internalization and degradation of the receptor rather than in transmitting a positive intracellular signal. If, on the other hand, the tyrosine kinase activity is involved in initiating intracellular events, it is likely that this will involve phosphorylation of intracellular proteins in addition to autophosphorylation. Two approaches have been taken to investigate intracellular substrates. The first has been to evaluate the ability of the PDGF-stimulated kinase to phosphorylate exogenously added purified substrates in vitro. Among the substrates tested, only histones (Ek and Heldin, 1982) and a synthetic peptide with an amino acid sequence comparable to the site of phosphorylation of the Rous sarcoma virus transforming protein (Pike *et al.*, 1983) are phosphorylated. The second approach has been to examine the intracellular proteins phosphorylated *in vivo* in response to addition of PDGF. For these experiments, the intracellular ATP pools of cells are preequilibrated by preincubation with $^{32}PO_4$. PDGF is added at 37°C for a short period (1–60 min) and intracellular phosphoproteins analyzed by two-dimensional gel electrophoresis. Such studies have shown (Cooper *et al.*, 1982) that PDGF does stimulate phosphorylation on tyrosine of a specific subset of intracellular proteins, including a pair of related 45,000-dalton proteins, a pair of related 43,000-dalton proteins, and a 42,000-dalton protein. The pattern is similar, but not identical to that seen when the cells are stimulated with epidermal growth factor. The current challenge will be to determine whether these proteins are phorphorylated directly by the PDGF receptor or by other kinases activated as a result of PDGF binding, and to determine the role of these proteins and the effect of phosphorylation on their activity.

VII. SUBCELLULAR LOCALIZATION AND FATE OF BOUND PDGF

The subcellular localization of bound PDGF has been followed morphologically by electron microscopic autoradiography of bound ^{125}I-PDGF (Nilsson *et al.*, 1983a; Rosenfeld *et al.*, 1985) and electron microscopy of bound PDGF–colloidal gold complexes (Rosenfeld *et al.*, 1985). These studies show that PDGF binding at 4–12°C is to cell surface sites, accounting for the ability of acidified saline to dissociate ^{125}I-PDGF bound under these conditions. Coated membrane areas bind more PDGF per unit membrane area than do uncoated areas. To what extent this preferential localization in coated areas reflects preconcentration of PDGF receptors or a ligand-induced movement of occupied receptors into such regions is not known. However, the concentration of bound PDGF in coated regions is increased during the first few minutes after warming to 37°C, suggesting that at least some induced concentration can occur (Rosenfeld *et al.*, 1985). Within 5 min after warming to 37°C, PDGF is found in intracellular vesicles and by 15 min is concentrated in lysosomes.

The morphological studies are supported by measurements of the kinetics of binding and degradation. When cells are incubated with ^{125}I-PDGF at 37°C, the total amount of cell-associated radioactivity reaches a maximum by 20 min to 1 hr, depending on the concentration of [125]-PDGF used, then declines. Concomitant with the decline in cell-associated radioactivity, TCA-soluble radioactivity appears in the incubation medium, suggesting degradation of the added ^{125}I-PDGF, and the ability of the cultures to bind freshly added ^{125}I-PDGF decreases to as little as 20% of binding to naive cultures (Heldin *et al.*, 1981, 1982; Bowen-Pope and Ross, 1982; Huang *et al.*, 1982).

The kinetics of degradation has also been determined by preincubating the cells with ^{125}I-PDGF at 4°C to permit binding but not processing. After warming to 37°C, cell surface-bound ^{125}I-PDGF is internalized with a $t_{1/2} = 15$ min (Rosenfeld *et al.*, 1985), and TCA-soluble radioactivity appears in the incubation medium with a $t_{1/2}$ of 60–90 min (Huang *et al.*, 1983; Nilsson *et al.*, 1983a; Rosenfeld *et al.*, 1985).

This pattern is comparable to that seen with other hormone ligands, e.g., EGF and insulin, and has been interpreted as reflecting the internalization of ligand–receptor complexes followed by degradation of both ligand and receptor in the lysosomes (Kaplan, 1981). A lysosomal site of ^{125}I-PDGF degradation is supported by the ability of inhibitors of lysosomal degradation, e.g., primary amines, which are thought to act by increasing the pH of the lysosomal compartment, to inhibit the appearance of TCA-soluble radioactivity in the incubation medium (Heldin *et al.*, 1982). Although degradation of the PDGF receptor itself has not been demonstrated, such degradation is strongly suggested by the decrease in binding sites which accompanies ^{125}I-PDGF degradation and by the requirement for protein synthesis to restore the number of binding sites to naive levels after down-regulation by exposure to PDGF (Heldin *et al.*, 1982).

REFERENCES

Adamson, E. D., and Rees, A. R. (1981). Epidermal growth factor receptors. *Mol. Cell. Biochem.* **34**, 129–152.

Antoniades, H. N., Stathakos, D., and Scher, C. D. (1975). Isolation of a cationic polypeptide from human serum that stimulates proliferation of 3T3 cells. *Proc. Natl. Acad. Sci. U.S.A.* **72**, 2635–2639.

Antoniades, H. N., Scher, C. D., and Stiles, C. D. (1979). Purification of human platelet-derived growth factor. *Proc. Natl. Acad. Sci. U.S.A.* **76**, 1809–1813.

Balk, S. D. (1971). Calcium as a regulator of the proliferation of normal, but not of transformed, chicken fibroblasts in a plasma-containing medium. *Proc. Natl. Acad. Sci. U.S.A.* **68**, 271–275.

Bernstein, L. R., Antoniades, H. N., and Zetter, B. R. (1982). Migration of cultured vascular cells in response to plasma and platelet-derived factors. *J. Cell Sci.* **56**, 71–82.

Berry, C. L., Looker, T., and Germain, J. (1972). The growth and development of the rat aorta: I. Morphological aspects. *J. Anat.* **113**, 1–16.

Bowen-Pope, D. F., and Ross, R. (1982). Platelet-derived growth factor: II. Specific binding to cultured cells. *J. Biol. Chem.* **257**, 5161–5171.

Bowen-Pope, D. F., and Ross, R. (1983). Is epidermal growth factor present in human blood? Alteration of EGF binding specificity in the radioreceptor assay. *Biochem. Biophys. Res. Commun.* **114**, 1036–1041.

Bowen-Pope, D. F., and Ross, R. (1985). The platelet-derived growth factor receptor. *In* "Methods in Enzymology" (L. Birnbaumer and B. W. O'Malley, eds.), Vol. 109, pp. 69–100. Academic Press, New York.

Bowen-Pope, D. F., and Rubin, H. (1983). Growth stimulatory precipitates of Ca^{2+} and pyrophosphate. *J. Cell Physiol.* **117**, 51–61.

Bowen-Pope, D. F., DiCorleto, P. E., and Ross, R. (1983). Interactions between receptors for platelet-derived growth factor and epidermal growth factor. *J. Cell Biol.* **96**, 679–683.

Bowen-Pope, D. F., Vogel, A., and Ross, R. (1984a). Production of platelet-derived growth factor-like molecules and reduced expression of platelet-derived growth factor receptors accompanies transformation by a wide spectrum of agents. *Proc. Natl. Acad. Sci. U.S.A.* **81**, 2396–2400.

Bowen-Pope, D. F., Malpass, T. W., Foster, D. M., and Ross, R. (1984b). Platelet-derived growth factor *in vivo*: Levels, activity, and rate of clearance. *Blood.* **64**, 458–469.

Brenner, E. G., Hakomori, S-i., Bowen-Pope, D. F., Raines, E., and Ross, R. (1984). Ganglioside-mediated modulation of cell growth, growth factor binding, and receptor phosphorylation. *J. Biol. Chem.* **259**, 6818–6825.

Burke, J. M. (1982). Cultured retinal glial cells are insensitive to platelet-derived growth factor. *Exp. Eye Res.* **35**, 663–669.

Butler-Gralla, E., and Herschman, H. (1981). Variants of 3T3 cells lacking mitogenic response to the tumor promoter tetradecanoylphorbol-acetate. *J. Cell. Physiol.* **107**, 59–67.

Canalis, E. (1981). Effect of platelet-derived growth factor on DNA and protein synthesis in cultured rat calvaria. *Metab., Clin. Exp.* **30**, 970–975.

Collins, M. K. L., Sinnett-Smith, J. W., and Rozengurt, E. (1983). Platelet-derived growth factor treatment decreases the affinity of the epidermal growth factor receptors of Swiss 3T3 cells. *J. Biol. Chem.* **258**, 11689–11693.

Cooper, J. A., Bowen-Pope, D. F., Raines, E., Ross, R., and Hunter, T. (1982). Similar effects of platelet-derived growth factor and epidermal growth factor on the phosphorylation of tyrosine in cellular proteins. *Cell (Cambridge, Mass.)* **31**, 263–273.

Cuatrecasas, P., and Hollenberg, M. D. (1976). Membrane receptors and hormone action. *Adv. Protein Chem.* **30**, 251–451.

Deuel, T. F., Huang, J. S., Proffitt, R. T., Baenziger, J. U., Chang, D., and Kennedy, B. B. (1981). Human platelet-derived growth factor—purification and resolution into two active protein fractions. *J. Biol. Chem.* **256**, 8896–8899.

Deuel, T. F., Senior, R. M., Huang, J. S., and Griffin, G. L. (1982). Chemotaxis of monocytes and neutrophils to platelet-derived growth factor. *J. Clin. Invest.* **69**, 1046–1049.

Deuel, T. F., Huang, J. S., Huang, S. S., Stroobant, P., and Waterfield, M. D. (1983). Expression of a platelet-derived growth factor-like protein in simian sarcoma virus transformed cells. *Science* **221**, 1348–1350.

Dicker, P., and Rozengurt, E. (1981). Stimulation of DNA synthesis by transient exposure of cell cultures to TPA or polypeptide mitogens: Induction of competence or incomplete removal. *J. Cell. Physiol.* **109**, 99–109.

Dicker, P., Pohjanpelto, P., Pettican, P., and Rozengurt, E. (1981). Similarities between fibroblast-derived growth factor and platelet-derived growth factor. *Exp. Cell Res* **135**, 221–227.

DiCorleto, P. E., and Bowen-Pope, D. F. (1983). Cultured endothelial cells produce a platelet-derived growth factor-like protein. *Prot. Natl. Acad. Sci. U.S.A.* **80,** 1919–1923.

Ek, B., and Heldin, C. A. (1982). Characterization of tyrosine-specific kinase activity in human fibroblast membranes stimulated by platelet-derived growth factor. *J. Biol. Chem.* **257,** 10486–10492.

Ek, B., Westermark, B., Wasteson, A., and Heldin, C. H. (1982). Stimulation of tyrosine-specific phosphorylation by platelet-derived growth factor. *Nature (London)* **295,** 419–420.

Fabricant, R. N., DeLarco, J. E., and Todaro, G. J. (1977). Nerve growth factor receptors on human melanoma cells in culture. *Proc. Natl. Acad. Sci. U.S.A.* **74,** 565–569.

Gerrity, R. G., and Cliff, W. J. (1975). The aortic tunica media of the developing rat: I. Quantitative stereologic and biochemical analysis. *Lab. Invest.* **32,** 585–600.

Glenn, K. C., Bowen-Pope, D. F., and Ross, R. (1982). Platelet-derived growth factor: III. Identification of a platelet-derived growth factor receptor by affinity labeling. *J. Biol. Chem.* **257,** 5172–5176.

Grossberg, A. L. (1977). Some chemical bases for biological specificity as exemplified by hapten–antibody interactions. In "Cell Interaction in Differentiation" (M. Karkinen-Jääskelainen, L. Saxén, and L. Weiss, eds.), pp. 291–309. Academic Press, New York.

Gudas, L. J., Singh, J. P., and Stiles, C. D. (1983). Secretion of growth regulatory molecules by teratocarcinoma stem cells. *Cold Spring Harbor Conf. Cell Proliferation* **10,** 229–236.

Hakamori, S-i. (1975). Structures and organization of cell surface glycolipids. Dependency on cell growth and malignant transformation. *Biochim. Biophys. Acta.* **417,** 55–89.

Haudenschild, C. C., Zahniser, D., Folkman, J., and Klagsbrun, M. (1976). Human vascular endothelial cells in culture. Lack of response to serum growth factors. *Exp. Cell Res.* **98,** 175–183.

Heldin, C.-H., Wasteson, A., and Westermark, B. (1977). Partial purification and characterization of platelet factors stimulating the multiplication of normal human glial cells. *Exp. Cell Res.* **109,** 429–437.

Heldin, C.-H., Westermark, B., and Wasteson, A. (1979a). Platelet-derived growth factor: Purification and partial characterization. *Proc. Natl. Acad. Sci. U.S.A.* **76,** 3722–3726.

Heldin, C.-H., Westermark, B., and Wasteson, A. (1979b). Purification and characterization of human growth factors. *Cold Spring Harbor Conf. Cell Proliferation* **6,** 17–31.

Heldin, C.-H., Westermark, B., and Wasteson, A. (1980). Chemical and biological properties of a growth factor from human-cultured osteosarcoma cells: Resemblance with platelet-derived growth factor. *J. Cell. Physiol.* **105,** 235–246.

Heldin, C.-H., Westermark, B., and Wasteson, A. (1981). Specific receptors for platelet-derived growth factor on cells derived from connective tissue and glia. *Proc. Natl. Acad. Sci. U.S.A.* **78,** 3664–3668.

Heldin, C.-H., Wasteson, A., and Westermark, B. (1982). Interaction of platelet-derived growth factor with its fibroblast receptor: Demonstration of ligand degradation and receptor modulation. *J. Biol. Chem.* **257,** 4216–4221.

Heldin, C.-H., Ek, B., and Ronnstrand, L. (1983). Characterization of the receptor for platelet-derived growth factor on human fibroblasts. Demonstration of an intimate relationship with a 185,000 dalton substrate for the PDGF-stimulated kinase. *J. Biol. Chem.* **258,** 10054–10061.

Holley, R. W., Armour, R., Baldwin, J. H., Brown, K. D., and Yeh, Y. C. (1977). Density-dependent regulation of growth of BSC-1 cells in cell culture: Control of growth by serum factors. *Proc. Natl. Acad. Sci. U.S.A.* **74**, 5046–5050.

Huang, J. S., Huang, S. S., Kennedy, B., and Deuel, T. F. (1982). Platelet-derived growth factor: Specific binding to target cells. *J. Biol. Chem.* **257**, 8130–8136.

Huang, J. S., Huang, S. S., and Deuel, T. F. (1983). Human platelet-derived growth factor: Radioimmunoassay and discovery of a specific plasma-binding protein. *J. Cell Biol.* **97**, 383–388.

Huang, J. S., Huang, S. S., and Deuel, T. F. (1984). Specific covalent binding of platelet-derived growth factor to human plasma alpha$_2$-macroglobulin. *Proc. Natl. Acad. Sci. U.S.A.* **81**, 342–346.

Kaplan, J. (1981). Polypeptide-binding membrane receptors: Analysis and classification. *Science* **212**, 14–20.

King, G. L., Kahn, C. R., and Heldin, C. H. (1983). Sharing of biological effect and receptors between guinea pig insulin and platelet-derived growth factor. *Proc. Natl. Acad. Sci. U.S.A.* **80**, 1308–1312.

Kohler, N., and Lipton, A. (1974). Platelets as a source of fibroblast growth-promoting activity. *Exp. Cell Res.* **87**, 297–301.

Lemke, G. E., and Brokes, J. (1983). Glial growth factor: A mitogenic polypeptide of the brain and pituitary. *Fed. Proc. Fed. Am. Soc. Exp. Biol.* **42**, 2627–2629.

Lim, R. W., and Hauschka, S. D. (1984). A rapid decrease in epidermal growth factor-binding capacity accompanies the terminal differentiation of mouse myoblasts *in vitro*. *J. Cell Biol.* **98**, 739–747.

Looker, T., and Berry, C. L. (1972). The growth and development of the rat aorta. II. Changes in nucleic acid and scleroprotein content. *J. Anat.* **113**, 17–34.

Maroudas, N. G. (1974). Short-range diffusion gradients. *Cell (Cambridge, Mass.)* **3**, 217–219.

Martin, G. R. (1980). Teratocarcinomas and mammalian embryogenesis. *Science* **209**, 768–776.

Nilsson, J., Thyberg, J., Heldin, C.-H., Wasteson, A., and Westermark, B. (1983a). Surface binding and internalization of platelet-derived growth factor in human fibroblasts. *Proc. Natl. Acad. Sci. U.S.A.* **80**, 5592–5596.

Nilsson, J., Ksiazek, T., Heldin, C.-H., and Thyberg, J. (1983b). Demonstration of stimulatory effects of platelet-derived growth factor on cultivated rat arterial smooth muscle cells. *Exp. Cell Res.* **145**, 231–237.

Nishimura, J., Huang, J. S., and Deuel, T. F. (1982). Platelet-derived growth factor stimulates tyrosine-specific protein kinase activity in Swiss mouse 3T3 cell membranes. *Proc. Natl. Acad. Sci. U.S.A.* **79**, 4303–4307.

Nister, M., Heldin, C.-H., Wasteson, A., and Westermark, B. (1982). A platelet-derived growth factor analog produced by a human clonal glioma cell line. *Ann. N.Y. Acad. Sci.* **397**, 25–33.

Pike, L. J., Bowen-Pope, D. F., Ross, R., and Krebs, E. G. (1983). Characterization of platelet-derived growth factor-stimulated phosphorylation in cell membranes. *J. Biol. Chem.* **258**, 9383–9390.

Pledger, W. J., Stiles, C. D., Antoniades, H. N., and Scher, C. D. (1977). Induction of DNA synthesis in BALB/c 3T3 cells by serum components: Re-evaluation of the commitment process. *Proc. Natl. Acad. Sci. U.S.A.* **74**, 4481–4485.

Pledger, W. J., Stiles, C. D., Antoniades, H. N., and Scher, C. D. (1978). An ordered sequence of events is required before BALB/c 3T3 cells become committed to DNA synthesis. *Proc. Natl. Acad. Sci. U.S.A.* **75**, 2839–2843.

Pratt, R. M., and Pastan, I. (1978). Decreased binding of epidermal growth factor to BALB/c 3T3 mutant cells defective in glycoprotein synthesis. *Nature (London)* **222**, 68–70.

Pruss, R. M., and Herschman, H. R. (1977). Variants of 3T3 cells lacking mitogenic response to epidermal growth factor. *Proc. Natl. Acad. Sci. U.S.A.* **74**, 3918–3921.

Raines, E. W., and Ross, R. (1982). Platelet-derived growth factor: I. High yield purification and evidence for multiple forms. *J. Biol. Chem.* **257**, 5154–5160.

Raines, E. W., Bowen-Pope, D. F., and Ross, R. (1984). Plasma binding proteins for platelet-growth factor that inhibit its binding to cell-surface receptors. *Proc. Natl. Acad. Sci. U.S.A.* **81**, 3424–3428.

Rizzino, A., and Bowen-Pope, D. F. (1985). Production of PDGF-like factors by embryonal carcinoma cells and response to PDGF by endoderm-like cells. *Dev. Biol.* (in press).

Rosenfeld, M. E., Bowen-Pope, D. F., and Ross, R. (1984). Platelet-derived growth factor: Morphologic and biochemical studies of binding, internalization, and degradation *in vitro*. *J. Cell. Physiol.* **121**, 263–274.

Rosenfeld, M., Keating, A., Bowen-Pope, D. F., Singer, J. W., and Ross, R. (1985). Responsiveness of the *in vitro* hematopoietic microenvironment to platelet-derived growth factor. *J. Leukemia Res.* (in press).

Ross, R., and Glomset, J. (1976). The pathogenesis of atherosclerosis. *N. Engl. J. Med.* **295**, 369–377, 420–425.

Ross, R., Glomset, J., Kariya, B., and Harker, L. (1974). Platelet-dependent serum factor that stimulates the proliferation of arterial smooth muscle cell *in vitro*. *Proc. Natl. Acad. Sci. U.S.A.* **71**, 1207–1210.

Ross, R., Vogel, A., Davies, P., Raines, E., Kariya, B., Rivest, M. J., Gustafson, C., and Glomset, J. (1979). The platelet-derived growth factor and plasma control cell proliferation. *Cold Spring Harbor Conf. Cell Proliferation* **6**, 3–16.

Rutherford, R. B., and Ross, R. (1976). Platelet factors stimulate fibroblasts and smooth muscle cells quiescent in plasma serum to proliferate. *J. Cell. Biol.* **69**, 196–203.

Sefton, B. M., and Hunter, T. (1984). Tyrosine protein kinases. *In* "Advances in Cyclic Nucleotide and Protein Phosphorylation Research" (P. Greengard and G. A. Robison, eds.) pp 195–226. Raven Press, New York.

Seifert, R. S., Schwartz, S. M., and Bowen-Pope, D. F. (1984). Developmentally regulated production of platelet-derived growth factor-like molecules. *Nature (London)* **311**, 669–671.

Singh, J. P., Chaikin, M. A., and Stiles, C. D. (1982). Phylogenetic analysis of platelet-derived growth factor by radioreceptor assay. *J. Cell Biol.* **95**, 667–671.

Singh, J. P., Chaikin, M. A., Pledger, W. J., Scher, C. D., and Stiles, C. D. (1983). Persistence of the mitogenic response to platelet-derived growth factor (competence) does not reflect a long-term interaction between the growth factor and the target cell. *J. Cell Biol.* **96**, 1497–1502.

Slayback, J. R. B., and Cheung, L. W. Y. (1977). *Comparative effects of human platelet growth factor on the growth and morphology* of human fetal and adult diploid fibroblasts. *Exp. Cell Res.* **110**, 462–466.

Smith, J. C., Singh, J. P., Lillquist, J. S., Goon, D. S., and Stiles, C. D. (1982). Growth factors adherent to cell substrate are mitogenically active *in situ*. *Nature (London)* **296**, 154–156.

Sporn, M. B., and Todaro, G. J. (1980). Autocrine secretion and malignant transformation of cells. *N. Engl. J. Med.* **303**, 878–880.

Striker, G. E., Soderland, C., Bowen-Pope, D. F., Gown, A. M., Schmer, G., Johnson, A.,

Luchtel, D., Ross, R., and Striker, L. J. (1984). Isolation, characterization, and propagation *in vitro* of human glomerular endothelial cells. *J. Exp. Med.* **160**, 325–328.

Tashjian, A. H., Jr., Hohman, E. L., Antoniades, H. N., and Levine, L. (1982). Platelet-derived growth factor stimulates bone resorption via a prostaglandin-mediated mechanism. *Endocrinology (Baltimore)* **111**, 118–124.

Van Obberghen-Schilling, E., Perez-Rodriguez, R., and Pouyssegur, J. (1982). Hirudin, a probe to analyze the growth-promoting activity of thrombin in fibroblasts: Re-evaluation of the temporal action of competence factors. *Biochem. Biophys. Res. Commun.* **106**, 79–86.

Vogel, A., Ross, R., and Raines, E. (1980). Role of serum in density-dependent inhibition of growth cells in culture: Platelet-derived growth factor is the major serum determinant of saturation density. *J. Cell Biol.* **85**, 377–385.

Wall, R. T., Harker, L. A., Quadracci, L. J., and Striker, G. E. (1978). Factors influencing endothelial cell proliferation *in vitro*. *J. Cell. Physiol.* **96**, 203–214.

Weinstein, R., Stemerman, M. B., and Maciag, T. (1981). Hormonal requirements for growth of arterial smooth muscle cells *in vitro*: An endocrine approach to atherosclerosis. *Science* **212**, 818–820.

Westermark, B., and Wasteson, A. (1976). A platelet factor stimulating human normal glial cells. *Exp. Cell Res.* **98**, 170–174.

Williams, L. T., Tremble, P., and Antoniades, H. N. (1982). Platelet-derived growth factor binds specifically to receptors on vascular smooth muscle cells and the binding becomes nondissociable. *Proc. Natl. Acad. Sci. U.S.A.* **79**, 5867–5870.

Williams, L. T., Antoniades, H. N., and Goetzl, E. J. (1983). Platelet-derived growth factor stimulates mouse 3T3 cell mitogenesis and leukocyte chemotaxis through different structural determinants. *J. Clin. Invest.* **72**, 1759–1763.

Witte, L. D., Kaplan, K. L., Nossel, H. L., Lages, B. A., Weiss, H. F., and Goodman, D. S. (1978). Studies of the release from human platelets of the growth factor for cultured human arterial smooth muscle cells. *Circ. Res.* **42**, 402–409.

IV

Transduction Mechanisms

11

The Role of Tyrosine Protein Kinases in the Action of Growth Factors

BARTHOLOMEW M. SEFTON

Molecular Biology and Virology Laboratory
Salk Institute
San Diego, California

I. INTRODUCTION AND BACKGROUND

The Discovery of Tyrosine Protein Kinases

Protein kinases with a specificity for tyrosine were unknown until such activity was detected in association with the transforming proteins of three tumor viruses: polyomavirus (Eckhart *et al.*, 1979), Abelson murine leukemia virus (Witte *et al.*, 1980), and Rous sarcoma virus (RSV) (Hunter and Sefton, 1980). It is now known that this activity is an inherent property of the transforming proteins of both

315

Abelson virus and RSV (Wang *et al.*, 1982; Gilmer and Erikson, 1981; McGrath and Levinson, 1982) and that it plays a fundamental role in cellular transformation (Sefton *et al.*, 1980b). In contrast, the tyrosine protein kinase activity associated with the transforming protein of polyomavirus, which may be equally important in cellular transformation, has been found to arise from a cellular enzyme which is bound to the viral protein (Courtneidge and Smith, 1983).

When the nature of the activity of these transforming proteins was discovered, it was immediately apparent that even normal uninfected cells might contain tyrosine protein kinases. This was clear because (1) the transforming gene of RSV, termed either *src* or v-*src*, was known to have originated by transduction of cellular chromosomal DNA by a relatively benign ancestral virus (Stehelin *et al.*, 1976), and (2) the cellular form of this gene, termed c-*src*, was known to encode a protein kinase (Oppermann *et al.*, 1979) with almost complete sequence identity with the viral polypeptide (Sefton *et al.*, 1980a). The crucial question was whether this activity was expressed *in vivo*. The discovery of phosphotyrosine in protein in normal uninfected cells revealed that either the product of the c-*src* gene or some other as yet unknown normal cellular tyrosine protein kinase was active *in vivo* (Hunter and Sefton, 1980). Since these original observations, it has become clear that the vertebrate genome encodes a remarkably large family of protein kinases with a specificity for tyrosine (Sefton and Hunter, 1984) and that the products of these genes are likely to play a crucial role in the control of cell multiplication in normal cells.

Fibroblasts transformed by RSV differ from normal fibroblasts in several significant respects. First, their growth is not inhibited at high cell density or in media which contain low concentrations of growth factors. Second, they are more rounded and noticeably less adherent than normal cells. Finally, they metabolize glucose abnormally. It is almost certain that all these changes result from the phosphorylation of cellular polypeptides by p60[src]. Transformation results, therefore, from either the excessive or untimely phosphorylation of cellular proteins whose activity or function is normally modulated by phosphorylation on tyrosine or the *de novo* phosphorylation of proteins which normally are not regulated by tyrosine protein kinases.

If the processes affected by transformation by RSV are normally regulated, in a benign manner, by cellular tyrosine protein kinases, identification of the substrates of p60[src] in a transformed cell should reveal simultaneously the substrates of crucial protein kinases in normal cells. Although there is little information available yet as to whether cell shape and glucose metabolism are modulated in uninfected cells by tyrosine protein kinases, there is now ample reason to believe that tyrosine protein kinases do in fact mediate the action of a diverse group of mitogenic compounds in normal cells and that growth regulation is lost after infection by RSV because of the intervention of the viral kinase in normal regulatory pathways.

The first suggestion that this might be the case came from the characterization of the receptor for epidermal growth factor (EGF; Chapter 7). Stanley Cohen and colleagues had found that this receptor possessed protein kinase activity which was stimulated by EGF *in vitro* (Carpenter *et al.*, 1978, 1979). As was the case with $p60^{src}$ (Collett and Erikson, 1978), this protein kinase activity was originally thought to have a specificity for threonine (Carpenter *et al.*, 1978). The realization that phosphotyrosine cannot be distinguished from phosphothreonine by the traditional phosphoamino acid analysis (electrophoresis at pH 1.9), prompted a reexamination of the amino acid specificity of the receptor and the discovery that the EGF receptor was a tyrosine protein kinase (Ushiro and Cohen, 1980). As is the case with $p60^{src}$, it is now clear that this protein kinase activity is an inherent property of the receptor (Cohen *et al.*, 1982, Buhrow *et al.*, 1982) and that the receptor manifests this activity *in vivo* (Hunter and Cooper, 1981).

The EGF receptor was just the tip of an iceberg. The cell surface receptors for platelet-derived growth factor (PDGF; see Chapter 10), insulin, and insulin-like growth factor I/somatomedin C (IGF; see Chapter 4) all have an associated tyrosine protein kinase activity (Ek *et al.*, 1982; Nishimura *et al.*, 1982; Kasuga *et al.*, 1982; Jacobs *et al.*, 1983; Rubin *et al.*, 1983). Indeed, in the case of the insulin receptor and the IGF I receptor, this activity is known to be an intrinsic property of the 95,000-dalton subunit of the tetrameric receptor (Shia and Pilch, 1983; van Obberghen *et al.*, 1983; Roth and Cassell, 1983).

II. SUBSTRATES OF THE GROWTH FACTOR RECEPTORS

Fundamental to an understanding of the role of protein phosphorylation in the relay of the mitogenic signal is knowledge of the primary polypeptide substrates of the growth factor receptor-associated protein kinase activities. Identification of these proteins is, however, no small task. The fact that they necessarily will contain elevated levels of phosphotyrosine after addition of the growth factor both aids and hinders their detection. The abundance of phosphotyrosine in protein, even in the most unusual cases, is vanishingly small. It normally constitutes only 0.03% of the acid-stable phosphoamino acids in total cellular protein (Hunter and Sefton, 1980) and has almost never been found to be present at a level in excess of 1% (Hunter and Cooper, 1981). As a result, the background of proteins phosphorylated on tyrosine by irrelevant tyrosine protein kinases is low. For example, it is likely that 9 of every 10 cellular proteins phosphorylated on tyrosine in cells transformed by RSV are phosphorylated as a direct result of the activity of the viral transforming protein (Sefton et al., 1980b). Nevertheless, the low absolute abundance of phosphotyrosine in cells treated with mitogens makes detection of proteins which contain it difficult, since they are obscured inevitably by the very much more abundant cellular proteins which contain exclusively phosphoserine or phosphothreonine.

III. METHODS OF DETECTION

All procedures employed to date to detect substrates of growth factor receptors have involved pre-labeling of cells with $^{32}P_i$ and comparison of cellular phosphoproteins from treated and untreated cells by gel electrophoresis. Inspection of autoradiograms of cell lysates fractionated by simple one-dimensional SDS–polyacrylamide gel electrophoresis is useless in this undertaking, since only the most abundant phosphoserine-containing polypeptides are visible. Indeed, inspection of autoradiograms of cell lysates fractionated by two-dimensional gel electrophoresis is only slightly more useful. It allows detection of one primary substrate of a growth factor receptor kinase, p36.

Three approaches have been taken to identify additional substrates. The first takes advantage of the fact that phosphotyrosine is much more stable to alkali than is either RNA or phosphoserine. Treatment of two-dimensional gels of ^{32}P-labeled cellular protein with 1 M KOH for 2 hr at 55°C hydrolyzes most of the RNA—which is found smeared throughout the gel—and dephosphorylates much of the phosphoserine in protein, without reducing the label in most phosphotyrosine-containing proteins significantly (Cheng and Chen, 1981; Cooper and Hunter, 1981a). Remarkably, the acrylamide gel retains sufficient physical integrity that significant diffusion of polypeptides does not occur. This procedure has revealed a small number of substrates of the receptors for both EGF and PDGF (Hunter and Cooper, 1981; Gilmore and Martin, 1983; Cooper et al., 1982, 1984).

Weber and colleagues (Martinez et al., 1982; Nakamura et al., 1983; Bishop et al., 1983) have looked for the substrates of growth factor receptors by fractionating proteins from mitogen-treated cells by one-dimensional SDS–polyacrylamide gel electrophoresis, cutting the gel into as many as 120 slices, and performing phosphoamino acid analysis of the proteins eluted from each slice. This approach is much more laborious than two-dimensional gel electrophoresis and offers significantly less sensitivity and resolution. It has, nevertheless, the decided advantage that it yields quantitative data as to the relative amount of phosphotyrosine in substrates of a specific size and should in theory detect all fairly abundant substrates. Their approach has revealed the apparent importance of the substrate now known as p42.

The third approach is also the newest. Alonzo Ross and Ray Frackelton have prepared antibodies to an analogue of phosphotyrosine, aminobenzylphosphonic acid, and have used these to isolate phosphotyrosine-containing proteins by affinity chromatography (Ross et al., 1981; Frackelton et al., 1983). The procedure has the potential to allow the detection of heretofore unsuspected substrates of tyrosine protein kinases, although such has not yet been achieved.

IV. STUDIES WITH CELL CULTURES

The first evidence that the tyrosine kinase activity associated with the EGF receptor in vitro was not an artifact came from the observation that the level of phosphotyrosine in total cellular protein increased severalfold within minutes of the addition of EGF to growing human A-431 cells (Hunter and Cooper, 1981). This cell line was used deliberately for this experiment because these cells express on their surfaces approximately 10- to 20-fold more EGF receptor molecules than do typical fibroblastic cells (Fabricant et al., 1977). In fact, significant increases in the abundance of phosphotyrosine in total cell protein in response to EGF are difficult to detect in cells other than A-431 cells (Hunter and Cooper, 1981; Cooper et al., 1982). However, this apparent insensitivity of other cell lines is only a reflection of their lower levels of the EGF receptor and not an indication that EGF does not stimulate tyrosine protein kinase activity.

The increase in the total abundance of phosphotyrosine in A-431 cells suggested, but did not prove, that the receptor was stimulated by EGF to phosphorylate cellular proteins. The reason for uncertainty is that the receptor is itself phosphorylated on tyrosine upon exposure to EGF (L. E. King et al., 1980; Hunter and Cooper, 1981), and in all probability a significant fraction of the increase in the level of total phosphotyrosine reflects only the apparent autophosphorylation of the receptor (Frackelton et al., 1983). Nevertheless, at least two cellular proteins other than the receptor are obviously phosphorylated on tyrosine in response to EGF in A-431 cells (Hunter and Cooper, 1981; Erikson et al., 1981). Analysis of alkali-treated two-dimensional gels of total cellular protein revealed the increased phosphorylation on tyrosine of a protein of 36,000 daltons and of one of 81,000 daltons (Hunter and Cooper, 1981). These two proteins are now known as p36 and p81, respectively.

The 36,000-dalton phosphoprotein was found to be identical to the major cellular substrate of most of the viral transforming tyrosine protein kinases (Radke et al., 1980; Hunter and Cooper, 1981; Erikson et al., 1981). In addition, it was found to be phosphorylated at exactly the same tyrosine residue in EGF-treated A-431 cells and in RSV-infected A-431 cells (Cooper and Hunter, 1981c). This protein appears to have a structural role in cells. It is found associated with the inner face of the plasma membrane of many types of cells and in the brush border of the intestine (Cooper and Hunter, 1982; Amini and Kaji, 1983; Courtneidge et al., 1983; Greenberg and Edelman, 1983; Lehto et al., 1983; Nigg et al., 1983; Gould et al., 1984). There is no evidence to date that its phosphorylation affects its function, although it can be hypothesized that phosphorylation interferes with its normal structural role and contributes to the alterations in cell shape that accompany cellular transformation.

p81 remained poorly characterized until very recently. It is now clear that it is identical to an 80,000-dalton protein isolated from the brush border of chicken intestinal cells and that it too has a structural role in cells (Bretscher, 1983; K. Gould, personal communication). In particular, it is concentrated in regions rich in microvilli (Bretscher, 1983). Again, whether the phosphorylation of p81 alters its function is not yet known. This protein is also the substrate of at least two viral transforming tyrosine protein kinases, p60^{src} (K. Gould, personal communication), and the p85$^{gag-fes}$ protein of Snyder–Theilen feline sarcoma virus (Hunter and Cooper, 1983).

An important question is whether the phosphorylation of either of these abundant cellular proteins plays a crucial role in mediating the action of EGF. The answer is probably no. A-431 cells are not stimulated to grow by the doses of EGF which are necessary to induce the phosphorylation of p36 and p81. Instead, their growth is arrested (Gill and Lazar, 1981). The analysis of EGF-dependent protein phosphorylation in cells which do reinitiate growth in response to EGF has revealed a very different picture to that seen with A-431 cells (Cooper et al., 1982, 1984). The phosphorylation of p81 is never observed and the phosphorylation of p36 is seen only in some cell sublines of mouse 3T3 cells (Cooper et al., 1982). It is very unlikely, therefore, that the phosphorylation of either of these proteins is necessary for the reinitiation of growth. p36 and p81 are in all probability occasional substrates of the EGF receptor which are phosphorylated noticeably in A-431 cells because of the abnormal level of the receptor.

V. p42

In contrast to what is found with A-431 cells, the phosphorylation on tyrosine of proteins of 42,000 and 45,000 daltons is stimulated by EGF in a variety of cells which resume growth when exposed to EGF (Cooper et al., 1982, 1984; Nakamura et al., 1983). The best studied of these are a pair of related proteins, p42A and p42B, with A indicating the more acidic of the pair and B the more basic. These two proteins are phosphorylated on tyrosine in response to EGF in both mouse cells and chick cells and can, in retrospect, be detected as very minor species in EGF-treated A-431 cells (Cooper and Hunter, 1981c). They are of particular interest because they were found earlier to be phosphorylated on tyrosine in chick cells transformed by RSV (Cooper and Hunter, 1981a). p42A and p42B are, in fact, the only two proteins identified to date which are phosphorylated on tyrosine in both RSV-transformed chick cells and EGF-treated chick cells. They fulfill therefore one of the criteria for cellular proteins whose phosphorylation on tyrosine might be responsible for the loss of growth regulation in RSV-transformed cells.

Like EGF, PDGF stimulates the phosphorylation of its presumptive receptor on tyrosine when added to membrane preparations from responsive cells (Ek et al., 1982; Nishimura et al., 1982). This reaction appears to be indicative of the activity of the receptor in living cells. Treatment of mouse and chick cells with PDGF also stimulates the phosphorylation of p42A and p42B (Cooper et al., 1984; Nakamura et al., 1983) and, in certain cells, the phosphorylation of proteins of 41,000 and 45,000 daltons (Cooper et al., 1982). In light of the activity of the PDGF receptor in vitro, it is reasonable to conclude that these proteins are primary substrates of the PDGF receptor. However, this is in no sense certain.

A surprisingly large number of quite unrelated mitogenic compounds also stimulate the phosphorylation of p42A and p42B on tyrosine. These include the tumor promoter tetradecanoylphorbol acetate (TPA) (Bishop et al., 1983; Gilmore and Martin, 1983; Cooper et al.,

1984) and the mitogenic protease trypsin (Cooper et al., 1984). There are three possible explanations for the ability of these latter two agents to mimic EGF and PDGF. One is that they have specific receptors which also possess tyrosine protein kinase activity and that p42A and p42B are also primary substrates of these receptors. A second is that these mitogenic agents stimulate directly the receptors for EGF or PDGF and that these receptors in turn phosphorylate p42A and p42B. A third is that p42A and p42B are in all cases substrates of an as yet unknown tyrosine protein kinase whose activity is stimulated by the receptors for a number of growth factors. One reason to consider the third possibility seriously is that the presumptive receptor for TPA, protein kinase C (Castagna et al., 1982; Niedel et al., 1983), is an enzyme with a specificity for serine and threonine, but not for tyrosine (Nishizuka, 1983). If the action of TPA is in fact mediated only by protein kinase C, then the phosphorylation of p42A and p42B which is stimulated by TPA must occur as the consequence of the action of a tyrosine kinase which is regulated by protein kinase C.

As yet, too little is known about p42A and p42B. The proteins appear to be of low abundance and located in the cytosol. Since they become phosphorylated rapidly upon exposure of cells to essentially all mitogenic agents, it is not unreasonable to conclude that their phosphorylation on tyrosine plays a role in the decision of cells to multiply in response to growth-stimulatory compounds. The unrestrained replication of chick cells transformed by viruses which induce the phosphorylation of p42A and p42B on tyrosine could therefore arise from the chronic, unregulated phosphorylation of p42A and p42B by the viral transforming protein. However, this tidy explanation cannot hold for virally transformed mouse cells. Although p42A and p42B are phosphorylated on tyrosine in mouse cells in response to exposure to a number of different mitogens (Cooper et al., 1982; Nakamura et al., 1983), they are not phosphorylated detectably in transformed mouse cells (Cooper and Hunter, 1981b). The loss of growth regulation in the transformed mouse cells cannot therefore arise from the chronic phosphorylation of this protein, and this raises the possibility that the phosphorylation of p42A and p42B may not be, in fact, involved in the abrogation of growth control in transformed chick cells either.

The cellular substrates of the insulin receptor and the IGF I receptor have not yet been identified. Given the generality of p42 as a substrate of protein kinases activated by growth factors, it would be surprising if p42 were not one of the proteins which becomes phosphorylated on tyrosine in cells treated with insulin and IGF I.

VI. IMPORTANT QUESTIONS

A. Do Important Substrates of Growth Factor Receptor Tyrosine Kinases Remain to Be Found?

p42A and p42B are the most obvious substrates of tyrosine kinases which are responsive to mitogenic compounds when either chick cells or mouse cells are analyzed on alkali-treated two-dimensional gels (Cooper et al., 1982, 1984). Although this procedure offers great senitivity, it will not detect proteins which contain alkali-labile phosphate on tyrosine—such as is the case with the p50 substrate of p60^{src} (Hunter and Sefton, 1980)—or proteins which are refractory to isoelectric focusing, such as the growth factor receptors themselves. Nevertheless, there is good reason to think that p42A and p42B are prominent substrates. Measurements of protein phosphorylation by phosphoamino acid analysis of every fraction from a one-dimensional SDS–polyacrylamide gel analysis of proteins phosphorylated in response to mitogens reveals increased phosphotyrosine only in proteins of 40,000–45,000 daltons (Nakamura et al., 1983; Bishop et al., 1983), a size range which must include both p42A and p42B. This result suggests at face value that other abundant substrates are unlikely. However, this conclusion is not certain. This method of analysis detects little increased phosphotyrosine in the receptors for either EGF or PDGF in stimulated cells (Nakamura et al., 1983). In contrast, analysis of phosphotyrosine-containing proteins by affinity chromatography with anti-phosphotyrosine antibodies indicates that the predominant phosphotyrosine-containing protein in EGF-treated cells is the receptor itself (Frackelton et al., 1983). Part of the discrepancy probably derives from the fact that large proteins like the EGF receptor are the most difficult to detect by gel analysis because they elute inefficiently from gel slices. Resolution of these apparently contradictory results is clearly in order.

B. Do Growth Factor Receptors Have Functions Other Than Protein Phosphorylation?

There is little doubt that the receptors for EGF, PDGF, IGF I, and insulin possess intrinsic tyrosine protein kinase activity which is regulated by the peptide ligand. The receptors for EGF and insulin have both been shown to be labeled specifically by reactive ATP analogues and hence to possess ATP binding sites (Buhrow et al., 1982; Shia and Pilch, 1983; Roth and Cassell, 1983; van Obberghen et al., 1983). In addition, the deduced amino acid sequences of the EGF and insulin receptors have extensive homology with both the viral tyrosine protein kinases and the cyclic AMP-dependent protein kinase (Ullrich *et al.*, 1984; 1985). Nevertheless, this does not prove that the protein kinase activity is either sufficient or essential for transmission of the signal for cell replication. There are, in fact, two observations which are inconsistent with the idea that the protein kinase activity of the EGF receptor is sufficient to induce a mitogenic response.

First, EGF which has been cleaved with cyanogen bromide is reported to retain the ability to bind to the EGF receptor, albeit with a 10-fold reduced affinity, and to mimic intact EGF in many ways, including the stimulation of the phosphorylation in membrane preparations from human A-431 cells (Schreiber *et al.*, 1981; Yarden *et al.*, 1982; see Chapter 13). Nevertheless, this preparation of cleaved EGF is reported to be essentially inactive when assayed for its ability to stimulate increased thymidine incorporation in either human fibroblasts or mouse 3T3 cells (Schreiber *et al.*, 1981; Yarden *et al.*, 1982).

Secondly, although the degree of stimulation of phosphorylation of cellular protein by EGF in whole A431 cells correlates well with binding, being half-maximal when half the receptors are occupied (Hunter and Cooper, 1981; Cooper *et al.*, 1983), EGF stimulates growth in normal human fibroblasts at concentrations which are an order of magnitude lower than those necessary to cause half-maximal occupancy of its receptor (Hollenberg and Cuatrecasas, 1975). The good correlation between receptor occupancy and protein phosphorylation in A-431 cells and the disparity between receptor occupancy and growth stimulation in human fibroblasts raises the possibility that EGF can stimulate cell division at concentrations at which it has no measurable effect on the phosphorylation of cellular proteins. This possibility is difficult to evaluate rigorously, however, because the cells in which it is possible to measure the stimulation of phosphorylation of cellular proteins most accurately, A431 cells, are not stimulated to grow by EGF, and the cells with which it is possible to measure a mitogenic response to EGF, normal human fibroblasts, show only a slight stimulation of protein phosphorylation in response to EGF.

Unlike what is found for EGF, there is a very good correlation between the extent of binding of PDGF, IGF I and insulin to their receptors and the response of cells. When studied with intact cells, half-maximal stimulation of growth, half-maximal stimulation of phosphorylation, and half-maximal binding all occur at the same concentration of ^{125}I-PDGF (Bowen-Pope and Ross, 1982; Cooper *et al.*, 1983).

Similarly, the degree of stimulation of both cell division and phosphorylation in membranes appear to parallel closely the extent of binding of IGF I (Zapf et al., 1978; Rubin et al., 1983). Further, the ability of insulin to stimulate phosphorylation in membranes correlates well with the ability of the hormone to elicit biological effects. Strikingly, both occur when only a small fraction of the available receptors are occupied (Gammeltoft and Gliemann, 1973; Zick et al., 1983). Taken together, these observations buttress the notion that the ability of these three peptides to stimulate protein phosphorylation is essential to their biological activity.

C. Other Peptide Hormone Receptors

Three of the peptide hormones whose receptors have an associated tyrosine protein kinase activity, EGF, PDGF, and IGF I, have clear-cut mitogenic activity. It is quite possible that other, and perhaps all, receptors for peptides which stimulate cell multiplication will be found to have an associated tyrosine protein kinase activity.

Insulin is somewhat different. Although the binding of insulin to its receptor is apparently mitogenic in a few types of cells (Koontz and Iwahasi, 1980), the mitogenic effect of insulin on most cells appears to arise from a low-affinity interaction with the receptor for IGF I (G. L. King et al., 1980). The predominant effects of insulin are metabolic. If insulin action is initiated entirely by the tyrosine protein kinase activity of the receptor, it could be that, as is the case with cell multiplication, important and as yet unappreciated aspects of cellular metabolism are controlled by tyrosine protein kinases.

Finally, it should be kept in mind that the extent of protein phosphorylation can be affected by regulating the activity of either protein kinases or protein phosphatases. Examples of peptides which increase the extent of protein phosphorylation on tyrosine by inhibiting a tyrosine protein phosphatase, rather than by stimulating a tyrosine protein kinase, may well be found.

D. Growth Factors, Growth Factor Receptors, and Oncogenes

Both viral oncogenes and peptide growth factors stimulate cell multiplication. Because viral oncogenes invariably represent transduced cellular genes, the possibility that they are derived from those cellular genes which encode growth factors, growth factor receptors, and the proteins which mediate growth factor action in the cell has received considerable attention.

Recent findings have been shown this notion to have a sound basis in fact. First, it has been discovered that the *sis* oncogene encodes one of the two subunits of PDGF (Doolittle *et al.*, 1983; Waterfield *et al.*, 1983). It is probable that *sis*-mediated transformation results, at least in part, from the chronic production of a PDGF and that it may be mediated by the PDGF receptor. Secondly, the *erbB* oncogene is now known to be a fragment of the gene for the EGF receptor (Downward *et al.* 1984; Ullrich *et al.*, 1984). Transformation in this case is likely to result from the unregulated tyrosine protein kinase activity of a truncated EGF receptor.

Do all oncogenes encode growth factors or receptors? Almost certainly not. The products of the *myc*, *myb*, and *fos* oncogenes are found exclusively in the nucleus (Abrams *et al.*, 1982; Klempnauer *et al.*, 1984; Curran *et al.*, 1984). Additionally, even a number of the oncogene products that have extensive amino acid sequence homology with the EGF receptor, such as p60^{src}, are not cell surface proteins. The evidence is compelling that p60^{src} is located entirely in the cytoplasm of transformed cells. This transforming protein may well function like the cytoplasmic catalytic domain of a growth factor receptor, but is probably not itself a receptor.

Given that the genes for both PDGF and the EGF receptor can function as oncogenes, it is not at all unreasonable to suspect that the genes for both the PDGF receptor and for EGF may also have oncogenic activity. Indeed, the genes for all mitogenic peptides and their receptors may eventually be implicated in malignant transformation. The study of growth factors and their receptors should continue to teach us much about oncogenes.

REFERENCES

Abrams, H. D., Rohrschneider, L. R., and Eisenman, R. N. (1982). Nuclear location of the putative transforming protein of avian myelocytomatosis virus. *Cell* **29**, 427–439.
Amini, S., and Kaji, A. (1983). Association of pp36, a phosphorylated form of the presumed target of the *src* protein of Rous sarcoma virus, with the membrane of chicken cells transformed by Rous sarcoma virus. *Proc. Natl. Acad. Sci. U.S.A.* **80**, 960–964.
Bishop, R., Martinez, R., Nakamura, K. D., and Weber, M. J. (1983). A tumor promoter stimulates phosphorylation on tyrosine. *Biochem. Biophys. Res. Commun.* **115**, 536–543.
Bowen-Pope, D. F., and Ross, R. (1982). Platelet-derived growth factor. II. Specific binding to cultured cells. *J. Biol. Chem.* **257**, 5161–5171.

Bretscher, A. (1983). Purification of an 80,000 dalton protein that is a component of the isolated microvillus cytoskeleton, and its localization in nonmuscle cells. *J. Cell Biol.* **97**, 425–432.

Buhrow, S. A., Cohen, S., and Stavos, J. V. (1982). Affinity labeling of the protein kinase associated with the epidermal growth factor receptor in membrane vesicles from A-431 cells. *J. Biol. Chem.* **257**, 4019–4022.

Carpenter, G., King, L., and Cohen, S. (1978). Epidermal growth factor stimulates phosphorylation in membrane preparations *in vitro*. *Nature (London)* **276**, 409–410.

Carpenter, G., King, L., and Cohen, S. (1979). Rapid enhancement of protein phosphorylation in A-431 cell membrane preparations by epidermal growth factor. *J. Biol. Chem.* **254**, 4884–4891.

Castagna, M., Takai, Y., Kaibuchi, K., Sano, K., Kikkawa, U., and Nishizuka, Y. (1982). Direct activation of calcium-activated, phospholipid-dependent protein kinase by tumor-promoting phorbol esters. *J. Biol. Chem.* **257**, 7847–7851.

Cheng, Y.-S., E., and Chen, L. B. (1981). Detection of phosphotyrosine-containing 36,000 dalton protein in the framework of cells transformed with Rous sarcoma virus. *Proc. Natl. Acad. Sci. U.S.A.* **78**, 2388–2392.

Cohen, S., Ushiro, H., Stoschek, C., and Chinkers, M. (1982). A native 170,000 epidermal growth factor receptor–kinase complex from shed plasma membrane vesicles. *J. Biol. Chem.* **257**, 1523–1531.

Collett, M. S., and Erikson, R. L. (1978). Protein kinase activity associated with the avian sarcoma virus *src* gene product. *Proc. Natl. Acad. Sci. U.S.A.* **75**, 2021–2024.

Cooper, J. A., and Hunter, T. (1981a). Changes in protein phosphorylation in Rous sarcoma virus transformed chicken embryo cells. *Mol. Cell. Biol.* **1**, 165–178.

Cooper, J. A., and Hunter, T. (1981b). Four different classes of retroviruses induce phosphorylation of tyrosines present in similar cellular proteins. *Mol. Cell Biol.* **1**, 394–407.

Cooper, J. A., and Hunter, T. (1981c). Similarities and differences between the effects of epidermal growth factor and Rous sarcoma virus. *J. Cell Biol.* **91**, 878–883.

Cooper, J. A., and Hunter, T. (1982). Discrete primary locations of a tyrosine protein kinase and of three proteins that contain phosphotyrosine in virally transformed chick fibroblasts. *J. Cell Biol.* **94**, 287–296.

Cooper, J. A., Bowen-Pope, D., Raines, E., Ross, R., and Hunter, T. (1982). Similar effects of platelet-derived growth factor and epidermal growth factor on the phosphorylation of tyrosine in cellular proteins. *Cell (Cambridge, Mass.)* **31**, 263–273.

Cooper, J. A., Scolnick, E. M., Ozanne, B., and Hunter, T. (1983). EGF receptor metabolism and protein kinase activity in human A-431 cells infected with Snyder–Theilin feline sarcoma virus or Harvey or Kirsten murine sarcoma viruses. *J. Virol.* **48**, 752–764.

Cooper, J. A., Sefton, B. M., and Hunter, T. (1984). Diverse mitogenic agents induce the phosphorylation of two related 42,000 dalton proteins on tyrosine in quiescent chick cells. *Mol. Cell. Biol.* **4**, 30–37.

Courtneidge, S. A., and Smith, A. E. (1983). Polyomavirus transforming protein associates with the product of the c-src cellular gene. *Nature (London)* **303**, 435–439.

Courtneidge, S. A., Ralston, R., Alitalo, K., and Bishop, J. M. (1983). The subcellular location of an abundant substrate (p36) for tyrosine–specific protein kinases. *Mol. Cell. Biol.* **3**, 340–350.

Curran, T., Miller, A. D., Zokas, L., and Verma, I. M. (1984). Viral and cellular *fos* proteins: A comparative analysis. *Cell* **36**, 259–268.

Doolittle, R. F., Hunkapillar, M. W., Hood, L. E., Devare, S. G., Robbins, K. C., Aaronson,

S. A., and Antoniades, H. N. (1983). Simian sarcoma virus onc gene, v-sis, is derived from the gene (or genes) encoding a platelet-derived growth factor. *Science* **221**, 275–277.

Downward, J., Yarden, Y., Mayes, E., Scrace, G., Totty, N., Stockwell, P., Ullrich, A., Schlessinger, J., and Waterfield, M. D. (1984). Close similarity of epidermal growth factor receptor and v-erb-B oncogene protein sequences. *Nature (London)* **307**, 521–527.

Eckhart, W., Hutchinson, M. A., and Hunter, T. (1979). An activity phosphorylating tyrosine in polyoma T antigen immunoprecipitates. *Cell (Cambridge, Mass.)* **18**, 925–933.

Ek, B., Westermark, B., Wasteson, A., and Heldin, C.-H. (1982). Stimulation of tyrosine-specific phosphorylation by platelet-derived growth factor. *Nature (London)* **295**, 419–420.

Erikson, E., Shealy, D. J., and Erikson, R. L. (1981). Evidence that viral transforming gene products and epidermal growth factor stimulate phosphorylation of the same cellular protein with similar specificity. *J. Biol. Chem.* **256**, 11381–11384.

Fabricant, R. N., DeLarco, J. E., and Todaro, G. J. (1977). Nerve growth factor receptors on human melanoma cells in culture. *Proc. Natl. Acad. Sci. U.S.A.* **74**, 565–569.

Frackelton, A. R., Ross, A. H., and Eisen, H. N. (1983). Characterization and use of monoclonal antibodies for isolation of phosphotyrosyl proteins from retrovirus-transformed and growth factor-stimulated cells. *Mol. Cell. Biol.* **3**, 1343–1352.

Gammeltoft, S., and Gliemann, J. (1973). Binding and degradation of ^{125}I-labeled insulin by isolated rat fat cells. *Biochim. Biophys. Acta* **320**, 16–32.

Gill, G. N., and Lazar, C. S. (1981). Increased phosphotyrosine content and inhibition of proliferation in EGF-treated A-431 cells. *Nature (London)* **293**, 305–307.

Gilmer, T. M., and Erikson, R. L. (1981). Rous sarcoma virus transforming protein, p60^{src}, expressed in *E. coli*, functions as a protein kinase. *Nature (London)* **294**, 771–773,

Gilmore, T., and Martin, G. S. (1983). Phorbol ester and diacylglycerol induce protein phosphorylation at tyrosine. *Nature (London)* **306**, 487–490.

Gould, K. L., Cooper, J. A., and Hunter, T. (1984). The 46,000 dalton tyrosine protein kinase substrate is widespread, whereas the 36,000 dalton substrate is only expressed at high level in certain rodent tissues. *J. Cell Biol.* **98**, 487–497.

Greenberg, M. E., and Edelman, G. M. (1983). The 34Da pp60^{src} substrate is located at the inner face of the plasma membrane. *Cell (Cambridge, Mass)* **33**, 767–779.

Hollenberg, M. D., and Cuatrecasas, P. (1975). Insulin and epidermal growth factor. Human fibroblast receptors related to deoxyribonucleic acid synthesis and amino acid uptake. *J. Biol. Chem.* **250**, 3845–3853.

Hunter, T., and Cooper, J. A. (1981). Epidermal growth factor induces rapid tyrosine phosphorylation of proteins in A-431 human tumor cells. *Cell (Cambridge, Mass.)* **24**, 741–752.

Hunter, T., and Cooper, J. A. (1983). The role of tyrosine phosphorylation in malignant transformation and in cellular growth control. *Prog. Nucleic Acid Res. Mol. Biol.* **29**, 221–232.

Hunter, T., and Sefton, B. M. (1980). The transforming gene product of Rous sarcoma virus phosphorylates tyrosine. *Proc. Natl. Acad. Sci. U.S.A.* **77**, 1311–1315.

Jacobs, S., Kull, F. C., Earp, H. S., Svoboda, M. E., Van Wyck, J. J., and Cuatrecasas, P. (1983). Somatomedin-C stimulates the phosphorylation of the β-subunit of its own receptor. *J. Biol. Chem.* **258**, 9581–9584.

Kasuga, M., Zick, Y., Blithe, D. L., Crettaz, M., and Kahn, C. R. (1982). Insulin stimulates tyrosine phosphorylation of the insulin receptor in a cell-free system. *Nature (London)* **298**, 667–669.

King, G. L., Kahn, C. R., Rechler, M. M., and Nissley, S. P. (1980). Direct demonstration of separate receptors for growth and metabolic activities of insulin and multiplication-stimulating activity (an insulin-like growth factor) using antibodies to the insulin receptor. *J. Clin. Invest.* **66**, 130–140.

King, L. E., Carpenter, G., and Cohen, S. (1980). Characterization by electrophoresis of epidermal growth factor stimulated phosphorylation using A-431 membranes. *Biochemistry* **19**, 1524–1528.

Klempnauer, K.-H., Symonds, G., Evan, G. I., and Bishop, J. M. (1984). Subcellular locations of proteins encoded by oncogenes of avian myeleoblastosis virus and avian leukemia virus E26 and by the chicken c-myb gene. *Cell* **37**, 537–547.

Koontz, J. W., and Iwahasi, M. (1980). Insulin as a potent, specific growth factor in a rat hepatoma cell line. *Science* **211**, 947–949.

Lehto, V.-P., Virtanen, I., Paasivuo, R., Ralston, R., and Alitalo, K. (1983). The p36 substrate of tyrosine-specific protein kineases co-localizes with non-erythrocyte α-spectrin antigen, p230, in surface lamina of cultured fibroblasts. *EMBO J.* **2**, 1701–1705.

McGrath, J. P., and Levinson, A. D. (1982). Bacterial expression of an enzymatically active protein encoded by RSV src gene. *Nature (London)* **295**, 423–425.

Martinez, R., Nakamura, K. D., and Weber, M. J. (1982). Identification of phosphotyrosine-containing proteins in untransformed and Rous sarcoma virus-transformed chicken embryo cells. *Mol. Cell. Biol.* **2**, 653–665.

Nakamura, K. D., Martinez, R., and Weber, M. J. (1983). Tyrosine phosphorylation of specific proteins following mitogen stimulation of chicken embryo fibroblasts. *Mol. Cell. Biol.* **3**, 380–390.

Niedel, J. E., Kuhn, L. J., and Vandenbark, G. R. (1983). Phorbol diester receptor copurifies with protein kinase C. *Proc. Natl. Acad. Sci. U.S.A.* **80**, 36–40.

Nigg, E. A., Cooper, J. A., and Hunter, T. (1983). Immunofluorescent localization of a 39,000 dalton substrate of tyrosine protein kinases to the cytoplasmic surface of the plasma membrane. *J. Cell Biol.* **97**, 1601–1609.

Nishimura, J., Huang, J. S., and Deuel, T. F. (1982). Platelet-derived growth factor stimulates tyrosine-specific protein kinase activity in Swiss mouse 3T3 cell membranes. *Proc. Natl. Acad. Sci. U.S.A.* **79**, 4303–4307.

Nishizuka, Y. (1983). Phospholipid degradation and signal translation for protein phosphorylation. *Trends Biochem. Sci.* **8**, 13–16.

Oppermann, H., Levinson, A. D., Levintow, L., Varmus, H. E., and Bishop, J. M. (1979). Uninfected vertebrate cells contain a protein that is closely related to the product of the avian sarcoma virus transforming gene (src). *Proc. Natl. Acad. Sci. U.S.A.* **76**, 1804–1808.

Radke, K., Gilmore, T., and Martin, G. S. (1980). Transformation by Rous sarcoma virus: A cellular substrate for transformation-specific protein phosphorylation contains phosphotyrosine. *Cell (Cambridge, Mass.)* **21**, 821–828.

Ross, A. H., Baltimore, D., and Eisen, H. (1981). Phosphotyrosine containing proteins isolated by affinity chromatography with antibodies to a synthetic hapten. *Nature (London)* **294**, 654–656.

Roth, R. A., and Cassell, D. J. (1983). Insulin receptor: Evidence that it is a protein kinase. *Science* **219**, 299–301.

Rubin, J. B., Shia, M. A., and Pilch, P. F. (1983). Stimulation of tyrosine-specific phosphorylation in vitro by insulin-like growth factor I. *Nature (London)* **305**, 438–440.

Schreiber, A. B., Yarden, Y., and Schlessinger, J. (1981). A non-mitogenic analogue of epidermal growth factor enhances the phosphorylation of endogenous membrane proteins. *Biochem. Biophys. Res. Commun.* **101**, 517–523.

Sefton, B. M., and Hunter, T. (1984). Tyrosine protein kinases. *Adv. Cyclic Nucleotide Res.* **18**, 195–226.

Sefton, B. M., Hunter, T., and Beemon, K. (1980a). The relationship of the polypeptide products of the transforming gene of Rous sarcoma virus and the homologous gene of vertebrates. *Proc. Natl. Acad. Sci. U.S.A.* **77**, 2059–2063.

Sefton, B. M., Hunter, T., Beemon, K., and Eckhart, W. (1980b). Phosphorylation of tyrosine is essential for cellular transformation by Rous sarcoma virus. *Cell (Cambridge, Mass.)* **20**, 807–816.

Shia, M. A., and Pilch, P. F. (1983). The β subunit of the insulin receptor is an insulin-activated protein kinase. *Biochemistry* **22**, 717–721.

Stehelin, D., Varmus, H. E., Bishop, J. M., and Vogt, P. K. (1976). DNA related to the transforming gene(s) of avian sarcoma viruses is present in normal avian DNA. *Nature (London)* **260**, 170–173.

Ullrich, A., Coussens, L., Hayflick, J. S., Dull, T. J., Gray, A., Tam, A. W., Lee, J., Yarden, Y., Liberman, T. A., Schlessinger, J., Downward, J., Mayes, E. L. V., Waterfield, M. D., Whittle, M., and Seeburg, P. H. (1984). Human epidermal growth factor receptor, cDNA sequence, and aberrant expression of the amplified gene in A431 epidermoid carcinoma cells. *Nature (London)* **309**, 418–425.

Ullrich, A., Bell, J. R., Chen, E. Y., Herrara, R., Petruzzelli, L. M., Dull, T. J., Gray, A., Coussens, L., Liao, Y.-C., Tsubokawa, M., Mason, A., Seeburg, P. H., Grunfeld, C., Rosen, O. M., and Ramachandran, J. (1985). Human insulin receptor and its relationship to the tyrosine kinase family of oncogenes. *Nature (London)* **313**, 756–761.

Ushiro, H., and Cohen, S. (1980). Identification of phosphotyrosine as a product of epidermal growth factor-activated protein kinase in A-431 cell membrane. *J. Biol. Chem.* **255**, 8363–8365.

van Obberghen, E., Rossi, B., Kowalski, A., Gazzano, H., and Ponzio, G. (1983). Receptor-mediated phosphorylation of the hepatic insulin receptor: Evidence that the M_r 95,000 receptor subunit is its own kinase. *Proc. Natl. Acad. Sci. U.S.A.* **80**, 945–949.

Wang, J. Y. J., Queen, C., and Baltimore, D. (1982). Expression of an Abelson murine leukemia virus-encoded protein in *Escherichia coli* causes extensive phosphorylation of tyrosine residues. *J. Biol. Chem.* **257**, 13181–13184.

Waterfield, M. D., Scrace, G. T., Whittle, N., Stroobant, P. Johnsson, A., Wasteson, A., Westermark, B., Heldin, C.-H., Huang, J. S., and Deuel, T. F. (1983). Platelet-derived growth factor is structurally related to the putative transforming protein p28[sis] of simian sarcoma virus. *Nature (London)* **304**, 35–39.

Witte, O. N., Dasgupta, A., and Baltimore, D. (1980). Abelson murine leukemia virus protein is phosphorylated *in vitro* to form phosphotyrosine. *Nature (London)* **283**, 826–831.

Yarden, Y., Schreiber, A. B., and Schlessinger, J. (1982). A nomitogenic analogue of epidermal growth factor induces early responses mediated by epidermal growth factor. *J. Cell Biol.* **92**, 687–693.

Zapf, J., Schoenle, E., and Froesch, E. R. (1978). Insulin-like growth factors I and II: Some biological actions and receptor binding characteristics of two purified constituents of nonsuppressible insulin-like activity of human serum. *Eur. J. Biochem.* **87**, 285–296.

Zick, Y., Kasuga, M., Kahn, C. R., and Roth, J. (1983). Characterization of insulin-mediated phosphorylation of the insulin receptor in a cell-free system. *J. Biol. Chem.* **258**, 75–80.

12

The Control of Cell Proliferation by Calcium, Ca²⁺-Calmodulin, and Cyclic AMP

JAMES F. WHITFIELD, ALTON L. BOYNTON,[1] R. H. RIXON, AND T. YOUDALE

Cell Physiology Group
Division of Biological Sciences
National Research Council of Canada
Ottawa, Ontario, Canada

> "'Tis better to have mused and missed
> than never to have mused at all."
> (With Apologies to Tennyson)

[1]Present address: Cancer Center of Hawaii, University of Hawaii, Honolulu, Hawaii 96813.

331

I. INTRODUCTION

Sometime during the earliest, pre-DNA or "vital mud" stage of their evolution (Cairns-Smith, 1982), organisms learned to store energy in the form of ion gradients which would later be harnessed to energize, feed, move, and reproduce the infinitely more sophisticated organisms of the future (Wilson and Lin, 1980). Even the creation of the eukaryotic nucleus (about 1.5×10^9 years ago) by surrounding the prokaryotic chromosome with a membrane and shifting the attachment sites for the choromosomal replication origin and the multienzyme replication complexes from the plasma membrane to the intranuclear protein lamina and matrix may have been done to isolate and protect the chromosome and the replication machinery from deformation and disruption by the Ca^{2+}-driven cytoplasmic streaming and movement of secretory vesicles to and from the cell membrane during endocytosis/exocytosis (Cavalier-Smith, 1981). Calcium also came to be used to trigger events leading to chromosome replication, to move the chromosomes during mitosis, and finally to divide the cell into two daughters (MacManus *et al.*, 1982; Whitfield, 1980, 1982; Whitfield *et al.*, 1979, 1980, 1982).

Cyclic AMP seems to have been invented by prokaryotes primarily to control the transcription of certain so-called catabolite gene operons (Ullmann and Danchin, 1983). Thus, when bound to a basic protein known as CAP (i.e., catabolite activating protein), the cyclic nucleotide promotes transcription of these genes by increasing the binding of RNA polymerase to their promoter regions, by opposing transcription termination, and possibly by overcoming the transcription-inhibiting action of a catabolite modulator protein (Ullmann and Danchin, 1983). The cyclic nucleotide's role expanded with the advent of the eukaryotes. The cyclic AMP-binding CAP became the regulatory subunits of the multipurpose cyclic AMP-dependent protein kinases (Weber *et al.*, 1982). When released by cyclic AMP from their inhibitory attachment to the cyclic AMP-binding regulatory subunits, these enzymes' catalytic subunits can stimulate a multitude of processes, among which is still the amplification of mRNA production (see review by Walker, 1983). According to the latest assessment of the evidence by Boynton and Whitfield (1983), bursts of adenylate cyclase activity and a couple of transient cyclic AMP surges stimulate essential prereplicative and premitotic events in the cycles of all kinds of eukaryotic cells ranging from free-living diatoms and yeasts to the specialized, proliferatively restrained hepatocyte.

The means by which such ancient and accomplished signal transducers and cascade triggerers as Ca^{2+} and cyclic AMP cooperate to drive the all-important prereplicative events in the cell cycle must be understood, because it is in this mysterious prereplicative "black box" (Whitfield, 1982) where lie the master proliferative control mechanisms and the keys to understanding how each newborn cell decides to proliferate or differentiate further, and the solution to the puzzle of the relentless expansion of cancer cell populations.

II. CALCIUM, THE TRANSITORY TRIGGERER

Cells use Ca^{2+} to trigger a vast variety of processes such as ciliary and flagellar beating, contraction, cytoplasmic streaming, endocytosis of things such as receptor–ligand complexes, exocytosis (secretion) of various products, locomotion, phosphorylation of nuclear and plasma membrane proteins, chromosome replication, and cytokinesis (Boynton and Whitfield, 1983; Cheung, 1980, 1982, 1983; Hepler and Wolniak, 1983; MacManus and Whitfield, 1981; MacManus et al., 1982; Rubin, 1982; Salisbury et al., 1983; Sikorska et al., 1983; Takai et al., 1982; Whitfield, 1982; Whitfield et al., 1982). Ca^{2+}, when combined with its intracellular receptor/signal transducer calmodulin, also controls the size and duration of cyclic AMP surges by stimulating adenylate cyclase and then, some time later (depending on the location and rate of diffusion of the Ca^{2+}-calmodulin complexes), activating cyclic nucleotide phosphodiesterase (Boynton and Whitfield, 1983; Klee et al., 1980).

Cells made Ca^{2+} into a potent triggerer of cascading events simply by pumping the ion out of their cytosols to create an extremely steep, inwardly directed $\Delta\mu_{Ca^{2+}}$ ($pCa^{2+}_{ext} \cong 3.0$; $pCa^{2+}_{int} \cong 7.0$), which certain agonists could then use to produce generalized or localized internal Ca^{2+} surges by opening specific receptor-linked or potential-operated channels in the plasma membrane, by activating membrane "porters," or by ferrying the ion across the plasma membrane. The size and nature of the cell's response to this internal Ca^{2+} surge depends on its intracellular location, the number and intracellular distribution of Ca^{2+} receptor/enzyme complexes formed during the surge, and the types and locations of enzymes and motive or structural components susceptible to, and available for, activation or modification by these complexes.

An extracellular or plasma membrane bound Ca^{2+}-dependent enzyme ("exoenzyme"), such as phospholipase A_2 which is part of a multipurpose signaling mechanism, is selectively activated by the relatively slow process of fitting Ca^{2+} into a hole in the molecule having the same radius as Ca^{2+} (Levine and Williams, 1982). By contrast, calcium's intracellular functions require much more rapid association–dissociation steps and greater binding affinities ($\ll 10^6$) (Levine and Williams, 1982). Account must be taken of the fact that Ca^{2+} is a dangerous signaling tool which, if its level inside the cell should exceed 10^{-5} M because of a damaged plasma membrane or malfunctioning ATP-producing or Ca^{2+} pumping systems, can cause internal chaos and kill the cell by further damaging the plasma membrane, disrupting lysosomal membranes, and destroying vital organelles such as mitochondria through a runaway phospholipase cascade (Schanne et al., 1979; Shier and Du Bourdieu, 1982; Trump et al., 1981). To prevent this, the cell uses some of the cascade's endoperoxide products (e.g., PGG_2, PGH_2) to stimulate the synthesis of cyclic GMP which inhibits phospholipase action and the release of diacylglycerols and arachidonic acid (Nishizuka, 1983a,b). Another solution to this problem was the development of internal Ca^{2+} receptors which are both ion buffers and enzyme activators. Consequently, long before the separation of the animal and plant lineages, eukaryotic cells had invented the small Ca^{2+} receptor/activator protein calmodulin, which is present in relatively large amounts inside the cell, has a high affinity for Ca^{2+} (log K_b ~ 7), assumes one or more enzyme activating configurations when it combines with one to four Ca^{2+}, and enables Ca^{2+} to selectively trigger certain cascades of events without having to reach dangerously high levels to do so (Cheung, 1980; Klee et al., 1980; Levine and Williams, 1982). Ca^{2+}-calmodulin complexes have the added safety feature of stimulating membrane pumps to remove excess Ca^{2+} in order to terminate the signal and prevent lethal runaways.

III. THE CELL CYCLE—AN OVERVIEW

To really understand how Ca^{2+} and cyclic AMP regulate cell proliferation we must first really understand the cell cycle. Unfortunately, this is far in the future, and all we can do now is briefly summarize the cycle, identify the pivotal transitions caused by these cascade triggerers and transcription amplifiers, and present some ideas on how they might do it.

The life cycle of most, though not all, eukaryotic cells can be divided into four phases (Figs. 1–4). It starts with the cell's birth when the parent divides. The newborn cell may decide, on prompting from sig-

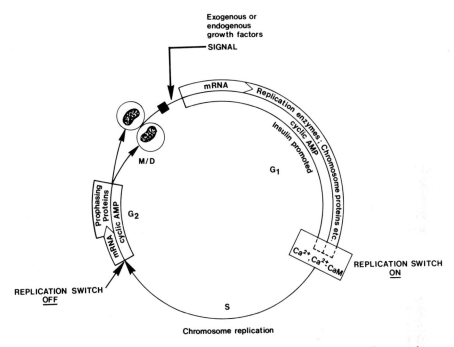

Fig. 1. The type 1 prereplicative period of continuously cycling cells, in culture dishes or tissues such as bone marrow and intestine, consists entirely of the basic G_1 events.

nals from its niche, not to initiate a cycle, but to switch from the proliferation mode and reversibly or terminally differentiate instead. If it decides to cycle, it first passes through a prereplicative period containing the G_1 black box (Figs. 1–4; Whitfield, 1982), the machinations within which commit the cell to a round of chromosome replication. The cell then goes through the S phase, during which it replicates its chromosomes, then switches off its replication machinery, pauses more or less briefly in the G_2 phase to prepare for mitosis, and finally divides into two new individuals at the end of the mitosis/division phase.

Progression through the cell cycle is driven by a string of gene activations, brief surges of regulators such as Ca^{2+}, Ca^{2+}-calmodulin, and cyclic AMP, bursts of different enzyme activities, cytoskeletal restructuring, and the transient unmasking or insertion into the plasma membrane of receptors for external growth factors (Boynton and Whitfield, 1983; Hochhauser et al., 1981; Leffert et al., 1979; see Chapter 5). Consequently, cellular morphology, the types, densities, binding, and

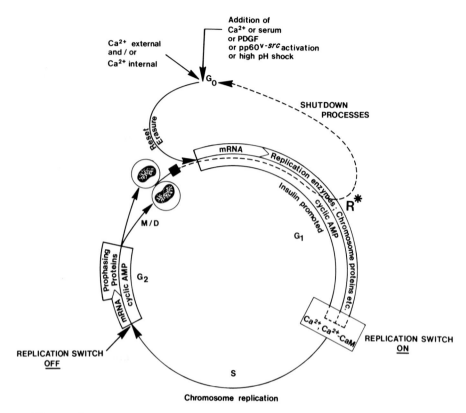

Fig. 2. The type 2 prereplicative period consists of a preliminary reset or competence establishment phase followed by a basic G_1 phase. It occurs in cultured cells (and perhaps cells in tumors) which have previously been set back from a type 1 G_1 phase into a quiescent G_0 state by lack of serum or a critical nutrient. R* is the restriction point of Pardee *et al.* (1982). Once R* has been passed, neither serum deprivation nor the protein synthesis inhibitor cycloheximide can prevent the initiation of chromosome replication. Abbreviations are Ca^{2+}-CaM, Ca^{2+}-calmodulin; pp60[v-src], the tyr-protein kinase product of the avian sarcoma virus.

down-regulation characteristics of surface receptors, and the cell's enzyme profile change during progression through the cycle. In other words, the cell in S phase is different from what it was in early G_1 phase and what it will be during mitosis/division. Because of this progressively changing structural and functional composition, regulators such as cyclic AMP will promote cell cycle transit when generated at the right times in the cycle, but will block transit if made to appear in abnormal amounts at other times (Boynton and Whitfield, 1983).

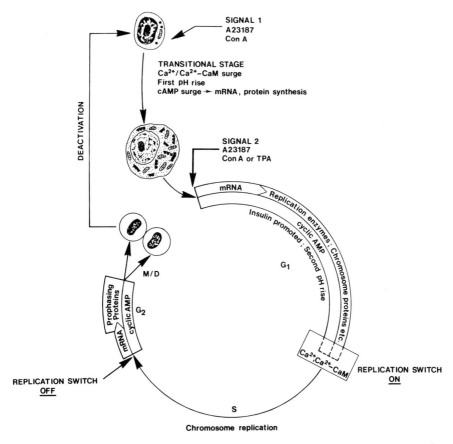

Fig. 3. The very long type 3 prereplicative period, which consists of a transitional phase followed by a basic G_1 phase, is driven by two signals and occurs in nearly dormant cells such as small lymphocytes. Abbreviations are Ca^{2+}-CaM, Ca^{2+}- calmodulin; Con A, concanavalin A; TPA, 12-O-tetradecanoylphorbol-13-acetate.

IV. THE PREREPLICATIVE PERIOD

A. Overview

During the prereplicative period, which may consist wholly or only partly of the basic G_1 phase (Figs. 1–4), there are gene activations, sometimes activation of untranslated "dowry" messenger RNA carried over from the parent, and cascades of enzyme activations and structural changes which lead to the production of active DNA-replicating en-

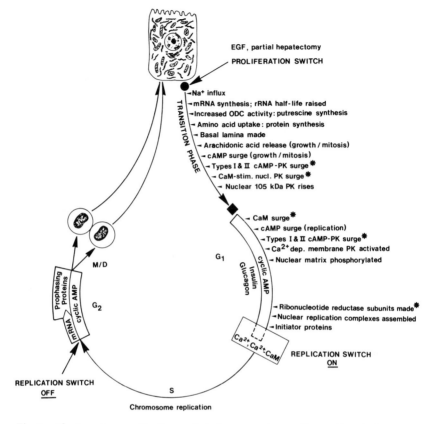

Fig. 4. The type 4 prereplicative period of regenerating rat liver cells, which consists of a preliminary transitional phase followed by a basic G_1 phase, is triggered by EGF or partial hepatectomy and is then driven by glucagon and insulin. Abbreviations are Ca^{2+}-CaM, Ca^{2+}-calmodulin; cAMP-pK, cyclic AMP-dependent protein kinase; CaM-stim.nucl.pK, the Ca^{2+}-calmodulin stimulable nuclear protein kinase of Sikorska *et al.* (1980); Nuclear 105 kilodalton pK, the 105-kilodalton protein kinase of Sikorska *et al.* (1983); ODC, ornithine decarboxylase; (*) indicates an event which does not occur in the liver cells remaining after partial hepatectomy in rats thyroparathyroidectomized 72 hr previously.

zymes, initiator protein(s), the five histones and other chromosomal proteins, nuclear matrix proteins, and the many other things needed for a round of chromosome replication (Boynton and Cohen, 1983; Das, 1980; Floros *et al.*, 1978; Gutowski and Cohen, 1983; Hochhauser *et al.*, 1981; Jazwinski *et al.*, 1976; Klein, 1982; Mercer *et al.*, 1982; Rao *et al.*, 1982; Whitson, 1980).

Cells are most sensitive to their microenvironment and exogenous factors and hormones during the prereplicative period. This has evolved from the need for multicellular organisms to selectively control and coordinate proliferative activity in their tissues and from the basic requirement that the cell be prepared to run through the chromosome replication and mitosis programs without interruption and without interference from external regulators. Only in this way can the cell avoid the disruption and death which would result from stopping or being stopped in the middle of the S phase or mitosis.

To understand the roles of Ca^{2+} and cyclic AMP in prereplicative development, it is essential to realize that there are four different kinds of prereplicative period (Boynton and Whitfield, 1983; Whitfield, 1982). The role of Ca^{2+}_{ext}, Ca^{2+}-calmodulin, and cyclic AMP during the types 1, 2, and 3 prereplicative periods will only briefly be discussed due to space limitations and will be more thoroughly reviewed for the type 4 prereplicative period.

B. Calcium and Cyclic AMP in Type 1 Prereplicative Periods

The type 1 prereplicative period is the classical G_1 phase containing the basic black box bursts of gene activities and enzyme cascades which are common to all four types of prereplicative period (Figs. 1–4; Whitfield, 1982). Started in newborn cells by a signal from external growth factors or hormones or from internal growth factors [e.g., the oncofetal growth factors of DeLarcro and Todaro (1978) and Swierenga et al. (1978)], it proceeds without interruption to the initiation of chromosome replication. This is the prereplicative period of the rapidly cycling amplifying progenitor and precursor cells in bone marrow and intestine in vivo, the continuously cycling cells of exponentially growing cultures, and the cells in synchronized cultures (Fig. 1).

Extracellular calcium involvement in the type 1 prereplicative phase is illustrated by simple Ca^{2+}_{ext} deprivation [incubation in low (0.02 mM) Ca^{2+} medium serum for 24–48 hr], the result of which is a reduction of DNA synthetic activity which is rapidly reversed by Ca^{2+} (1.25 mM) readdition (Boynton et al., 1981; Whitfield et al., 1982). The Ca^{2+}_{ext}-deprived cells that remain in late G_1 as long as 48 hr may be "treadmilling" like the BALB/c 3T3 mouse cells mentioned by Pardee et al. (1982), which are held in late G_1 phase by a concentration of cycloheximide just high enough to counterbalance degradative processes and keep a labile regulator protein(s) at a subthreshold level, but not high enough to set the cell back into a G_0 state by completely suppressing the regulator's synthesis and allowing its complete degradation. Alternatively, Ca_{ext}^{2+}-deprived cells may be trapped in late G_1 phase because accumulated replication enzymes or their subunits and replication initiator protein(s) cannot get into the nucleus to start chromosome replication. However, with prolonged Ca_{ext}^{2+} deprivation the accumulated components are lost and with them the cell's ability to respond quickly to Ca_{ext}^{2+}, and a slowly responding subpopulation accumulates (see Chapter 5). Thus, like serum-deprived cells, they are eventually set back into a quiescent G_0 state and must restart the entire string of prereplicative events in order to replicate their chromosomes.

Involvement of Ca^{2+}-calmodulin in late G_1 events is suggested by the facts that synchronously cycling CHO cells double their calmodulin content during a brief period just before and shortly after the onset of DNA synthesis (though this might just be the time when these cells make most of their proteins), and Ca^{2+}-calmodulin blockers (the naphthalenesulfonamides W7 and W13) delay entry of these cells into the S phase (Chafouleas and Means, 1982; Hidaka *et al.*, 1987); that several Ca^{2+}-calmodulin blockers (W7, R24571, trifluoperazine) prevent the avian sarcoma virus's pp60^{v-src} tyr-protein kinase from triggering DNA synthesis in *ts*LA23-NRK cells trapped in late G_1 phase by external Ca^{2+} deprivation; and that Ca^{2+}-calmodulin itself triggers prompt DNA synthetic responses from Ca_{ext}^{2+}-deprived *ts*LA23-NRK cells (at the nonpermissive 40°C), planarian cells (*Polycelis tenuis*), and T51B rat liver cells (Boynton *et al.*, 1980a; Martelly *et al.*, 1983). It must be borne in mind that all of these Ca^{2+}-calmodulin blockers also inhibit protein kinase C (Kuo *et al.*, 1985), which could be wholly or partly responsible for their blockage of late G_1 events. However, according to Tanaka *et al.* (1982), the naphthalenesulfonamides seem to inhibit Ca^{2+}-calmodulin-dependent processes much more selectively than protein kinase C.

There may be two kinds of Ca^{2+}-dependent process operating during the G_1 phase, one of which has a permissive need for Ca_{ext}^{2+}-dependent protein kinase C activity, and the other of which is triggered by a transient surge of internal Ca^{2+}-calmodulin complexes (Whitfield *et al.*, 1982). The former may have been separated from the latter by the Ca_{ext}^{2+}-bypassing, protein kinase C-stimulating TPA and the avian sarcoma virus's pp60^{v-src} tyr-protein kinase, both of which stimulate cells to proliferate indefinitely in severely Ca^{2+}-deficient medium without reducing their susceptibility to late G_1 arrest by Ca^{2+}-calmodulin blockers (Durkin *et al.*, 1983; Jones *et al.*, 1982). Thus, the insensitivity of the proliferative activity of all kinds of neoplastic cell to drastic Ca_{ext}^{2+} deprivation might be due to a change in the Ca^{2+}-binding regulatory part of the protein kinase C molecule (Kishimoto *et al.*, 1983; see Boynton *et al.*, this volume).

It should be noted that Ca^{2+}-calmodulin function does not involve transcriptional or translational events because the Ca^{2+}-calmodulin-dependent late G_1 phase of serum or $pp60^{v\text{-}src}$ tyr-protein kinase-stimulated NRK cells is insensitive to actinomycin D and cycloheximide (Durkin and Whitfield, 1984).

Some nonneoplastic epithelial lining cells, such as the basal cells in primary cultures of newborn mouse epidermis and human foreskin keratinocytes, proliferate, but do not terminally differentiate into keratinous squames when cultivated in medium having a Ca^{2+} concentration between 0.01 mM and 0.10 mM (Hennings *et al.*, 1980), which is too low for the optimal proliferation of nonneoplastic parenchymal epithelial cells such as neonatal rat hepatocytes (Armato *et al.*, 1983). Rat esophageal epithelial cells also proliferate maximally in the presence of 0.1 mM Ca^{2+}, but they cannot proliferate in the presence of anything less than 0.01 mM Ca^{2+} (Babcock *et al.*, 1983). Raising the Ca^{2+}_{ext} concentration to 0.3 or 1.0 mM encourages the esophageal cells to contact one another, establish desmosomes at the points of contact, stop proliferating, and terminally differentiate into keratinous squames (Babcock *et al.*, 1983). Raising the Ca^{2+}_{ext} concentration to 1.0, 1.4, or 1.9 mM causes murine epidermal basal cells or human foreskin keratinocytes to stop cycling sometime during the next 24 to 36 hr, an event which is preceded by the establishment of cell-to-cell contacts, the appearance of desmosomes, in the case of the foreskin cells expulsion of involucrin-containing cells from the basal layer, and is followed by the terminal differentiation of the cells into cornified, keratinized squames (Hennings *et al.*, 1980). The importance of cell-to-cell contact for hypersensitivity to Ca^{2+}_{ext} and the triggering of terminal differentiation of lining epithelial cells is demonstrated by the fact that noncontacting human bronchial epithelial cells in sparse cultures proliferate maximally when Ca^{2+}_{ext} is between 0.1 and 1.0 mM, but contacting cells in crowded cultures (5000 cells/cm²) are hypersensitive to Ca^{2+}_{ext} and stop proliferating within 20 hr after Ca^{2+}_{ext} is raised from 0.1 to 1.0 mM (Lechner, 1982).

The hypersensitivity of contacting lining epithelial cells to Ca^{2+}_{ext} might be explained by the same loss of internal Ca^{2+} buffering capacity that has been cited as the reason for the secretion of the Ca^{2+}-mobilizing parthyroid hormone being inhibited rather than stimulated by Ca^{2+}_{ext} (Rubin, 1982). Thus, contacting epidermal basal cells may reduce the number and/or activity of the Ca^{2+} pumps in their plasma membranes in order to enable the internal Ca^{2+} concentration to rise to a level high enough to trigger Ca^{2+}-dependent keratinizing mechanisms and stop such unnecessary things as chromosome replication and division. Therefore, to keep these lining cells cycling in culture, either the cell density must be adjusted to prevent cell contacts or Ca^{2+}_{ext} must be lowered enough to avoid overwhelming the cells' limited internal Ca^{2+} buffering capacity (even 0.01 mM Ca^{2+}_{ext} is 10–1000 times greater than the internal Ca^{2+} concentration) and initiating the terminal differentiation program, while still being high enough for the small internal Ca^{2+} surges and the generation of the Ca^{2+}-calmodulin complexes needed to drive late G_1 events.

Cyclic AMP and cyclic AMP-dependent protein kinases also trigger critical events in late G_1 (Fig. 1; Boynton and Whitfield, 1983; Whitfield, 1982). Thus, in a wide variety of continuously cycling cells [e.g., diatoms (*Cylindrotheca fusiformis*), budding yeast (*Saccharomyces cerevisiae*), Chinese hamster CHO cells, HeLa human carcinoma cells, human RPMI 8866 lymphosarcoma cells], there is a short burst of adenylate cyclase activity, a transient reduction of cyclic nucleotide phosphodiesterase activity, and a resulting transient cyclic AMP surge which peaks in middle or late G_1 phase and usually subsides just before the onset of chromosome replication (Fig. 5A; Boynton and Whitfield, 1983; Sullivan and Volcani, 1981). Inhibiting the burst of adenylate cyclase activity and preventing the cyclic AMP surge in *S. cerevisiae* cells with the tridecapeptide α mating pheromone stops G_1 transit at the "start" step where the *cdc* 28 gene product would normally start the syntheses of RNA and protein species and the series of gene activations needed to initiate DNA replication (Elliott and McLaughlin, 1983). *Saccharomyces cerevisiae* mutants with defective adenylate cyclase can advance beyond the start step to initiate chromosome replication and budding only in medium containing cyclic AMP (Goodenough and Thorner, 1983). Moreover, exposure to external cyclic AMP (which binds to the regulatory subunits of cyclic AMP-dependent protein kinases in the plasma membrane), to any one of many adenylate cyclase stimulators (e.g., Ca^{2+}_{ext}, calcitonin, cholera toxin, epinephrine, glucagon, parathyroid hormone), or to a cyclic nucleotide phosphodiesterase inhibitor (e.g., caffeine, 3-isobutyl-1-methylxanthine) triggers DNA synthesis in BALB/c 3T3 mouse cells, rat thymic lymphoblasts, and rat T51B cells which have been trapped in late G_1 phase by Ca^{2+}_{ext} deprivation (Boynton and Whitfield, 1983). The importance of cyclic AMP-dependent protein kinase activity is demonstrated by the fact that *S. cerevisiae* mutants producing a cyclic AMP-dependent protein kinase, which is relatively insensitive to activation by cyclic AMP,

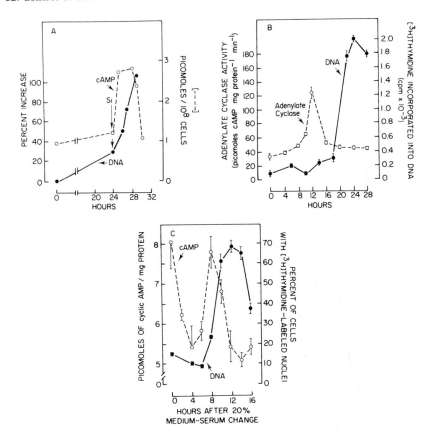

Fig. 5. A short burst of adenylate cyclase activity and a transient cyclic AMP surge occur just before the initiation of DNA replication in such different cells as (A) diatoms (*Cylindrotheca fusiformis*), first rendered quiescent (except for a slow accumulation of plastid and mitochondrial DNA), by silicon deprivation and then stimulated by silicon addition; (B) rat aortic smooth muscle cells, first rendered quiescent by incubation in medium supplemented with platelet-poor plasma and then stimulated by adding platelet extract (i.e., crude PDGF) at time 0; (C) T51B rat liver cells rendered quiescent by serum deprivation and then stimulated by fresh fetal bovine serum at time 0. The culture procedures and methods of measuring cyclic AMP level and DNA content of the diatoms in (A) are described by Sullivan and Volcani (1981). The culture procedures and method of measuring adenylate cyclase activity in the rat vascular smooth muscle cells in (B) have been described by Franks *et al.* (1984). The autoradiographic and culture procedures and the method of measuring the cyclic AMP content of the T51B rat liver cells in (C) have been described by Boynton and Whitfield (1979). The data for the curves in (A) and (B) were generously provided by B. Volcani (Scripps Oceanographic Institution) and D. J. Franks (Clinical Research Institute of Montréal).

do not advance beyond the G_1 start step unless enough cyclic nucleotide is added to the medium to compensate for this reduced sensitivity (Matsumoto *et al.*, 1983). It is further demonstrated by the fact that another mutation, which permanently derepresses this protein kinase activity by reducing the production of the inhibitory cyclic AMP-binding regulatory subunits, enables both adenylate cyclase-deficient mutants and the mutants with cyclic AMP-insensitive protein kinase to advance beyond the start step and proliferate without help from external cyclic AMP (Matsumoto *et al.*, 1983). In higher eukaryotes, it seems that it is the type II cyclic AMP-dependent protein kinase which is specially involved in the late G_1 events. Thus, synchronously cycling Chinese hamster CHO cells make this enzyme specifically in late G_1 phase (see review by Boynton and Whitfield, 1983); exogenous type II holoenzyme, but not type I holoenzyme, triggers DNA synthesis in Ca^{2+}-deprived T51B rat liver cells and enables them to proliferate in Ca^{2+}-deficient medium (Boynton *et al.*, 1981); and the stimulation of DNA synthesis in T47D human mammary cancer cells by calcitonin is preceded by a burst of adenylate cyclase activity and an almost immediate activation of the type II isoenzyme with little change in the type I isoenzyme (Ng *et al.*, 1983).

C. Calcium and Cyclic AMP in Type 2 Prereplicative Periods

The type 2 prereplicative period is much longer than the type 1 prereplicative period. It is the most studied and the easiest to study of the four types of prereplicative period, but it may be the least relevant to the *in vivo* proliferation of normal and neoplastic cells because it follows the stimulation of quiescent cells in variously starved or confluent cell cultures in plastic petri dishes (Fig. 2; Boynton and Whitfield, 1983; Whitfield, 1982). However, it may also be the type of prereplicative period which follows the stimulation of tumor cells rendered quiescent by outgrowing their blood supply.

A newborn cell starting out on a normal type 1 prereplicative program begins activating genes and accumulating things which commit it to a round of chromosome replication when they reach critical levels (Pardee *et al.*, 1982). If the medium lacks essential growth factors or contains the transcription inhibitor actinomycin D or a protein synthesis inhibitor such as cycloheximide at the time of the cell's birth, the cell cannot accumulate enough components to reach this critical commitment point, the so-called restriction point. However, if this point has already been passed before the withdrawal of growth factor(s) or addition of actinomycin D or cycloheximide, the cell will complete G_1 development and start replicating its chromosomes (Durkin and Whitfield, 1984; Pardee et al., 1982; Streumer-Svoboda et al., 1982). If the cell is caught before the restriction point, it will stop advancing toward the S phase and, bearing marks of its aborted G_1 transit [e.g., monophophorylated H1 histones, ciliated centrioles, elevated levels of calmodulin and cyclic AMP (Boynton and Whitfield, 1983; Chafouleas and Means, 1982; Hochhauser et al., 1982)], it will "park" itself indefinitely in a so-called G_0 state (Fig. 2). Short-lived proliferogenic components [e.g., the p53 protein (Klein, 1982)] are prevented from accumulating by having their synthesis reduced and their degradation enhanced (Pardee et al., 1982; Schneiderman et al., 1971). To enter the G_0 state of quiescence, the cell loses an internal signal mediator/gene activator, inactivates the pivotal eIF-2 polypeptide chain initiation factor (which reduces or stops all protein synthesis), disaggregates polyribosomes, accumulates inactive 80 S ribosomes, increases proteolysis, degrades already accumulated labile components, and lowers its ribosomal RNA content to a minimum level (Austin and Kay, 1982; Ballard et al., 1980; Darzynkiewicz and Traganos, 1982; Stiles, 1983; Van Venrooij et al., 1972; Warburton and Poole, 1977).

Ca_{ext}^{2+} is sometimes needed, while an internal Ca^{2+} surge (resulting from Ca_{ext}^{2+} influx or mobilization of internally accumulated calcium) may always be needed to trigger type 2 prereplicative development (Boynton and Whitfield, 1976, 1984; Damluji and Riley, 1979; Dulbecco and Elkington, 1975; Tupper et al., 1980; Whitfield et al., 1982), and Ca_{ext}^{2+} appears always to be needed for late G_1/S transition of non-neoplastic cells (Armato et al., 1983; Boynton et al., 1976; Engström et al., 1982; Hazelton et al., 1979; Swierenga et al., 1978). In addition, Ca^{2+}-calmodulin function is also required for G_1/S transition during the type 2 prereplicative period (Fig. 6; Durkin et al., 1983).

Since cyclic AMP triggers critical events near the end of the basic type 1 (i.e., G_1) prereplicative period, it should do the same near the end of the G_1 portion of the type 2 prereplicative period (Fig. 2). Indeed, a burst of adenylate cyclase activity and a transient cyclic AMP surge, which peak before the initiation of DNA synthesis, occur later in the type 2 prereplicative periods of BALB/c 3T3 mouse cells (Boynton and Whitfield, 1983), rat vascular smooth muscle cells (Fig. 5B; Franks et al., 1984), and T51B rat liver cells (Fig. 5C; Boynton and Whitfield, 1979, 1983), induced by PDGF or serum. The cause of this burst of adenylate cyclase activity is unknown. It might be due to the transient production or unmasking of cell surface receptors for an adenylate cyclase-stimulating external growth factor or hormone (e.g., β-adrenergic catecholamines, glucagon), an internal surge of adenylate cyclase-stimulating prostaglandins, or a cholera toxin-like factor which stimulates the enzyme by (mono) ADP-ribosylating the enzyme's GTP-binding N_s subunit (Van Heyningen, 1982).

D. Calcium and Cyclic AMP in Type 3 Prereplicative Periods

The third type of prereplicative period is the very long one ($\sim$50 hr) following the activation of nearly dormant and proliferatively inactive cells such as small lymphocytes (Fig. 3).

The small lymphocyte consists of a functionally repressed, primary gene transcript-wasting nucleus containing highly condensed chromatin and a drastically reduced nuclear matrix, which is surrounded by a thin rim of cytoplasm containing a poorly developed Golgi apparatus, few mitochondria, many monoribosomes, few or no polyribosomes, and a reduced capacity for forming protein chain initiation complexes. Its proliferative engine is not idling—it and the cell cycle genes are turned off. The prereplicative period following the activation of this cell by a mitogen or proliferogen such as concanavalin A or phytohemagglutinin owes its great length to the extensive construction needed to bring the cell to the threshold of the G_1 phase. Thus, there is an initial period during which the chromatin decondenses and early genes are activated, the ubiquitous p53 protein kinase appears (Klein, 1982), the interchromatin nuclear matrix expands greatly to process primary gene transcripts, transcript wastage stops, the ribosomes are made and transported into the cytoplasm where the ability to form protein chain initiation complexes is improved, a large new protein synthetic apparatus is set up, and lots of mitochondria are made to fuel it. Receptors for 1,25-dihydroxyvitamin D_3 also appear (Provvedini *et al.*, 1983), an event which may be important for proliferation because this hormone controls the levels of the cyclic AMP-dependent protein kinases and the transient surges in these levels during the prereplicative development of regenerating liver cells (Sikorska *et al.*, 1983; Whitfield *et al.*, 1984).

When its new synthetic machinery is in place, the cell is ready to implement the basic G_1 program for activating the genes coding for chromosomal proteins, DNA synthetic enzymes, and replication initiator protein(s) (Gutowski and Cohen, 1983), raising the internal cyclic AMP and pH levels to maximize the production and translation of the mRNA for these enzymes and proteins, and finally transporting these components into the nucleus to trigger chromosome replication.

The results of experiments on blood, lymph node, and splenic lymphocytes from cattle, humans, mice, and pigs indicate that Ca^{2+} is involved in the initiation of type 3 prereplicative development, a process requiring contact with a suitable antigen, plant lectin (e.g., concanavalin A, phytohemagglutinin), or cationophore (e.g., A23187, quin 2) for about 6 hr, during which time there is a transient (0–6 hr), small (0.1–1.0 μM) rise in the internal Ca^{2+} level and a transient (0–6 hr) rise in the internal pH from about 7.18 to about 7.4 (Fig. 3; Gerson, 1982; Mastro and Smith, 1982). These cells need Ca^{2+}_{ext} again between 19 and 47 hr to start chromosome replication (Mastro and Smith, 1982). Lowering the Ca^{2+}_{ext} concentration at either one of these times prevents the completion of prereplicative development and the initiation of chromosome replication (Fig. 3; Hesketh et al., 1982; Mastro and Smith, 1982).

Since Ca^{2+} and Ca^{2+}-calmodulin trigger the basic G_1 events, they should trigger the same events in the G_1 portion of the type 3 prereplicative period (Fig. 3). Indeed, according to Mastro and Smith (1982), a second Ca^{2+}_{ext}-dependent signal must be given to bovine lymph node lymphocytes between 19 and 47 hr after exposure to concanavalin A or cationophore A23187 to keep them advancing toward the S phase. The ability of a Ca^{2+}-calmodulin blocker such as trifluoperazine to prevent completely the initiation of DNA synthesis in phytohemagglutinin-activated human peripheral lymphocytes when added as late as 24 hr after the mitogen (MacNeil et al., 1982) suggests an involvement of Ca^{2+}-calmodulin in this second Ca^{2+}-dependent period.

Cyclic AMP may be involved along with Ca^{2+} in the early stage of the type 3 prereplicative period. Thus, there is some evidence of an early burst of adenylate cyclase activity, a transient cyclic AMP surge, and a specific activation of type 1 cyclic AMP-dependent protein kinase (see Boynton and Whitfield, 1983, for references). Since cyclic AMP and its dependent protein kinases trigger critical events at the end of the basic G_1 period in many other cells, including lymphoid cells such as human RPMI 8866 cells and rat thymic lymphoblasts (Boynton and Whitfield, 1983), they should also be involved in the G_1 portion of the small lymphocyte's type 3 prereplicative period. Indeed, there is a large cyclic AMP surge in the second Ca^{2+}-dependent (and high internal pH) stage of the mouse spleen lymphocyte's prereplicative development which appears to be linked to the initiation of DNA synthesis and begins 10 hr after exposure to mitogen, peaks around 20 hr, and then subsides before DNA synthetic activity peaks at 50 hr (Foker et al., 1979; Whitfield et al., 1980).

E. Calcium and Cyclic AMP in Type 4
Prereplicative Periods

Secretory cells such as hepatocytes and parotid gland acinar cells in the nutritionally complete, homeostatically maintained conditions in an adult rat are not proliferatively dormant like small lymphocytes or idling in G_0 like serum-starved cells in culture. They are amply endowed with protein-synthesizing machinery and the mitochondria to fuel it, but they are variously programmed to make, secrete, detoxify, and dispose of many things rather than to use this machinery to grow, replicate their chromosomes, and divide. Despite the often intense protein synthetic activity, such cells do not grow and divide because their cell cycle genes are either totally repressed or function only sporadically and randomly at the "noise" level (Walker and Whitfield, 1984). Most of the protein they make is exported and the rest is used to replace components or is degraded. However, these cells can be made to proliferate by injecting certain proliferogens or functionally overloading them by partially removing or damaging a large part of the organ.

When an actively functioning hepatocyte or parotid acinar cell is proliferatively activated, it must increase its already substantial protein-making capability in order to continue functioning while producing the many short-lived things needed for one or two rounds of chromosome replication and division. Therefore, the type 4 prereplicative period, like the type 3 period, has an initial transitional phase during which the cell prepares to enter the basic G_1 phase by activating the transcription of cell cycle genes, increasing amino acid uptake, greatly stepping up ribosome accumulation, reducing autophagocytosis and protein degradation, activating ornithine decarboxylase, briefly increasing the level of a nuclear Ca^{2+}-calmodulin-stimulable protein kinase, transiently increasing for the first time the cytoplasmic levels of types I and II cyclic AMP-dependent protein kinase holoenzyme, and laying down additional basal lamina (Fig. 4; Boynton *et al.*, 1980b; Gospodarowicz and Tauber, 1980; Sikorska *et al.*, 1983; Walker and Whitfield, 1984; Whitfield, 1980; Whitfield *et al.*, 1980, 1982, 1985). When the cell's synthetic capability has reached its new peak (usually by about 10 hr), it is ready to enter the G_1 phase where it activates the genes coding for DNA-replicating enzymes, chromosomal proteins, and some oncofetal genes such as c-*myc*, c-*ras*, and c-*src* (Figs. 4 and 6; Fausto *et al.*, 1982; Goyette *et al.*, 1983; see Chapter 1).

While virtually nothing is known about the factors controlling the prepreplicative development of parotid acinar cells (except that they

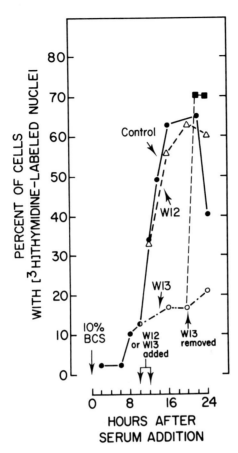

Fig. 6. A selective Ca^{2+}-calmodulin blocker like the naphthalenesulfonamide W13 (but not the much less active control naphthalenesulfonamide, W12) traps T51B rat liver cells in confluent cultures at the end of the G_1 phase of their serum-triggered type 2 prereplicative period. W13 or W12 (10^{-5} mole/liter) was added twice, at 10 and 12 hr, after serum addition and then removed (by discarding the culture medium and replacing it with inhibitor-free medium) at 20 hr. The almost immediate initiation of DNA synthesis after W13 removal demonstrates the lateness of the prereplicative block as well as the lateness of the requirement for Ca^{2+}-calmodulin activity. Culture and autoradiographic procedures are described in Boynton and Whitfield (1979) and Boynton et al. (1982a). The points are the means of the values in four cultures. The coefficients of variation of these means were less than 10%.

can be proliferatively activated by β-adrenergic catacholamines such as isoproterenol), it is known that hepatocytes, like lymphocytes, need two signals (between 0 and 3 hr and again between 12 and 20 hr) to send them through their type 4 prereplicative period. The starting signal can be given (certainly *in vitro* and perhaps *in vivo*) to hepatocytes by a brief (3-hr) exposure to EGF (Armato *et al.*, 1983; Leffert, 1983; Leffert *et al.*, 1979). The second signal is generated by the adenylate cyclase-stimulating glucagon and the cell-"energizing" (Moore, 1983) insulin, which together promote, but cannot start prereplicative development; i.e., they amplify mRNA production and translation, but do not activate proliferogenic genes (Armato *et al.*, 1983; de Hemptinne and Leffert, 1983; Leffert, 1982; Leffert *et al.*, 1983).

The first signal requires external Na^+ and causes a membrane-depolarizing Na^+ influx (Leffert, 1982; Leffert *et al.*, 1983; see Chapter 13). It also triggers (probably via an Na^+/H^+ exchange mechanism) a substantial, possibly insulin-dependent (Moore, 1983), internal pH rise (e.g., from 6.68 to 6.90). This pH rise [like the one in the fertilized sea urchin egg (Epel, 1982; Steinhardt, 1982)] might trigger synthesis of early proteins such as those needed to start prereplicative development in isoproterenal-activated parotid cells (Whitfield, 1980).

Ca^{2+}_{ext} seems not to be required for the first signal because dropping it prevents the completion, but not the initiation, of prereplicative development of neonatal rat hepatocytes *in vitro*, adult rat liver cells *in vivo*, and rat parotid acinar cells *in vivo* (Armato *et al.*, 1983; Rixon and Whitfield, 1976; Tsang *et al.*, 1981; Whitfield, 1980). Thus, EGF-activated, glucagon/insulin-promoted neonatal rat hepatocytes approach the end of their prereplicative development in Ca^{2+}-deficient medium, but they cannot replicate their chromosomes until more Ca^{2+} is added to the medium (Armato *et al.*, 1983). Similarly, hepatocytes activated by HPX (i.e., partial hepatectomy) in the hypocalcemic rat 72 hr after TPTX (i.e., thyroparathyroidectomy) increase messenger and ribosomal RNA synthesis, generate the first burst of cyclic AMP synthesis, and initiate some later events such as the activation of a Ca^{2+}_{ext}-dependent membrane protein kinase (which would be expected because the plasma Ca^{2+} level in the TPTX rat is still 0.5 mM) (Fig. 4). However, they do not generate an early burst of Ca^{2+}-calmodulin stimulable nuclear protein kinase activity, and the severe 1,25-dihydroxyvitamin D_3 deficiency which develops in TPTX rats prevents both the early and the late surges of types I and II cyclic AMP-dependent protein kinase holoenzyme (Fig. 4; Sikorska *et al.*, 1983; Whitfield *et al.*, 1985). They also do not make the L_1 (CDP-specific, substrate-binding effector) and L_2 (nonheme iron-containing) subunits of ribonucleotide reductase; they do not increase the cytoplasmic and nuclear ribonucleotide reductase holoenzyme activity; and they do not make the deoxyribonucleotides needed to replicate DNA (Figs. 4 and 7; Whitfield, 1980; Whitfield et al., 1979; Youdale et al., 1982).

Raising the blood Ca^{2+} concentration in hypocalcemic rats (TPTX 24 hr before HPX) by intrajugular infusion of $CaCl_2$ solution for only the first 8 hr after HPX does not enable the remaining hepatocytes to start replicating their chromosomes, but just one intraperitoneal $CaCl_2$ injection as late as 12 hr after HPX enables them to start replicating their chromosomes at the normal time (Rixon and Whitfield, 1976; Rixon et al., 1979). It should be noted, however, that intraperitoneal injection of a whole cocktail of Ca^{2+}-homeostatic hormones (calcitonin, parathyroid hormone, 1,25-dihydroxyvitamin D_3) is needed to enable hepatocytes to initiate chromosome replication at the normal time in rats thyroparathyroidectomized 72 hr before HPX (Rixon et al., 1979).

Another indication of the unimportance of Ca^{2+}_{ext} for the first signal is the fact that the prereplicative development of parotid acinar cells is not triggered by α-adrenergic catecholamines that cause Ca^{2+}_{ext} influx, but is triggered by β-adrenergic catecholamines that can mobilize internal Ca^{2+} but do not stimulate Ca^{2+}_{ext} influx (Whitfield, 1980; Whitfield et al., 1982).

Although Ca^{2+}_{ext} may not be important for the first signal, a Ca^{2+}-dependent phospholipase cascade (which could be triggered by an internal Ca^{2+} release) does occur along with the first cyclic AMP surge during the first 2 or 3 hr of the hepatocyte's prereplication period. The arachidonic acid liberated by this cascade and prostaglandins made from it are not involved in the events leading up to chromosome replication, but they provide something hepatocytes need to initiate mitosis after they have finished replicating their chromosomes (Rixon and Whitfield, 1982). If a phospholipase cascade also is involved in starting parotid gland acinar cell proliferation, it does not include increased phospholipase C activity and phosphatidylinositol turnover because this turnover is stimulated in these cells by the nonmitogenic α-adrenergic catecholamines, but not by the mitogenic β-adrenergic catecholamines (Michell, 1978).

The role of cyclic AMP in the first signal is uncertain. Both the brief, but large cyclic AMP surge which peaks sharply around 15 min in isoproterenol-activated parotid cells and the smaller, but much more prolonged surge which peaks between 2 and 4 hr in HPX-activated liver cells are not needed for the initiation of chromosome replication (MacManus et al., 1973); Tsang et al., 1980; Whitfield, 1980). However, the early surge in HPX-activated liver cells appears to be needed for the much later initiation of mitosis after chromosome replication because it coincides with the early mitosis-related, Ca^{2+}-dependent phospholipase-arachidonate cascade (Rixon and Whitfield, 1982) and the drug propranolol, which eliminates it without affecting the initiation of DNA synthesis (MacManus *et al.*, 1973; see Chapter 16), reduces the postreplicative flow of cells into mitosis, an effect which can be reversed by injecting cyclic AMP (but not 2',3'-cyclic AMP) into the propranolol-treated animal around 90 min after HPX (Rixon and Whitfield, unpublished observations).

At this point, we must show how insulin might work with Ca^{2+} and cyclic AMP to prime and promote later prereplicative development of cells such as those in regenerating rat liver. Although the level of circulating insulin actually drops after partial hepatectomy, the remaining proliferatively activated liver cells still manage to use the hormone to raise their protein synthetic and growth capabilities by putting more receptors out on their surfaces and by decreasing the hormone's degradation or endocytosis which increases the effectiveness of the available hormone by increasing the time it spends on its receptors (de Hemptinne and Leffert, 1983; Leffert *et al.*, 1979). Insulin seems to "energize" the cell by stimulating membrane Na^+,K^+-ATPase pumps, which increases Na^+-dependent amino acid uptake by transport system A by raising the Na^+ gradient ($\Delta\mu_{Na}$) across the plasma membrane (Moore, 1983). Since the methionine of the protein chain initiator complexes is transported by system A, this will maximize the translation of cyclic AMP-amplified mRNAs. The hormone also selectively stimulates the translation of mRNAs for ribosomal proteins (Austin and Kay, 1982). It then provides for the fueling of the $\Delta\mu_{Na}$-energized cell by stimulating the transfer of internal glucose transporters to the plasma membrane (Moore, 1983). It also stimulates an Na^+/H^+ exchange mechanism (Moore, 1983), which increases amino acid transport by system A, increases protein synthesis, and reduces protein degradation by raising the intracellular pH. Finally, raising the internal pH sensitizes Ca^{2+}-dependent mechanisms to Ca^{2+} by reducing the competition between H^+ and Ca^{2+} for binding sites on proteins, an effect which might explain the ability of insulin to trigger DNA replication in T51B rat liver cells trapped in late G_1 phase by Ca^{2+}_{ext} deprivation (A. L. Boynton and J. F. Whitfield, unpublished observation).

Since Ca^{2+} and Ca^{2+}-calmodulin complexes trigger some of the basic G_1 events, they should also trigger events in the G_1 part of the type 4 prereplicative period. As stated, adding enough Ca^{2+} to the medium late in the prereplicative period enables Ca^{2+}_{ext}-deprived, EGF-activated neonatal hepatocytes *in vitro* and HPX-activated hepatocytes in the hypocalcemic TPTX rat to finish their prereplicative development and start replicating their chromosomes (Armato *et al.*, 1983; Rixon and Whitfield, 1976; Rixon *et al.*, 1979). The possibility of calmodulin being involved in the prereplicative development of HPX-activated rat hepatocytes is suggested by a transient rise in the soluble (i.e., not sedimentable by centrifugation of liver homogenates at 105,000 g) calmodulin level which peaks broadly between 8 and 14 hr after HPX and subsides before the cells start replicating their chromosomes (MacManus *et al.*, 1981). The soluble (but not the total) calmodulin level peaks sharply 2 hr before the onset of DNA synthesis in parotid cells activated in the rat by injection of 150 mg of *dl*-isoproterenol/kg of body weight (Campos-Gonzalez *et al.*, 1984). While we do not know whether the prolonged surge of soluble calmodulin in hepatocytes is needed for prereplicative development, the late prereplicative surge of soluble calmodulin in parotid cells *cannot* be needed to start chromosome replication simply because it does not occur before the onset of DNA synthesis in parotid cells activated by a much lower dose (e.g., 25 mg/kg of body weight) of isoproterenol (Campos-Gonzalez *et al.*, 1984). However, this does not mean that calcicalmodulin activity is unnecessary because in other cells, such as tsLA23-NRK cells and T51B rat liver cells, calcicalmodulin activity is essential for the initiation of chromosome replication, but in the former cells there is no prereplicative calmodulin surge, while there is a surge in the latter cells (Fig. 6; Durkin *et al.*, 1983).

Neoplastic transformation of hepatocytes (and many other rodent and human cells) results in a two- to fourfold increase in their total calmodulin content as well as the appearance of another, smaller (11.7 kilodaltons) calcium-binding protein called oncomodulin which has not been found in any proliferatively active or quiescent nonneoplastic adult tissue (MacManus and Whitfield, 1983; Whitfield *et al.*, 1982). In the normal rat, oncomodulin normally appears only briefly in visceral yolk sac and placental cells during the thirteenth and fourteenth days of gestation and then disappears from the fetus and its membranes by the nineteenth and twentieth days, while still lingering on in the soon-to-be-discarded placenta (Brewer and MacManus, 1983). Although oncomodulin can stimulate cyclic nucleotide phospodiesterase and trigger DNA synthesis in Ca^{2+}-deprived nonneoplastic T51B rat liver cells like its larger cousin calmodulin, it is not involved in normal proliferation like calmodulin because it does not appear even briefly in proliferating normal adult cells, and it is not essential for the deregulation proliferation of neoplastic cells because about 15% of all neoplastic cells do not make it (Boynton *et al.*, 1982; MacManus and Whitfield, 1983).

It may be that a burst of Ca^{2+}/phospholipid-dependent (Ca^{2+}-calmodulin/cyclic AMP-independent) protein kinase activity might be needed to activate final G_1 events. Indeed, inhibition of this, or a related enzyme could be the reason for the as yet inexplicable partial inhibition of the second cyclic AMP surge and the nearly complete inhibition of the initiation of DNA synthesis in HPX-activated rat liver cells by a later injection of an α-adrenergic drug such as phenoxybenzamine or phentolamine (MacManus *et al.*, 1973), which are both now known to be protein kinase C inhibitors (Nishizuka, 1983a,b; Takai *et al.*, 1982). Therefore, the Ca^{2+}-dependent (Ca^{2+}-calmodulin/cyclic AMP-independent), protein kinase C-like membrane protein kinase, which phosphorylates certain (35, 48, and 100 kilodaltons) plasma membrane proteins and is switched on briefly and specifically in the second half of the HPX-activated liver cell's prereplicative period, might be the target of these drugs (MacManus and Whitfield, 1981).

Cyclic AMP and cyclic AMP-dependent protein kinases may amplify the second signal because there is a second, small (yet potent because of its cascade-triggering capability) cyclic AMP surge during the G_1 portion of the prereplicative period, which both isoproterenol-activated parotid cells and regenerating liver cells need to initiate chromosome replication (Boynton and Whitfield, 1983; MacManus *et al.*, 1973; Tsang *et al.*, 1980). In regenerating liver cells, this second surge is accompanied by second, 1,25-dihydroxyvitamin D_3-dependent transient rises in the levels of the two cyclic AMP-dependent protein kinases (Sikorska *et al.*, 1983; Whitfield *et al.*, 1984). This G_1 cyclic AMP surge may promote the production of DNA replication enzymes because cyclic AMP is known to stimulate gene expression possibly by forming ternary complexes with type II cyclic AMP-dependent protein kinase holoenzyme which is specifically concentrated in liver nuclei (Boynton and Whitfield, 1983; Cho-Chung, 1980; Severin and Nesterova, 1982; Whitfield *et al.*, 1984). Moreover, regenerating liver cells and EGF-stimulated liver cells *in vitro* need the well-known cyclic AMP elevator and liver enzyme inducer (Walker, 1983), glucagon, to transit the second half of their prereplicative period (de Hemptinne and Leffert, 1983; Leffert *et al.*, 1983; see Chapter 4). Indeed, when the levels of cyclic AMP and the two cyclic AMP-dependent protein kinases peak for the second time in regenerating liver cells, the normally inactive oncofetal c-*ras* gene is transcribed (Goyette *et al.*, 1984), synthesis of the L_1 and L_2 subunits of ribonucleotide reductase begins in the cytoplasm, and the level of L_1

subunits sharply rises in the nucleus (Fig. 6). At first glance, it seems unlikely that cyclic AMP could be the stimulator of ribonucleotide reductase synthesis because TPTX stops the synthesis of the enzyme's subunits (Fig. 7) without stopping the second post-HPX cyclic AMP surge (Fig. 4; Whitfield et al., 1979). However, this cyclic AMP surge in the TPTX animal is probably impotent because of the failure of the levels of the cyclic AMP-dependent protein kinase holoenzymes to rise after HPX in the TPTX animal (Sikorska et al., 1983; Whitfield et al., 1984). If the cyclic nucleotide is involved in the enzyme's synthesis, it must only start the synthesis of the subunits without affecting their subsequent accumulation of assembly into active holoenzyme because both the cytoplasmic L_1 subunit level and the cytoplasmic holoenzyme activity continue to rise long after the cyclic AMP surge has subsided and even after the cells have finished replicating their DNA (Fig. 7).

The cause of the second, or G_1, cyclic AMP surge is unknown, but since it occurs even in free-living organisms such as diatoms and yeasts, it seems that it must be triggered internally by some evolutionarily conserved mechanism (Boynton and Whitfield, 1983). However, the proliferation of liver cells seems to be limited by the need for glucagon to generate the G_1 cyclic AMP surge because blockage of the DNA synthetic response to HPX by diverting portal blood from the liver remnant through a portal-vena caval shunt can be overcome by injecting glucagon 8–10 hr after HPX and not earlier (de Hemptinne and Leffert, 1983), and primary rat hepatocytes in serum-free medium need glucagon, dibutyryl cyclic AMP, or the cyclic AMP-elevating cyclic nucleotide phosphodiesterase inhibitor 3-isobutyl-1-methylxanthine late in the prereplicative period to initiate DNA synthesis (Leffert et al., 1979; Hasegawa et al., 1980), The maximum utilization of the glucagon/cyclic AMP-amplified later prereplicative mRNAs requires that insulin be available to raise the cellular $\Delta\mu_{Na}$ and an internal pH high enough to maximize protein synthesis, minimize protein degradation, and maximally sensitize Ca^{2+}-dependent activation mechanisms to Ca^{2+} (Austin and Kay, 1982; Leffert et al., 1983; Moore, 1983).

V. CONCLUSIONS

We are gradually beginning to understand the roles of Ca^{2+}, Ca^{2+}-calmodulin, and cyclic AMP in the cell cycle. A round of chromosome replication and division is triggered in a cell in one of four possible proliferative states (newborn-cycling, quiescent-G_0, quiescent-dormant, quiescent-actively functioning) by a signal from an appropriate activator or mitogen, and chromosome replication and mitosis begin sometime

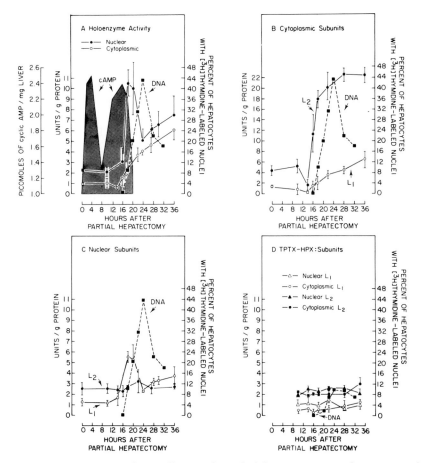

Fig. 7. As regenerating liver cells near the end of their type 4 prereplicative period, they generate the second, or G_1, cyclic AMP surge [the universal G_1 cyclic AMP surge identified by Boynton and Whitfield (1983)] and start making the separately inactive L_1 (CDP-specific, substrate-binding effector) and L_2 (nonheme iron-containing) subunits of ribonucleotide reductase (Youdale *et al.*, 1982) and assembling them into active holoenzyme. (A) The changes in cytoplasmic (○) and nuclear (●) ribonucleotide reductase holoenzyme activity (measured by the reduction of cytidine diphosphate to deoxycytidine diphosphate) and the temporal relations between these changes, the cyclic AMP surge, and the onset of nuclear DNA replication. The at first sight surprising transience of the burst of nuclear holoenzyme activity despite the rising DNA synthetic activity is likely due to the sequestration of nuclear holoenzyme in substrate-channeling, multienzyme replication complexes which renders it inaccessible to assay using exogenous cytidine diphosphate (CDP) (Prem ver Reddy and Pardee, 1982). (B) The levels of cytoplasmic L_1 and L_2 subunits start rising at very different rates and in both cases independently from DNA synthetic activity. The L_1 level is measured by mixing cytoplasmic extracts with an excess of a standard preparation of L_2 subunits; the amount

later as the results of a string of gene activations and the actions of short-lived products of these genes which cause the accumulation of DNA-replicating enzymes and their subunits, chromosomal proteins, replication initiator protein(s), and the prophasing phosphoprotein(s) (Figs. 1–4), which have not been discussed here for want of space. The cyclic AMP surges preceding chromosome replication, and entry into mitosis (Figs. 1–4) may be used by the cell to stimulate the transcription of these "cell cycle" genes as well as directly to activate some of the enzymes and proteins [e.g., the prophasing protein(s)] produced from these transcripts. The translation of the cyclic AMP-amplified mRNA population into proteins is probably maximized and the degradation of these new proteins is probably minimized by insulin (or functionally similar hormones such as the somatomedins), the essential universal cellular energizer which acts by increasing the Na^+ gradient across the plasma membrane, raising internal pH by stimulating Na^+/H^+ exchange, and transferring glucose transporters from internal membranes to the plasma membrane. Finally, internal Ca^{2+} surges (with their effectiveness increased by the insulin-induced higher pH) and the resulting internal Ca^{2+}-calmodulin surges and bursts of Ca^{2+}-phospholipid-dependent protein kinase C activity may directly or indirectly stimulate the transcription of early genes, stimulate translation of these gene transcripts, activate some of the products of these genes, and finally trigger the entry of replication enzymes and the replication initiator protein(s) into the late G_1 nucleus. Although the later stages of the cell cycle could not be discussed because of space limitations, it should be noted that internal Ca^{2+} and Ca^{2+}-calmodulin surges also mobilize tubulin from the cytoskeleton to form the spindle, operate the spindle mechanism, and finally trigger cytoplasmic division (see Chapter 16).

---

of CDP-reducing activity is then a function of the level of L_1 which combines with L_2 to form active holoenzyme. Conversely, the L_2 level is measured with a standard preparation of L_1 subunits. (C) The transient burst of nuclear holoenzyme activity upon the onset nuclear DNA replication is due to a sharp squirt of L_1 subunits into the nucleus where they combine with a relatively high and unchanging complement of L_2 subunits to form active holoenzyme. The at first surprising transience of the nuclear L_1 surge in the face of rising nuclear DNA synthetic activity could be due to the incorporation of these subunits into multienzyme replication complexes which would render them inaccessible for assay using exogenous CDP (Prem ver Reddy and Pardee, 1982). (D) Thyroparathyroidectomy (TPTX) 72 hr before partial hepatectomy prevents the regenerating livers from initiating DNA synthesis. It also prevents increases in L_1 and L_2 subunits and holoenzyme activity. Experimental procedures may be found in Rixon and Whitfield (1982), Sikorska et al. (1983), and Youdale et al. (1982). The points are means ±SEM of at least eight determinations. (○), Cytoplasmic L_1; (●), Cytoplasmic L_2; (△), nuclear L_1; (▲), nuclear L_2.

ADDENDUM

Several pieces of evidence have appeared which suggest that the c-*ras* protein mimics the action of adenylate cyclase's GTP-binding N_s subunit. The importance of this finding stems from the fact that cells ranging from diatoms, yeast, planarian cells, as well as cells from more complex organisms such as mice, rats, and humans, all have a burst of adenylate cyclase activity and a transient cyclic AMP surge which precedes chromosome replication. We can now speculate that the cause of the replication-related burst of adenylate cyclase activity and the cyclic AMP surge may be the product of the c-*ras* gene. Indeed, there is a rise in c-*ras* transcripts at the time of the G_1 cyclic AMP transient in BALB/c 3T3 mouse cells and regenerating rat liver cells (Campisi *et al.*, 1984; Goyette *et al.*, 1984). Toda *et al.*, (1985) have found that the GTP-binding c-*ras* proteins of yeast (which are homologous to mammalian c-*ras* proteins) stimulate adenylate cyclase. Thus, cells may trigger the replication-related burst of adenylate cyclase activity by producing the ubiquitous, highly conserved GTP-binding c-*ras* proteins which act by mimicking the action of the cyclase's stimulatory GTP-binding N_s subunit. The importance of such a concept for the understanding of cell proliferation and neoplastic transformation by v-*ras* oncogenes is obvious.

ACKNOWLEDGMENTS

We thank Dr. Doug Franks of the Clinical Research Institute of Montréal (Montréal, Québec) and Dr. Ben Volcani of the Scripps Oncenographic Institution (La Jolla, California) for the curves of Fig. 5A and B, respectively. We also thank Judi Meredith for word processing. Finally, and as always, we must thank Dave Gillan for preparing the illustrations.

REFERENCES

Armato, U., Andreis, P. G., and Whitfield, J. F. (1983). The calcium dependence of the stimulation of neonatal rat hepatocyte DNA synthesis and division by epidermal growth factor, glucagon and insulin. *Chem.-Biol. Interact.* **45**, 203–222.

Austin, S. A., and Kay, J. E. (1982). Translational regulation of protein synthesis in eukaryotes. *Essays Biochem.* **18**, 79–120.

Babcock, M. S., Marino, M. R., Gunning, W. T., and Stoner, G. D. (1983). Clonal growth and serial propagation of rat esophageal epithelial cells. *In Vitro* **19**, 403–415.

Ballard, F. J., Knowles, S. E., Wong, S. S. C., Bodner, J. B., Wood, C. M., and Gunn, J. M. (1980). Inhibition of protein breakdown in cultured cells is a consistent response to growth factors. *FEBS Lett.* **114**, 202–219.

Boynton, A. L., and Whitfield, J. F. (1976). The different actions of normal and supranormal calcium concentrations on the proliferation of BALB/c 3T3 mouse cells. In Vitro 12, 479–484.

Boynton, A. L., and Whitfield, J. F. (1979). The cyclic AMP-dependent initiation of DNA synthesis by T51 rat liver epithelioid cells. J. Cell. Physiol. 101, 139–148.

Boynton, A. L., and Whitfield, J. F. (1983). The role of cyclic AMP in cell proliferation: A critical assessment of the evidence, Adv. Cyclic Nucleotide Res. 15, 193–294.

Boynton, A. L., and Whitfield, J. F. (1984). The effect of serum on the calcium content of BALB/c 3T3 mouse cells. Exp. Cell Res. 152, 266–269.

Boynton, A. L., Whitfield, J. F., and Isaacs, R. J. (1976). The different roles of serum and calcium in the control of proliferation of BALB/c 3T3 mouse cells. In Vitro 12, 31–36.

Boynton, A. L., Whitfield, J. F., and MacManus, J. P. (1980a). Calmodulin stimulates DNA synthesis by rat liver cells. Biochem. Biophys. Res. Commun. 95, 745–749.

Boynton, A. L., Whitfield, J. F., and Walker, P. R. (1980b). The possible roles of polyamines in prereplicative development and DNA synthesis: A critical assessment of the evidence. In "Polyamines in Biomedical Research" (J. M. Gaugas, ed.), pp. 63–80. Wiley, New York.

Boynton, A. L., Whitfield, J. F., MacManus, J. P., Armato, U., Tsang, B. K., and Jones, A. (1981). Involvement of cyclic AMP and cyclic AMP-dependent protein kinases in the initiation of DNA synthesis by rat liver cells. Exp. Cell Res. 135, 199–211.

Boynton, A. L., MacManus, J. P., and Whitfield, J. F. (1982b). Stimulation of liver cell DNA synthesis by oncomodulin, an MW 11,500 calcium-binding protein from hepatoma. Exp. Cell Res. 138, 454–458.

Boynton, A. L., Kleine, L. P., Durkin, J. P., Whitfield, J. F., and Jones, A. (1982a). Mediation by calcicalmodulin and cyclic AMP of tumor promoter-induced DNA synthesis in calcium-deprived rat liver cells, In "Ions, Cell Proliferation, and Cancer" (A. L. Boynton, W. L. McKeehan, and J. F. Whitfield, eds.), pp. 417–431. Academic Press, New York.

Brewer, L. M., and MacManus, J. P. (1983). Oncomodulin detection during embryonic development. J. Cell Biol. 97, 32a.

Cairns-Smith, A. G. (1982). "Genetic Takeover." Cambridge Univ. Press, London and New York.

Campisi, J., Gray, H. E., Pardee, A. B., Dean, M., and Sonnenshein, G. E. (1984). Cell cycle control of c-myc but not c-ras expression is lost following chemical transformation. Cell (Cambridge, Mass.) 36, 241–247.

Campos-Gonzalez, R., Whitfield, J. F., Boynton, A. L., MacManus, J. P., and Rixon, R. H. (1984). Prereplicative changes in the soluble calmodulin of isoproterenol-activated rat parotid glands. J. Cell. Physiol. 118, 257–261.

Cavalier-Smith, T. (1981). The origin and early evolution of the eukaryotic cell. Symp. Soc. Gen. Microbiol. 32, 33–84.

Chafouleas, J. G., and Means, A. R. (1982). Calmodulin is an important regulatory molecule in cell proliferation. In "Ions, Cell Proliferation, and Cancer" (A. L. Boynton, W. L. McKeehan, and J. F. Whitfield, eds.), pp. 449–464. Academic Press, New York.

Cheung, W. Y., ed. (1980). "Calcium and Cell Function," Vol. 1. Academic Press, New York.

Cheung, W. Y., ed. (1982). "Calcium and Cell Function," Vol. 2. Academic Press, New York.

Cheung, W. Y., ed. (1983). "Calcium and Cell Function," Vol. 3. Academic Press, New York.

Cho-Chung, Y. S. (1980). Cyclic AMP and its receptor protein in tumor growth regulation *in vivo. J. Cyclic Nucleotide Res.* **6**, 163–177.

Damluji, R., and Riley, P. A. (1979). Time dependency of a crucial effect of EGTA on DNA synthesis in cultures of Swiss 3T3 cells. *Exp. Cell Biol.* **47**, 446–453.

Darzynkiewicz, Z., and Traganos, F. (1982). RNA content and chromatin structure cycling in and noncycling cell populations studied by flow cytometry. *In* "Genetic Expression in the Cell Cycle" (G. M. Padilla and K. S. McCarty, Sr., eds.), pp. 103–128. Academic Press, New York.

Das, M. (1980). Mitogenic hormone-induced intracellular message: Assay and partial characterization of an activator of DNA replication induced by epidermal growth factor. *Proc. Natl. Acad. Sci. U.S.A.* **77**, 112–116.

de Hemptinne, B., and Leffert, H. L. (1983). Selective effects of portal blood diversion and glucagon on rat hepatocyte rates of S-phase entry and deoxyribonucleic acid synthesis. *Endocrinology (Baltimore)* **112**, 1224–1232.

DeLarco, J. E., and Todaro, G. J. (1978). Growth factors from murine sarcoma virus-transformed cells. *Proc. Natl. Acad. Sci. U.S.A.* **75**, 4001–4005.

Dulbecco, R., and Elkington, J. (1975). Induction of growth in resting fibroblastic cell cultures by Ca^{2+}. *Proc. Natl. Acad. Sci. U.S.A.* **72**, 1584–1588.

Durkin, J. P., and Whitfield, J. F. (1984). Partial characterization of the mitogenic action of pp60[v-src], the oncogenic protein product of the *src* gene of avian sarcoma virus. *J. Cell. Physiol.* **120**, 135–145.

Durkin, J. P., Whitfield, J. F., and MacManus, J. P. (1983). The role of calmodulin in the proliferation of transformed and phenotypically normal ts ASV-infected rat cells. *J. Cell. Physiol.* **115**, 313–319.

Elliott, S. G., and McLaughlin, C. S. (1983). The yeast cell cycle: Coordination of growth and division rates. *Prog. Nucleic Acid Res. Mol. Biol.* **28**, 143–176.

Engström, W., Zetterberg, A., and Auer, G. (1982). Calcium, phosphate and cell proliferation. *In* "Ions, Cell Proliferation, and Cancer" (A. L. Boynton, W. L. MacKeehan, and J. F. Whitfield, eds.), pp. 341–357. Academic Press, New York.

Epel, D. (1982). The cascade of events initiated by rises in cytosolic Ca^{2+} and pH following fertilization in sea urchin eggs. *In* "Ions, Cell Proliferation, and Cancer" (A. L. Boynton, W. L. McKeehan, and J. F. Whitfield, eds.), pp. 327–339. Academic Press, New York.

Fausto, N., Goyette, M., Petropoulos, C., and Shank, P. (1982). Oncogene expression during liver regeneration. *J. Cell Biol.* **95**, 476a.

Floros, J., Chang, H., and Baserga, R. (1978). Stimulated DNA synthesis in frog nuclei by cytoplasmic extracts of temperature-sensitive mammalian cells. *Science* **201**, 651–652.

Foker, J. E., Malkinson, A. M., Sheppard, J. R., and Wang, T. (1979). Studies of cAMP metabolism in proliferating lymphocytes. *In* "Molecular Basis of Immune Cell Function" (J. G. Kaplan, ed.), pp. 57–74. Elsevier/North-Holland, New York.

Franks, D. J., Plamondon, J., and Hamet, P. (1984). An increase in adenylate cyclase activity precedes DNA synthesis in cultured vascular smooth muscle cells. *J. Cell. Physiol.* **119**, 41–45.

Gerson, D. F. (1982). The relation between intracellular pH and DNA synthesis rate in proliferating lymphocytes. *In* "Intracellular pH: Its Measurement, Regulation, and Utilization in Cellular Functions" (R. Nuccitelli and D. W. Deamer, eds.), pp. 375–383. Alan R. Liss, Inc., New York.

Goodenough, V. W., and Thorner, J. (1983). Sexual differentiation and mating strategies in the yeast *Saccharomyces* and in the green alga *Chlamydomonas. In* "Cell Interactions and Development: Molecular Mechanisms" (K. M. Yamada, ed.), pp. 29–75. Wiley, New York.

Gospodarowicz, D., and Tauber, J.-P. (1980). Growth factors and the extracellular matrix. *Endoc. Rev.* **1**, 201–227.

Goyette, M., Petropoulous, C. J., Shank, P. R., and Fausto, N. (1984). Regulated transcription of c-Ki-*ras* and c-*myc* during compensatory growth of rat liver. *Mol. Cell Biol.* **4**, 1493–1498.

Gutowski, J. K., and Cohen, S. (1983). Induction of DNA synthesis in isolated nuclei by cytoplasmic factors from spontaneously proliferating and mitogen-activated lymphoid cells. *Cell. Immunol.* **75**, 300–311.

Hasegawa, K., Ohtake, H., and Koga, M. (1980). Induction of rat hepatocyte DNA synthesis in primary monolayer culture. *Nippon Seikigaku Zasshi* **42**, 224.

Hazelton, B., Mitchell, B., and Tupper, J. (1979). Calcium, magnesium, and growth control in the WI-38 human fibroblast cell. *J. Cell Biol.* **83**, 487–498.

Hennings, H., Michael, D., Cheng, C., Steinert, P., Holbrook, K., and Yuspa, S. H. (1980). Calcium regulation of growth and differentiation of mouse epidermal cells in culture. *Cell (Cambridge, Mass.)* **19**, 245–254.

Hepler, P. K., and Wolniak, S. M. (1983). Membranous compartments and ionic transients in the mitotic apparatus. *In* "Spatial Organization of Eukaryotic Cells" (J. R. McIntosh, ed.), pp. 93–112. Alan R. Liss, Inc., New York.

Hesketh, T. R., Smith, G. A., and Metcalfe, J. C. (1982). Calcium and lymphocyte activation. *In* "Ions, Cell Proliferation and Cancer" (A. L. Boynton, W. L. McKeehan, and J. F. Whitfield, eds.), pp. 397–415. Academic Press, New York.

Hidaka, H., Sasaki, Y., Tanaka, T., Endo, T., Ohno, S., Fujii, Y., and Nagato, T. (1981). N-(6-aminohexyl)-5-chloro-1-naphthalene sulfonamide, a calmodulin antagonist inhibits cell proliferation. *Proc. Natl. Acad. Sci. U.S.A.* **78**, 4354–4357.

Hochhauser, S. J., Stein, J., and Stein, G. S. (1981). Gene expression and cell cycle regulation. *Int. Rev. Cytol.* **71**, 95–243.

Jazwinski, S. M., Wang, J. L., and Edelman, G. M. (1976). Initiation of replication in chromosomal DNA induced by extracts from proliferating cells. *Proc. Natl. Acad. Sci. U.S.A.* **73**, 2231–2235.

Jones, A., Boynton, A. L., MacManus, J. P., and Whitfield, J. F. (1982). Ca-calmodulin mediates the DNA-synthetic response of calcium-deprived liver cells to the tumor promoter TPA. *Exp. Cell Res.* **138**, 87–93.

Kishimoto, A., Kajikawa, N., Shiota, M., and Nishizuka, Y. (1983). Proteolytic activation of calcium-activated, phospholipid-dependent protein kinase by calcium-dependent neutral protease. *J. Biol. Chem.* **258**, 1156–1164.

Klee, C. B., Crouch, T. H., and Richman, P. G. (1980). Calmodulin. *Annu. Rev. Biochem.* **49**, 409–515.

Klein, G., ed. (1982). "The Transformation-Associated Cellular p53 Protein," Adv. Viral Oncol., Vol. 2. Raven Press, New York.

Kuo, J. F., Mazzei, G. J., Shatzman, R. C., and Turner, R. S. (1985). The phospholipid/ Ca^{2+}-dependent protein phosphorylation system. *In* "Ions, Membranes, and the Electrochemical Control of Cell Function" (A. Pilla and A. L. Boynton, eds.), Verlag-Chemie, New York (in press).

Lechner, J. F. (1982). Effect of calcium on growth and differentiation of normal human bronchial epithelial cells. *In* "Ions, Cell Proliferation, and Cancer," Book of Abstracts (A. L. Boynton, W. L. McKeehan, and J. F. Whitfield, eds.), p. 62. W. Alton Jones Cell Science Center, Lake Placid, New York.

Leffert, H. L. (1982). Monovalent cations, cell proliferation and cancer. *In* "Ions, Cell

Proliferation, and Cancer" (A. L. Boynton, W. L. MacKeehan, and J. F. Whitfield, eds.), pp. 93–102. Academic Press, New York.

Leffert, H. L., Koch, K. S., Moran, T., and Rubalcava, B. (1979). Hormonal control of rat liver regeneration. *Gastroenterology* **76**, 1470–1482.

Leffert, H. L., Koch, K. S., Lad, P. J., de Hemptinne, B., and Skelly, H. (1983). Glucagon and liver regeneration. *Handb. Exp. Pharmacol.* **66**, Pt. 1, 453–484.

Levine, B. A., and Williams, R. J. P. (1982). The chemistry of calcium ion and its biological relevance. *In* "The Role of Calcium in Biological Systems" (L. J. Anghileri and A. M. Tuffet-Anghileri, eds.), pp. 3–26. CRC Press, Boca Raton, Florida.

MacManus, J. P., and Whitfield, J. F. (1981). Stimulation of autophosphorylation of liver cell membrane proteins by calcium and partial hepatectomy. *J. Cell. Physiol.* **106**, 33–40.

MacManus, J. P., and Whitfield, J. F. (1983). Oncomodulin—A calcium-binding protein from hepatoma. *In* "Calcium and Cell Function" (W. Y. Cheung, ed.), Vol. 4, pp. 411–440. Academic Press, New York.

MacManus, J. P., Braceland, B. M., Youdale, T., and Whitfield, J. F. (1973). Adrenergic antagonists and a possible link between the increase in cyclic adenosing $3',5'$-monophosphate in DNA synthesis during liver regeneration. *J. Cell. Physiol.* **82**, 157–164.

MacManus, J. P., Braceland, B. M., Rixon, R. H., Whitfield, J. F., and Morris, H. P. (1981). An increase in calmodulin during growth of normal and cancerous liver *in vivo*. *FEBS Lett.* **133**, 90–102.

MacManus, J. P., Boynton, A. L., and Whitfield, J. F. (1982). The role of calcium in the control of cell reproduction. *In* "The Role of Calcium in Biological Systems" (L. J. Anghileri and A. M. Tuffet-Anghileri, eds.), Vol. 1, pp. 147–164. CRC Press, Boca Raton, Florida.

MacNeil, S., Walker, S. W., Brown, B. L., and Tomlinson, S. (1982). Evidence that calmodulin may be involved in phytohemagglutinin-stimulated lymphocyte division. *Biosci. Rep.* **2**, 891–897.

Martelly, I., Molla, A., Thomasset, M., and Le Moigne, A. (1983). Planarian regeneration: *In vivo* and *in vitro* effects of calcium and calmodulin on DNA synthesis. *Cell Differ.* **13**, 25–34.

Mastro, A. M., and Smith, M. C. (1982). Two signal, calcium-dependent lymphocyte activation by an ionophore, A23187, and a comitogenic phorbol ester, TPA. *J. Cell Biol.* **95**, 21a.

Matsumoto, K., Uno, I., and Ishikawa, T. (1983). Initiation of meiosis in yeast mutants defective in adenylate cyclase and cyclic AMP-dependent protein kinase. *Cell (Cambridge, Mass.)* **32**, 417–423.

Mercer, W. E., Nelson, D., DeLeo, A. B., Old, L. J., and Baserga, R. (1982). Microinjection of monoclonal antibody to protein p53 inhibits serum-induced DNA synthesis in 3T3 cells. *Proc. Natl. Acad. Sci. U.S.A.* **79**, 6309–6312.

Michell, R. H. (1978). Inositol phospholipids and cell surface receptor function. *Biochim. Biophys. Acta* **415**, 81–147.

Moore, R. D. (1983). Effects of insulin upon ion transport. *Biochim. Biophys. Acta* **737**, 1–49.

Ng, K. W., Livesay, S. A., Larkins, R. G., and Martin, T. J. (1983). Calcitonin effects on growth and on selective activation of type II isoenzyme of cyclic adenosine $3',5'$-monophosphate-dependent protein kinase in T47D human breast cancer cells. *Cancer Res.* **43**, 794–800.

Nishizuka, Y. (1983a). Phospholipid degradation and signal translation for protein phosphorylation. *Trends Biochem. Sci.* **8**, 13–16.

Nishizuka, Y. (1983b). A receptor-linked cascade of phospholipid turnover in hormone action. In "Endocrinology" (K. Shizume, H. Imura, and N. Shimizu, eds.), pp. 15–24. Excerpta Medica, Amsterdam.

Pardee, A. B., Campisi, J., and Croy, R. G. (1982). Differences in growth regulation of normal and tumor cells. *Ann. N.Y. Acad. Sci.* **397**, 121–129.

Prem ver Reddy, G., and Pardee, A. B. (1982). Coupled ribonucleoside diphosphate reduction, channeling and incorporation into DNA of mammalian cells. *J. Biol. Chem.* **257**, 12526–12531.

Provvedini, D. M., Tsoukas, C. D., Deftos, L. J., and Manolagas, S. C. (1983). 1,25-Dihydroxyvitamin D_3 receptors in human leukocytes. *Science* **221**, 1181–1183.

Rao, P. N., Johnson, R. T., and Sperling, K., eds. (1982). "Premature Chromosome Condensation: Application in Basic, Clinical and Mutation Research." Academic Press, New York.

Rixon, R. H., and Whitfield, J. F. (1976). The control of liver regeneration by parathyroid hormone and calcium. *J. Cell. Physiol.* **87**, 147–156.

Rixon, R. H., and Whitfield, J. F. (1982). An early mitosis-determining event in regenerating rat liver and its possible mediation by prostaglandins or thromboxane. *J. Cell. Physiol.* **113**, 281–288.

Rixon, R. H., MacManus, J. P., and MacManus, J. P. (1979). The control of liver regeneration by calcitonin, parathyroid hormone and 1α,25-dihydroxycholecalciferol. *Mol. Cell. Endocrinol.* **15**, 79–89.

Rubin, R. P. (1982). "Calcium and Cellular Secretion." Plenum, New York.

Salisbury, J. L., Condeelis, J. S., and Satir, P. (1983). Receptor-mediated endocytosis: Machinery and regulation of the clathrin-coated vesicle pathway. *Int. Rev. Exp. Pathol.* **24**, 1–62.

Schanne, F. A. Y., Kane, A. B., Young, E. E., and Furber, J. L. (1979). Calcium dependence of toxic cell death: A final common pathway. *Science* **206**, 700–702.

Schneiderman, M. H., Dewey, W. C., and Highfield, D. P. (1971). Inhibition of DNA synthesis in synchronized Chinese hamster cells treated in G_1 with cycloheximide. *Exp. Cell Res.* **67**, 147–155.

Severin, E. S., and Nesterova, M. V. (1982). An effect of cyclic AMP-dependent protein kinases on gene expression. *Adv. Enzyme Regul.* **20**, 167–193.

Shier, W. T., and Du Bourdieu, D. J. (1982). Role of phospholipid hydrolysis in the mechanism of toxic cell death by calcium and inophore A23187. *Biochem. Biophys. Res. Commun.* **109**, 106–112.

Sikorska, M., MacManus, J. P., Walker, P. R., and Whitfield, J. F. (1980). The protein kinases of rat liver nuclei. *Biochem. Biophys. Res. Commun.* **93**, 1196–1203.

Sikorska, M., Whitfield, J. F., and Rixon, R. H. (1983). The effects of thyroparathyroidectomy and 1,25-dihydroxyvitamin D_3 on changes in the activities of some cytoplasmic and nuclear protein kinases during liver regeneration. *J. Cell. Physiol.* **115**, 297–304.

Steinhardt, R. A. (1982). Ionic logic in activation of the cell cycle. In "Ions, Cell Proliferation, and Cancer" (A. L. Boynton, W. L. McKeehan, and J. F. Whitfield, eds.), pp. 311–325. Academic Press, New York.

Stiles, C. D. (1983). The molecular biology of platelet-derived growth factor. *Cell (Cambridge, Mass.)* **33**, 653–655.

Streumer-Svoboda, Z., Wiegant, F. A. C., van Dongen, A. A. M. S., and van Wijk, R.

(1982). Variations in some molecular events during the early phases of the Reuber H35 hepatoma cycle. *Biochimie* **64**, 411–418.

Sullivan, C. W., and Volcani, B. E. (1981). Silicon in the cellular metabolism of diatoms. *In* "Silicon and Siliceous Structures in Biological Systems" (T. L. Simpson and B. E. Volcani, eds.), pp. 15–42. Springer-Verlag, Berlin and New York.

Swierenga, S. H. H., Whitfield, J. F., and Boynton, A. L. (1978). Age-related and carcinogen-induced alterations of the extracellular growth factor requirements for cell proliferation *in vitro*. *J. Cell Physiol.* **94**, 171–180.

Takai, Y., Kishimoto, A., and Nishizuka, Y. (1982). Calcium and phospholipid turnover as transmembrane signaling for protein phosphorylation. *In* "Calcium and Cell Function" (W. Y. Cheung, ed.), Vol. 2, pp. 385–412. Academic Press, New York.

Tanaka, T., Ohmura, T., Yamakado, T., and Hidaka, H. (1982). Two types of calcium-dependent protein phosphorylations modulated by calmodulin antagonists. Naphthalenesulfonamide derivatives. *Mol. Pharmacol.* **22**, 408–412.

Toda, T., Uno, I., Ishikawa, T., Powers, S., Kataoka, T., Brook, B., Cameron, S., Broach, J., Matsumoto, K., and Wigler, M. (1985). In yeast, RAS proteins are controlling elements of adenylate cyclase. *Cell (Cambridge, Mass.)* **40**, 27–36.

Trump, B. J., Berezesky, I. K., and Osomio-Vargas, A. R. (1981). Cell death and the disease process. The role of calcium. *In* "Cell Death in Biology and Pathology" (I. D. Bowen and R. A. Lockshin, eds.), pp. 209–242. Chapman & Hall, London.

Tsang, B. K., Rixon, R. H., and Whitfield, J. F. (1980). A possible role of cyclic AMP in the initiation of DNA synthesis by isoproterenol-activated parotid gland cells. *J. Cell. Physiol.* **102**, 19–26.

Tsang, B. K., Whitfield, J. F., and Rixon, R. H. (1981). The reversible inability of isoproterenol-activated parotid gland cells to initiate DNA synthesis in the hypocalcemic, thyroparathyroidectomized rat. *J. Cell. Physiol.* **107**, 41–46.

Tupper, J. T., Kaufman, L., and Bodine, P. V. (1980). Related effects of calcium and serum on the G_1 phase of the human WI-38 fibroblast. *J. Cell. Physiol.* **104**, 97–103.

Ullman, A., and Danchin, A. (1983). The role of cyclic AMP in bacteria. *Adv. Cyclic Nucleotide Res.* **15**, 1–53.

Van Heyningen, S. (1982). Cholera toxin. *Biosci. Rep.* **2**, 135–146.

Van Venrooij, W. J. W., Henshaw, E. C., and Hirsch, C. A. (1972). Effects of deprival of glucose or individual amino acids on polyribosome distribution and rate of protein synthesis in cultured mammalian cells. *Biochim. Biophys. Acta* **259**, 127–137.

Walker, P. R. (1983). "The Molecular Biology of Enzyme Synthesis." Wiley, New York.

Walker, P. R., and Whitfield, J. F. (1984). Colchicine prevents the translation of mRNA molecules transcribed immediately after proliferative activation of hepatocytes in regenerating rat liver. *J. Cell. Physiol.* **118**, 179–186.

Warburton, M. J., and Poole, B. (1977). Effect of medium composition on protein degradation and DNA synthesis in rat embryo fibroblasts. *Proc. Natl. Acad. Sci. U.S.A.* **74**, 2427–2431.

Weber, I. T., Takio, K., Titani, K., and Steitz, T. A. (1982). The cAMP binding domains of the regulatory subunit of cAMP-dependent protein kinase and the catabolite gene activator protein are homologous. *Proc. Natl. Acad. Sci. U.S.A.* **79**, 7679–7683.

Whitfield, J. F. (1980). Adrenergic agents, calcium ions, and cyclic nucleotides in the control of cell proliferation. *Handb. Exp. Pharmacol.* **54**, Pt. 1, 267–317.

Whitfield, J. F. (1982). The roles of calcium and magnesium in cell proliferation. *In* "Ions, Cell Proliferation, and Cancer" (A. L. Boynton, W. L. McKeehan, and J. F. Whitfield, eds.), pp. 283–294. Academic Press, New York.

Whitfield, J. F., Boynton, A. L., MacManus, J. P., Sikorska, M., and Tsang, B. K. (1979).

The regulation of cell proliferation by calcium and cyclic AMP. *Mol. Cell. Biochem.* **27**, 155–179.

Whitfield, J. F., Boynton, A. L., MacManus, J. P., Rixon, R. H., Sikorska, M., Tsang, B., Walker, P. R., and Swierenga, S. H. H. (1980). The roles of calcium and cyclic AMP in cell proliferation. *Ann. N.Y. Acad. Sci.* **339**, 216–240.

Whitfield, J. F., MacManus, J. P., Boynton, A. L., Durkin, J., and Jones, A. (1982). Futures of calcium, calcium-binding proteins, cyclic AMP and protein kinases in the quest for an understanding of cell proliferation and cancer. *In* "Functional Regulation at the Cellular and Molecular Levels" (R. A. Corradino, ed.), pp. 61–87. Elsevier/North-Holland, New York.

Whitfield, J. F., Rixon, R. H., and Sikorska, M. (1985). Calcium, cyclic AMP and protein kinases combine to trigger chromosome replication in regenerating liver. *In* "Ions, Membranes, and the Electrochemical Control of Cell Function" (A. Pilla and A. L. Boynton, eds.). Verlag-Chemie, New York (in press).

Whitson, G. L., ed. (1980). "Nuclear-Cytoplasmic Interactions in the Cell Cycle." Academic Press, New York.

Wilson, T. H., and Lin, E. C. C. (1980). Evolution of membrane bioenergetics. *J. Supramol. Struct.* **13**, 421–446.

Youdale, T., MacManus, J. P., and Whitfield, J. F. (1982). Rat liver ribonucleotide reductase: Separation, purification, and properties of two nonidentical subunits. *Can. J. Biochem.* **60**, 463–470.

13

Growth Regulation by Sodium Ion Influxes

HYAM L. LEFFERT* AND KATHERINE S. KOCH[†,1]

*Department of Medicine
Division of Pharmacology
and
†Department of Biology
University of California, San Diego
La Jolla, California

[1]Present address: Department of Medicine, Division of Pharmacology, University of California, San Diego, La Jolla, California 92093.

CONTROL OF ANIMAL CELL PROLIFERATION,
VOLUME I

I. INTRODUCTION

In 1976, a paper from David Epel's laboratory appeared suggesting that events leading to the initiation of DNA synthesis in fertilized sea urchin eggs start with a transient increase in the rate of exchange of extracellular sodium ions for intracellular protons across the cell surface (Johnson et al., 1976). Soon thereafter, in an apparently unrelated series of experiments, Lionel Jaffe and his colleagues reported that artificially applied cathodal currents stimulated amphibian limb regeneration (Borgens et al., 1977a). Similar electrical currents were generated spontaneously as inwardly directed Na^+ fluxes at cut regenerating stumps (Borgens et al., 1977b). Despite their differences, proliferation was blocked in both systems by amiloride, a reversible inhibitor of passive Na^+ influx. A common pattern emerged when these observations were extended to mammalian cells from findings that amiloride-sensitive Na^+ influxes were stimulated by peptide hormones required for rat liver regeneration (Koch and Leffert, 1979a). Futhermore, these fluxes seemed necessary for several cellular functions that preceded the initiation of rat hepatocyte proliferation in vitro (Koch and Leffert, 1979a,b, 1980; Leffert and Koch, 1980, 1982a,b; Leffert et al., 1979).

Since these observations were made, some 40 laboratories have begun to explore this phenomenon—mitogen-induced sodium influx—as it relates to the control of animal cell proliferation and differentiation (Leffert, 1980, 1982). This chapter will summarize approaches and experimental systems used to study these fluxes. After reviewing pertinent phenomenology, attempts will be made to evaluate evidence for or against the concept that such fluxes control growth, summarize biochemical properties of Na^+ flux-mediating structures and the ways in which their activation might occur, and outline cellular changes that appear linked to them.

II. EXPERIMENTAL ASPECTS

A. Biological

Tissue culture systems have been used by most investigators because they provide controlled conditions without complex host variables. Proliferative transitions in culture usually are studied by adding growth factors to quiescent cell populations (G_0). Then the kinetics of cellular entry into S phase (DNA synthesis) and mitosis are monitored. These transitions are measured by labeling with tritiated thymidine ($[^3H]dT$), followed by radioautography, to determine the cellular fraction synthesizing DNA (the labeling index), by staining cells with a fluorescent DNA-binding dye (e.g., mithramycin) and examining them with a flow microfluorimeter to determine the cell cycle distribution ($G_{0,1}$, S, and G_2 + M), by determining, from [^{3}H]dT pulse and density labeling studies, rates of isotope incorporation into semiconservatively replicated DNA or, from Colcimid-treated cells followed by radio-autography, the mitotic labeling index, and by simply counting the cells. Under appropriate conditions, the kinetics of these processes in vitro closely simulate physiological ones in vivo (Leffert et al., 1982a,b, 1983).

Quiescent cultures (or tissues) generally contain small proportions of S phase (0.05–5%) and mitotic cells (0.001–0.1%). After exposure to growth-promoting conditions, a lengthy interval of time passes (the prereplicative phase) before new cells initiate DNA synthesis. Growth factors exert their effects at both early and late periods in this pre-replicative phase (Leffert and Koch, 1977). Depending upon the depth of quiescence (defined by the labeling index), this interval usually lasts 10–16 hr. It is here during its early period that mitogen-mediated Na$^+$ influxes, which we have termed "Signal 1," are thought to control growth (Koch and Leffert, 1979a, 1980; Leffert and Koch, 1977, 1980, 1982a,b).

It is not always preferable or possible to work under quiescent conditions. In some situations in culture, drug-synchronized cells have been used (Panet et al., 1983; Warley et al., 1983) or, in studies with trans-plantable tumors, constitutively proliferating cells were analyzed (Smith et al., 1978). Results from non-steady-state conditions can be informative, but are more difficult to interpret.

B. Physicochemical

Several methods have been used to measure Na$^+$ fluxes across cell membranes and Na$^+$ contents inside cells (Boynton et al., 1982).

Radionuclide uptake (^{22}NaCl) and flame photometry are standard procedures. Rapid (seconds) and precise (nanomoles) measurements can be made as a function of time during proliferative transitions. Ion influxes are determined with the Na$^+$,K$^+$-ATPase pump fully inhibited by ouabain. Under these conditions, net influxes can be monitored, since both active Na$^+$ efflux and Na$^+$: Na$^+$ exchange are negligible (Villereal and Owen, 1982; Leffert and Koch, 1982a). Atomic absorption spectrometry also has been used to measure cellular contents of several mono- and divalent cations simultaneously (Sanui and Rubin, 1982). These approaches, while relatively straightforward, are limited to the extent that they cannot distinguish "free" from "bound" Na$^+$ (hence, they are inadequate for measurements of intracellular Na$^+$ concentrations), identify Na^+ compartments inside cells, or resolve fast ionic conductances on the scale of single channels.

Thus, single-cell microelectrode analysis, nuclear magnetic resonance spectroscopy (NMR), and energy-dispersive X-ray microprobe analysis are becoming increasingly important for studying the concentration and distribution of intracellular Na^+ ions, expecially in situations where Na^+ influxes might alter these parameters. These analytical tools are elaborate and invasive, but they overcome some of the limitations mentioned above. For example, classical electrophysiological procedures have been used successfully to measure passive Na^+ influxes in single neuroblastoma cells (Moolenaar et al., 1981a). In addition, Na^+-selective liquid ion exhanger microelectrodes have provided measurements of Na^+ activities in single intestinal epithelial cells of Necturus maculosa (Hudson and Schultz, 1984). Although it is unclear if impaled cells in either of these preparations survive long enough, it would be interesting to see if their proliferative transitions could be measured under appropriate mitogenic conditions to prove directly that cells in which early passive Na^+ influxes occur are, in fact, destined to initiate DNA synthesis (see Section III,C). A less invasive NMR method, employing dysprosium[III]tripolyphosphate as a paramagnetic shift reagent for $^{23}Na^+$, has been developed to measure free Na^+ (Gupta and Gupta, 1982). An advantage of this technique is its applicability to living cells. Sodium compartmentation also has been measured in a variety of proliferating and quiescent cells with X-ray microprobes (Smith et al., 1978; Cameron et al., 1980; Zs.-Nagy et al., 1981,1983; Lukács et al., 1983; Pieri et al., 1983; Warley et al., 1983). This technique eliminates ion redistribution artifacts and permits quantitative estimates of nuclear and cytoplasmic Na^+ contents of unfixed nonnecrotic tissue.

Recently, patch-clamp electrophysiological methods have been used to monitor single-channel K^+ conductances in human lymphocytes (Matteson and Deutsch, 1984) exposed to mitogens like phytohemagglutinin (PHA; De Coursey et al., 1984). These techniques might eventually be adapted to monitor single-channel Na^+ conductances.

III. PHENOMENOLOGY OF GROWTH-RELATED SODIUM INFLUXES

A. Growth Factor Activation

Table I lists the kinds of growth stimuli that activate Na^+ influxes in quiescent cells.

These stimuli are remarkably diverse. Among them are sperm; a large number of naturally occurring peptides and hormones delivered to cells by endocrine [EGF, vasopressin, insulin, angiotensin, and progesterone (Morill *et al.*, 1983,1984; Cameron *et al.*, 1984)], paracrine (FGF, PDGF, NGF, bradykinin, EGF), and, perhaps, autocrine routes (FGF, PDGF); purified (thrombin), unpurified serum, and partially characterized blood-borne macromolecular proteins (HGF; see Fig. 1); plant lectins [concanavalin A (Con A) and PHA]; anti-immunoglobulin antibodies; a tumor promoter [12-O-tetradecanoylphorbol-13-acetate (TPA)]; a chemical carcinogen (1,2-dimethylhydrazine); cathodal electrical currents; and 67% hepatectomy.

Thus far, no naturally occurring growth factor acting in G_0 and/or early G_1 has been found that does not also stimulate an amiloride-sensitive Na^+ influx. These observations suggest that unsuspected substances might behave like growth factors if they activate passive Na^+ influxes under appropriate conditions. This prediction has been fulfilled from studies with nonphysiological substances like the diuretic furosemide (Koch and Leffert, 1979a), naturally occurring factors like the bee venom peptide mellitin or other peptides like vasopressin (see Table I) or bombesin (Rozengurt and Smith, 1983), and ions like sodium orthovanadate (Smith, 1983; Cassel *et al.*, 1984).

Most growth factors listed in Table I work only upon cells bearing specific cell surface receptors. For example, EGF stimulates Na^+ influxes and proliferation in many cells of endodermal origin, like hepatocytes (see Fig. 2). In contrast, PDGF (see Lipton, Chapter 6) and FGF preferentially activate proliferation and Na^+ influxes in cells of mesodermal origin which have receptors to these peptides (e.g., fibroblasts). Hepatocytes display EGF but not PDGF or FGF receptors (see Bowen-Pope *et al.*, Chapter 10 and Gospodarowicz, Chapter 3). Hence, the latter two peptides activate neither fluxes nor proliferation in hepatocytes cultured under chemically defined conditions (Koch and Leffert, 1979a; Leffert *et al.*, 1979).

It seems, therefore, that some degree of proliferative specificity is mediated through the operation of Na^+ influx systems. However, other growth factor determinants of proliferative specificity exist (Leffert and Koch, 1977). For example, cells bearing receptors that mediate adenylate cyclase activation by peptides like glucagon (see Beckner *et al.*, Chapter 9), such as hepatocytes and kidney epithelia (Taub *et al.*, 1984), are stimulated to grow optimally only when exposed to both Na^+ influx and adenylate cyclase-activating mitogens. Glucagon does not activate Na^+ influxes in hepatocytes (Fig. 2) or proliferation of cells that lack glucagon receptors.

Dose–response curves delineating initial rates of activation of Na^+

TABLE I

Rapid Sodium Ion Flux-Related Events Stimulated by Growth Factors and Inhibited by Amiloride[a]

Factor/ stimulus	Cell system	Na+ influx	Na+ pump activation	Na+ : H+ exchange and ↑ pH$_{in}$	Na+ : Ca^{2+} exchange	↑Cell Na+ content	References
Sperm	Sea urchin egg	+	+	+	U	+	Johnson et al., 1976; Shen and Steinhardt, 1978; Epel, 1980; Johnson and Epel, 1981; Steinhardt, 1982; Whitaker and Steinhardt, 1982; Girard et al., 1982
	Medaka blastula	+	U	U	U	U	Lum et al., 1983
	Rabbit preimplantation embryo	+	U	U	U	U	Benos, 1983
EGF	Rat hepatocyte	+	+	+	–	–	Koch and Leffert, 1979a,b, 1980; Hasegawa et al., 1980; Leffert and Koch, 1980, 1982a,b; Fehlmann et al., 1981a; Koch et al., 1984
	Mouse fibroblast	+	+	+	U	+	Rozengurt and Mendoza, 1980; Burns and Rozengurt, 1984
	Human fibroblast	+	+	+	+/–	+	Villereal, 1981; Villereal and Owen, 1982; Moolenaar et al., 1982a,b; W. H. Moolenaar (personal communication, 1984)

	Cell type					Reference
	Monkey kidney (BSC-1)	+	+	+	U	Rothenberg et al., 1982
	Human epidermoid carcinoma (A-431)	+	U	+	–	Rothenberg et al., 1983a,b
FGF	Rat pheochromocytoma (PC12)	+	+	+	U	Boonstra et al., 1983
	Hamster fibroblast	+	+	+	U	Pouysségur et al., 1982a
PDGF	Hamster fibroblast	+	+	+	U	Pouysségur et al., 1982a; Chambard et al., 1983
	Mouse fibroblast	+	+	+	+	Schuldiner and Rozengurt, 1982; Cassel et al., 1983; Burns and Rozengurt, 1983
NGF	Rat PC12	+	+	+	U	Boonstra et al., 1983
Vasopressin	Mouse fibroblast	+	+	+	+	Mendoza et al., 1980b; Rozengurt, 1982; Burns and Rozengurt, 1983
	Monkey BSC-1	+	U	U	+	Walsh-Reitz and Toback, 1983
Insulin	Rat hepatocyte	+	+	–	–	Leffert and Koch, Fig. 2 (this chapter); Fehlmann and Freychet, 1981; H. L. Leffert and R. Grosse (unpublished results); Reyes and Benos, 1983
	Frog skeletal muscle	+	+	+	U	Moore et al., 1982
	Xenopus oocyte	+	U	+	U	Morrill et al., 1983, 1984

(continued)

373

TABLE I (Continued)

Factor/ stimulus	Cell system	Na⁺ influx	Na⁺ pump activation	Na⁺:H⁺ exchange and ↑ pH$_{in}$	Na⁺:Ca²⁺ exchange	↑Cell Na⁺ content	References
Angiotensin	Rat smooth muscle	+	+	U	U	U	Smith and Brock, 1983
Bradykinin	Human fibroblast	+	U	U	U	U	Owen and Villereal, 1983a
Mellitin	Mouse fibroblast	+	+	U	U	U	Gelehter and Rozengurt, 1980; Rozengurt et al., 1981
Thrombin	Human fibroblast	+	U	U	U	U	Vicentini et al., 1984
	Hamster fibroblast	+	+	+	–	U	Pouysségur et al., 1982a,b; L'Allemain et al., 1984a,b
	Mouse fibroblast	+	+	U	U	U	Stiernberg et al., 1983, 1984
HGF	Rat hepatocyte	+	U	U	U	U	Leffert and Koch (Fig. 1, this chapter)
Serum	Mouse fibroblast	+	+	+	U	+	Smith and Rozengurt, 1978a; Mendoza et al., 1980a; Burns and Rozengurt, 1983; Frelin et al., 1983
	Human fibroblast	+	+	+	U	U	Pouysségur et al., 1980, 1982a
	Mouse neuroblastoma	+	+	+	–	U	Moolenaar et al., 1981a,b, 1982a,b, 1983
	Hamster fibroblast	+	+	U	–	U	Pouysségur et al., 1983a; J. Pouysségur (personal communication, 1984)
	Monkey BSC-1	+	U	+	U	+	Rothenberg et al., 1982

Agent	Cell/tissue					References
	Dog kidney (MDCK)	+	+	U	U	Reznik et al., 1983
	Mouse glioma–neuroblastoma hybrid	+	+	U	U	O'Donnell and Villereal, 1982; Benos and Sapirstein, 1983
Plant lectins						
Con A, PHA	Mouse T lymphocyte	+	+	U	+	Averdunk and Lauf, 1975; Averdunk, 1976; Averdunk and Gunther, 1980; Gerson, 1982
Con A, PHA	Human lymphocyte	+	−	−	+	Segel et al., 1979; Kaplan and Owens, 1980; Deutsch et al., 1981, 1984; Deutsch and Price, 1982a,b
Anti-Ig antisera	Mouse T lymphocyte	U	+	U	U	Averdunk and Lauf, 1975
	Human B leukemia	+	+	U	U	Heikkilä et al., 1983a,b
Phorbol esters						
TPA	Mouse fibroblast	+	+	U	U	Dicker and Rozengurt, 1981; Burns and Rozengurt, 1983
1,2-Dimethylhydrazine	Rat distal colon	+	U	U	U	Davies et al., 1983
Cathodal currents	Regenerating frog limb	+	U	U	U	Borgens et al., 1977a,b; Jaffe, 1980
Partial hepatectomy	Rat liver	+	+	U	+	Koch and Leffert, 1979a; Leffert and Koch, 1979; Hasegawa and Koga, 1979

[a] A "+" or "—" indicates the occurrence or nonoccurrence of the event, respectively; a "+/−" indicates conflicting results; "U" indicates that information is currently unavailable. EGF, FGF, PDGF, NGF, "HGF," Con A, PHA, TPA, and PMA are epidermal growth factor, fibroblast growth factor, platelet-derived growth factor, nerve growth factor, "hepatocyte growth factor," concanavalin A, phytohemagglutinin, and 12-O-tetra-decanoylphorbol-13-acetate, respectively. See text for details.

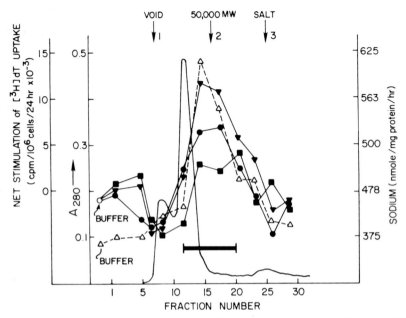

Fig. 1. Cofractionation of rat serum [³H]dT and Na⁺ uptake-stimulating activity ("HGF," hepatocyte growth factor) by gel filtration. An ammonium sulfate fraction of rat serum (proteins soluble at 40% salt saturation) was chromatographed over Sephacryl-200 in phosphate-buffered saline (pH 7.4). The absorbances ($A_{280 \text{ nm}}$) of eluted proteins were followed by a Uvicord monitor (———). Bioassays for growth-promoting activity (net stimulation of [³H]dT uptake) in 10 (■), 20 (●), and 40 μl (▼) fraction/ml of medium were performed on stationary phase hepatocytes cultured in plastic Costar wells (Koch et al., 1982) under standard DNA synthesis reinitiation conditions (Koch and Leffert, 1979a). [³H]dT uptake in control cultures (fresh medium, plus column buffer, plus 10 ng insulin, glucagon, EGF/ml) was 22,000 cpm/10⁶ cells/24 hr. Bioassays for Na⁺ uptake-stimulating activity were performed with stationary phase hepatocytes cultured in 35-mm dishes in order to obtain enough culture sample for cation analysis by flame photometry (Leffert and Koch, 1982a). Fractions for testing were added ("15%" v/v equivalent volumes of rat serum) to fresh chemically defined medium containing 1 mM ouabain at 37°C. One hour later, the fluids were aspirated and the dishes washed six times with 2 ml of 0.1 M MgCl₂ at 4°C. The monolayers (≈350,000 cells/dish) were dissolved in 0.5 ml of a solution of 15 mM LiCl₂ and 1% toluene. Sodium contents were determined by flame photometry. Each point is the average of 4 dishes from two separate experiments (△). Arrows (1, 2, 3) indicate elution positions of molecular weight marker proteins chromatographed separately and of salt. The peak of Na⁺ uptake stimulation was abolished if amiloride (0.4 mM) was present during the bioassay incubation (data not shown).

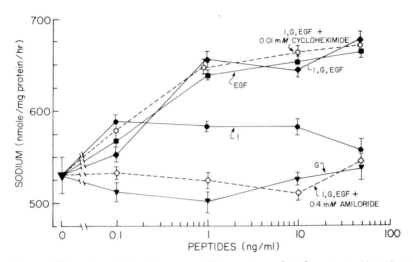

Fig. 2. Effects of peptide mitogens on Na^+ contents of rat hepatocytes in primary culture. Twelve-day-old stationary cultures (~500,000 cells/dish) were shifted into fresh 1 mM ouabain-supplemented chemically defined media without (▼) or with EGF (■), I (insulin) (●), G (glucagon) (▾), or a mixture of the three peptides at the concentrations given on the x axis [without (◆) or with amiloride (◇) or cycloheximide (◯)]. Cell Na^+ contents were determined 1 hr later [values are averages ±SEM (N = 13–16)] by flame photometry (see legend, Fig. 1).

influx are hyperbolic. One example, illustrated for EGF with cultured hepatocytes (Fig. 2), shows an $EC_{50} \simeq 1 \times 10^{-10}$ M, similar to the $K_{D[app]}$ of ^{125}I-labeled EGF for binding to its 170,000-dalton hepatocyte receptor (Fehlmann et al., 1981a; see Herschman, Chapter 7). Another curve, reported for vasopressin with cultured BSC-1 kidney cells, shows an apparent $EC_{50} \simeq 1 \times 10^{-11}$ M (Walsh-Reitz and Toback, 1983). In both cases, high concentrations of peptides ($\geq 10^{-7}$M) inhibit cell proliferation (EGF) or show reduced efficacy (vasopressin) without comparable effects upon Na^+ influxes. The underlying causes and significance of these observations are unclear (see Section III,C).

B. Properties

Mitogen-stimulated Na^+ influx rates range from 5 to 20 nmols $Na^+/10^6$ cells/min (Koch and Leffert, 1979a; Rozengurt and Mendoza, 1980; Villereal, 1981; Moolenaar et al., 1981a,b; Pouysségur et al., 1982a; Mummery et al., 1982). These rates represent 1.5- to 5.0-fold elevations above basal unstimulated rates and, with respect to $[Na^+]_{out}$, probably reflect increases in the V_{max} for transport (Smith and Rozengurt, 1978b). Problems encountered in their quantitation, owing to nonuniform cell surface geometry, have been discussed elsewhere (Adam et al., 1982).

The stimulated fluxes do not require exogenous Ca^{2+} (Rothenberg et al., 1983a,b; Frelin et al., 1983), apart from fluxes stimulated by tumor promoters like phorbol 12-myristate 13-acetate (Whitely et al., 1984), nor do they require extracellular amino acids cotransported by A system carriers or new protein synthesis (see Fig. 2; Leffert and Koch, 1982a; Rozengurt, 1982; Lubin et al., 1982).

Mitogen-activated Na^+ influxes are amiloride sensitive. They are not inhibited by fast Na^+ channel blockers like tetrodotoxin (Koch and Leffert, 1979a; Pouysségur et al., 1980; Moolenaar et al., 1981a; see Section V,B). The extracellular concentrations of amiloride that half-maximally inhibit mitogen-inhibited Na^+ influxes (ID_{50}) vary from 20 to 400 μM (Koch and Leffert, 1979a, 1980; O'Donnell et al., 1983; Frelin et al., 1983; Benos and Sapirstein, 1983; L'Allemain et al., 1984a; Cameron et al., 1984). The reasons for this variation are unclear; the data come from different biological systems under different experimental conditions. Nonetheless, amiloride's effects are nontoxic and reversible for 24–36 hr. Amiloride is not metabolized by animal cells, so its effects are not produced by breakdown products.

C. Relationships to Proliferation

Amiloride-sensitive Na^+ influxes have been observed in virtually all quiescent animal cell systems when they are stimulated to grow. As Table I shows, this property exists among phyla separated through evolution over millions of years. It applies to animal cells originating from all embryonic germ layers in different phases of their organism's life cycle, and it encompasses widely separated and closely related species. For these reasons, it is reasonable to envision these fluxes as fundamental properties of proliferative activation in animal cells. The extent to which this process operates in prokaryotic cells or in plant cells has not yet been clarified, although monovalent cation fluxes have been implicated as regulators of bacterial (Lubin, 1982; Zilberstein et al., 1982) and plant cell proliferation (Leigh and Wyn Jones, 1984).

Mitogen-activated Na^+ influxes are rapid "burst-like" events, occurring within seconds after mitogen treatment and persisting at activated levels for at least 2–60 min in the presence of mitogen (Johnson et al., 1976; Koch and Leffert, 1979a; Fehlmann et al., 1981b; Deutsch et al., 1981; Moolenaar et al., 1981a, 1982a,b; Rozengurt, 1982; Rothenberg et al., 1982; Pouysségur et al., 1982a; Villereal and Owen, 1982; Cassel et al., 1983; Frelin et al., 1983; Benos and Sapirstein, 1983; Felber and

Brand, 1983). Although elevations of Na^+,K^+-ATPase pump activity near or at the G_1/S boundary have been interpreted as evidence for a second burst (or late-G_1 phase) of amiloride-sensitive Na^+ influx (Mummery et al., 1982; Stiernberg et al., 1983), other explanations could account for these results (see Sections IV,B and VI,B).

Direct correlations exist between elevated [³H]dT labeling indices and initial Na^+ influx rates (Koch and Leffert, 1980; Leffert and Koch, 1982a; Owen and Villereal, 1983a). Mitogen concentrations that are required to half-maximally stimulate initiation of DNA synthesis also half-maximally stimulate amiloride-sensitive Na^+ influxes under appropriate conditions (Leffert, 1982; Leffert and Koch, 1982b). Exceptions to this behaviour might reflect, on the one hand, different structural aspects of ligand–receptor interactions in different cells (Walsh-Reitz and Toback, 1983) or, on the other, different modes of rapidly terminating the signal—an issue which others have begun to consider with various Na^+ flux stimulators in fibroblast systems (Owen and Villereal, 1983b; Collins and Rozengurt, 1984).

Individual growth factors that are relatively more effective in stimulating DNA synthesis usually are more effective Na^+ influx activators. For example, EGF stimulates Na^+ influx more effectively than insulin (see Fig. 2). A similar trend is found with respect to EGF's ability to stimulate initiation of adult hepatocyte DNA synthesis (Koch and Leffert, 1979a). Figure 2 also shows that EGF and insulin individually stimulate Na^+ influxes in cultured hepatocytes at levels of peptide below those that, in combination, are required to stimulate DNA synthesis initiation optimally (see also Fehlmann and Freychet, 1981). These discrepancies are not understood at this time. They might arise from marked differences in the intervals of time that peptides must interact with cells in order to generate responses—seconds (for flux measurements) as opposed to hours (for growth measurements), or additional processes activated by peptides (that are necessary to stimulate DNA synthesis) might require greater fractional occupancy of the requisite receptors. This too might underlie shifts in the dose–response curves.

In studies with nonquiescent systems—for example, growing fibroblasts (Adam et al., 1982; Panet et al., 1983) or developing liver (Bellemann, 1981)—direct correlations are reported or implicated, respectively, for rates of passive Na^+ influx and cell proliferation. Detailed experiments with the fibroblast systems reveal that rates of passive Na^+ influx also decrease as cell density increases (Adam et al., 1982). The mechanisms responsible for these changes are unknown. Mitogen receptor display, intercellular contacts, growth-regulatory conditioning factors produced by cells, and/or rates of mitogen and mitogen receptor degradation—variables which are known to affect proliferative responses and to be dependent upon cell density in some systems—might be contributory.

When the kinetics of initiation of cell proliferation are monitored in quiescent systems, amiloride blocks mitogenesis by delaying the time when new cells start to synthesize DNA [the onset time, designated S_t (de Hemptinne and Leffert, 1983)], but not the rates at which committed cells enter S phase (Koch and Leffert, 1979a, 1980; Rozengurt and Mendoza, 1980; Mummery et al., 1982). Once cells have begun to replicate DNA, their susceptibility to amiloride's growth-inhibitory effects is markedly diminished (Koch and Leffert, 1979a; Mummery et al., 1982). Whether this stems from completion of early prereplicative Na^+ influx-dependent processes, from desensitization of Na^+ influx systems, and/or from an inability of amiloride to bind to structures where it exerts its specific effects in preactivated cells is unclear. The interpretation of these results is complicated because amiloride is transported into most cells and exerts nonspecific effects (see Section IV,B).

D. Relationships to Differentiation

Differentiated functions tend to be more efficiently expressed by quiescent cells and less efficiently expressed by chronically proliferating cells. If Na^+ influxes are required to initiate proliferation, one would expect to find relatively low influx rates in differentiated or differentiating cell systems. According to this coupling hypothesis, conditions that inhibit or limit passive Na^+ influxes should stimulate or facilitate differentiation (Leffert and Koch, 1979). Since Na^+ influxes stimulate Na^+,K^+-ATPase pump activity (see Section VI,B,2), a corollary of this hypothesis is that reductions in Na^+,K^+-ATPase-driven ion fluxes correlate with or promote differentiation.

In support of this hypothesis, amiloride potentiates the induction of granulocyte differentiation by dimethyl sulfoxide (DMSO) in a promyelocytic cell line (Carlson et al., 1984). Blockade of plasma membrane Na^+ fluxes may account for these findings. In a different system where proliferating neuroblastoma cells are induced to differentiate by serum growth factor withdrawal, an inverse relationship exists between enhanced differentiation (measured by neurite outgrowth) and EGF receptor number per cell. Consequently, EGF's capacity to activate Na^+ influxes and reinitiate proliferation in such differentiated neurons is markedly reduced (Mummery et al., 1983).

In support of the corollary of this hypothesis, agents like DMSO or ouabain that inhibit uphill K^+ influx mediated by Na^+,K^+-ATPase induce differentiation in several model systems (Kaplan, 1978). This is exemplified by the induction of hemoglobin synthesis in Friend erythroleukemia cells (Bernstein et al., 1976), increased surface IgM expression in a pre-B cell tumor line (Rosoff and Cantley, 1983), and granulocyte differentiation in a human leukemia cell line (Gargus et al., 1983). It is unclear whether decreases of $[K^+]_{in}$ and/or increases of $[Na^+]_{in}$, two potential consequences of diminished pump activity, contribute to differentiation. However, since ouabain does not block amiloride-sensitive Na^+ influxes (Moolenaar et al., 1981a), its effects upon them must be indirect. Alternatively, agents like DMSO might alter Na^+ influxes directly by shifting mitogen receptor conformations into inactive modes (see Herschman, Chapter 7).

Several findings conflict with this hypothesis. For example, elevated amiloride-sensitive Na^+ influxes induced by LPS (lipopolysaccharide) or TPA precede differentiation (surface IgM expression) in a pre-B cell tumor line (Rosoff and Cantley, 1983; Rosoff et al., 1984). An earlier report (Levenson et al., 1980) showing that amiloride blocked differentiation of a mouse erythroleukemia cell line supports this view. But the proposed blockade mechanism in this latter system, an inhibition of plasma membrane Na^+/Ca^{2+} exchange (Smith et al., 1982) as opposed to an inhibition of protein synthesis (see Section IV,B), has not been clarified. Hypertonic conditions also stimulate amiloride-sensitive Na^+ influxes (Whitely et al., 1984) and the expression of specific differentiated functions in other systems, for example, viral protein synthesis (Koch et al., 1980; Garry et al., 1982), globin mRNA transcription (Groudine and Weintraub, 1982), and type III collagen synthesis (Hata et al., 1983). The mechanisms underlying these observations are unknown.

An apparent contradiction to the corollary of this hypothesis has been reported by Hennings et al., (1983). They have observed that ouabain blocks Ca^{2+}-induced differentiation of epidermal cells into keratinocytes. In contrast to the findings of Whitfield et al. (see Chapter 12), however, epidermal cells grow better when external Ca^{2+} levels are ≤ 0.1 mM. Were such cells to possess a growth-regulatory cell surface Na^+/Ca^{2+} exchange system poised by ion gradients generated by the Na^+ pump, this contradiction might be explained.

IV. EVIDENCE FOR CAUSE AND EFFECT

A. Dependence on External Sodium

If Na^+ influxes regulate cell growth, the stimulation of cell proliferation should depend upon external levels of Na^+ ions. This dependence has been observed in different quiescent cell culture systems (Smith and Rozengurt, 1978a; Koch and Leffert, 1979a). The dose–response curves are similar and suggest that Na^+ entry is carrier mediated: Stimulated [^{3}H]dT uptake saturates at $\simeq 100$ mM NaCl, and half-maximal responses occur at 60–70 mM NaCl with proliferative unresponsiveness at ≤ 20 mM NaCl (see Section V,A). Growing cell populations also show $[Na^+]_{out}$ optima. For example, 180 and 179 mM NaCl have been reported for cultured BSC-1 kidney cells (Toback, 1980) and for secondary chick embryo cells, respectively (Garry et al., 1981).

These observations suggest but do not prove that Na^+ influxes are necessary for initiating DNA synthesis. Although these results probably do not depend upon the counterions employed, the interpretations of proliferative dependence in these types of studies are complicated further from observations that the chemical nature of the osmotic substitute for Na^+ may itself affect growth (Balk and Polimeni, 1982; Lubin, 1982; Burns and Rozengurt, 1984). Alternatively, Na^+ dependence might reflect cell surface ion binding, as discussed elsewhere (Leffert and Koch, 1982a).

B. Pharmacologic Blockade

If mitogen-activated Na^+ influxes are necessary to initiate cell proliferation, their inhibition should block (or delay) the initiation of DNA synthesis.

Studies along these lines have exploited amiloride, a specific inhibitor of passive Na^+ influx (Benos, 1982). Amiloride inhibits DNA synthesis reinitiation with ID_{50} values of 0.02–0.4 mM. These concentrations are similar to those which half-maximally block mitogen-activated Na^+ influx (see Section III,B). For the most part, Na^+ influxes are insensitive to amiloride in quiescent cell populations (basal states) (Moolenaar et al., 1981a). Partial sensitivity has been seen in cultured hepatocytes and glioma–neuroblastoma hybrid cells (Koch and Leffert, 1979a; O'Donnell and Villereal, 1982; O'Donnell et al., 1983), but the proportion of G_0 cells in these systems may have been lower than others. Amiloride's growth-inhibitory effects are reversible and are manifested during the prereplicative phase. As noted (Section III,C), in contrast to inhibitors of protein synthesis, amiloride delays the onset of DNA replication but does not block DNA synthesis once it has started.

Amiloride does not block peptide mitogen binding to or the clustering and internalization of cell surface receptors (Koch and Leffert, 1979a; Moolenaar et al., 1982a). However, its actions are not exerted solely at the cell surface (its presumptive binding site for Na^+ influx blockade). This has become increasingly apparent from the fact that amiloride is transported by a saturable carrier (for hepatocytes, the $K_{m[app]} \simeq 0.02$ mM, and the $V_{max} \simeq 1.43$ nmol/10^6 cells/min) into and is concentrated by animal cells (Leffert et al., 1982c; Smith et al., 1982; Moolenaar et al., 1982b). Under these conditions, amiloride directly inhibits protein synthesis (Lubin et al., 1982; Leffert et al., 1982c), cyclic AMP-dependent protein kinases (Ralph et al., 1982; Holland et al., 1983), and mitochondrial functions (Taub, 1978; Stiernberg et al., 1983), and it stimulates protein phosphatases (Le Cam et al., 1982). Whether the hepatocyte amiloride "carrier" is identical to the putative Na^+/H^+ exchange porter is not yet clear (see Section V,A,1).

Amiloride's effects on protein synthesis do not interfere with mitogen-activated Na^+ influx (Leffert and Koch, 1982a). But the drug's side effects might interfere with the initiation of DNA synthesis, especially with respect to proliferative requirements of protein synthesis and cyclic AMP-dependent phosphorylation. This problem has led to studies with amiloride analogues as well as to nonpharmacologic approaches.

The results of studies with analogues from three different laboratories are promising but inconclusive. For example, Cameron et al. (1982, 1984) report that benzamil inhibits progesterone-induced germinal vesicle breakdown (GVBD) in Xenopus oocytes (ID$_{50} \simeq 2$ mM). Unlike amiloride, benzamil apparently does not penetrate the egg and, as revealed by electron microscopic radioautographic studies using [^{3}H]benzamil, reacts solely with the egg cell surface. However, no correlations were reported between rank orders of potency with respect to analogue inhibition of passive Na^+ influx and GVBD (Cameron et al., 1984).

Similar analogues were tested for their ability to inhibit $^{22}Na^+$ influxes and DNA synthesis in human and rodent fibroblasts. In one case, trends were obtained in equivalent rank orders of potencies for both responses (fluxes and proliferation) with one interesting exception: Two functionally distinct benzamil analogues were found (O'Donnell et al., 1983). Thus, benzamil inhibited both $^{22}Na^+$ influx and initiation of DNA synthesis (ID$_{50} \simeq 15$ and 18 µM, respectively), whereas 1-µM levels of modified benzamil (replacement of the 5-amino group with an H^+) inhibited DNA synthesis 50% without inhibiting Na^+ influx. However, no drug uptake studies were reported, therefore permeability differences might also account for these anomalous results. In another case (using Chinese hamster lung fibroblasts), significant correlations among rank orders of inhibition of $^{22}Na^+$ influxes (ID$_{50} \simeq 0.04$–3000 µM) and DNA synthesis (ID$_{50} \simeq 0.04$–900 µM) were observed; amiloride analogues containing protonated guanidino moieties were more potent than nonprotonated congeners (L'Allemain et al., 1984a). These results were not obtained in normal media, but instead by growing cells in media containing low Na^+ levels (25–50 mM) without bicarbonate (buffered with HEPES). Apparently, these conditions were necessary to permit amiloride binding to specific membrane sites as well as to restrict intracellular regulation of intracellular pH (pH_{in}) to an Na^+- and H^+-dependent mode (see Sections IV,D and V,A,1). Additional experiments performed with cell-free protein-synthesizing reticulocyte lysates revealed that much higher concentrations of different analogues (e.g., 5-N,N-dimethylamiloride or 5-N,N-diethylamiloride) were needed to block protein synthesis directly ($ID_{50} \simeq 0.4$ mM) in contrast to functions of intact cells ($ID_{50} \simeq 0.15$ or 0.04 μM, respectively). However, no concomitant studies were performed with living cells to rule out differential drug uptake. And, in the latter two studies, drug exposure intervals were not examined to eliminate direct effects on cells replicating DNA.

Other drugs that stimulate [monensin (Smith and Rozengurt, 1978a; Koch and Leffert, 1979a)] or inhibit [indomethacin, aspirin, and calmodulin antagonists such as trifluoperazine (Owen and Villereal, 1982, 1983c)] $^{22}Na^+$ influxes have been tested for their effects on mitogen-activated DNA synthesis. The interpretations of results of these experiments are equivocal. For example, monensin, a neutral ionophore that activates Na^+/H^+ exchange, blocks proliferation (Koch and Leffert, 1979a) whereas it stimulates the Na^+ pump (Smith and Rozengurt, 1978a) and increases intracellular pH (Koch and Leffert, 1980; Leffert and Koch, 1982a,b) as early-acting mitogens do. However, monensin perturbs many cellular processes, possibly because it interacts extensively with all cellular membranes. And, unlike early-acting mitogens, monensin may not exert pleiotypic effects at the mitogen receptor level (see Section IV,E). The proliferative and $^{22}Na^+$ influx inhibitory effects of indomethacin and aspirin are provocative (the different drugs show equivalent ID_{50} values for each response), but the specificity of these drugs also is questionable owing to the high concentrations needed to obtain biological effects (millimolar range). A similar problem of interpretation exists with regard to trifluoperazine and its congeners.

Since mitogen-activated $^{22}Na^+$ influxes stimulate K^+ influxes via the Na^+ pump, it might be expected (if pump activation also is needed to induce DNA synthesis) that ouabain, a putative specific inhibitor of Na^+,K^+-ATPase, inhibits the induction of cell proliferation (see Section VI,B,2). Findings of this kind, first reported with lymphocytes by

Quastel and Kaplan (1970), have been obtained in virtually all quiescent cell systems examined (Table I; for one exception, see Section V,B). Ouabain, like amiloride, blocks only during the prereplicative phase and not afterward. Depending upon cell species, the ID_{50} for proliferation and the $K_{i[app]}$ for Na^+,K^+-ATPase activity in membrane preparations from stimulated cells are in close agreement.

Two cases of "dissociation" of Na^+ pump-mediated K^+ influxes from the cell cycle have been reported. The interpretation of the results in one case (Tupper and Zografos, 1978) is complicated from the experimental design (mitogen deprivation as opposed to mitogen addition). In the other study, ouabain blocked $^{86}Rb^+$ influxes in quiescent cultures but not initiation of DNA synthesis (Frantz et al., 1981). This probably occurred because intracellular K^+ levels already had exceeded threshold concentrations needed for the initiation of DNA synthesis (Lopez-Rivas et al., 1982; Lubin, 1982). Quiescent cell systems with high basal levels of $[K^+]_{in}$ often appear refractory to ouabain's antiproliferative effects, a phenomenon which Lubin (1982) has termed the "K^+ sensitivity shift."

Ouabain permeates cells and, in rodent systems, very high ouabain concentrations ($K_i \simeq 0.2$ mM) are necessary to block growth and pump activity. Therefore, ouabain might exert nonspecific detergent-like effects as well as unknown intracellular effects. Attempts to circumvent this problem are being made (see Section IV,E).

C. Chemically Modified Mitogens

Another approach to testing the hypothesis that Na^+ influxes are necessary for initiating DNA synthesis is to modify early-acting mitogens so that they are more or less capable of enhancing these fluxes and then determine their effects on proliferation. In principle, there are five kinds of modifications that might be made.

One of these would be an alteration that renders the mitogen fully defective in both its capacity to activate influxes and proliferation. This has not yet been achieved. A provocative attempt along these lines (a second category) has been made by L'Allemain et al. (1984a). They have shown that PMS-thrombin (phenylmethanesulfonyl-conjugated α-thrombin) behaves like a partial agonist with respect to influx (Na^+/H^+ exchange) and proliferative responsiveness (both were 10% of normal). No evidence of significant contamination of PMS-thrombin with unmodified mitogen was obtained. Along similar lines, all of the available non-mitogenic derivatives of human α-thrombin (γ-, nitro-α-, or the diisopropylphospho-α- forms) at low concentration are incapable of stimulating ouabain-sensitive ^{86}Rb$^+$ influxes (although each form binds to thrombin receptors), in contrast to α-thrombin, the natural mitogen, which is capable (Stiernberg et al., 1984). Construction of more active agonists (a third category) has not yet been achieved.

Two other kinds of modifications might involve mitogens that are either defective with respect to flux activation, but are proliferation competent (these have not yet been constructed nor have they been identified in nature), or that activate Na$^+$ influxes but not proliferation. Alterations of the latter kind also have not yet been reported, although there are claims that modified EGF (CNBr-EGF; Yarden et al., 1982) and thrombin (specifically, high concentrations of either the γ- or nitro-α-form; Stiernberg et al., 1983) are capable of activating ouabain-sensitive Rb$^+$ influxes but not DNA synthesis. In both instances, no direct measurements of ^{22}Na$^+$ influxes were made. Such experiments are necessary, since pump activation can occur by mechanisms that do not necessarily involve Na$^+$ influx (Hudson and Schultz, 1984; Serpersu and Tsong, 1984; Schenk et al., 1984).

Apart from the fact that the results with cyanogen bromide-cleaved EGF have not been confirmed (J. Pouysségur or W. Moolenaar, personal communications, 1984), the possibility remains that chemically modified EGF or thrombin preparations are contaminated with unmodified mitogen and that activation of early events like accelerated Na$^+$,K$^+$-ATPase pumping requires lower EC$_{50}$ levels (Leffert and Koch, 1982a). In addition, when using "mitogens" in either of the latter two categories, the possibilities cannot be excluded that amiloride-sensitive Na$^+$ influxes are insufficient (as opposed to unnecessary) for proliferation, as discussed elsewhere (Koch and Leffert, 1979a,b, 1980; Leffert and Koch, 1982a), or that Na$^+$ influxes fit into a sequential physiological cascade of early prereplicative events that modified mitogens might be "bypassing" because of altered interactions with cellular receptors.

D. Genetic Analysis

If Na$^+$ influxes are necessary for proliferation, one should be able to select for cellular mutations (or variants) in systems mediating these fluxes which confer altered proliferative properties upon these cells.

Three approaches have been reported. One involves the selection of amiloride-resistant mutants of MDCK cells (Taub, 1978). Mutant cells have decreases in both the K_m and V_{max} for ^{22}Na$^+$ influx with respect to wild-type cells (Taub and Saier, 1981). Although one of the mutant cell lines (Am^{r}22) has a generation time almost 50% lower than the wild type, no additional growth control studies have been performed to relate this difference to altered Na$^+$ influx. In addition, other metabolic defects were noted in the four variant lines, raising the possibility of secondary mutations.

A second approach, using human fibroblasts, has been to select for cell variants in low Na^+ media (50 mM) that survive lethal doses of H^+ in culture media (pH $\simeq$ 5.5) lacking HCO_3^- and substituted with organic buffers (L'Allemain *et al.*, 1984b; Pouysségur *et al.*, 1984). The rationale of this selection scheme is based upon the premise that only those cells which are defective in Na^+ influx coupled to H^+ efflux (Na^+/H^+ antiport—see Section V,A,1) should survive the stress of environmental acidification. When these variant cells are "challenged" with an acid load by shifting them into culture media at low pH (≤ 7.2), they grow poorly (their pH_{in} levels are 0.2–0.3 units more acidic than wild-type cells), and they fail to demonstrate effective Na^+/H^+ exchange in response to mitogens such as α-thrombin and insulin.

These findings support the hypothesis that under stringent culture conditions, Na^+ influxes are causally related to proliferation. [These results should not be confused with the conflicting reports of initiation of DNA synthesis in quiescent cells by brief exposures to alkaline media (Zetterberg and Engström, 1981; Burroni and Ceccarini, 1984).] When HCO_3^- is added to the culture media (pH 8.0–8.3) of these variant cells to raise their pH_{in}, normal proliferation occurs, but Na^+/H^+ exchange remains defective. This suggests that such variant cells contain an additional membrane "pH stat," perhaps an HCO_3^-/Cl^- exchanger. Two pH regulating systems may coexist in kidney medullary cell membranes (Kinsella and Aronson, 1980) and in unfertilized and fertilized sea urchin eggs (Johnson and Epel, 1981). However, HCO_3^-/Cl^- exchange is insensitive to amiloride and available evidence limits its role, if any, in mitogen activation of growth (Pouysségur *et al.*, 1984).

The third approach has been to analyze Na^+ fluxes and contents in fibroblasts transformed with DNA or RNA tumor viruses (Mendoza *et al.*, 1980a; Garry *et al.*, 1981; Moyer *et al.*, 1982). Both wild-type viruses as well as viruses bearing temperature-sensitive mutations in their genes coding for transforming proteins have been studied. While interesting correlations have been reported [e.g., decreases or increases in amiloride-sensitive Na^+ influxes at nonpermissive (growth inhibitory) or permissive (growth stimulatory) temperatures, respectively], no definitive conclusions have been reached.

E. Immunochemical Probes

Several approaches employing monospecific polyclonal or monoclonal antibodies are being tried in attempts to circumvent artifacts associated with pharmacologic blockade. For example, antibodies that block Na^+ influx and the initiation of DNA synthesis would lend support to the Na^+ influx hypothesis. To our knowledge, such antibodies have not yet been obtained. An inherent problem preparing them arises from difficulties in isolating appropriate immunogens—for example, membrane proteins mediating Na^+ influx (see Section V,A,1). One way this problem might be solved (although it biases the experimental outcome) is to select for monoclonal antibodies, against membrane proteins, that protect cells from cytotoxic H^+ loads (J. Pouysségur or W. Moolenaar, personal communications, 1984).

Two other classes of probes have begun to yield interesting but inconclusive results. One involves a panel containing anticatalytic mouse monoclonal antibodies to rat Na^+,K^+-ATPase (Schenk and Leffert, 1983). Since early mitogen-activated Na^+ influxes activate the Na^+ pump and since pump activation and the initiation of DNA synthesis are blocked by ouabain, it is reasonable to expect that specific anticatalytic antibodies might also block proliferation.

Figure 3 shows the results of preliminary experiments from our laboratory along these lines (D. B. Schenk, H. Skelly and H. L. Leffert, unpublished results). Two responses were measured after shifting primary rat hepatocyte cultures into growth promoting media: (1) initiation of DNA synthesis and (2) cell Na^+ contents. The two monoclonal antibodies tested, 9-A5 and 9-B1, are potent inhibitors of Na^+,K^+-ATPase in broken cell preparations ($K_{[app]} \simeq 30$ and 600 nM, respectively), and both antibodies bind to distinct sites on the pump's α subunit (Schenk and Leffert, 1983; Schenk et al., 1984). Antibody 9-A5 showed no biological effects on living hepatocytes (it did not block DNA synthesis and it did not increase cell Na^+ contents). In contrast, antibody 9-B1 blocked [³H]dT incorporation and increased cell Na^+ contents. Antibody 9-A5 (an IgG_1) has recently been shown to recognize a single intracellular epitope (Schenk et al., 1984; Leffert et al., 1985; Farley and Ochoa, 1985) which probably explains 9-A5's biological inactivity upon intact cells. This antibody has proved useful for other structural, quantitative, and localization studies of Na^+ pumps in quiescent and proliferating tissues (Schenk and Leffert, 1983; Schenk et al., 1984; Hubert et al., 1985; Leffert et al., 1985). Antibody 9-B1 has posed technical problems because of chemical instability (its anticatalytic effects are lost after prolonged storage at $-20°$ or $4°C$). Whether this is due to 9-B1's primary or secondary structure is unclear. Nonetheless, further studies seem warranted with this approach, especially those geared toward selection of externally directed antipump antibodies.

A second class of probes involves anti-growth factor receptor anti-

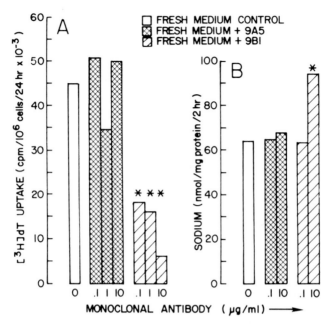

Fig. 3. Effects of monoclonal antibodies on [³H]dT uptake and cell Na⁺ contents in primary adult hepatocyte cultures. Thirteen-day-old cultures (≈500,000 cells/dish) were shifted into fresh chemically defined media without or with the concentrations of purified monoclonal antibodies shown (Schenk and Leffert, 1983). After 24 hr, DNA synthesis was measured (A) or, after 1 hr, cell Na⁺ contents were determined (B) as described above. The (*) indicates values significantly different from controls (N = 12–18; $p \ll 0.05$). Unstimulated basal [³H]dT uptake rates were 15,000 cpm/10⁶ cells/24 hr.

bodies. One might expect to generate within this group of antibodies at least five types with properties that fall into categories such as those enumerated above for chemically modified growth factors. W. H. Moolenaar (personal communication, 1984) has recently produced one such mouse monoclonal antibody ("IgG-2Eg") that competes with ¹²⁵I-labeled EGF for specific binding to the EGF receptor of human A-431 fibroblasts. This antibody lacks Na⁺ influx, Na⁺/H⁺ exchange, and DNA synthesis stimulating activity. But it perturbs the EGF receptor because it stimulates the phosphorylation of tyrosine in cellular proteins (from [γ-³²P]ATP) which is thought to be at least one of the catalytic functions of the EGF receptor in living cells (Carpenter, 1984). In contrast, another mitogenic anti-EGF receptor mouse monoclonal antibody has been produced (Schreiber et al., 1981) that in its divalent form ("2G2-IgM") induces EGF receptor "microclustering" (Schreiber et al., 1983). This phenomenon has been proposed to be a necessary event for induction of DNA synthesis. When monovalent Fab fragments of 2G2-

IgM were tested in fibroblasts for their biological properties, the fragments stimulated several early events including ouabain-sensitive $^{86}Rb^+$ fluxes (and, although not directly examined, presumably Na^+ influxes), but neither receptor microclustering nor DNA synthesis (see also Gospodarowicz, Chapter 3). These experiments, if confirmed, suggest that Na^+ influxes are mediated by unclustered receptors and are necessary but insufficient to initiate cell proliferation. These studies do not clarify the issue of how "early" receptor-mediated events such as Na^+ influxes and tyrosine kinase activation are linked. This possibility is suggested from recent findings that antibodies against SV40 transforming viral T antigens (which have associated tyrosine kinase activity) block ouabain-sensitive $^{86}Rb^+$ influxes in mouse fibroblasts (Hand et al., 1983).

F. Natural Inhibitors

There is mounting evidence that many cells produce (and secrete) molecules that inhibit their own proliferation ("chalones"; for review, see Leffert and Koch, 1977). It might be predicted that at least two classes of chalones exist: (1) those acting early and (2) those acting late in the prereplicative phase (Koch and Leffert, 1979a; Leffert and Koch, 1980). One logical test of this hypothesis is to isolate and identify chalones that specifically block Na^+ influxes (Leffert, 1982). An interesting result along these lines was reported recently by Walsh-Reitz et al. (1984) using a partially purified protein that is secreted by and inhibits proliferation of BSC-1 kidney cells in culture. A rough proportionality was observed between the ability of this protein [which may be chemically similar to β-TGF (Tucker et al., 1984)] to block EGF-activated elevations of cellular Na^+ contents and EGF-stimulated DNA synthesis in quiescent cultures. Other reports of early-acting chalone-like substances have appeared, for example, the inhibition of initiation of proliferation of fetal rat hepatocytes by very low-density lipoproteins or by acidic phospholipids and/or gangliosides associated with these particles (Leffert and Weinstein, 1976). Their effects on mitogen-activated Na^+ influxes are unknown.

V. FLUX-MEDIATING STRUCTURES

The plasma membrane harbors a complex molecular network that mediates Na^+ transport. These molecules operate directly as porters, channels, or pumps, or indirectly as ion transport modulators. The critical questions are these: What flux-mediating structures are stimulated by early-acting mitogens; what are their molecular and biophysical properties; and how do mitogens activate them (Leffert, 1982)? Current research has focused upon their phenomenological characterization from kinetic studies.

A. Porters

1. Antiport

Sodium influxes stimulated by growth factors seem to involve molecular entities—"Na^+/H^+ antiporters"—which exchange extracellular Na^+ ions for intracellular protons. The growth-controlling role of this membrane process was first suggested from studies of the fertilization of sea urchin eggs (Johnson et al., 1976) where it was shown by both chemical (pH titration and weak inorganic acid uptake) and single-cell microelectrode measurements that elevations of egg cytoplasmic pH (pH_{in}) of 0.1–0.3 pH units were necessary to initiate DNA synthesis (see Table I and Nuccitelli and Deamer, 1982; for an opposing view from studies of fertilized eggs injected with acidic buffers, see Jaffe, 1980). Cytoplasmic alkalinization and DNA synthesis were blocked when eggs were fertilized in isoosmotic Na^+-free seawater or when sperm-induced Na^+ influxes were specifically inhibited by amiloride. These observations have been confirmed (Girard et al., 1982; Lum et al., 1983; Benos, 1983), although species differences exist (Epel, 1980).

Kinetic studies of Na^+/H^+ exchange have been performed in cellular and subcellular systems derived from muscle (Aickin and Thomas, 1977; Moore et al., 1982), kidney (Kinsella and Aronson, 1980, 1981; Taub and Saier, 1979; Rindler and Saier, 1981; La Belle and Lee, 1982), liver (Koch and Leffert, 1980; Arias and Forgac, 1983; Kashiwagura et al., 1984), nerve (Moolenaar et al., 1981a,b, 1982a,b, 1983), lymphoid (Gerson, 1982; Deutsch et al., 1984), and connective tissue (Pouysségur et al., 1982a,b, 1984; Paris and Pouysségur, 1983,1984; L'Allemain et al., 1984a,b; Cassel et al., 1983, 1984; Rothenberg et al,, 1983a,b; Schuldiner and Rozengurt, 1982; Burns and Rozengurt, 1983, 1984; Frelin et al., 1983).

The properties of Na^+/H^+ exchange are comparable among different cell types. Antiport is Na^+ ion selective ($Na^+ \simeq Li^+ >> K^+$); neither choline$^+$ nor K^+ substitute for Na^+. Monovalent cation exchange is electroneutral because it occurs with a fixed stoichiometry, probably 1:1. Exchange is driven physiologically by energy derived from an inwardly directed Na^+ gradient ($[Na^+]_{out} > [Na^+]_{in}$). This gradient is actively maintained by the Na^+ pump (Na^+,K^+-ATPase), but neither ouabain nor changes in $\Delta\Psi$ (membrane potential) nor ATP depletion directly inhibit Na^+/H^+ exchange. Amiloride (and its congeners) are, thus far, the only known pharmacologic compounds that specifically and directly inhibit this antiporter.

In studies with isolated plasma membrane vesicles (Kinsella and Aronson, 1980, 1981), with reconstituted phospholipid proteoliposomes (La Belle and Lee, 1982), or with intact living cells, it has been shown that a membrane protein (or proteins) and sealed membranes are required for Na^+/H^+ antiport, that the exchange process is reversible (inwardly directed Na^+ gradients stimulate uphill H^+ effluxes, whereas outwardly directed H^+ gradients stimulate uphill Na^+ influxes), and that amiloride (or its congeners) blocks Na^+/H^+ exchange by competing with Na^+ ($K_{i[app]} \simeq 1$–$40~\mu M$) at a saturable extracellular Na^+ binding site ($K_{m[app]}^{Na^+} \simeq 40$ mM). By contrast, in cell-free systems, amiloride's blockade of Na^+-dependent H^+ efflux appears to be noncompetitive with respect to a distinct intracellular H^+ "binding site." Inwardly directed anion gradients (Cl^-, SO_4^{2-}, or SCN^-) exert no detectable effects on Na^+/H^+ exchange (Kinsella and Aronson, 1980). This observation suggests that if additional electroneutral plasma membrane ion exchange systems exist that regulate pH_{in}, such as HCO_3^-/Cl^- porters, their rates of functioning must be slow with respect to Na^+/H^+ exchange (Pouysségur et al., 1984).

Recently, single-channel electrophysiological measurements made upon putative solubilized Na^+/H^+ antiporters (from A6 glial cells) reconstituted into planar lipid bilayers have suggested that the Na^+ conductance of one antiporter is 4–80 pS (at $[Na^+]_{out} = 200$ mM), that Na^+ conductance is voltage independent, and that amiloride blocks Na^+ conductance by decreasing the fraction of porters operating in the "open state" (Sariban-Sohraby et al., 1984). The relevance of model systems like these to Na^+/H^+ exchange in living cells is unclear, especially in view of the high Na^+ conductances observed.

There are at least three somewhat overlapping hypotheses to explain how growth factors might activate Na^+/H^+ exchange (Leffert, 1982; see Section V,C).

One hypothesis is that early-acting mitogens stimulate increases in the levels of free $[Ca^{2+}]_{in}$. This could occur in various ways: directly, by mitogenic activation of plasma membrane Ca^{2+} influx coupled to allosteric stimulation by Ca^{2+} of Na^+/H^+ antiporters (Taub and Saier, 1979) or indirectly, by perturbation of intracellular Ca^{2+} sequestration sites to "release" bound Ca^{2+} for interaction with calmodulin and

rons (in primary cultures of dissociated spinal cord) are stimulated to proliferate by veratridine, a voltage-sensitive Na$^+$ channel agonist (Cone and Cone, 1976). These findings were attributed to sustained $\Delta\Psi$ depolarization (Cone, 1980; see Section VI,A,1). They might reflect a specific property of CNS neurons since in fibroblast systems that contain fast Na$^+$ channels, neither mitogenic effects with specific agonists nor antiproliferative effects with specific antagonists (like tetrodotoxin) were obtained (Pouysségur et al., 1980).

C. Other Membrane Proteins

Could sodium influx mediating structures be growth factors, growth factor receptors, or other protein modulators (Leffert, 1982)?

If growth factor and antiporter were identical, it is difficult to envision how structurally different mitogens [e.g., EGF (single-chain, 6000-dalton acidic polypeptide), PDGF (double-chain, 31000-dalton basic polypeptide), insulin (double-chain, 6000-dalton acidic polypeptide), IGF I (single-chain, 7600-dalton basic polypeptide, and phorbol esters (complex lipids)] stimulate Na$^+$ influx with virtually identical kinetic properties.

Alternatively, if receptor and antiporter were identical, one might expect to see large-scale structural similarities among different mitogen receptors. Here again, except for insulin and IGF I (see Moses and Pilistine, Chapter 4), the receptors to several early-acting mitogens are structurally distinct [EGF (single-chain, 170,000-dalton glycoprotein), PDGF (see Bowen-Pope et al., Chapter 10), and phorbol esters (protein kinase C; Nishizuka, 1984)]. For these reasons, although direct evidence is lacking, and although we have assumed that extensive primary sequence homologies alone determine ion transporting function [which may not necessarily be true (Kaiser and Kézdy, 1984)], it seems more likely that mitogen–receptor interactions indirectly activate Na$^+$/H$^+$ exchangers. Whether microclustering of mitogen receptors (or aggregation) might confer upon them Na$^+$/H$^+$ exchange properties is currently unknown.

What other membrane proteins (or processes) might be involved? One candidate might be cell surface phospholipase C. It catalyzes the formation of InsP$_3$ from phosphatidylinositol 4,5-diphosphate (Majerus et al., 1984). InsP$_3$ may rapidly initiate a cascade of Ca^{2+}-dependent biochemical changes culminating in accelerated Na$^+$ influx (Vicentini et al., 1984). Counterarguments make this possibility, however, of questionable biological significance (see Section V,A,1). Similar arguments would apply to hypothetical receptor-mediated stimulation of InsP$_3$

formation if, as a consequence of mitogen activation, receptor kinases function in ways akin to *src* kinases (Sugimoto *et al.*, 1984). [In this regard, despite their large-scale structural differences, receptors to PDGF, EGF, insulin, and IGF I contain intrinsic tyrosine-specific kinase activity (see Sefton, Chapter 11) and, therefore, presumably a similar catalytic sequence.] Whether the putative Na^+/H^+ antiporter becomes activated if phosphorylated on specific tyrosine residues (or if methylated at specific sites) is currently unknown.

VI. FUNCTIONAL CHANGES LINKED TO SODIUM INFLUXES

A variety of ways exist in which mitogen-activated Na^+ influxes might change cellular function. With respect to growth control, it is unclear if one alteration is more crucial than another. We briefly describe the changes currently thought to be relevant.

A. Direct

1. Plasma Membrane Potential

Despite Cone's reports (1980; Cone and Cone, 1976) and *in vivo* evidence of decreased $\Delta\Psi$ in rat hepatomas (Binggeli and Cameron, 1980), there is no firm evidence yet to link (or not to link) mitogen-induced $\Delta\Psi$ depolarization in G_0/G_1 with the initiation of DNA synthesis. Deutsch and Price (1982a) have argued that early $\Delta\Psi$ depolarization is not required since succinyl-Con A stimulates lymphocyte DNA synthesis without eliciting such changes. A similar conclusion has been reached from results of single-cell microelectrode measurements in cultured fiboblasts (Moolenaar *et al.*, 1981a,b) and kidney cells (Rothenberg *et al.*, 1982) treated with serum or EGF. In the lymphocyte studies, however, the possibility was not eliminated that chemically modified lectins might produce alterations "bypassing" earlier steps in the normal activation sequence. And, although early $\Delta\Psi$ depolarization occurred in the nonlymphatic culture systems, it was not amiloride sensitive. In addition, no evidence was presented to show that "clamping" $\Delta\Psi$ or blocking depolarization would also block initiation of growth.

Other investigators also have observed rapid growth factor-induced $\Delta\Psi$ depolarization [e.g., in B or T lymphocytes, stimulated by LPS or Con A, respectively (Kiefer *et al.*, 1980; Felber and Brand, 1983), in hepatocytes (Koch and Leffert, 1980) and in neuroblastoma cells (O'Donnell and Villereal, 1982)]. With regard to hepatocytes and neuroblastoma cells, it appears that such $\Delta\Psi$ depolarizations are amiloride sensitive. In addition, in fibroblast cultures, growth factor-depleted serum fails to depolarize cells (Moolenaar et al., 1979). Therefore, barring measurement artifacts arising from the use of microelectrodes or radiolabeled lipophilic cations, it remains possible that mitogen-induced $\Delta\Psi$ depolarization (mediated by Na^+ influx) is necessary for initiation of DNA synthesis.

Membrane potential depolarization would be expected to rapidly perturb membrane function. Mechanisms involved could include surface electrophoresis of membrane proteins (Jaffe, 1977, 1980; Edidin et al., 1980), alterations in A system transport [e.g., in short-term experiments, EGF acutely inhibits glucagon-stimulated [^{14}C]AIB uptake in isolated hepatocytes (Morin et al., 1981)], or alterations in junctional communications between cells (Loewenstein, 1979).

2. Intracellular Sodium Content

Several groups using X-ray microprobes have observed that different tumor cell nuclei and cytoplasms contain elevated Na^+ levels (see Section II,B). These changes are not seen with other elements such as K^+, Mg^{2+}, phosphorus, and sulfur. In general, tumor Na^+ levels are three- to fourfold higher than Na^+ levels in normal quiescent tissue (≈100 mmol/kg dry weight for both compartments). Normal proliferating tissues have intermediate levels, elevated $\approx$ twofold. The mechanisms responsible for these changes are unknown. Sodium influxes have been implicated from reports in which amiloride was found to reduce tumor nuclear Na^+ levels (Sparks et al., 1983). But defective Na^+ pumping might also account for these alterations.

Little is known about how the contents of intracellular Na^+ compartments are regulated. Whether nuclear Na^+ content might affect gene expression at the levels of chromatin and DNA structure or transcription and replication of RNA and DNA, respectively, is unclear at this time. The role that compartmented Na^+ ions might also play in controlling cytoplasmic volume, and consequently, rates of molecular diffusion, also is unknown (Mastro and Keith, 1984). It is interesting to consider all of these parameters as potentially regulatory, especially if one attempts to account for ways in which mitogens might rapidly activate nuclear processes.

3. Sodium Activity

To our knowledge, direct measurements of intracellular Na^+ activities in mitogen-treated cells have not yet been made (see Section II,B). It is not known, therefore, whether mitogen-activated Na^+ influxes raise cytoplasmic or nuclear Na^+ concentrations. Such information is critical to understanding functional linkage, especially as Na^+ ion concentrations may regulate cell volume, ultrastructure (Mastro and Keith, 1984), and protein synthesis at posttranslational (Koch et al., 1980) or translational levels (Piatiagorski et al., 1980).

In several systems, Li^+ ions promote cell proliferation (Levitt and Quesenberry, 1980; Toback, 1980; Rybak and Stockdale, 1981; Tomooka et al., 1983). It has not yet been established definitively if Li^+ acts as an Na^+ substitute in these situations. In adrenal glomerulosa cells, Li^+ blocks incorporation of $^{32}P_i$ into phosphatidylinositol and promotes a concomitant accumulation of $^{32}P_i$ into phosphatidic acid (Balla et al., 1984). Whether these biochemical changes account for lithium's (or sodium's) mitogenic properties is unknown (see Section V,A,1).

B. Indirect

1. Intracellular pH

Except for studies by Deutsch et al. (1984) using mixed populations of human lymphocytes, Na^+ influx-dependent amiloride-sensitive elevations of pH_{in} during G_0/G_1 have been observed in 21 different kinds of mitogen-activated systems (see Table I). The reasons for the discrepant findings are unclear (Deutsch et al. argue that they may be methodological), especially since lectin-activated increases in lymphocyte pH_{in} have been reported by others (Gerson, 1982; Grinstein et al., 1983).

Elevated pH_{in} would be expected to affect many enzymatic pathways and cascades. For example, hepatocyte proliferation in vivo (stimulated by 67% hepatectomy) and in vitro (stimulated by EGF, insulin, and glucagon) is accompanied by increased lactate gluconeogenesis (Brown et al., 1983; Lad et al., 1984). As suggested earlier from studies of perfused liver (Friedmann, 1972) and confirmed recently with isolated hepatocytes (Kashiwagura et al., 1984), hepatocyte gluconeogenesis rates are stimulated by Na^+-dependent increases in pH_{in}. Several other key regulatory enzymes, including DNA polymerases, show significant changes in their rates of activity at higher pH optima (summarized by Gerson, 1982). Although protein subunit polymerization (Epel, 1980) and bioenergetic rates (Mitchell, 1979) are exquisitely sensitive to changes of pH_{in}, their roles, if any, in the initiation of DNA synthesis and mitosis remain to be rigorously proved.

2. Intracellular Potassium

Table I shows that mitogen-activated Na^+ influxes are associated with rapid and transient activation of the Na^+ pump. Sodium pump activation in cultured cells often, but not always (see Leffert and Koch, 1982a; Lubin, 1982), generates transient increases in cellular K^+ content and in membrane premeability to K^+ (Boonstra et al., 1981). Whether these changes reflect increases in overall or compartmented ion activities in this "first phase" of pump activation, and whether such changes influence membrane processes like receptor-mediated endocytosis in which both K^+ and Na^+ ions are implicated (Samuelson et al., 1984), is unknown.

Late in G_1, a "second phase" of activation of preexisting Na^+ pumps occurs. This activation is not solely substrate dependent, since plasma membranes isolated from such cells show an increased Na^+,K^+-ATPase activity (Mummery et al., 1981; Schenk et al., 1984). This activation phase is associated with net increases in $[K^+]_{in}$ and $\Delta\Psi$, at least in regenerating mouse and rat liver, as determined from single-cell microelectrode measurements (Wondergem and Harder, 1980; Wondergem, 1982; de Hemptinne et al., 1982). It is not yet clear whether changes in the structures of the membrane and/or the pump itself account for the increased turnover activity of the enzyme (Schenk et al., 1984).

The late G_1 phase is associated with increases in A system amino acid transport (Leffert and Koch, 1977) which is markedly sensitive to inhibitors to protein synthesis (Leffert and Koch, 1979; Fehlmann et al., 1981b). Both processes depend upon the Na^+ pump, indirectly (energy for transport) and directly [intracellular K^+ is required for protein chain elongation (Cahn and Lubin, 1978)]. However, evidence that elevated rates of A system transport and protein chain elongation are necessary for $G_1 \rightarrow S$ transit is lacking (see Section IV,E).

3. Intracellular Calcium

Elevations of free $[Ca^{2+}]_{in}$ generally, but not always (see Section V,A,1), occur early in G_0/G_1. Therefore, biochemical changes mediated by Ca^{2+}-dependent and calmodulin-activated enzymes might be involved (see Whitfield et al., Chapter 12).

Despite earlier reports (Lowe et al., 1976), the available evidence does not appear to support a direct role for Na^+ influx or Na^+/H^+ exchanges as an activator of increases in free $[Ca^{2+}]_{in}$ (Johnson et al., 1976; see Section V,A,1). Na^+/Ca^{2+} exchange also does not appear to be a "universal" mechanism by which mitogens regulate free $[Ca^{2+}]_{in}$.

This type of exchange seems to be a specialized property of excitable membranes, not a general property of nucleated animal cell membranes, since it is absent from epithelial (Koch et al., 1984), lymphoid (Deutsch and Price, 1982b), and most fibroblast cells in culture (Rothenberg et al., 1983a,b; Moolenaar et al., 1984). In those cases where mitogens stimulate such increases, the elevated levels of Ca^{2+} may instead act synergistically with increased pH_{in} to stimulate protein synthesis (Zucker et al., 1978) or inhibit protein degradation.

4. Amino Acid Transport

As discussed (Sections V,A, VI,A,1, and VI,B,2), there are at least two ways in which amiloride-sensitive Na^+ influxes in late G_1 might be linked to amino acid cotransport through the A system. First, Na^+ influxes might activate genes coding for or processes stimulating the synthesis of new porters. Second, Na^+ influxes might lead to chronic activation of the Na^+ pump, causing sustained $\Delta\Psi$ hyperpolarization which energizes the transport system.

5. Phosphorylation

Mitogen-activated Na^+ influxes have been linked to cyclic AMP-dependent protein kinase phosphorylation of ribosomal protein S6 (Pouysségur et al., 1982b). This finding is probably artifactual, since amiloride is taken up by cells. Under these conditions, S6 phosphorylation is blocked because either amiloride acts as a competitive inhibitor with respect to ATP in the kinase reaction (Holland et al., 1983) or the drug activates an intracellular phosphatase (see Section IV,B).

Many growth factors listed in Table I activate membrane receptor kinases that phosphorylate protein substances on tyrosine residues (e.g., PDGF, EGF, insulin, IGF I), or they facilitate Ca^{2+}-dependent activation of protein kinase C, which phosphorylates its protein substrates on serine and threonine residues in a cyclic AMP-independent fashion (Nishizuka, 1984). No available evidence links any of these (or other) phosphorylations to mitogen-activated Na^+ influxes.

VII. CONCLUSIONS

We have tried to place into perspective observations that implicate transient influxes of Na^+ ions as "early" prereplicative regulators of the initiation of DNA synthesis. We have not dealt here with the plethora of outstanding growth control issues (Leffert and Koch, 1977; Leffert,

1982). Clearly, other growth factors that do not activate Na$^+$ influxes are necessary for optimal proliferative responses (Leffert and Koch, 1982a,b; Leffert et al., 1982a,b, 1983). Accordingly, Na$^+$ influxes might be viewed as necessary but insufficient to fully stimulate cell proliferation (Leffert and Koch, 1977; Koch and Leffert, 1979a).

Although rigorous proof of cause and effect has not yet been obtained, the available evidence supports the Na$^+$ influx hypothesis. Work is needed to define and simplify its exact details, particularly with regard to issues of functional linkage. Until this is done, the question of whether abnormal growth control is a consequence of the defective regulation of such fluxes cannot be answered (Leffert, 1982).

ACKNOWLEDGMENTS

This work was supported in part by grants from the USPHS (AM28215 and AM28392). We thank Hal Skelly for technical assistance, Yedy Israel for helpful comments, and Sandy Dutky for computerized word processing.

REFERENCES

Adam, G., Kleuser, B., Seher, J. P., and Ullrich, S. (1982). Relation between K$^+$, Na$^+$, Ca^{2+}, and proliferation of normal and transformed 3T3 mouse cells. In "Ions, Cell Proliferation and Cancer" (A. L. Boynton, W. L. MeKeehan, and J. F. Whitfield, eds.), pp. 219–244. Academic Press, New York.

Aickin, C. C., and Thomas, R. C. (1977). An investigation of the ionic mechanism of intracellular pH regulation in mouse soleus muscle fibers. J. Physiol. (London) 273, 295–316.

Amsler, K., Donahue, J., Slayman, C. W., and Adelberg, E. A. (1983). The effect of growth state on K$^+$ efflux pathways in Swiss 3T3K fibroblasts. J. Gen. Physiol. 82, 16a.

Arias, I. M., and Forgac, M. (1983). Hepatocyte sinusoidal membrane vesicles contain a Na$^+$/H$^+$ antiport which regulates intracellular pH. Hepatology 3, 872.

Averdunk, R. (1976). Early changes of "leak flux" and the cation content of lymphocytes by concanavalin A. Biochem. Biophys. Res. Commun. 70, 101–109.

Averdunk, R., and Gunther, T. (1980). Effect of concanavalin A on intracellular K$^+$ and Na$^+$ concentration and K$^+$ transport of human lymphocytes. Immunobiology 157, 132–144.

Averdunk, R., and Lauf, P. K. (1975). Effects of mitogens on sodium–potassium transport, ^{3}H-ouabain binding, and adenosine triphosphatase activity in lymphocytes. Exp. Cell Res. 93, 331–342.

Balk, S. D., and Polimeni, P. I. (1982). Effect of reduction of culture medium sodium, using different sodium chloride substitutes, on the proliferation of normal and Rous sarcoma virus-infected chicken fibroblasts. J. Cell. Physiol. 112, 251–256.

Balla, T., Enyedi, P., Hunyady, L., and Spät, A. (1984). Effects of lithium on angiotensin-stimulated phosphatidylinositol turnover and aldosterone production in adrenal glomerulosa cells: A possible causal relationship. FEBS Lett. 171, 179–182.

Bellemann, P. (1981). Amino acid transport and rubidium-ion uptake in monolayer cultures of hepatocytes from neonatal rats. *Biochem. J.* **198**, 475–483.

Benos, D. J. (1982). Amiloride: A molecular probe of sodium transport in tissues and cells. *Am. J. Physiol.* **242**, C131–C145.

Benos, D. J. (1983). Developmental aspects of sodium-dependent transport processes of preimplantation rabbit embryos. *J. Gen. Physiol.* **82**, 8a.

Benos, D. J., and Sapirstein, V. S. (1983). Characteristics of an amiloride-sensitive sodium entry pathway in cultured rodent glial and neuroblastoma cells. *J. Cell. Physiol.* **116**, 213–220.

Bernstein, A., Hunt, D. M., Crichley, V., and Mak, T. W. (1976). Induction by ouabain of hemoglobin synthesis in cultured Friend erythroleukemic cells. *Cell (Cambridge, Mass.)* **9**, 375–381.

Binggeli, R., and Cameron, I. L. (1980). Cellular potentials of normal and cancerous fibroblasts and hepatocytes. *Cancer Res.* **40**, 1830–1835.

Boonstra, J., Mummery, C. L., Tertoolen, L. G. J., van der Saag, P. T., and de Laat, S. (1981). Cation transport and growth regulation in neuroblastoma cells. Modulations of K^+ transport and electrical membrane properties during the cell cycle. *J. Cell. Physiol.* **107**, 75–83.

Boonstra, J., Moolenaar, W. H., Harrison, P. H., Moed, P., van der Saag, P. T., and de Laat, S. (1983). Ionic responses and growth stimulation induced by nerve growth factor and epidermal growth factor in rat pheochromocytoma (PC12) cells. *J. Cell Biol.* **97**, 92–98.

Borgens, R. B., Vanable, J. W., and Jaffe, L. F. (1977a). Bioelectricity and regeneration. 1. Initiation of frog limb regeneration by minute currents. *J. Exp. Zool.* **200**, 403–416.

Borgens, R. B., Vanable, J. W., and Jaffe, L. F. (1977b). Bioelectricity and regeneration: Large currents leave the stumps of regenerating newt limbs. *Proc. Natl. Acad. Sci. U.S.A.* **74**, 4528–4532.

Boynton, A. L., McKeehan, W. L., and Whitfield, J. F., eds. (1982). "Ions, Cell Proliferation, and Cancer," pp. 1–551. Academic Press, New York.

Brown, J. W., Lad, P. J., Skelly, H., Koch, K. S., Lin, M., and Leffert, H. L. (1983). Expression of differentiated function by hepatocytes in primary culture: Variable effects of glucagon and insulin on gluconeogenesis during cell growth. *Differentiation* **25**, 176–184.

Burgess, G. M., Godfrey, P. P., McKinney, J. S., Berridge, M. J., Irvine, R. F., and Putney, J. W. (1984). The second messenger linking receptor activation to internal Ca^{2+} release in liver. *Nature (London)* **309**, 63–66.

Burns, C. P., and Rozengurt, E. (1983). Serum, platelet-derived growth factor, vasopressin and phorbol esters increase intracellular pH in Swiss 3T3 cells. *Biochem. Biophys. Res. Commun.* **116**, 931–938.

Burns, C. P., and Rozengurt, E. (1984). Extracellular Na^+ and initiation of DNA synthesis: Role of intracellular pH and K^+. *J. Cell Biol.* **98**, 1082–1089.

Burroni, D., and Ceccarini, C. (1984). The effect of alkaline pH on the cell growth of six different mammalian cells in tissue culture. *Exp. Cell Res.* **150**, 505–508.

Cahn, F., and Lubin, M. (1978). Inhibition of elongation steps of protein synthesis at reduced potassium concentrations in reticulocytes and reticulocyte lysate. *J. Biol. Chem.* **253**, 7798–7803.

Cameron, I. L., Smith, N. K. R., Pool, T. B., and Sparks, R. L. (1980). Intracellular concentration of sodium and other elements as related to mitogenesis and oncogenesis *in vivo*. *Cancer Res.* **40**, 1493–1500.

Cameron, I. L., Hunter, K. E., and Cragoe, E. J. (1982). A stage-specific inhibitory effect of benzamil on *Xenopus* oocyte maturation located at the cell surface. *Exp. Cell Res.* **139**, 455–457.

Cameron, I. L., Lum, J. B., and Cragoe, E. J. (1984). Amiloride and other blockers of electrolyte flux as inhibitors of progesterone-stimulated meiotic maturation in *Xenopus* oocytes. *Cell Tissue Kinet.* **17**, 161–169.

Carlson, J., Dorey, F., Cragoe, E. and Koeffler, H. P. (1984). Amiloride potentiation of differentiation of human promyelocytic cell line HL-60. *JNCI, J. Natl. Cancer Inst.* **72**, 13–17.

Carpenter, G. (1984). Properties of the receptor for epidermal growth factor. *Cell (Cambridge, Mass.)* **37**, 357–358.

Cassel, D., Rothenberg, P., Zhuang, Y. X., Deuel, T. F., and Glaser, L. (1983). Platelet-derived growth factor stimulates Na^+/H^+ exchange and induces cytoplasmic alkalinization in NR6 cells. *Proc. Natl. Acad. Sci. U.S.A.* **80**, 6224–6228.

Cassel, D., Zhuang, Y. X., and Glaser, L. (1984). Vanadate stimulates Na^+/H^+ exchange activity in A-431 cells *Biochem. Biophys. Res. Commun.* **118**, 675–681.

Chambard, J. C., Franchi, A., Le Cam, A., and Pouysségur, J. (1983). Growth factor-stimulated protein phosphorylation in G_0/G_1-arrested fibroblasts: Two distinct classes of growth factors with potentiating effects. *J. Biol. Chem.* **258**, 1706–1713.

Collins, M. K. L., and Rozengurt, E. (1984). Homologous and heterologous mitogenic desensitization of Swiss 3T3 cells to phorbol esters and vasopressin: Role of receptor and post-receptor steps. *J. Cell. Physiol.* **118**, 133–142.

Cone, C. D. (1980). Ionically mediated induction of mitogenesis in CNS neurons. *Ann. N.Y. Acad. Sci.* **339**, 115–131.

Cone, C. D., and Cone, C. M. (1976). Induction of mitosis in mature neurons in central nervous system by sustained depolarization. *Science* **192**, 155–158.

Crane, R. K., and Dorando, F. C. (1980). On the mechanism of Na^+-dependent glucose transport. *Ann. N.Y. Acad. Sci.* **339**, 46–52.

Davies, R. J., Weidema, W. F., Palmer, L., and De Cosse, J. J. (1983). Increased sodium absorption by the distal colon in 1,2-dimethylhydrazine-treated Sprague–Dawley rat. *Surg. Forum* **34**, 183–185.

De Coursey, T. E., Chandy, K. G., Gupta, S., and Cahalan, M. D. (1984). Voltage-gated K^+ channels in human T-lymphocytes: A role in mitogenesis? *Nature (London)* **307**, 465–468.

de Hemptinne, B., and Leffert, H. L. (1983). Selective effects of portal blood diversion and of glucagon on rat hepatocyte rates of S-phase entry and DNA synthesis. *Endocrinology (Baltimore)* **112**, 1224–1232.

de Hemptinne, B., Leffert, H. L., and Lambotte, L. (1982). Hyperpolarization after partial hepatectomy. *Arch. Int. Physiol. Biochim.* **90**, P59–P60.

Deutsch, C., and Price, M. (1982a). Role of extracellular Na^+ and K^+ in lymphocyte activation. *J. Cell. Physiol.* **113**, 73–79.

Deutsch, C., and Price, M. (1982b). Cell calcium in human peripheral blood lymphocytes and the effect of mitogen. *Biochim. Biophys. Acta* **687**, 211–218.

Deutsch, C., Price, M., and Johansson, C. (1981). A sodium requirement for mitogen-induced proliferation in human peripheral blood lymphocytes. *Exp. Cell Res.* **136**, 359–369.

Deutsch, C., Taylor, J. S., and Price, M. (1984). pH homeostasis in human lymphocytes: Modulation by ions and mitogen. *J. Cell Biol.* **98**, 885–894.

Dicker, P., and Rozengurt, E. (1981). Phorbol ester stimulation of Na^+ influx and Na-K pump activity in Swiss 3T3 cells. *Biochem. Biophys. Res. Commun.* **100**, 433–441.

Edidin, M., Wei, T., and Holmberg, S. (1980). The role of membrane potential in determining rates of lateral diffusion in the plasma membrane of mammalian cells. *Ann. N.Y. Acad. Sci.* **339**, 1–7.

Epel, D. (1980). Ionic triggers in the fertilization of sea urchin eggs. *Ann. N.Y. Acad. Sci.* **339**, 74–85.

Farley, R., and Ochoa, G. T. (1985). Location of the major antibody binding sites on the α-subunit of dog kidney (Na+,K+)-ATPase. Submitted for publication.

Fehlmann, M., and Freychet, P. (1981). Insulin and glucagon stimulation of (Na+,K+)-ATPase transport activity in isolated rat hepatocytes. *J. Biol. Chem.* **236**, 7449–7453.

Fehlmann, M., Canivet, B., and Freychet, P. (1981a). Epidermal growth factor stimulates monovalent cation transport in isolated rat hepatocytes. *Biochem. Biophys. Res. Commun.* **100**, 254–260.

Fehlmann, M., Samson, M., Koch, K. S., Leffert, H. L., and Freychet, P. (1981b). The effect of amiloride on hormonal regulation of amino acid transport in isolated and cultured adult rat hepatocytes. *Biochim. Biophys. Acta* **642**, 88–95.

Felber, S. M., and Brand, M. D. (1983). Concanavalin A causes an increase in sodium permeability and intracellular sodium content of pig lymphocytes. *Biochem. J.* **210**, 893–897.

Frantz, C. N., Nathan, D. G., and Scher, C. D. (1981). Intracellular univalent cations and the regulation of the BALB/c-3T3 cell cycle. *J. Cell Biol.* **88**, 51–56.

Frelin, C., Vigne, P., and Lazdunski, M. (1983). The amiloride-sensitive Na+/H+ antiport in 3T3 fibroblasts. Characterization and stimulation by serum. *J. Biol. Chem.* **258**, 6272–6276.

Friedmann, N. (1972). Effects of glucagon and cyclic AMP on ion fluxes in the perfused liver. *Biochim. Biophys. Acta* **274**, 214–225.

Gargus, J. J., Adelberg, E. A., and Slayman, C. W. (1983). Coordinated changes in potassium fluxes as early events in the differentiation of the human promyelocyte line HL60. *J. Gen. Physiol.* **82**, 6a.

Garry, R. F., Moyer, M. P., Bishop, J. M., Moyer, R. C., and Waite, M. R. F. (1981). Transformation parameters induced in chick cells by incubation in media of altered NaCl concentration. *Virology* **111**, 427–439.

Garry, R. F., Ulug, E. T., and Bose, H. R. (1982). Membrane mediated alterations of intracellular Na+ and K+ in lytic virus-infected and retrovirus-transformed cells. *Biosci. Rep.* **2**, 617–623.

Gelehter, T. D., and Rozengurt, E. (1980). Stimulation of monovalent ion fluxes and DNA synthesis in 3T3 cells by mellitin and vasopressin is not mediated by phospholipid deacylation. *Biochem. Biophys. Res. Commun.* **97**, 716–724.

Gerson, D. F. (1982). The relation between intracellular pH and DNA synthesis rate in proliferating lymphocytes. *In* "Intracellular pH: Its Measurement, Regulation, and Utilization in Cellular Functions" (R. Nuccitelli and D. W. Deamer, eds.), pp. 375–383. Alan R. Liss, Inc., New York.

Girard, J. P., Payan, P., and Sardet, C. (1982). Changes in intracellular cations following fertilization of sea urchin eggs. *Exp. Cell Res.* **142**, 215–221.

Grinstein, S., Cohen, S., and Rothstein, A. (1983). Cytoplasmic pH regulation in thymocytes by an amiloride-sensitive Na+/H+ antiport. *J. Gen. Physiol.* **82**, 17a–18a.

Groudine, M., and Weintraub, H. (1982). Propagation of globin DNase I-hypersensitive sites in absence of factors required for induction: A possible mechanism for determination. *Cell (Cambridge, Mass.)* **30**, 131–139.

Gupta, R. K., and Gupta, P. (1982). Direct observation of resolved resonances from intra- and extracellular sodium-23 ions in NMR studies of intact cells and tissues using dysprosium (III) tripolyphosphate as paramagnetic shift reagent. *J. Magn. Reson.* **47**, 344–350.

Hand, R., Malnasi, L., and Heyman, M. (1983). Inhibition of $^{86}Rb^+$ uptake by SV40 anti-T-antiserum in simian virus-40-transformed mouse and human cells. *Cell Tissue Kinet.* **16**, 523–524.

Hasegawa, K., and Koga, M. (1979). Relationship between the induction of DNA synthesis and changes in the concentration of intracellular Na^+ in the liver of intact rats. *J. Physiol. Soc. Jpn.* **41**, 282.

Hasegawa, K., Namai, K., and Koga, M. (1980). Induction of DNA synthesis in adult rat hepatocytes cultured in a serum-free medium. *Biochem. Biophys. Res. Commun.* **95**, 243–249.

Hata, R., Sunada, H., and Nagai, Y. (1983). Sodium ion modulates collagen types in human fibroblastic cells in culture. *Biochem. Biophys. Res. Commun.* **117**, 313–318.

Heikkilä, R., Iversen, J. G., and Godal, T. (1983a). Early, anti-immunoglobulin induced events prior to Na^+-K^+ pump activation: An analysis in a monoclonal human B-lymphoma cell population. *J. Cell Physiol.* **117**, 1–8.

Heikkilä, R., Iversen, J. G., and Godal, T. (1983b). Amiloride inhibits anti-Ig induction of proliferation in a human B-lymphoma cell population. *Exp. Cell Res.* **149**, 299–302.

Hennings, H., Holbrook, K. A., and Yuspa, S. H. (1983). Potassium mediation of calcium-induced terminal differentiation of epidermal cells in culture. *J. Invest. Dermatol.* **81**, 50s–55s.

Holland, R., Woodgett, J. R., and Hardie, D. G. (1983). Evidence that amiloride antagonizes insulin-stimulated protein phosphorylation by inhibiting protein kinase activity. *FEBS Lett.* **154**, 269–273.

Hubert, J. J., Schenk, D. B., and Leffert, H. L. (1985). Purification and structural studies of the α-subunit of rat hepatic (Na^+,K^+)-ATPase. Submitted for publication.

Hudson, R. L., and Schultz, S. G. (1984). Sodium-coupled sugar transport: Effects on cellular sodium activities and sodium pump activity. *Science* **224**, 1237–1239.

Jaffe, L. F. (1977). Electrophoresis along cell membranes. *Nature (London)* **265**, 600–602.

Jaffe, L. F. (1980). Calcium explosions as triggers of development. *Ann. N.Y. Acad. Sci.* **339**, 86–101.

Jayme, D. W., Slayman, C. W., and Adelberg, E. A. (1984). Furosemide-sensitive potassium efflux in cultured mouse fibroblasts. *J. Cell. Physiol.* **120**, 41–48.

Johnson, C. H., and Epel, D. (1981). Intracellular pH of sea urchin eggs measured by dimethyloxazolidinedione (DMO) method. *J. Cell Biol.* **89**, 284–291.

Johnson, J. D., Epel, D., and Paul, M. (1976). Intracellular pH and activation of sea urchin eggs after fertilization. *Nature (London)* **262**, 661–664.

Kaiser, E. T., and Kézdy, F. J. (1984). Amphiphilic secondary structure: Design of peptide hormones. *Science* **223**, 249–255.

Kaplan, J. G. (1978). Membrane cation transport and the control of proliferation of mammalian cells. *Annu. Rev. Physiol.* **40**, 19–41.

Kaplan, J. G., and Owens, T. (1980). Activation of lymphocytes of man and mouse: Monovalent cation fluxes. *Ann. N.Y. Acad. Sci.* **339**, 191–200.

Kashiwagura, T., Deutsch, C. J., Taylor, J., Erecinska, M., and Wilson, D. F. (1984). Dependence of gluconeogenesis, urea synthesis and energy metabolism of hepatocytes on intracellular pH. *J. Biol. Chem.* **259**, 237–242.

Kiefer, H., Blume, A. J., and Kaback, H. R. (1980). Membrane potential changes during mitogenic stimulation of mouse spleen lymphocytes. *Proc. Natl. Acad. Sci. U.S.A.* **77**, 2200–2204.

Kinsella, J. L., and Aronson, P. S. (1980). Properties of the Na^+-H^+ exchanger in renal microvillus membrane vesicles. *Am. J. Physiol.* **238**, F461–F469.

Kinsella, J. L., and Aronson, P. S. (1981). Interaction of NH_4^+ and Li^+ with the renal microvillus membrane Na^+-H^+ exchanger. *Am. J. Physiol.* **241**, C220–C226.

Koch, K. S., and Leffert, H. L. (1979a). Increased sodium ion influx is necessary to initiate rat hepatocyte proliferation. *Cell (Cambridge, Mass.)* **18**, 153–163.

Koch, K. S., and Leffert, H. L. (1979b). Ionic landmarks along the mitogenic route. *Nature (London)* **279**, 104–105.

Koch, K. S., and Leffert, H. L. (1980). Growth control of differentiated adult rat hepatocytes in primary culture. *Ann. N.Y. Acad. Sci.* **349**, 111–127.

Koch, G., Bilello, J. A., Kruppa, J., and Koch, F. (1980). Amplification of translational control by membrane-mediated events: A pleiotropic effect on cellular and viral gene expression. *Ann. N.Y. Acad. Sci.* **339**, 280–306.

Koch, K. S., Shapiro, P., Skelly, H., and Leffert, H. L. (1982). Rat hepatocyte proliferation is stimulated by insulin-like peptides in defined medium. *Biochem. Biophsy. Res. Commun.* **109**, 1054–1060.

Koch, K. S., Grosse, R., Skelly, H., and Leffert, H. L. (1984). Initiation of cultured rat hepatocyte proliferation does not involve sodium-dependent calcium fluxes across the plasma membrane. *Cell Biol. Int. Rep.* **8**, 309–316.

La Belle, E. F., and Lee, S. O. (1982). Solubilization and reconstitution of an amiloride-inhibited sodium transporter from rabbit kidney medulla. *Biochemistry* **21**, 2693–2697.

Lad, P. J., Blásquez, E., Lin, M., and Leffert, H. L. (1984). Changes in adenylate cyclase and phosphodiesterase activities during the growth cycle of adult rat hepatocytes in primary culture. *Arch. Biochem. Biophys.* **232**, 679–684.

L'Allemain, G., Franchi, A., Cragoe, E., and Pouysségur, J. (1984a). Blockade of the Na^+/H^+ antiport abolishes growth factor-induced DNA synthesis in fibroblasts. Structure activity relationships in the amiloride series. *J. Biol. Chem.* **259**, 4313–4319.

L'Allemain, G., Paris, S., and Pouysségur, J. (1984b). Growth factor action and intra-cellular pH regulation in fibroblasts. Evidence for a major role of the Na^+/H^+ antiport. *J. Biol. Chem.* **259**, 5809–5815.

Le Cam, A., Auberger, P., and Samson, M. (1982). Insulin enhances protein phosphoryla-tion in isolated hepatocytes by inhibiting an amiloride-sensitive phosphatase. *Biochem. Biophys. Res. Commun.* **106**, 1062–1070.

Leffert, H. L., ed. (1980). "Growth Regulation by Ion Fluxes," pp. 1–335. N.Y. Acad. Sci. New York.

Leffert, H. L. (1982). Monovalent cations, cell proliferation and cancer: An overview. *In* "Ions, Cell Proliferation, and Cancer" (A. L. Boynton, W. L. McKeehan, and J. F. Whitfield, eds.), pp. 93–102. Academic Press, New York.

Leffert, H. L., and Koch, K. S. (1977). Control of animal cell proliferation. *In* "Growth, Nutrition and Metabolism of Cells in Culture" (G. H. Rothblat and V. J. Cristofalo, eds.), vol. 3, pp. 225–294. Academic Press, New York.

Leffert, H. L., and Koch, K. S. (1979). Regulation of growth of hepatocytes by sodium ions. *Prog. Liver Dis.* **6**, 123–134.

Leffert, H. L., and Koch, K. S. (1980). Ionic events at the membrane initiate rat liver regeneration. *Ann. N.Y. Acad. Sci.* **339**, 201–215.

Leffert, H. L., and Koch, K. S. (1982a). Monovalent cations and the control of hepatocyte proliferation in chemically defined medium. *In* "Ions, Cell Proliferation, and Can-cer" (A. L. Boynton, W. L. McKeehan, and J. F. Whittfield, eds.), pp. 103–130. Academic Press, New York, New York.

Leffert, H. L., and Koch, K. S. (1982b). Hepatocyte growth regulation by hormones in chemically defined medium. A two-signal hypothesis. *Cold Spring Harbor Symp. Cell Proliferation.* **9**, 597–613.

Leffert, H. L., and Weinstein, D. B. (1976). Growth control of differentiated fetal rat hepatocytes in primary monolayer culture. IX. Specific inhibition of DNA synthesis initiation by very low density lipoprotein and possible significance to the problem of liver regeneration. *J. Cell Biol.* **70**, 20–32.

Leffert, H. L., Koch, K. S., Moran, T., and Rubalcava, B. (1979). Hormonal control of rat liver regeneration. *Gastroenterology* **76**, 1470–1482.

Leffert, H. L., Koch, K. S., Lad, P. J., Skelly, H., and de Hemptinne, B. (1982a). Hepatocyte growth factors. *In* "Hepatology" (D. Zakim and T. D. Boyer, eds.), pp. 64–75. Saunders, Philadelphia, Pennsylvania.

Leffert, H. L., Koch, K. S., Lad, P. J., Skelly, H., and de Hemptinne, B. (1982b). Hepatocyte regeneration, replication, and differentiation. *In* "Pathobiology of the Liver" (I. Arias, H. Popper, D. Schacter, and D. Shafritz, eds.), pp. 601–614. Raven Press, New York.

Leffert, H. L., Koch, K. S., Fehlmann, M., Heiser, W., Lad. P. J., and Skelly, H. (1982c). Amiloride blocks cell-free protein synthesis at levels attained inside cultured hepatocytes. *Biochem. Biophys. Res. Commun.* **108**, 738–745.

Leffert, H. L., Koch, K. S., Lad, P. J., de Hemptinne, B., and Skelly, H. (1983). Glucagon and liver regeneration. *Handb. Exp. Pharmacol.* **66**, 453–484.

Leffert, H. L., Schenk, D. B., Hubert, J. J., Skelly, H., Schumacher, M., Ariyasu, R., Ellisman, M., Koch, K. S., and Keller, G. (1985). Hepatic (Na^+,K^+)-ATPase: A current view of its structure, function and localization in rat liver as revealed by studies with monoclonal antibodies. *Hepatology* **5**, 501–507.

Leigh, R. A., and Wyn Jones, R. G. (1984). A hypothesis relating critical potassium concentrations for growth to the distribution and functions of this ion in the plant cell. *New Phytol.* **97**, 1–13.

Levenson, R., Housman, D., and Cantley, L. (1980). Amiloride inhibits murine erythroleukemia cell differentiation: Evidence for a Ca^{2+} requirement for commitment. *Proc. Natl. Acad. U.S.A.* **77**, 5948–5952.

Levitt, L. J., and Quesenberry, P. J. (1980). The effect of lithium on murine hematopoiesis in a liquid culture system. *N. Engl. J. Med.* **302**, 713–719.

Loewenstein, W. R. (1979). Junctional intercellular communication and the control of cell growth. *Biochim. Biophys. Acta* **560**, 1–66.

Lopez-Rivas, A., Adelberg, E. A., and Rozengurt, E. (1982). Intracellular K^+ and the mitogenic response of 3T3 cells to peptide factors in serum-free medium. *Proc. Natl. Acad. Sci. U.S.A.* **79**, 6275–6279.

Lowe, D. A., Richardson, B. P., Taylor, P., and Donatsch, P. (1976). Increasing intracellular sodium triggers calcium release from bound pools. *Nature (London)* **260**, 337–338.

Lubin, M. (1982). The potassium sensitivity shift and other matters. *In* "Ions, Cell Proliferation, and Cancer" (A. L. Boynton, W. L. McKeehan, and J. F. Whitfield, eds.), pp. 131–150. Academic Press, New York.

Lubin, M., Cahn, F., and Coutermarsh, B. A. (1982). Amiloride, protein synthesis, and activation of quiescent cells. *J. Cell. Physiol.* **113**, 247–251.

Lukács, G. L., Zs.-Nagy, I., Lustyik, Gy., and Balázs, Gy. (1983). Microfluorimetric and X-ray microanalytic studies on the DNA content and $Na^+:K^+$ ratios of the cell nuclei in various types of thyroid tumors. *Cancer Res. Clin. Oncol.* **105**, 280–284.

Lum, J. B., Lawrence, W. C., and Cameron, I. L. (1983). A blastula specific requirement for Na^+ influx into the blastocoel for continued cell proliferation. *Cell Tissue Kinet.* **16**, 523.

McRoberts, J. A., Erlinger, S., Rindler, M. J., and Saier, M. H. (1982). Furosemide-sensitive salt transport in the Madin-Darby canine kidney cell line. Evidence for the cotransport of Na^+, K^+ and Cl^-. *J. Biol. Chem.* **257**, 2260–2266.

Majerus, P. W., Neufeld, E. J., and Wilson, D. B. (1984). Production of phosphoinositide-deprived messengers. *Cell (Cambridge, Mass.)* **37**, 701–703.

Mastro, A. M., and Keith, A. D. (1984). Diffusion in the aqueous compartment. *J. Cell Biol.* **99**, 180s–187s.

Matteson, D. R., and Deutsch, C. (1984). K^+ channels in T lymphocytes: A patch clamp study using monoclonal antibody adhesion. *Nature (London)* **307**, 468–471.

Mendoza, S. A., Wigglesworth, N. M., Pohjanpelto, P., and Rozengurt, E. (1980a). Na^+ entry and Na^+, K^+ pump activity in murine, hamster, and human cells—effect of monensin, serum, platelet extract, and viral transformation. *J. Cell. Physiol.* **103**, 17–27.

Mendoza, S. A., Wigglesworth, N. M., and Rozengurt, E. (1980b). Vasopressin rapidly stimulates Na^+ entry and Na^+-K^+ pump activity in quiescent cultures of mouse 3T3 cells. *J. Cell. Physiol.* **105**, 153–162.

Mitchell, P. (1979). Keilin's respiratory chain concept and its chemiosmotic consequences. *Science* **206**, 1148–1159.

Moolenaar, W. H., de Laat, S. W., and van der Saag, P. T. (1979). Serum triggers a sequence of rapid ionic conductance changes in quiescent neuroblastoma cells. *Nature (London)* **279**, 721–723.

Moolenaar, W. H., Mummery, C. L., van der Saag, P. T., and de Laat, S. W. (1981a). Rapid ionic events and the initiation of growth in serum-stimulated neuroblastoma cells. *Cell (Cambridge, Mass.)* **23**, 789–798.

Moolenaar, W. H., Boonstra, J., van der Saag, P. T., and de Laat, S. W. (1981b). Sodium/proton exchange in mouse neuroblastoma cells. *J. Biol. Chem.* **256**, 12883–12887.

Moolenaar, W. H., Yarden, Y., de Laat, S. W., and Schlessinger, J. (1982a). Epidermal growth factor induces electrically silent Na^+ influx in human fibroblasts. *J. Biol. Chem.* **257**, 8502–8506.

Moolenaar, W. H., de Laat, S. W., Mummery, C. L., and van der Saag, P. T. (1982b). Na^+/H^+ exchange in the action of growth factors. *In* "Ions, Cell Proliferation and Cancer" (A. L. Boynton, W. L. McKeehan, and J. F. Whitfield, eds.), pp. 151–162. Academic Press, New York.

Moolenaar, W. H., Tertoolen, L. G. J., Tsien, R. Y., van der Saag, P. T., and de Laat, S. W. (1983b). Na^+/H^+ exchange: Cytoplasmic pH and the action of growth factors in human fibroblasts. *Nature (London)* **304**, 645–648.

Moolenaar, W. H., Tertoolen, L. G. J., and de Laat, S. W. (1984). Growth factors immediately raise cytoplasmic free Ca^{2+} in human fibroblasts. *J. Biol. Chem.* **259**, 8066–8069.

Moore, R. D., Fidelman, M. L., Hansen, J. C., and Otis, J. N. (1982). The role of intracellular pH in insulin action. *In* "Intracellular pH: Its measurement, Regulation, and Utilization in Cellular Functions" (R. Nuccitelli and D. W. Deamer, eds.), pp. 386–416. Alan R. Liss, Inc., New York.

Morin, D., Forest, C., and Fehlmann, M. (1981). EGF inhibits glucagon stimulation of amino acid transport in primary cultures of adult rat hepatocytes. *FEBS Lett.* **127**, 109–111.

Morrill, G. A., Kostellow, A. B., Weinstein, S. P., and Gupta, R. K. (1983). NMR and electrophysiological studies of insulin action on cation regulation and endocytosis in the amphibian oocyte: Possible role of membrane recycling in the meiotic divisions. *Physiol. Chem. Phys. Med. NMR* **15**, 357–362.

Morrill, G. A., Kostellow, A. B., Mahajan, S., and Gupta, R. K. (1984). Role of calcium in regulating intracellular pH following the stepwise release of the metabolic blocks at first meiotic prophase and second meiotic metaphase in amphibian oocytes. *Biochim. Biophys. Acta* **804**, 107–117.

Moyer, M. P., Moyer, R. C., and Waite, M. R. F. (1982). A survey of intracellular Na$^+$ and K$^+$ of various normal, transformed and tumor cells. *J. Cell Physiol.* **113**, 129–133.

Mummery, C. L., Boonstra, J., van der Saag, P. T., and de Laat, S. W. (1981). Modulation of functional and optimal (Na$^+$,K$^+$)-ATPase activity during the cell cycle of neuroblastoma cells. *J. Cell. Physiol.* **107**, 1–9.

Mummery, C. L., Boonstra, J., van der Saag, P. T., and de Laat, S. W. (1982). Modulation of Na$^+$ transport during the cell cycle of neuroblastoma cells. *J. Cell. Physiol.* **112**, 27–34.

Mummery, C. L., van der Saag, P. T., and de Laat, S. W. (1983). Loss of EGF binding and cation transport response during differentiation of mouse neuroblastoma cells. *J. Cell Biochem.* **21**, 63–75.

Nishizuka, Y. (1984). Turnover of inositol phospholipids and signal transduction. *Science* **225**, 1365–1370.

Nuccitelli, R., and Deamer, D. W., eds. (1982). "Intracellular pH: Its Measurement, Regulation, and Utilization in Cellular Functions." Alan R. Liss, Inc., New York.

O'Donnell, M. E., and Villereal, M. L. (1982). Membrane potential and sodium flux in neuroblastoma × glioma hybrid cells: Effects of amiloride and serum. *J. Cell. Physiol.* **113**, 405–412.

O'Donnell, M. E., Cragoe, E., and Villereal, M. L. (1983). Inhibition of Na$^+$ influx and DNA synthesis in human fibroblasts and neuroblastoma–glioma hybrid cells by amiloride analogs. *J. Pharmacol. Exp. Ther.* **226**, 369–372.

Owen, N. E., and Villereal, M. L. (1982). Evidence for a role of calmodulin in serum stimulation of Na$^+$ influx in human fibroblasts. *Proc. Natl. Acad. Sci. U.S.A.* **79**, 3537–3541.

Owen, N. E., and Villereal, M. L. (1983a). Lys-bradykinin stimulates Na$^+$ influx and DNA synthesis in cultured human fibroblasts. *Cell (Cambridge, Mass.)* **32**, 979–985.

Owen, N. E., and Villereal, M. L. (1983b). Mechanism of desensitization of serum-stimulated Na$^+$ influx in cultured human fibroblasts. *J. Cell Biol.* **97**, 341a.

Owen, N. E., and Villereal, M. L. (1983c). Na$^+$ influx and cell growth in cultured human fibroblasts. *Exp. Cell Res.* **143**, 37–46.

Panet, R., Fromer, I., and Alayoff, A. (1983). Rb$^+$ influxes differentiate between growth arrest of cells by different agents. *J. Membr. Biol.* **75**, 219–224.

Paris, S., and Pouysségur, J. (1983). Biochemical characterization of the amiloride-sensitive Na$^+$/H$^+$ antiport in Chinese hamster lung fibroblasts. *J. Biol. Chem.* **258**, 3503–3508.

Paris, S., and Pouysségur, J. (1984). Growth factors activate the Na$^+$/H$^+$ antiporter in quiescent fibroblasts by increasing its affinity for intracellular H$^+$. *J. Biol. Chem.* **259**, 10989–10994.

Piatigorsky, J., Shinohara, T., Bhat, S. P., Reszelbach, R., Jones, R. E., and Sullivan, M. A. (1980). Correlated changes in δ-crystalline synthesis and ion concentrations in the embryonic chick lens: Summary, current experiments and speculations. *Ann. N.Y. Acad. Sci.* **339**, 265–279.

Pieri, C., Giuli, C., and Bertoni-Freddari, C. (1983). X-ray microanalysis of monovalent electrolyte contents of quiescent, proliferating as well as tumor rat hepatocytes. *Carcinogenesis* **4**, 1577–1581.

Pouysségur, J., Jacques, Y., and Lazdunski, M. (1980). Identification of a tetrodotoxin-sensitive Na$^+$ channel in a variety of fibroblast lines. *Nature (London)* **286**, 162–164.

Pouysségur, J., Paris, S., and Chambard, J. C. (1982a). Na$^+$, K$^+$, H$^+$, and protein phosphorylation in the growth factor induced G_0/G_1 transition in fibroblasts. *In* "Ions, Cell Proliferation, and Cancer" (A. L. Boynton, W. L. Mckeehan, and J. F. Whitfield, eds.), pp. 205–218. Academic Press, New York.

Pouysségur, J., Chambard, J. C., Franchi, A., Paris, S., and van Obberghen-Schilling, E. (1982b). Growth-factor activation of an amiloride-sensitive Na$^+$/H$^+$ exchange system in quiescent fibroblasts: Coupling to ribosomal protein S6 phosphorylation. *Proc. Natl. Acad. Sci. U.S.A.* **79**, 3935–3939.

Pouysségur, J., Sardet, C., Franchi, A., L'Allemain, G., and Paris, S. (1984). A specific mutation abolishing Na$^+$/H$^+$ antiport activity in hamster fibroblasts precludes growth at neutral and acidic pH. *Proc. Natl. Acad. Sci. U.S.A.* **81**, 4833–4837.

Quastel, M. R., and Kaplan, J. G. (1970). Early stimulation of potassium uptake in lymphocytes treated with PHA. *Exp. Cell Res.* **53**, 203–233.

Ralph, R. K., Smart, J., Wojcik, S. J., and McQuillan, J. (1982). Inhibition of mouse mastocytoma protein kinases by amiloride. *Biochem. Biophys. Res. Commun.* **104**, 1054–1059.

Reyes, J., and Benos, D. J. (1983). Sodium transport in isolated mammalian hepatocytes. *In* "Isolation, Characterization and Use of Hepatocytes" (R. A. Harris and N. W. Cornell, eds.), pp. 239–244. Am. Elsevier, New York.

Reznik, V. M., Villela, J., and Mendoza, S. A. (1983). Serum stimulates Na$^+$ entry and the Na$^+$-K$^+$ pump in quiescent cultures of epithelial cells (MDCK). *J. Cell. Physiol.* **117**, 211–214.

Rindler, M. J., and Saier, M. H. (1981). Evidence for Na$^+$/H$^+$ antiport in cultured dog kidney cells (MDCK). *J. Biol. Chem.* **256**, 10820–10825.

Rosoff, P. M., and Cantley, L. C. (1983). Increasing the intracellular Na$^+$ concentration induces differentiation in a pre-B lymphocyte cell line. *Proc. Natl. Acad. Sci. U.S.A.* **80**, 7547–7550.

Rosoff, P. M., Stein, L. F., and Cantley, L. C. (1984). Phorbol esters induce differentiation in a pre-B-lymphocyte cell line by enhancing Na$^+$/H$^+$ exchange. *J. Biol. Chem.* **259**, 7056–7060.

Rothenberg, P., Reuss, L., and Glaser, L. (1982). Serum and epidermal growth factor transiently depolarize quiescent BSC-1 epithelial cells. *Proc. Natl. Acad. Sci. U.S.A.* **79**, 7783–7787.

Rothenberg, P., Glaser, L., Schlesinger, P., and Cassel, D. (1983a). Epidermal growth factor stimulates amiloride-sensitive ^{22}Na$^+$ uptake in A-431 cells. Evidence for Na$^+$/H$^+$ exchange. *J. Biol. Chem.* **258**, 4883–4889.

Rothenberg, P., Glaser, L., Schlesinger, P., and Cassel, D. (1983b). Activation of Na$^+$/H$^+$ exchange by epidermal growth factor elevates intracellular pH in A-431 cells. *J. Biol. Chem.* **258**, 12644–12653.

Rozengurt, E. (1982). Monovalent ion fluxes, cyclic nucleotides, and the stimulation of DNA synthesis in quiescent cells. *In* "Ions, Cell Proliferation, and Cancer" (A. L. Boynton, W. L. McKeehan, and J. F. Whitfield, eds.), pp. 259–281. Academic Press, New York.

Rozengurt, E., and Mendoza, S. (1980). Monovalent ion fluxes and the control of cell proliferation in cultured fibroblasts. *Ann. N.Y. Acad. Sci.* **339**, 175–190.

Rozengurt, E., and Smith, J. S. (1983). Bombesin stimulation of DNA synthesis and cell division in cultures of Swiss 3T3 cells. *Proc. Natl. Acad. Sci. U.S.A.* **80**, 2936–2940.

Rozengurt, E., Gelehrter, T. D., Legg, A., and Pettican, P. (1981). Mellitin stimulates Na$^+$

entry, Na^+-K^+ pump activity and DNA synthesis in quiescent cultures of mouse cells. *Cell (Cambridge, Mass.)* **23**, 781–788.

Rybak, S. M., and Stockdale, F. E. (1981). Growth effects of lithium chloride in BALB/c 3T3 fibroblasts and Madin-Darby canine kidney epithelial cells. *Exp. Cell Res.* **136**, 263–270.

Samuelson, A. C., Withers, D. M., Stockert, R. J., and Wolkoff, A. W. (1984). Role of monovalent cations in receptor-mediated endocytosis. *Hepatology* **4**, 1075.

Sanui, H., and Rubin, H. (1982). Atomic absorption measurement of cations in cultured cells. In "Ions, Cell Proliferation, and Cancer" (A. L. Boynton, W. L. McKeehan, and J. F. Whitfield, eds.), pp. 41–52. Academic Press, New York.

Sariban-Sohraby, S., Latorre, R., Burg, M., Olans, L., and Benos, D. (1984). Amiloride-sensitive epithelial Na^+ channels reconstituted into planar lipid bilayer membranes. *Nature (London)* **308**, 80–82.

Schenk, D. B., and Leffert, H. L. (1983). Monoclonal antibodies to rat (Na^+,K^+)-ATPase block enzymatic activity. *Proc. Natl. Acad. Sci. U.S.A.* **80**, 5281–5285.

Schenk, D. B., Hubert, J. J., and Leffert, H. L. (1984). Use of a monoclonal antibody to quantify (Na^+,K^+)-ATPase activity and sites in normal and regenerating rat liver. *J. Biol. Chem.* **259**, 14941–14951.

Schreiber, A. B., Lax, I., Yarden, Y., Eshhar, Z., and Schlessinger, J. (1981). Monoclonal antibodies against receptor for epidermal growth factor induce early and delayed effects of epidermal growth factor. *Proc. Natl. Acad. Sci. U.S.A.* **78**, 7535–7539.

Schreiber, A. B., Libermann, T. A., Lax, I., Yarden, Y., and Schessinger, J. (1983). Biological role of epidermal growth factor-receptor clustering. Investigation with monoclonal anti-receptor antibodies. *J. Biol. Chem.* **258**, 846–853.

Schuldiner, S., and Rozengurt, E. (1982). Na^+/H^+ antiport in Swiss 3T3 cells: Mitogenic stimulation leads to cytoplasmic alkalinization. *Proc. Natl. Acad. Sci. U.S.A.* **79**, 7778–7782.

Segel, G. B., Simon, W., and Lichtman, M. A. (1979). Regulation of sodium and potassium transport in phytohemagglutinin-stimulated human blood lymphocytes. *J. Clin. Invest.* **64**, 834–841.

Serpersu, E. H., and Tsong, T. Y. (1984). Activation of electrogenic Rb^+ transport of (Na^+,K^+)-ATPase by an electric field. *J. Biol. Chem.* **259**, 7155–7162.

Shen, S. S., and Steinhardt, R. A. (1978). Direct measurement of intracellular pH during metabolic derepression of the sea urchin egg. *Nature (London)* **272**, 253–254.

Shier, W. T. (1979). Activation of high levels of endogenous phospholipase A_2 in cultured cells. *Proc. Natl. Acad. Sci. U.S.A.* **76**, 195–199.

Smith, J. B. (1983). Vanadium ions stimulate DNA synthesis in Swiss mouse 3T3 and 3T6 cells. *Proc. Natl. Acad. Sci. U.S.A.* **80**, 6162–6166.

Smith, J. B., and Brock, T. A. (1983). Analysis of angiotensin-stimulated sodium transport in cultured smooth muscle cells from rat aorta. *J. Cell. Physiol.* **114**, 284–290.

Smith, J. B., and Rozengurt, E. (1978a). Serum stimulates the Na^+,K^+ pump in quiescent fibroblasts by increasing Na^+ entry. *Proc. Natl. Acad. Sci. U.S.A.* **75**, 5560–5564.

Smith, J. B., and Rozengurt, E. (1978b). Lithium transport by fibroblastic mouse cells: Characterization and stimulation by serum and growth factors in quiescent cultures. *J. Cell. Physiol.* **97**, 441–450.

Smith, N. R., Sparks, R. L., Pool, T. B., and Cameron, I. L. (1978). Differences in the intracellular concentration of elements in normal and cancerous liver cells as determined by X-ray microanalysis. *Cancer Res.* **38**, 1952–1959.

Smith, R.L., Macara, I. G., Levenson, R., Housman, D., and Cantley, L. (1982). Evidence that a Na^+/Ca^{2+} antiport system regulates murine erythroleukemia cell differentiation. *J. Biol. Chem.* **257**, 773–780.

Sohraby, S. S., Burg, M., Weismann, W. P., Chiang, P. K., and Johnson, J. P. (1984). Transport into A6 apical membrane vesicles: Possible mode of aldosterone action. *Science* **225**, 745–746.

Sparks, R. L., Pool, T. B., Smith, N. K. R., and Cameron, I. L. (1983). Effects of amiloride on tumor growth and intracellular element content of tumor cells *in vivo*. *Cancer Res.* **43**, 73–77.

Steinhardt, R. A. (1982). Ionic logic in activation of the cell cycle. *In* "Ions, Cell Proliferation, and Cancer" (A. L. Boynton, W. L. McKeehan, and J. F. Whitfield, eds.), pp. 311–325. Academic Press, New York.

Stiernberg, J., La Belle, E. F., and Carney, D. H. (1983). Demonstration of a late amiloride-sensitive event as a necessary step in initiation of DNA synthesis by thrombin. *J. Cell. Physiol.* **117**, 272–281.

Stiernberg, J., Carney, D. H., Fenton, J. W., and La Belle, E. F. (1984). Initiation of DNA synthesis by human thrombin: Relationships between receptor binding, enzymic activity, and stimulation of $^{86}Rb^+$ influx. *J. Cell. Physiol.* **120**, 289–295.

Sugimoto, Y., Whitman, M., Cantley, L. C., and Erikson, R. L. (1984). Evidence that the Rous sarcoma virus transforming gene product phosphorylates phosphatidylinositol and diacylglycerol. *Proc. Natl. Acad. Sci. U.S.A.* **81**, 2117–2121.

Taub, M. (1978). Isolation of amiloride-resistant clones from dog kidney epithelial cells. *Somatic Cell Genet.* **4**, 609–616.

Taub, M., and Saier, M. H. (1979). Regulation of $^{22}Na^+$ transport by calcium in an established kidney epithelial cell line. *J. Biol. Chem.* **254**, 11440–11444.

Taub, M., and Saier, M. H. (1981). Amiloride-resistant Madin-Darby canine kidney (MDCK) cells exhibit decreased cation transport. *J. Cell. Physiol.* **106**, 191–199.

Taub, M., Devis, P. E., and Grohol, S. H. (1984). PGE_1-independent MDCK cells have elevated intracellular cAMP but retain the growth stimulatory effects of glucagon and epidermal growth factor in serum-free medium. *J. Cell. Physiol.* **120**, 19–28.

Toback, F. G. (1980). Induction of growth in kidney epithelial cells in culture by Na^+. *Proc. Natl. Acad. Sci. U.S.A.* **77**, 6654–6656.

Tomooka, Y., Imagawa, W., Nandi, S., and Bern, H. A. (1983). Growth effect of lithium on mouse mammary epithelial cells in serum-free collagen gel culture. *J. Cell. Physiol.* **117**, 290–296.

Tucker, R. F., Shipley, G. D., Moses, H. L., and Holley, R. W. (1984). Growth inhibitor from BSC-1 cells closely related to platelet type β transforming growth factor. *Science* **226**, 705–707.

Tucker, R. W., and Fay, F. S. (1984). Growth factors stimulate increases in free Ca^{2+} as measured by quin 2 fluorescence in single living BALB/c 3T3 cells. *J. Cell Biol.* **99**, 272a.

Tupper, J. T., and Zografos, L. (1978). Effect of imposed serum deprivation on growth of the mouse 3T3 cell. Dissociation from changes in potassium ion transport as measured from $^{86}Rb^+$ ion uptake. *Biochem. J.* **174**, 1063–1065.

Tupper, J. T., Del Rosso, M., Hazelton, B., and Zorgniotti, F. (1978). Serum-stimulated changes in calcium transport and distribution in mouse 3T3 cells and their modification by dibutyryl cAMP. *J. Cell. Physiol.* **95**, 71–84.

Vicentini, L. M., Miller, R. J., and Villereal, M. L. (1984). Evidence for a role of phospholipase activity in the serum stimulation of Na^+ influx in human fibroblasts. *J. Biol. Chem.* **259**, 6912–6919.

Villereal, M. L. (1981). Sodium fluxes in human fibroblasts: Effect of serum, Ca^{2+}, and amiloride. *J. Cell. Physiol.* **107**, 359–369.

Villereal, M. L., and Owen, N. E. (1982). The involvement of Ca^{2+} in the serum stimulation of Na^+ influx in human fibroblasts. In "Ions, Cell Proliferation, and Cancer" (A. L. Boynton, W. L. McKeehan, and J. F. Whitfield, eds.) pp. 245–257. Academic Press, New York.

Walsh-Reitz, M. M., and Toback, F. G. (1983). Vasopressin stimulates growth of renal epithelial cells in culture. *Am. J. Physiol.* **245**, C365–C370.

Walsh-Reitz, M. M., Toback, F. G., and Holley, R. W. (1984). Cell growth and net Na^+ flux are inhibited by a protein produced by kidney epithelial cells in culture. *Proc. Natl. Acad. Sci. U.S.A.* **81**, 793–796.

Warley, A., Stephen, J., Hockaday, A., and Appleton, T. C. (1983). X-Ray microanalysis of HeLa S3 cells. *J. Cell Sci.* **62**, 339–350.

Whitaker, M. J., and Steinhardt, R. A. (1982). Ionic regulation of egg activation. *Q. Rev. Biophys.* **15**, 593–666.

Whiteley, B., Cassel, D., Zhuang, Y. X., and Glaser, L. (1984). Tumor promoter phorbol 12-myristate 13-acetate inhibits mitogen-stimulated Na^+/H^+ exchange in human epidermoid carcinoma A-431 cells. *J. Cell Biol.* **99**, 1162–1166.

Wondergem, R. (1982). Intracellular potassium activity during liver regeneration. In "Ions, Cell Proliferation, and Cancer" (A. L. Boynton, W. L. McKeehan, and J. F. Whitfield, eds.) pp. 175–186. Academic Press, New York.

Wondergem, R., and Harder, D. R. (1980). Membrane potential measurements during rat liver regeneration. *J. Cell. Physiol.* **102**, 193–197.

Yarden, Y., Schreiber, A. B., and Schlessinger, J. (1982). A nonmitogenic analogue of epidermal growth factor induces early responses mediated by epidermal growth factor. *J. Cell Biol.* **92**, 687–693.

Zetterberg, A., and Engström, W. (1981). Mitogenic effect of alkaline pH on quiescent, serum-starved cells. *Proc. Natl. Acad. Sci. U.S.A.* **78**, 4334–4338.

Zilberstein, D., Agmon, V., Schuldiner, S., and Padan, E. (1982). The sodium/proton antiporter is part of the pH homeostasis mechanism in *Escherichia coli*. *J. Biol. Chem.* **257**, 3687–3691.

Zs.-Nagy, I., Lustyik, G., Zs.-Nagy, V., Zarandi, B., and Bertoni-Freddari, C. (1981). Intracellular $Na^+:K^+$ ratios in human cancer cells as revealed by energy dispersive X-ray microanalysis. *J. Cell Biol.* **90**, 769–777.

Zs.-Nagy, I., Lustyik, G., Lukács, G., Zs.-Nagy, V., and Balázs, G. (1983). Correlation of malignancy with the intracellular $Na^+:K^+$ ratio in human thyroid tumors. *Cancer Res.* **43**, 5395–5402.

Zucker, R. S., Steinhardt, R. A., and Winkler, M. M. (1978). Intracellular calcium release and the mechanisms of parthenogenetic activation of the sea urchin egg. *Dev. Biol.* **65**, 285–295.

V

Regulation

14

Structural Heterogeneity of Duplex DNA

CHARLES K. SINGLETON[1]

Department of Bacteriology
University of Wisconsin
Madison, Wisconsin

I. INTRODUCTION

DNA is a thin, extremely long polymer of nucleotides consisting of two antiparallel strands with an outer sugar–phosphate backbone and nitrogenous bases, hydrogen-bonded in a complementary manner and stacked as base pairs down the center of the molecule. The two strands are in a helical configuration. A commonly held view is that the geometric characteristics of the helix, the backbone conformation, and the spatial relationship of the base pairs to the rest of the molecule are in fairly good agreement to the B DNA structure first proposed by Watson and Crick (1953); thus, the conformation up and down the double helix is regarded as fairly monotonic. To a first approximation, such a picture of the DNA molecule is a legitimate one. However, as will be detailed, a very active area of biochemical and biophysical research is quickly

[1]Present address: Department of Molecular Biology, Vanderbilt University, Nashville, Tennessee 37235.

CONTROL OF ANIMAL CELL PROLIFERATION,
VOLUME I

dispelling the notion of a DNA molecule possessing a monotonous, uniform conformation. Instead, this work is demonstrating that conformational differences, both subtle and others more dramatic, can exist within a single DNA molecule. These differences in spatial geometry are dependent not only on the environment of the DNA, but also on the particular sequence of bases along the DNA polymer.

Early studies with simple DNA polymers, both homo- and heteropolymers, convincingly demonstrated sequence-dependent conformations in that polymers composed of one to four different bases showed different conformational characteristics (Wells *et al.*, 1980; Leslie *et al.*, 1980). Recently, a renewed interest in the sequence-dependent conformational properties of duplex DNA has occurred, being brought about by new findings in this area as well as new and improved methodologies for answering questions in this field. One finding which has renewed interest is that of left-handed helical DNA, a novel structure previously given little thought. Improved methodologies include technical advances in areas as diverse as chemical synthesis of nucleic acids, NMR and Raman spectroscopy, single-crystal X-ray diffraction studies, molecular cloning, sequencing, and enzymology of nucleic acids metabolizing proteins. Elegant combinations of these techniques have given rise to new insights into DNA polymorphism at much more subtle (propellar twist, base pair roll, twist angle) and dramatic (left-handed helices, cruciforms, alternating backbones) levels than previously existed.

This chapter serves as a general review of DNA structure with emphasis on environmental- and sequence-dependent conformational properties. It is not intended to be comprehensive, yet at least some mention of the many recent interesting findings is strived for. Other recent reviews concerning various aspects of DNA structure are mentioned at the appropriate places within the text.

II. RIGHT-HANDED DUPLEX DNA

A. Solid State

1. Fiber Diffraction Analysis

Almost all of the conformational families of duplex DNA have been defined initially by interpretation of diffraction patterns of DNA fibers along with proposed detailed models consistent with the interpretations (Leslie *et al.*, 1980; reviewed in Wells *et al.*, 1980, and Zimmerman, 1982). Three forms are generally found, A, B, and C, dependent on salt concentration and type as well as the relative humidity. The conformational characteristics for these DNA forms have been extensively considered (Leslie et al., 1980) and a brief listing can be found in Table I. Studies using homopolymers and di- and trinucleotide repeating polymers show that different sequences adopt different conformations within the three families (Leslie et al., 1980); within a family, the dihedral angles characterizing the helix conformation remain within the same local minima. A given sequence can also adopt more than one of these forms, depending upon the specific environmental conditions. Other specific allomorphs can be assumed given the appropriate conditions: B', C', C", D, E, and S (left-handed helix). Some of these structures are peculiar to specific sequences (Leslie et al., 1980). A recent paper (Mahendrasigam et al., 1983) describes conditions for which a single polymer, poly(dA-dT)·poly(dA-dT), assumes the A, B, C, or D form. The conclusion from most of these studies clearly indicates that a variety of secondary structures can be adopted by duplex DNA, depending upon both the DNA environment and the particular base sequence.

Two recent fiber analyses have given insights into finer structural polymorphisms. The first involves a conformation the authors term "wrinkled" DNA and appears to be an intrinsic feature of alternating purine/pyrimidine sequences (Arnott et al., 1983b). The structure was seen with poly(dG-dC)·poly(dG-dC), poly(dI-dC)·poly(dI-dC), and poly (dA-dT)·poly(dA-dT) when in an overall B-type conformation for the first polymer and a D-type for the second and third. The purine–3',5'-pyrimidine (Pu–Pyr) steps assume the same conformation as smooth B DNA, whereas the Pyr–Pu steps have a different conformation at C-3'–O-3'. This leads to a distinctive change in the phosphate group orienta-

TABLE I

Helical Parameters of Duplex DNA from Fiber Diffraction Analysis[a]

DNA form	Residues/turn	Rise/residue (Å)	Rotation/residue
A	11	2.56	32.7°
B	10	3.38	36.0°
C	9.3	3.32	38.6°

[a]Values obtained with permission from Leslie et al. (1980). With permission from Journal of Molecular Biology, Vol. 143, pp. 49–72. Copyright 1980 Academic Press Inc. (London) Ltd. Original references can be found therein.

tion at every other nucleotide unit, thus giving rise to a B or D dinucleotide repeat (five- or fourfold helix, respectively, of dinucleotides) (Arnott *et al.*, 1980). The overall appearance of the polymer is a wrinkled
version of a regular B or D DNA. The change in the sugar–phosphate
backbone leads to very little, if any, change in base pair stacking. However, the authors point out that potential interactions with N and O
atoms within the grooves would be altered. The wrinkled conformation
characterizes changes which, although alternating, are not as great as
the proposed "alternating B DNA" model for poly(dA-dT)·poly(dA-dT)
in solution (Klug *et al.*, 1979; Viswamitra *et al.*, 1982). This structure,
in which the sugar pucker alternates between a C-3'-endo and C-2'-
endo conformation, has been reviewed in detail elsewhere (Zimmerman, 1982; Wells *et al.*, 1980).

The second investigation concerns the duplex structure of
poly(dA)·poly(dT). Both strands assume a 10_1 helix with normal base
pairing, but one strand (probably the dA) has a C-3'-endo sugar conformation characteristic of A DNA while the other strand has a C-2'-endo
conformation typical of B DNA (Arnott *et al.*, 1983a). Base stacking
down both chains resembles that of B DNA, while the backbone,
glycosylic, and furanose conformations are chain dependent and typical of either A (dA strand) or B DNA (dT strand). A significant propeller twist of 19° is seen (dihedral angle between the base planes
within a base pair). Significant differences with respect to A and B
DNA are seen for the position and arrangement of the residues in relation to the helix axes.

2. Single-Crystal Diffraction

Although fiber diffraction analyses have proved to be invaluable concerning DNA duplex structure, the information produced from these
studies is limited by the inherent low resolution obtainable. Sequence-
dependent conformational heterogeneity has been found using simple
repeating sequences, but the conformations observed are averaged over
the entire sequence or polymer length. Thus, microheterogeneity of
structural differences between individual neighboring residues, if exis-
tent, is not found by this method of investigation. Determination of
sequence-dependent geometrical differences at the single nucleotide
level in complex sequence mixtures is an important goal of DNA structural studies. Another is being able to understand how a given sequence of base pairs, whether very few in number or more numerous,
will influence the configuration of that region and surrounding regions
of the DNA molecule. Single-crystal X-ray diffraction studies have the

potential high resolution to help achieve this goal. Indeed, as will be described and reviewed previously (Zimmerman, 1982; Dickerson et al., 1982), such studies are providing many exciting glimpses at attaining the above-mentioned goals.

a. B DNA

In 1980, Wing et al. presented a single-crystal X-ray diffraction study on a defined duplex DNA molecule of greater than one turn of helix, the self-complementary dodecamer d(CpGpCpGpApApTpTpCpGpCpG). Although at an early stage of refinement, the results indicated a Watson–Crick base-paired duplex with a 3.4 Å average rise per residue and an average of 10.1 residues per turn, i.e., a typical B DNA conformation. One of the first features clearly different from those predicted by fiber studies was a propeller twisting of the base pairs. This twisting, a nonclassical B-helix feature, results in a greater overlap between one base and the adjacent bases on the same strand. The enhanced stacking, predicted from energy calculations (Levitt, 1978), lends stability to the duplex structure. As will be discussed, propeller twisting may play a highly important role in producing sequence-specific conformations. As the analysis was continued and refined to atomic resolution, significant information concerning sequence-dependent polymorphism, DNA hydration, and helical bending became apparent (Zimmerman, 1982; Dickerson et al., 1982). These findings will be discussed.

Base-Specific Polymorphism.   Significant sequence-specific influences on the geometry of the individual residues of the dodecamer have been found (Drew et al., 1981; Dickerson and Drew, 1981a,b). Variations in sugar pucker were indicated by the representation of the C-2′-endo, C-1′-exo, O-1′-endo, and C-3′-endo (at the 3′ terminus) conformations. The C-1′–N glycosyl torsion angle (χ) and the main chain C-5′–C-4′–C-3′–O-3′ torsion angle (δ) were strongly correlated and both showed significant variation: As χ varied from $-140°$ to $90°$, δ varied from $80°$ to $160°$ (Dickerson and Drew, 1981b). These values for paired bases were anticorrelated such that glycosyl rotations were of opposite signs but equal magnitude. Furthermore, larger χ and δ values were associated with purines while relatively smaller values characterized pyrimidine residues. The rise per residue varied within the dodecamer showing a range of 3.14–3.31 Å. The number of base pairs per turn ranged from 8.05 to 11.18, thus spanning the values seen for D, B, and A DNA defined by fiber analysis. This range reflects a variation in rotation per residue of from 44.7° to 32.2° (Dickerson and Drew, 1981a). It was found that CpG steps showed a smaller helical twist angle than GpC steps. The central GAATTC gave an average of 9.8 base pairs/turn, similar to classical B DNA. Yet one base pair on either side showed A-like helix geometry with 10.8 and 11.1 residues per turn (Dickerson and Drew, 1981). Hence, these investigations dramatically reveal how, within a B DNA framework, base sequence can influence the helix structure in subtle but significant ways.

The conformation of a base pair can be characterized in relation to adjacent base pairs by the three parameters of twist, tilt, and roll (defined in Fig. 1); all three can and do show variations along the duplex molecule. To these, propeller twist, or the dihedral angle between base planes within a single base pair, can be added to further characterize the geometry of the bases. Variations in helical twist were previously described. Differences in roll angle lead to an opening up or closing of two adjacent base pair edges within the major and minor grooves in a correlated manner. For the dodecamer, it was found that Pyr–Pu steps roll in a way so as to open the base planes toward the minor groove, Pu–Pyr steps were opened to the major groove, and homopolymer steps

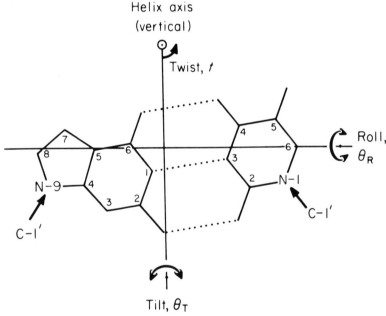

Fig. 1. The three rotational degrees of freedom of rigid base planes. Twist, t, is rotation about the helix axis. Tilt, θ_T, is rotation about the pseudodyad axis passing through the base plane. Roll, θ_R, is rotation about an axis in the plane of the bases perpendicular to the pseudodyad. From Dickerson and Drew (1981a). With permission from *Journal of Molecular Biology*, Vol. 149, pp. 761–786. Copyright 1981 Academic Press Inc. (London) Ltd.

were fairly resistant to roll in either direction (Dickerson and Drew, 1981a). Propeller twisting also demonstrated sequence-dependent values as seen by an average of 17.3° for A-T base pairs and 11.5° for G-C pairs. Such changes are reflected in a corresponding change in the minor groove width.

One of the fundamental goals of DNA structural studies is to determine what underlying causes give rise to sequence-specific polymorphism such as that amply described above for a duplex dodecamer. Given such an understanding of the basic principles involved, predictions and knowledge of conformational features could be attained simply from the sequence of the region of interest. Calladine (1982) has recently put forward a single, simple explanation for much of the variation found within the dodecamer. His proposal is based on the structural mechanics of a deformable elastic system and can account for the variations seen in the torsion angles δ and χ, roll, twist, and propeller twist as well as explain the anticorrelation principle discussed above (Fratini et al., 1982; Dickerson, 1983).

The proposal is that propeller twisting is a desirable feature, since the resulting enhanced base stacking up and down each strand leads to a more stable molecule. However, as shown in Fig. 2, propeller twisting also brings about a steric clash, and hence repulsive forces, between purine bases in consecutive base pairs but on opposite strands (Calladine, 1982). For Pyr–Pu steps, the purine position 6 side chains overlap in the major groove, whereas the position 2 side chains give rise to steric interference in the minor groove for Pu–Pyr steps. No clash is found for Pu–Pu or Pyr–Pyr steps. The steric interference can be relieved in four ways: a lessening of the propeller twisting, decreasing the local twist angle between base pairs, increasing the roll angle in the direction of the interference, or separating the purines by sliding the base pairs in a way to partially pull them out of the stack (Dickerson, 1983; Calladine, 1982). Hence, sequence-dependent clashes give rise to sequence-dependent variations in the above parameters.

Dickerson (1983) has recently put forward a quantitation of Calladine's proposal. Four sum functions for the involved parameters (helix twist, roll angle, torsion angle δ, and propeller twist) were described from which the expected heterogeneity of these parameters may be calculated from a given base sequence. His results showed that all four gave good predictions of the behavior of B DNA, only twist and roll functions were applicable to A DNA, and twist began to fail for A DNA of a RNA/DNA hybrid. The functions presented were in good agreement with the observed helix parameter variations (correlation coefficient >0.900).

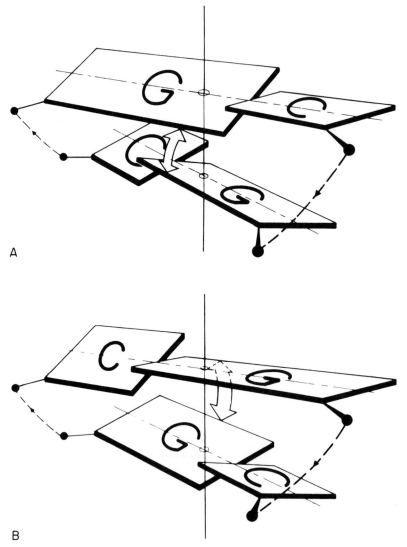

Fig. 2. (A) Schematic drawing of two successive base pairs of a B DNA double helix showing the way in which positive propeller twist induces a clash of purine ring edges (double-headed arrow) within the minor groove for a pyrimidine–purine sequence such as C-G. (B) Schematic drawing showing the way in which positive propeller twist induces a major groove clash of purines (double-headed arrow) for a purine–pyrimidine sequence such as G-C. From Dickerson (1983). With permission from *Journal of Molecular Biology*, Vol. 166, pp. 419–441. Copyright 1983 Academic Press Inc. (London) Ltd.

In summary, it seems that Calladine's proposal of purine–purine clashes can account for variations in many different parameters. The steric problems arise from propeller twisting which is desirable in order to obtain more efficient base–base overlap and stacking. It is significant that in all single-crystal studies to date, propeller twisting of 10–20° has been observed. In order to relieve the sequence-dependent clashes, changes in the twist, roll, propeller twist, and main chain and glycosyl torsion angle are introduced in a sequence-specific manner. As illustrated by Dickerson (1983), this simple steric interference can be quantitated so as to predict structure from base sequence.

b. A DNA

Diffraction studies on three different duplex oligomers have given information on the crystal structure of A DNA. One investigation concerned the tetramer, d(ICpCpGpG), which crystallized as a duplex molecule whose conformation was very similar to A DNA defined by fiber diffraction (Conner et al., 1982). The average values for several parameters were 10.7 residues per turn or a twist angle of 33.6° per residue, a 19° base tilt, and 2.3 Å rise per residue (compare to values given in Table I). The deep, narrow major groove and shallow minor groove of A DNA were observed with a somewhat smaller width than predicted from fibers for both grooves and a slightly deeper minor groove. The major difference from fiber A DNA was the significant 18° mean propeller twist, indicating as for B DNA the importance of this twisting in order to enhance base stacking along each strand.

Another duplex that has been examined is d(GpGpCpCpGpGpCpC) which existed in a modified A-type structure (Wang et al., 1982). The two base pairs at each end of the molecule were in a more or less A DNA conformation. However, the middle four base pairs had an alternating structure such that the third and fifth pairs possessed a B-like sugar pucker (C-2'-endo) and the fourth and sixth possessed A-like puckers (C-3'-endo). Other differences from the classical A form were a tilt of 12–14° (compared to 19°), and the base pairs were somewhat closer to the central axis. The stacking pattern showed some variation but was largely A DNA-like. There was not a uniform conformation for each base pair with respect to sugar conformations and torsion angles of the backbone, i.e., significant sequence-dependent polymorphism was observed.

Single-crystal studies on another self-complementary octamer, d(GpGpTpApTpApCpC), have shown this molecule to also adopt a duplex A DNA structure (Shakked et al., 1983). In contrast to the above study, the sugar–phosphate backbone was only minimally affected by base sequence while the base-stacking pattern revealed a large sequence-dependent variation. The very small variations in backbone and sugar conformations also contrast with the large degree of variance in these parameters for the B DNA dodecamer. The sugar geometry was essentially that of a C3′-endo conformation, the classical A DNA conformation. Helical twist variation was manifest in the 10 to 12 residues per turn values seen locally. The central T-A steps, with an average of 29.9°, had lower helical twists than did the A-T steps, with an average of 34.1°. Propeller twist, as in all other single-crystal analyses, was large and varied significantly over the range of 8.3−18.5°. The roll angles for all the base pairs (excluding the terminal pairs) were small (0.5−2.2°) except for the T-A pairs at position 3 and 6 with roll angles of 12° and 17°. Thus, the central TATA region possessed a large degree of conformational heterogeneity.

Finally, diffraction studies have been carried out on an RNA–DNA hybrid duplex, r(GpCpGp)d(TpApTpApCpGpC) (Wang et al., 1983). This molecule consists of a DNA core flanked by hybrid ends when in the duplex state. The entire molecule exists in an A form very close to the 11-fold RNA A helix. There is, however, propeller twisting (14° average) for each base pair. Also, the central TATA, as with the above octamer, showed variations in local conformations.

B. Solution Studies

Fiber and single-crystal diffraction investigations have provided and continue to provide a wealth of information concerning specific conformations available to duplex DNA molecules of various sequence. Are solid-state conformations a true reflection of solution conformations? This is a question of central importance in DNA structural investigations. The retention and/or alteration of these conformations when DNA is in solution are being vigorously pursued in a number of laboratories with increasing sophistication. Several recent findings from studies involving a variety of approaches and techniques will be described; the discussion is not intended to be an exhaustive coverage of this area of research.

1. Polymeric DNA

It is generally acknowledged that duplex DNA exists in a B-type conformation in low ionic strength solutions (Langridge et al., 1960). By varying the conditions of ionic strength, counterion species, and addition of dehydrating agents, conformational changes can be brought about giving rise to the other (A, C, Z, etc.) forms of DNA (reviewed in Well et al., 1980; Zimmerman, 1982).

Hydrodynamic and spectroscopic investigations have given more detailed information concerning solution conformations. Mitra et al. (1981) calculated magnetic shielding constants for poly(dG-dC)·poly(dG-dC) using ten DNA models from fiber and single-crystal diffraction data. Comparison to experimentally observed solution NMR results gave a good fit with the Arnott, Hukins B DNA of fiber studies (Arnott and Hukins, 1972) when the polymer was in low salt solutions. There was no evidence of large propelling twisting ($\leq 10°$) in contrast to single-crystal studies (Section II,A,2).

More recent ^{31}P (Patel et al., 1982a; Chen et al., 1983) and ^{13}C NMR (Chen et al., 1983) results have also indicated that this polymer adopts a regular B DNA conformation in low salt solutions. Yet in a solution of 2.5 M LiCl, poly(dG-dC)·poly(dG-dC) adopts an alternating B form as revealed by a double resonance for the phosphorus atoms of the phosphodiester backbone (Patel et al., 1982a). Similarly, both investigations found that the C-5 methylated derivative of this polymer was in an alternating B-type structure with a dinucleotide repeat. The alternating conformation may be similar to the wrinkled DNA conformation (Section II,A,1) seen for poly(dG-dC)·poly(dG-dC) from fiber diffraction analysis.

Other studies have been carried out concerning the propeller twisting of DNA in solution. Transient electric dichroism has been used to investigate the inclination of the base pairs to the helix axis. Hogan et al. (1978) and Wu et al. (1981) interpreted their data as suggesting a 17° average propeller twist for B DNA, and the latter workers also determined a small twisting for A DNA (0–7°). Thus for B DNA, the large propeller twisting is consistent with single-crystal diffraction data (Section II,A,2,a) whereas the low value for A DNA is not (Section II,A,2,b). However, Charney and Yamaoka (1982) have questioned the interpretations. They also find an approximate 70° angle between the DNA helix axis and direction of the 260 nm absorption band transition moment. They argue, though, that interpretation of these values in terms of DNA structure is equivocal and may not reflect propeller twisting.

Similar investigations have been conducted to estimate the average rise per residue for DNA in solution. Transient electric dichroism gave values of a 3.4 Å rise per residue for B DNA (Wu et al., 1981; Chen et al., 1982) and 2.8 Å for A DNA (Wu et al., 1981) (compare with values of Table I). A similar value of 3.34 Å per residue was found for B DNA based on rotational and translational frictional coefficients (Mandelkern *et al.*, 1981). The data of this study suggested that the rise per residue was not strictly constant but showed some sequence dependency.

The number of base pairs per turn of helix for DNA in solution has also been examined. These studies have been extensively reviewed (Zimmerman, 1982; Wells *et al.*, 1980) and will only briefly be mentioned. The results indicate a value of 10.6 ± 0.1 base pair per turn for random sequence B DNA, a value significantly greater than the 10 base pair value of fiber and crystal results (Strauss *et al.*, 1981; Peck and Wang, 1981; Rhodes and Klug, 1981). This conformational parameter showed sequence-dependent variation since poly(dA)·poly(dT) had 10.1 residues per turn, poly(dG)·poly(dC) gave 10.7 residues per turn, and poly(dA-dT)·poly(dA-dT) had a value of 10.5. It should also be pointed out that the helical twist (from which base pairs per turn is derived or vice versa) varies significantly with ionic strength, counterion species, solvent composition, and temperature (Lee *et al.*, 1981; reviewed in Bauer, 1978). The changes involved, however, are of a magnitude such that the helix type remains in the B family.

2. Defined Sequence Oligonucleotides

Several recent reports have appeared concerning solution conformations for duplex oligonucleotides. The combination of facile chemical synthesis of oligonucleotides and proton NMR methods for conformation determinations has proved to be a very powerful means of studying DNA structure [see Patel *et al.* (1982b) for an overview of existing NMR methodologies in relation to DNA structural studies]. Many of these investigations are preliminary and involve description of resonance assignments: d(CpApCpGpTpG) and d(GpTpGpCpApC), Tran-Dihn *et al.* (1983, also see references therein); d(ApTpApTpCpGpApTpApT), Feigon *et al.* (1982); d(CpCpApApGpCpTpTpGpG), Kan *et al.* (1982); d[Cp(Tp)$_5$G]·d[Cp(Ap)$_5$G], Pardi *et al.* (1981); d(GpGp^{m}Cp^{m}CpGpGpCpC), Sanderson *et al.* (1983); and d(G-C)$_5$, Reid *et al.* (1983). Complete assignment of proton resonances for many of these oligomers was accomplished which should allow for highly specific conformational studies to be carried out at the level of individual base pairs. Several of these investigators were able to conclude that the overall conformation was a B-type helix based mainly on the presence of C-2′-endo sugar conformations. For example, Chen *et al.* (1982) found that all sugar rings were predominately C-2′-endo within a range of graded variation of 66–87%. The refinement of the analysis of such hints at sequence-dependent conformations in solution should prove to be highly informative.

Solution studies on d(CpGpCpGpApApTpTpCpGpCpG), the dodeca-mer described in Section II,A,2,a, have also been carried out. Cleavage of the duplex dodecamer with DNase I revealed a specific cleavage pattern not predicted from DNase I base specificity (Lomonosoff et al., 1981). Instead, the pattern was closely correlated with the local helical twist angle between adjacent residues. The greater the twist angle [which ranged from 32° to 42° (Dickerson and Drew, 1981b)], the greater the apparent rate constant for digestion. The major conclusion was that the solution conformation of the dodecamer apparently was an accu-rate reflection of that defined by single-crystal diffraction analysis (Lomonosoff et al., 1981).

^{31}P and ^{1}H NMR have also been performed on this dodecamer (re-viewed in Patel et al., 1982a). The ^{31}P spectrum gave eight partially resolved resonances with the chemical shifts spread over a 0.45 ppm range (Patel et al., 1982c). This indicated a distribution of O-P torsion angles and/or O-P-O bond angles within the phosphate backbone, and thus is consistent with diffraction studies (Section II,A,2,a). The ^{1}H NMR data suggested that the duplex adopted a structure which exhib-ited twofold rotational symmetry. This is in contrast to the absence of such symmetry in the crystalline state, probably accounted for by the bending of the crystal molecule.

Predictions of sequence-dependent base-stacking properties arising from Calladine's proposal of steric clash (Section II,A,2,a) have been tested for a similar dodecamer, d(CpGpApTpTpApTpApApTpCpG), in solution. The predictions were that a 3.4 Å separation should occur for the homopolymer step dA-dA whereas a smaller separation for dT-dA steps and a larger one for dA-dT would be expected if propeller twist-ing and purine/purine cross-chain repulsion were occurring (Fig. 2; Patel et al., 1983). These predictions were confirmed by ^{1}H NMR deter-mination of interproton separations of the H-2s of the involved ade-nines. Thus, sequence-dependent base pair-stacking patterns induced by propeller twisting are found in solution as well as in crystals (Patel et al., 1983).

NMR investigations of the solution structures of three oligomers pos-sessing one non-self-complementary position have been undertaken. Patel et al. (1982d) found that G-T pairs formed in duplex molecules of d(CpGpTpGpApApTpTpCpGpCpG) with little disruption of the helical structure. A small decrease in stacking of the G-T pair with the adjacent pairs was observed as well as a possible change in the geometry of the two phosphodiester linkages. Calorimetric measurements of the en-thalpy of denaturation indicated essentially normal stacking (Patel et al., 1982d).

The "extra" deoxyadenosine residue in the molecule d(CpGpCpApGpApApTpTpCpGpCpG) was found to stack into the tridecamer duplex instead of looping out (Patel et al., 1982e; Pardi et al., 1982). The phosphodiester linkage opposite the nonpaired dA had atypical O-P-O bond angles and/or O-P torsion angles and was in an extended state. Calorimetric enthalpic measurements suggested that the extra dA stacking energy balanced the disruption of the opposite adjacent bases.

Finally, ^{1}H NMR has provided evidence of guanine–adenine pairing in d(CpCpApApGpApTpTpGpG) (Kan et al., 1983). The bases involved were in an anti conformation and thus the backbone in this region was probably disrupted to some degree.

III. LEFT-HANDED DUPLEX DNA

A. Solid State

1. Single-Crystal Diffraction

Some of the earliest single-crystal diffraction analyses of duplex oligonucleotides resulted in the startling conclusion that DNA can adopt a left-handed helix (reviewed in Zimmerman, 1982, and Dickerson et al., 1982). The oligonucleotides involved were a hexamer (Wang et al., 1979) and a tetramer (Drew et al., 1980) of the alternating sequence (dC-dG)$_n$, where n = 3 or 2. The description of left-handed helical DNA, or Z DNA (Wang et al., 1979), has evoked a tremendous amount of research on the occurrence, properties, and biological significance of this form of DNA.

The crystallographic work on Z DNA [the term will be used in a general way for left-handed structures similar to those defined by diffraction analyses, although it also signifies a specific left-handed conformation (Wang et al., 1979)] has shown that there exists several conformations which define a Z DNA family (Crawford et al., 1980; Wang et al., 1981; Drew and Dickerson, 1981). The general features of the Z DNA family (Fig. 3) are a left-handed antiparallel, Watson–Crick base-paired duplex with a dinucleotide repeat (Wang et al., 1979; Drew et al., 1980). There are 12 base pairs per helical turn with a helical rotation between repeating units of $-60°$. The base pairs are nearly perpendicular to the helix axis (tilt of 7–9°), and the average axial translation between residues is 3.7 Å or 3.8 Å. The dinucleotide repeat is made up of dC residues having a C-2′-endo sugar pucker and an anti-glycosylic conformation. The dG residues, however, have a C-3'-endo (Wang et al., 1979) or C-1'-exo (Drew et al., 1980) sugar pucker and a syn-glycosylic conformation. This alternation between residues results in a sugar–phosphate backbone which has a zigzag path around the helix (Fig. 3). A deep and narrow minor groove exists, but the "major groove" is actually a convex surface and not a groove.

The major conformational changes within the Z DNA family involve slight modifications in the phosphate geometry and the sugar pucker (Crawford et al., 1980; Wang et al., 1981; Drew and Dickerson, 1981).

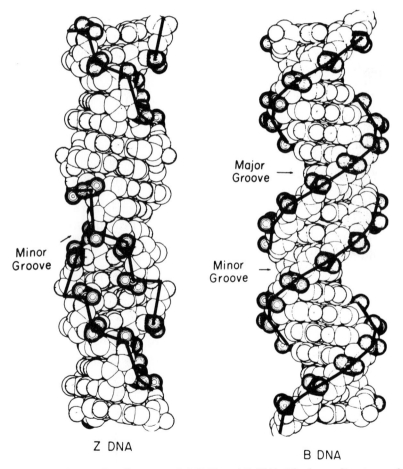

Z DNA

B DNA

Fig. 3. van der Waals side views of Z DNA and B DNA. The heavy line traces the phosphate-to-phosphate path around the helix molecule. Reprinted by permission from *Nature* Vol. 282, pp. 680–686. Copyright © 1979 Macmillan Journals Limited.

Many of these variations can be attributed, at least in part, to differences in associated counterions. Solution studies to be described below also indicate a wide range of structural variability within an overall Z framework. Drew and Dickerson (1981) have suggested that Z DNA possesses a greater degree of independent motion of phosphate groups, sugar residues, and base pairs than found in B DNA.

Crystallographic analysis has also been carried out on $(m^5dC-dG)_3$ (Fujii et al., 1982). The molecule adopted a left-handed duplex structure similar to the unmethylated hexamer. Some interatomic contacts were modified and slight changes in twist angles were found. The twist angle for C-G steps increased from $-8°$ to $-13°$ for the methylated hexamer, whereas that of G-C steps decreased from $-51°$ (unmethylated) to $-46°$ (methylated). The C-3'-endo sugar conformation of the dG residues was flattened when compared to the nonmethylated molecule. The positioning of the methyl groups will be discussed (Section III,B,3,b).

2. Fiber Diffraction

Fiber diffraction studies have followed the single-crystal results in demonstrating the existence of left-handed helices (Leslie et al., 1980; Arnott et al., 1980). Left-handed structures were found for poly(dG-dC)·poly(dG-dC), poly(dA-dC)·poly(dG-dT), and poly(dA-s^{4}dT)·poly-(dA-s^{4}dT) (Arnott et al., 1980), and it was suggested that any alternating purine–pyrimidine sequence can adopt a Z-type structure. The overall conformations consistent with the data showed a good correlation to the left-handed structures defined by single-crystal diffraction analysis.

B. Solution Studies

1. Defining Left-Handed DNA in Solution

In 1972, Pohl and Jovin observed a reversible, intramolecular isomerization for poly(dG-dC)·poly(dG-dC). The R to L transition, as termed by the authors, was brought about by high concentrations of sodium and magnesium salts. The highly cooperative R to L transition was characterized by a red shift in the UV spectrum of the polymer, a decrease in the A_{260}/A_{295} ratio, and a dramatic near inversion of the circular dichroism spectrum. When the crystallographic study of (dC-dG)$_3$ indicated the duplex oligomer was left-handed, the suggestion was made that the left-handed Z DNA structure corresponded to the high salt or L form of poly(dG-dC)·poly(dG-dC) described by Pohl and

Jovin (Wang et al., 1979). It was only a matter of time before this suggestion was corroborated and left-handed DNA was known to exist in solution as well as in the crystalline state.

The correlation between the crystalline and high salt solution structure involved several approaches. The investigations revealed, the high salt form of the alternating copolymer was similar in conformation to the crystalline duplex: The dG residues had a syn-glycosylic conformation (Patel et al., 1979, 1982a), the phosphodiester linkages alternated between two conformations (Patel et al., 1979), and the rise per residue was 3.7 Å (Wu et al., 1981). A more detailed account of this correlation can be found in Zimmerman (1982).

Klysik et al. (1981) carried these studies one step further by demonstrating that Z DNA could exist in solution when contiguous with B DNA. (dC-dG)$_n$, where n = 16,13, regions embedded within nonalternating sequences in restriction fragments and supercoiled plasmids were found to adopt a left-handed helix by a number of criteria.

2. Sequences That Form Left-Handed Helices

From fiber diffraction results it was suggested that any alternating purine–pyrimidine sequence can adopt a Z-type conformation (Arnott et al., 1980). The requirement for an alternating sequence is based on the alternating syn–anti glycosyl conformation of Z DNA. Although purines can adopt a syn conformation fairly readily, the O-2 carboynl oxygen of pyrimidines sterically interferes with the formation of a syn conformation. Thus, it is thought that every other residue must be a purine in order to attain the syn geometry.

Studies of various polymers in solution have demonstrated that many modified alternating dG-dC sequences can adopt a Z-type structure (Section III,B,3,b). Poly(dI-br^{5}dC)·poly(dI-br^{5}dC) has also been shown to adopt a left-handed helix (Patel et al., 1979; Hartmann et al., 1982) as well as (dC-dA)·(dT-dG) sequences (Wells et al., 1982; Vorlickova et al., 1982; Zimmer et al., 1982). It is probably not unreasonable to suggest that the detailed geometry of the left-handed helices formed by these various sequences is diverse and represents several different left-handed conformations, depending upon the particular sequence involved. However, poly(dA-dT)·poly(dA-dT) apparently cannot enter into a Z-type helix in solution (Quadrifoglio et al., 1982; Vorlickova et al., 1980). Wang et al. (1979) have argued that dA·dT substitutions for dG·dC pairs may result in a decreased ability to adopt a Z DNA structure. They propose that a hydrogen-bonded water bridge between the N-2 amino group of guanine and a phosphate oxygen atom, which is not possible with A-T pairs, exhibits a stabilizing influence on Z DNA.

This would be consistent with poly(dA-dT) not adopting a Z-type helix and formation of left-handed poly(dC-dA)·poly(dT-dG) requiring more rigorous conditions than poly(dG-dC) (Section III,B,3,a).

It is also of interest to know the minimum length requirements for Z DNA formation in solution. Quandrifoglio et al. (1981) determined that $(dG-dC)_3$ but not $(dG-dC)_2$ formed Z DNA. In contrast, Reid et al. (1983) were unable to demonstrate a Z-type helix for $(dG-dC)_5$. Ten base pairs of (dG-dC), the shortest stretch examined, when embedded within non-alternating sequences was shown to adopt a left-handed helical structure (Singleton et al., 1983a). Similarly, Nordheim and Rich (1983a) presented evidence indicating left-handed DNA formation for three stretches of alternating purine–pyrimidine sequences of eight base pairs (ATGTGTGT, GCATGCAT, and GCATGCAT) found in SV40 DNA (Section III,B,3,c).

Questions still remain regarding the efficiency of Z DNA formation for various combinations of alternating purine–pyrimidine sequences, interruptions by nonalternating base pairs, and the lengths of each. Nordheim et al. (1982) found that a pBR322 14 base pair sequence of alternating purine–pyrimidine residues with one base pair out of register adopted a left-handed helix. However, it was not known if the entire region or only the nine pairs of strictly alternating sequences were in a left-handed state. Singleton et al. (1983a) demonstrated that a four base pair region of nonalternating sequence, GATC, formed a left-handed helix when flanked by regions of 32 and 26 base pairs of left-handed (dG-dC). Other studies concerning mixtures of such sequences can be found in Quadrifoglio et al. (1982), Peck et al. (1982), and Kilpatrick et al. (1984). These studies have not reached a stage at which general conclusions concerning the effects of interruption of the alternating pattern can be drawn.

Finally, it has been demonstrated that form V DNA (annealed covalently closed, complementary single strands) possesses some regions in a left-handed state (Stettler et al., 1979; Lang et al., 1982; Pohl et al., 1982). The amount of left-handed DNA present was greater than the amount of alternating purine–pyrimidine sequences suggesting that nonalternating sequences were in a left-handed state.

3. Conditions for Obtaining Z DNA

a. Ionic Environment/Dehydrating Agents

The induction of the B to Z transition for poly(dG-dC) was originally brought about by high salt or ethanol concentrations. The midpoint of the cooperative transition occurred at 2.56 M NaCl, 1.67 M $NaClO_4$, and 0.66 M $MgCl_2$ (Pohl and Jovin, 1972) or at 50% ethanol (Pohl, 1976).

More recently, alternative conditions for the B → Z transition have been sought primarily to obtain conditions which were more physiological in nature. Several investigators showed that divalent metal ions in conjunction with dehydrating reagents acted synergistically in promoting the B → Z transition. Solutions of 0.4 mM $MgCl_2$/20% ethanol or 4 mM $MgCl_2$/10% ethanol (van de Sande and Jovin, 1982) and combinations of 0.2 mM $CoCl_2$ or 0.25 mM MnCl with 25% ethylene glycol or 15% ethanol (Zacharias et al., 1982) were found to be effective inducing conditions for Z DNA. Van de Sande et al. (1982) demonstrated that the chloride salts of Mn^{2+}, Co^{2+}, and Ni^{2+} at submillimolar concentrations were efficient at inducing Z DNA without the presence of ethanol. However, the forward reaction had a large activation energy and thus a heating step (>45°C) was required. Based on the circular dichroism end point spectra under the various conditions examined, it was concluded that a family of Z conformations occurs in solution, depending on the ionic environment as well as the presence or absence of various dehydrating reagents (Zacharias et al., 1982; van de Sande et al., 1982).

Behe and Felsenfeld (1981) observed that cobalt hexamine trichloride was a potent inducer of Z DNA. As little as 0.02 mM was required to bring about the B → Z transition. Millimolar levels of spermidine (Behe and Felsenfeld, 1981) and submillimolar amounts of spermine (Behe and Felsenfeld, 1981; Ivanov and Minyat, 1981) were also capable of inducing Z DNA formation.

Polyarginine interacts with poly(dG-dC) to facilitate the B → Z transition (Klevan and Shumaker, 1982). In the presence of 170 mM NaCl, Z DNA was formed at amino acid to nucleotide ratios of greater than 0.5. The results suggested that arginine-rich proteins might facilitate Z DNA formation.

The above studies were performed on the (dG-dC) polymer. Similar studies have been carried out on poly(dC-dA)·poly(dT-dG). Although this polymer does appear to adopt a Z-type structure in solution, the conditions required were much more drastic than those required for poly(dG-dC). Very high concentrations of cesium salts or ethanol solutions promoted the transition (Vorlickova et al., 1982; Zimmer et al., 1982). Many of the conditions found to promote Z DNA formation for poly(dG-dC) (Zacharias et al., 1982) were found to be ineffective for poly(dC-dA)·poly(dT-dG) (Wells et al., 1982). In order to obtain the left-handed state for this polymer, reaction with acetylaminofluorene and high salt solutions was required (Wells et al., 1982).

b. Covalent Modification/Complexation

Several modified polymers of (dG-dC) have been shown to adopt a Z DNA conformation more readily than the unmodified polymer. One of the most dramatic effects as well as perhaps the most physiologically significant modification is found with the 5-methylcytosine derivative of poly(dG-dC) (Behe and Felsenfeld, 1981). Conditions of 0.7 M NaCl, 0.6 mM MgCl$_2$, 20% ethanol, and 0.005 mM Co(NH$_3$)$_6$Cl$_3$ stabilized the Z structure for poly(dG-m^{5}dC). Based on crystallographic results, Fuji et al. (1982) suggest that the stabilization arises from the methyl groups existing in a "hydrophobic pocket" and thus being less accessible to solvent molecules for Z DNA relative to B DNA.

Klysik et al. (1983) investigated the effects of cytosine methylation on the B $\rightarrow$ Z transition for restriction fragments and plasmids containing segments of the methylated alternating copolymer. Although methylation lowered the salt concentration required for the transition, the salt requirements were still higher than the physiological range for NaCl and MgCl$_2$. Thus, it seems that the presence of junctions, which must exist between the B and Z segments, countered some of the enhancing effect of methylation when compared to the polymer (no contiguous B and Z) results. Methylation was found to contribute a favorable free-energy input of 6–8 kcal per (dC-dG) block whereas the unfavorable free energy per junction was 3–5 kcal (Klysik et al., 1983).

Other modifications which facilitate the B $\rightarrow$ Z transition for poly(dG-dC) are 5-bromocytosine (Malfoy et al., 1982), 7-methylguanine (Möller et al., 1981), and 8-bromoguanine (Lafer et al., 1981). Reaction at the C-8 positions of 6–9% of the guanine bases with the carcinogen, N-acetoxy-2-acetylaminofluorene, also promoted Z DNA formation (Sage and Leng, 1980). The polymer with about 20% modification existed as Z DNA in 1 mM phosphate buffer (Sage and Leng, 1981; Santella et al., 1981). Complexation of the polymer with the antitumor drug chlorodiethylenetriamino platinum(II) chloride at a bound Pt-to-base ratio of 0.12 stabilized the Z helix at 0.2 M NaCl (Malfoy et al., 1981).

In contrast, O-6 methylation of guanine brought about a requirement for more rigorous conditions for Z DNA induction (Kuzmich et al., 1983). Similarly, Nordheim et al. (1983) demonstrated that aflatoxin B$_1$ gave a strong inhibition of Z DNA formation, even at levels of binding of one molecule per 42 nucleotides. Previously, it was found that high concentrations of ethidium bromide inhibited left-handed DNA formation (Pohl et al., 1972).

c. Supercoiling

The torsional strain of negative supercoiling was shown to cause the B to Z transition for (dC-dG)$_n$·(dC-dG)$_n$ regions contained in recombinant plasmids (Peck et al., 1982; Singleton et al., 1982; Stirdivant et al.,

1982). The free energy of supercoil formation is unfavorable and increases quadratically with increasing supercoiling (Pulleyblank et al., 1975; Depew and Wang, 1975; reviewed in Wang, 1980, and Bauer, 1978). Thus, processes which remove negative supercoils (analogous to unwinding the two strands) obtain a favorable free-energy input from the loss of supercoil turns. Formation of left-handed DNA from right-handed DNA reduces the twist of the duplex molecule and hence results in a loss of negative supercoils in a topologically closed molecule. The favorable free energy derived from this loss compensates for the unfavorable free-energy difference between the B and Z states, and sufficient levels of superhelicity can drive the equilibrium to the Z, or left-handed, state.

The amount of negative supercoiling necessary to induce and stabilize Z DNA was found to increase with decreasing length of the (dC-dG) regions (Nordheim et al., 1982; Singleton et al., 1983). Superhelical densities in the range -0.03 to -0.07 were sufficient to induce the left-handed conformation for lengths of from 10 to 32 base pairs of CG. These studies have led to estimations for the free-energy difference between the B and Z states, ΔG_{BZ}, of about 0.45 kcal/mole base pair (Nordheim et al., 1982; Klysik et al., 1983; C. K. Singleton, unpublished results).

Negative supercoiling also was found to induce Z DNA in other alternating purine–pyrimidine sequences. The work of Nordheim and Rich (1983a), Nordheim et al. (1982), and Singleton et al. (1983), discussed in Section III,B,2, all involved supercoil-induced Z DNA in non-(dG-dC) sequences. Those non-(dG-dC) sequences which were shown to adopt a left-handed state required higher levels of negative superhelicity in order to attain the Z-like conformation. Similarly, regions of $(dC\text{-}dA)_n \cdot (dT\text{-}dG)_n$ were shown to undergo a supercoil-induced B to Z transition (Nordheim and Rich, 1983b; Haniford and Pulleyblank, 1983; Singleton et al., 1984). More supercoiling was required to bring about the transition than for the $(dC\text{-}dG)_n$ regions. The estimate of ΔG_{BZ} for the $(dC\text{-}dA)_n \cdot (dT\text{-}dG)_n$ sequences gave 0.70 kcal/mole base pair (Singleton et al., 1984), a value significantly higher than that for $(dC\text{-}dG)_n$ (see above).

4. Junctions

Many of the studies described above involve sequences which form a Z-type structure while embedded within and surrounded by sequences which retain the B DNA conformation. The juxtaposition of B and Z DNA necessitates junctional regions (Klysik et al., 1981; Wang et al.,

1979). To date, evidence from several sources suggests the junction region is small and resides mainly outside of the alternating residues (Stirdivant *et al.*, 1982; Peck *et al.*, 1982; Singleton *et al.*, 1982, 1983; Wartell *et al.*, 1982) (However, see Kilpatrick *et al.*, 1984.) The conformational features of the junction regions are of interest, since they represent distinct structures differing from the neighboring conformations.

Singleton *et al.* (1982) demonstrated that the B-Z junction regions were susceptible to cleavage by the single-strand specific S_1 nuclease. Conformational aberrations within these regions were also demonstrated by a loss of cleavage susceptibility to *Bam*HI when the recognition site for this endonuclease resided between B- and Z-type conformations (Singleton *et al.*, 1983). The junction regions showed conformational flexibility and heterogeneity in a sequence and supercoil-dependent manner (Singleton *et al.*, 1983, 1984). The heterogeneity was revealed by variations in susceptibility of different junctions to cleavage by single-strand specific nucleases. These investigations suggest that junction regions between B and Z DNA are conformationally heterogeneous and may have some single-stranded character.

Estimates of the junction free energy exist and lie in the range of 3–8 kcal/mole junction (Klysik *et al.*, 1983; Nordheim *et al.*, 1982; Singleton *et al.*, 1983a; Azorin *et al.*, 1983). Thus, formation of B-Z junctions is rather costly compared to the free energy between the B and Z states (see above). Azorin *et al.* (1983) demonstrated that the junctions were sensitive to the ionic environment such that they were more costly at higher salt concentrations.

IV. CRUCIFORMS AT INVERTED REPEATS

Inverted repeat sequences appear abundantly in what are thought to be biologically important regions—promoters, operators, transcription termination sites, replication origins, and others (Wells *et al.*, 1980; Müller and Fitch, 1982; Fitch, 1983; Day and Blake, 1982). Early on, it was recognized that inverted repeats have the potential to form a double hairpin or cruciform structure as well as the linear duplex form (Platt, 1955; Gierer, 1966). Only recently has direct evidence been obtained for the existence of the cruciform state. Gellert *et al.* (1978) and Mizuuchi *et al.* (1982) presented evidence for cruciform structures in a completely palindromic plasmid molecule. Lilley (1980, 1981) and Panayotatos and Wells (1981) found that single-strand specific nucleases cleaved within the centers (potential loops) of certain small inverted repeats of various plasmids. Cleavage was interpreted to be a result of recognition of the central loops by the nucleases when the inverted repeat adopted a cruciform structure. This interpretation has been verified by use of single-strand specific chemical agents (Lilley, 1983), supercoil relaxation studies (Lyamichev *et al.*, 1983; Courey and Wang, 1983; Gellert *et al.*, 1983), and psoralen–DNA cross-linking experiments (Sinden *et al.*, 1983).

All of these investigations have shown that the cruciform state at inverted repeats exists only in negatively supercoiled DNA. Supercoiling stabilizes cruciforms since cruciform extrusion results in an unwinding of the DNA strands, i.e., a loss of negative supercoil turns in a topologically closed molecule (see Section III,B,3,c). The detailed relationship between superhelicity and cruciform formation at several inverted repeats (from 20 to 68 base pairs) has been examined (Singleton and Wells, 1982; Sinden *et al.*, 1983; Singleton, 1983; Courey and Wang, 1983; Gellert *et al.*, 1983). The results demonstrate that cruciform induction by supercoiling is a highly cooperative process with each inverted repeat having a characteristic superhelical density at the transition midpoint (within the range of -0.04 to -0.08). Singleton (1983) found a sharp dependency on stem length for cruciform formation at a region with a constant loop size but variable stem size. A change in stem length from 13 to 10 base pairs resulted in no apparent change in stability, whereas a further reduction to 7 base pairs completely abolished the ability of the inverted repeat to adopt the cruciform state. However, not enough systematic studies exist so that rules can be proposed for correlations between stability, stem and loop size, and levels of supercoiling required for induction of the cruciform state.

The influence of salt concentration, salt type, and temperature on the supercoil induced formation of several cruciforms has been examined (Singleton, 1983). The general conclusions reached were that conditions which stabilize duplex DNA over single-stranded DNA (increasing salt, decreasing temperature) resulted in a relative destabilization of the cruciform state. Some evidence was obtained for stabilization of specific cruciforms by certain metal ions.

Examination of the kinetics of cruciform formation has shown the process to be slow and dependent on temperature and level of supercoiling (Courey and Wang, 1983; Gellert *et al.*, 1983; Sinden *et al.*, 1983). The resorption rate was also slow (Gellert *et al.*, 1983; Mizuuchi *et al.*, 1982). The results suggested that at physiological temperatures the cruciform state may be kinetically forbidden even though it is thermodynamically stable. Gellert *et al.* (1983) estimated the heat of activation for a 48 base pair potential cruciform at about 50 kcal/mol.

From the relationship between supercoiling and cruciform extrusion, estimates of the free energy of cruciform generation have been made. The unfavorable aspects of the cruciform state are the unpairing of several base pairs within the loop regions and any perturbations of duplex structure at the ends of the stem–junction region. Values of 17 and 18 kcal/mole were obtained for cruciforms with a potential four base loop of all A-T pairs (Courey and Wang, 1983) and two G-C and two A-T pairs (Gellert et al., 1983), respectively. Steric constraints impose a minimum of three unpaired bases within a loop, and thus the 17 kcal/mole value of Courey and Wang represents a near minimal estimate of the ΔG of cruciform generation. A value of 22 kcal/mole was obtained for a cruciform with a loop requiring disruption of four A-T and one G-C pair (Singleton and Wells, 1982). Finally, for a three base loop containing cruciform disrupting one A-T and two G-C pairs, an estimate of 20 kcal/mol was obtained (the stem of this cruciform may contain a C-T and A-G mispairing) (Singleton, 1983).

V. SUMMARY

Throughout this review, I have discussed recent investigations of duplex DNA structure with emphasis on environmental- and sequence-dependent conformational heterogeneity. The material covered readily demonstrates the abundance of DNA structural polymorphisms. Although much progress has been made and much has been learned in the past few years, many questions remain including those concerning the fundamental principles behind much of this polymorphism. No effort has been made herein to correlate DNA structure with its involvement in gene regulation and cell functioning. Studies concerning this correlation include investigations of transcription, replication, recombination, transposition, protein–nucleic acid interactions, chromatin structure, and others (see Chapter 13). A myriad of ways can be thought of, even with a dull imagination, for how the cellular machinery involved in these processes can capitalize on the types of polymorphism discussed. The work described here thus serves as a foundation for relating DNA structure with function as well as serving as a stepping stone toward the understanding of these processes.

ACKNOWLEDGMENTS

The author's work was supported in part by the American Cancer Society (PF-1904) and the National Institutes of Health (GM09260-02).

REFERENCES

Arnott, S., and Hukins, D. W. L. (1972). Optimized parameters for A-DNA and B-DNA. *Biochem. Biophys. Res. Commun.* **47**, 1504–1510.

Arnott, S., Chandrasekaren, R., Birdsall, D. L., Leslie, A. G. W., and Ratliff, R. L. (1980). Left-handed DNA helices. *Nature (London)* **283**, 743–745.

Arnott, S., Chandrasekaren, R., Hall, I. H., and Puigjaner, L. C. (1983a). Fiber diffraction of poly d(A)·d(T). *Nucleic Acids Res.* **11**, 4141–4155.

Arnott, S., Chandrasekaren, R., Puigjaner, L. C., Walker, J. K., Hall, I. H., Birdsall, D. L., and Ratliff, R. L. (1983b). Wrinkled DNA. *Nucleic Acids Res.* **11**, 1457–1474.

Azorin, F., Nordheim, A., and Rich, A. (1983). Formation of Z-DNA in negatively super-coiled plasmids is sensitive to small changes in the salt concentration within the physiological range. *EMBO J.* **2**, 649–555.

Bauer, W. R. (1978). Structure and reactions of closed duplex DNA. *Annu. Rev. Biophys. Bioeng.* **7**, 287–313.

Behe, M., and Felsenfeld, G. (1981). Effects of methylation on a synthetic polynucleotide: The B-Z transition in poly(dG-m^{5}dC)·poly(dG-m^{5}dC). *Proc. Natl. Acad. Sci. U.S.A.* **78**, 1619–1623.

Calladine, C. R. (1982). Mechanics of sequence-dependent stacking of bases in B-DNA. *J. Mol. Biol.* **161**, 343–352.

Charney, E., and Yamaoka, K. (1982). Electric dichroism of deoxyribonucleic acid in aqueous solutions: Electric field dependence. *Biochemistry* **21**, 834–842.

Chen, C. W., Cohen, J. S., and Behe, M. (1983). B to Z transition of double-stranded poly[deoxyguanylyl(3'-5')-5-methyldeoxycytidine] in solution by phosphorus-31 and carbon-13 nuclear magnetic resonance spectroscopy. *Biochemistry* **22**, 2136–2142.

Chen, H. H., Charney, E., and Rau, D. C. (1982). Length changes in solution accompany-ing the B-Z transition of poly(dG-m^{5}dC) induced by Co $(NH_3)_6^{3+}$. *Nucleic Acids Res.* **10**, 3561–3571.

Conner, B. N., Takano, T., Tanaka, S., Itakura, K., and Dickerson, R. E. (1982). The molecular structure of d(^{I}CpCpGpG), a fragment of right-handed double helical A-DNA. *Nature (London)* **295**, 294–299.

Courey, A. J., and Wang, J. C. (1983). Cruciform formation in a negatively supercoiled DNA may be kinetically forbidden under physiological conditions. *Cell (Cambridge, Mass.)* **33**, 817–829.

Crawford, J. K., Kolpak, F. L., Wang, A. H.-J., Quigley, G. J., van Boom, J. H., van der Marel, G., and Rich, A. (1980). *Proc. Natl. Acad. Sci. U.S.A.* **80**, 4015–4020.

Day, G. R., and Blake, R. D. (1982). Statistical significance of symmetrical and repetitive segments in DNA. *Nucleic Acids Res.* **10**, 8323–8339.

Depew, R. E., and Wang, J. C. (1975). Conformational fluctuations of DNA helix. *Proc. Natl. Acad. Sci. U.S.A.* **72**, 4275–4279.

Dickerson, R. E. (1983). Base sequence and helix structure variation in B and A DNA. *J. Mol. Biol.* **166**, 419–441.

Dickerson, R. E,, and Drew, H. R. (1981a). Structure of a B DNA dodecamer. II. Influence of base sequence on helix structure. *J. Mol. Biol.* **149**, 761–786.

Dickerson, R. E., and Drew, H. R. (1981b). Kinematic model for B-DNA. *Proc. Natl. Acad. Sci. U.S.A.* **78**, 7318–7322.

Dickerson, R. E., Drew, H. R., Conner, B. N., Wing, R. M., Fratini, A. V., and Kopka, M. L. (1982). The anatomy of A-, B-, and Z-DNA. *Science* 216, 475–485.

Drew, H. R., and Dickerson, R. E. (1981). Conformation and dynamics in a Z-DNA tetramer. *J. Mol. Biol.* **152**, 723–736.

Drew, H. R., Takano, T., Tanaka, S., Itakura, K., and Dickerson, R. E. (1980). High-salt d(CpGpCpG), a left-handed Z' DNA double helix. *Nature (London)* **286**, 567–572.

Drew, H. R., Wing, R. W., Takano, T., Broka, C., Tanaka, S., Itakura, K., and Dickerson, R. E. (1981). Structure of a B-DNA dodecamer: Conformation and dynamics. *Proc. Natl. Acad. Sci. U.S.A.* **78**, 2179–2183.

Feigon, J., Wright, J. M., Leupin, W., Denny, W. A., and Kearns, D. R. (1982). Use of two-dimensional NMR in the study of a double stranded DNA decamer. *J. Am. Chem. Soc.* **104**, 5540–5541.

Fitch, W. M. (1983). Calculating the expected frequencies of potential secondary structure in nucleic acids as a function of stem length, loop size, base composition, and nearest-neighbor frequencies. *Nucleic Acids Res.* **11**, 4655–4663.

Fratini, A. V., Kopka, M. L., Drew, H. R., and Dickerson, R. E. (1982). Reversible bending and helix geometry in a B-DNA dodecamer: CGCGAATT^{Br}CGCG. *J. Biol. Chem.* **257**, 14608–14707.

Fuji, S., Wang, A. H.-J., van der Marel, G., van Boom, J. H., and Rich, A. (1982). Molecular structure of $(m^5dC\text{-}dG)_3$: The role of the methyl group of 5-methylcytosine in stabilizing Z-DNA. *Nucleic Acids Res.* **10**, 7879–7892.

Gellert, M., Mizuuchi, K., O'Dea, M., Ohmori, H., and Tomizawa, J. (1978). DNA gyrase and supercoiling. *Cold Spring Harbor Symp. Quant. Biol.* **43**, 35–40.

Gellert, M., O'Dea, M. H., and Mizuuchi, K. (1983). Slow cruciform transitions in palindromic DNA. *Proc. Natl. Acad. Sci. U.S.A.* **80**, 5545–5549.

Gierer, A. (1966). Model for DNA and protein interaction and the function of the operator. *Nature (London)* **212**, 1480–1481.

Haniford, D. B., and Pulleyblank, D. E. (1983). Facile transition of poly[d(TG)·d(CA)] into a left-handed helix in physiological conditions. *Nature (London)* **302**, 632–634.

Hartmann, B., Pilet, J., Ptak, M., Ramstein, J., Malfoy, B., and Leng, M. (1982). The B → Z transition of poly(dI-br^{5}dC)·poly(dI-br^{5}dC). A quantitative description of the Z form dynamic structure. *Nucleic Acids Res.* **10**, 3261–3277.

Hogan, M., Dattagupta, N., and Crothers, D. M. (1978). Transient electric dichroism of rod-like DNA molecules. *Proc. Natl. Acad. Sci. U.S.A.* **75**, 195–199.

Ivanov, V. I., and Minyat, E. E. (1981). The transition between left- and right-handed forms of poly(dG-dC). *Nucleic Acids Res.* **9**, 4783–4798.

Kan, L.-S., Cheng, D. M., Jayaraman, K., Leutzinger, E. E., Miller, P. S., and Ts'o, P. O. P. (1982). Proton nuclear magnetic resonance study of a self-complementary decadeoxyribonucleotide, C-C-A-A-G-C-T-T-G-G-. *Biochemistry* **21**, 6723–6732.

Kan, L.-S., Chandrasegaran, S., Pulford, S. M., and Miller, P. S. (1983). Detection of a guanine-adenine base pair in a decadeoxyribonucleotide by proton magnetic resonance spectroscopy. *Proc. Natl. Acad. Sci. U.S.A.* **80**, 4263–4265.

Kilpatrick, M. W., Klysik, J., Singleton, C. K., Zarling, D. A., Jovin, T. M., Hanau, L. H., Erlanger, B. F., and Wells, R. D. (1984). Intervening sequences in human fetal globin genes adopt left-handed Z helices. *J. Biol Chem.* **259**, 7268–7274.

Klevan, L., and Shumaker, V. N. (1982). Stabilization of Z-DNA by polyarginine near physiological ionic strength. *Nucleic Acids Res.* **10**, 6809–6817.

Klug, A., Jack, A., Viswamitra, M. A., Kennard, O., Shakked, Z., and Steitz, T. A. (1979). A hypothesis on a specific sequence-dependent conformation of DNA and its relation to the binding of the *lac*-repressor protein. *J. Mol. Biol.* **131**, 669–680.

Klysik, J., Stirdivant, S. M., Larson, J. E., Hart, P. A., and Wells, R. D. (1981). Left-handed DNA in restriction fragments and recombinant plasmid. *Nature (London)* **290**, 672–677.

Klysik, J., Stirdivant, S. M., Singleton, C. K., Zacharias, W., and Wells, R. D. (1983).

Effects of 5-cytosine methylation on the B-Z transition in DNA restriction fragments and recombinant plasmids. *J. Mol. Biol.* **168**, 51–71.

Kuzmich, S., Marky, L. A., and Jones, R. A. (1983). Specifically alkylated DNA fragments. Synthesis and physical characterization of d[CGC(O⁶Me)GCG] and d(CGT(O⁶Me)GCG]. *Nucleic Acids Res.* **11**, 3393–3404.

Lafer, E. M., Möller, A., Nordheim, A., Stollar, B. D., and Rich, A. (1981). Antibodies specific for left-handed DNA. *Proc. Natl. Acad. Sci. U.S.A.* **78**, 3546–3550.

Lang, M. C., Malfoy, B., Freund, A. M., and Daune, M., and Leng, M. (1982). Visualization of Z sequences in form V of pBR322 by immuno-electron microscopy. *EMBO J.* **1**, 1149–1153.

Langridge, R., Marvin, D. A., Seeds, W. E., Wilson, H. R., Hooper, C. W., Wilkins, M. H. F., and Hamilton, C. D. (1960). The molecular configuration of deoxyribonucleic acid. *J. Mol. Biol.* **2**, 38–64.

Lee, C.-H., Mizusawa, H., and Kakefuda, T. (1981). Unwinding of double-stranded DNA helix by dehydration. *Proc. Natl. Acad. Sci. U.S.A.* **78**, 2838–2842.

Leslie, A. G., Arnott, S., Chandrasekaran, R., and Ratliff, R. L. (1980). Polymorphism of DNA double helices. *J. Mol. Biol.* **143**, 49–72.

Levitt, M. (1978). How many base pairs per turn does DNA have in solution and in chromatin? Some theoretical calculations. *Proc. Natl. Acad. Sci. U.S.A.* **75**, 640–644.

Lilley, D. M. J. (1980). The inverted repeat as a recognizable structural feature in super-coiled DNA molecules. *Proc. Natl. Acad. Sci. U.S.A.* **77**, 6468–6472.

Lilley, D. M. J. (1981). Hairpin-loop formation by inverted repeats in supercoiled DNA is a local and transmissible property. *Nucleic Acids Res.* **9**, 1271–1289.

Lilley, D.-M. J. (1983). Cruciform detection with bromoacetaldehyde. *Nucleic Acids Res.* **11**, 3097–3112.

Lomonossoff, G. P., Butler, G. J. G., and Klug, A. (1981). Sequence-dependent variation in the conformation of DNA. *J. Mol. Biol.* **149**, 745–760.

Lyamichev, V. I., Panyutin, I. G., and Frank-Kamenetskii, M. D. (1983). Evidence of cruciform structures in superhelical DNA provided by two-dimensional gel electrophoresis. *FEBS Lett.* **153**, 298–302.

Mahendrasigam, A., Rhodes, N. J., Goodwin, D. C., Nave, C., Pigram, W. J., Fuller, W., Brams, J., and Vergne, J. (1983). Conformational transitions in oriented fibers of the synthetic polynucleotide poly[d(A-T)]·poly[d(A-T)] double helix. *Nature (London)* **301**, 535–537.

Malfoy, B., Hartmann, B., and Leng, M. (1981). The B→Z transition of poly(dG-dC)·poly(dG-dC) modified by some platinum derivatives. *Nucleic Acids Res.* **9**, 5659–5669.

Malfoy, B., Rousseau, N., and Leng, M. (1982). Interaction between antibodies to Z-form deoxyribonucleic acid and double-stranded polynucleotides. *Biochemistry* **21**, 5463–5467.

Mandelkern, M., Elias, J. G., Eden, D., and Crothers, D. M. (1981). The dimensions of DNA in solution. *J. Mol. Biol.* **152**, 153–161.

Mitra, C. K., Sarma, M. H., and Sarma, R. H. (1981). Plasticity of the DNA double helix. *J. Am. Chem. Soc.* **103**, 6727–6737.

Mizuuchi, K., Mizuuchi, M., and Gellert, M. (1982). Cruciform structures in palindromic DNA are favored by DNA supercoiling. *J. Mol. Biol.* **156**, 229–243.

Möller, A., Nordheim, A., Nichols, S. R., and Rich, A. (1981). 7-Methylguanine in poly(dG-dC)·poly(dG-dC) facilitates Z-DNA formation. *Proc. Natl. Acad. Sci. U.S.A.* **78**, 4777–4781.

Müller, U. R., and Fitch, W. M. (1982). Evolutionary selection for perfect hairpin structures in viral DNAs. *Nature (London)* **298**, 582–585.

Nordheim, A., and Rich, A. (1983a). Negatively supercoiled simian virus 40 DNA contains Z-DNA segments within transcriptional enhancer sequences. *Nature (London)* **303**, 674–679.

Nordheim, A., and Rich, A. (1983b). The sequence $(dC-dA)_n \cdot (dG-dT)_n$ forms a left-handed Z-DNA in negatively supercoiled plasmids. *Proc. Natl. Acad. Sci. U.S.A.* **80**, 1821–1825.

Nordheim, A., Lafer, E. M., Peck, L. J., Wang, J. C., Stollar, B. D., and Rich, A. (1982). Negatively supercoiled plasmids contain left-handed Z-DNA segments as determined by specific antibody binding. *Cell (Cambridge, Mass.)* **31**, 309–318.

Nordheim, A., Hao, W. M., Wogan, G. N., and Rich, A. (1983). Salt-induced conversion of B-DNA to Z-DNA inhibited by aflatoxin B_1. *Science* **219**, 1434–1436.

Panayotatos, N., and Wells, R. D. (1981). Cruciform structures in supercoiled DNA. *Nature (London)* **289**, 466–470.

Pardi, A., Martin, F. H., and Tinoco, I. (1981). Comparative study of ribonucleotide, deoxyribonucleotide, and hybrid oligonucleotide helices by nuclear magnetic resonance. *Biochemistry* **20**, 3986–3996.

Pardi, A., Morden, K. M., Patel, D. J., and Tinoco, I. (1982). Kinetics for exchange of imino protons in the (C-G-C-G-A-A-T-T-C-G-C-G) double helix and in two similar helices that contain a GT base pair, d(C-G-T-G-A-A-T-T-C-G-C-G), and an extra adenine, d(C-G-C-A-G-A-A-T-T-C-G-C-G). *Biochemistry* **21**, 6567–6574.

Patel, D. J., Canuel, L. L., and Pohl, R. M. (1979). Alternating B-DNA conformation for the oligo(dG-dC) duplex in high salt solution. *Proc. Natl. Acad. Sci. U.S.A.* **76**, 2508–2511.

Patel, D. J., Kozlowski, S. A., Nordheim, A., and Rich, A. (1982a). Right-handed and left-handed DNA: Studies of B- and Z-DNA by using proton nuclear Overhauser effect and ^{31}P NMR. *Proc. Natl. Acad. Sci. U.S.A.* **79**, 1413–1417.

Patel, D. J., Pardi, A., and Itakura, K. (1982b). DNA conformation, dynamics, and interactions in solution. *Science* **216**, 581–590.

Patel, D. J., Kozlowski, S. A., Marky, L. A., Broka, C., Rice, J. A., Itakura, K., and Breslauer, K. J. (1982c). Premelting and melting transitions in the d(C-G-C-G-A-A-T-T-C-G-C-G) self-complementary duplex in solution. *Biochemistry* **21**, 428–436.

Patel, D. J., Kozlowski, S. A., Marky, L. A., Rice, J. A., Broka, C., Dallas, J., Itakura, K., and Breslauer, K. J. (1982d). Structure, dynamics, and energetics of deoxyguanosine·thymidine wobble base pair formation in the self-complementary d(C-G-T-G-A-A-T-T-C-G-C-G) duplex in solution. *Biochemistry* **21**, 437–444.

Patel, D. J., Kozlowski, S. A., Marky, L. A., Rice, J. A., Broka, C., Itakura, K., and Breslauer, K. J. (1982e). Extra adenosine stacks into the self-complementary d(C-G-C-A-G-A-A-T-T-C-G-C-G) duplex in solution. *Biochemistry* **21**, 445–451.

Patel, D. J., Kozlowski, S. A., and Bhatt, R. (1983). Sequence dependence of base pair stacking in right-handed DNA in solution: Proton nuclear Overhauser effect NMR measurements. *Proc. Natl. Acad. Sci. U.S.A.* **80**, 3908–3912.

Peck, L. J., and Wang, J. C. (1981). Sequence dependence of the helical repeat of DNA in solution. *Nature (London)* **292**, 375–378.

Peck, L. J., Nordheim, A., Rich, A., and Wang, J. C. (1982). Flipping of cloned $d(pCpG)_n \cdot d(pCpG)_n$ DNA sequences from right- to left-handed helical structure by salt, Co(III), or negative supercoiling. *Proc. Natl. Acad. Sci. U.S.A.* **79**, 4560–4564.

Platt, J. R. (1955). Possible separation of intertwined nucleic acid chains by transfer twist. *Proc. Natl. Acad. Sci. U.S.A.* **41**, 181–183.

Pohl, F. M. (1976). Polymorphism of a synthetic DNA in solution. *Nature (London)* **260**, 365–366.

Pohl, F. M., and Jovin, T. M. (1972). Salt-induced cooperative conformational change of a synthetic DNA: Equilibrium and kinetic studies with poly(dG-dC). *J. Mol. Biol.* **67**, 375–396.

Pohl, F. M., Jovin, T. M., Baehr, W., and Holbrook, J. J. (1972). Ethidium bromide as a cooperative effector of a DNA structure. *Proc. Natl. Acad. Sci. U.S.A.* **69**, 3805–3809.

Pohl, F. M., Thomae, R., and DiCapua, E. (1982). Antibodies to Z-DNA interact with form V DNA. *Nature (London)* **300**, 545–546.

Pulleyblank, D. E., Shure, M., Tang, D., Vinograd, J., and Vosberg, H. (1975). Action of nicking-closing enzyme on supercoiled and nonsupercoiled closed circular DNA: Formation of a Boltzmann distribution of topological isomers. *Proc. Natl. Acad. Sci. U.S.A.* **72**, 4280–4284.

Quadrifoglio, F., Manzini, G., Vasser, M., Dinkelspiel, K., and Crea, R. (1981). Conformational stability of alternating d(CG) oligomers in high salt solution. *Nucleic Acids Res.* **9**, 2195–2206.

Quadrifoglio, F., Manzini, G., Dinkelspiel, K., and Crea, R. (1982). Simultaneous stability of short alternating Z and B helices in synthetic DNA concatemers. *Nucleic Acids Res.* **10**, 3759–3768.

Reid, D. G., Salisbury, S. A., and Williams, D. H. (1983). A ^{1}H nOe and CD study of the salt concentration dependence of the structure of d(G-C)$_3$. *Nucleic Acids Res.* **11**, 3779–3793.

Rhodes, D., and Klug, A. (1981). Sequence-dependent helical periodicity of DNA. *Nature (London)* **292**, 378–380.

Sage, E., and Leng, M. (1980). Conformation of poly(dG-dC)·poly(dG-dC) modified by the carcinogens N-acetoxy-N-acetyl-2-aminofluorene and N-hydroxy-N-2-aminofluorene. *Proc. Natl. Acad. Sci. U.S.A.* **77**, 4597–4601.

Sage, E., and Leng, M. (1981). Conformational changes in poly(dG-dC)·poly(dG-dC) modified by the carcinogen N-acetoxy-N-acetyl-2-aminofluorene. *Nucleic Acids Res.* **9**, 1241–1250.

Sanderson, M. R., Mellema, J.-R., van der Marel, G. A., Wille, G., van Boom, J. H., and Altona, C. (1983). NMR on GGC^{m}C^{m}GGCC. *Nucleic Acids Res.* **11**, 3333–3346.

Santella, R. M., Grunberger, D., Weinstein, I. B., and Rich, A. (1981). Induction of the Z conformation in poly(dG-dC)·poly(dG-dC) by binding of N-2-acetylaminofluorene to guanine residues. *Proc. Natl. Acad. Sci. U.S.A.* **78**, 1451–1455.

Shakked, Z., Rabinovich, D., Kennard, O., Cruse, W. B. T., Salisbury, S. A., and Viswamitra, M. A, (1983). Sequence-dependent conformation of an A-DNA double helix: The crystal structure of the octamer d(G-G-T-A-T-A-C-C). *J. Mol. Biol.* **166**, 183–201.

Sinden, R. R., Broyles, S. S., and Pettijohn, D. E. (1983). Perfect palindromic *lac* operator DNA exists as a stable cruciform structure in supercoiled DNA *in vitro* but not *in vivo*. *Proc. Natl. Acad. Sci. U.S.A.* **80**, 1797–1801.

Singleton, C. K. (1983). Effects of salt, temperature, and stem length on supercoil induced formation of cruciforms. *J. Biol. Chem.* **258**, 7661–7668.

Singleton, C. K., and Wells, R. D. (1982). Relationship between superhelical density and cruciform formation in plasmid pVH51. *J. Biol. Chem.* **257**, 6292–6295.

Singleton, C. K., Klysik, J., Stirdivant, S. M., and Wells, R. D. (1982). Left-handed Z-DNA is induced by supercoiling in physiological ionic conditions. *Nature (London)* **299**, 312–316.

Singleton, C. K., Klysik, J., and Wells, R. D. (1983). Conformational flexibility of junctions between contiguous B- and Z-DNA in supercoiled plasmids. *Proc. Natl. Acad. Sci. U.S.A.* **80**, 2447–2451.

Singleton, C. K., Kilpatrick, M. W., and Wells, R. D. (1984). S_1 nuclease recognizes conformational junctions between left-handed helical $(dT-dG)_n \cdot (dC-dA)_n$ and contiguous right-handed sequences. *J. Biol. Chem.* **259**, 1963–1967.

Stettler, U. H., Weber, H., Koller, T., and Weissman, (1979). Preparation and characterization of form V DNA, the duplex DNA resulting from association of complementary, circular single-stranded DNA. *J. Mol. Biol.* **131**, 21–40.

Stirdivant, S. M., Klysik, J., and Wells, R. D. (1982). Energetic and structural interrelationship between DNA supercoiling and the right- to left-handed Z helix transition in recombinant plasmids. *J. Biol. Chem.* **257**, 10159–10165.

Strauss, F., Gaillerd, C., and Prunell, A. (1981). Helical periodicity of DNA, poly(dA)·poly(dT), and poly(dA-dT)·poly(dA-dT) in solution. *Eur. J. Biochem.* **118**, 215–222.

Tran-Dihn, S., Neuman, J.-M., Taboury, J., Huynh-Dihn, T., Renous, S., Genissiel, B., and Igolen, J. (1983). DNA fragment conformations. *Eur. J. Biochem.* **133**, 579–589.

van de Sande, J. H., and Jovin, T. M. (1982). Z*DNA, the left-handed helical form of poly[d(G-C)] in MgCl$_2$-ethanol, is biologically active. *EMBO J.* **1**, 115–120.

van de Sande, J. H., McIntosh, L. P., and Jovin, T. M. (1982). Mn^{2+} and other transition metals at low concentrations induce the right-to-left helical transformation of poly[d(G-C)]. *EMBO J.* **1**, 777–782.

Viswamitra, M. A., Shakked, Z., Jones, P. G., Sheldrick, G. M., Salisbury, S. A., and Kennard, O. (1982). Structure of the deoxytetranucleotide d-pApTpApT and a sequence-dependent model for poly(dA-dT). *Biopolymers* **21**, 513–533.

Vorlickova, M., Kypr, J., Kleinwachter, V., and Palacek, E. (1980). Salt-induced conformational changes of poly(dA-dT). *Nucleic Acids Res.* **8**, 3965–3973.

Vorlickova, M., Kypr, J., Stokrova, S., and Sponar, J. (1982). A Z-like form of poly(dA-dC)·poly(dG-dT) in solution? *Nucleic Acids Res.* **10**, 1071–1080.

Wang, A. H.-J., Quigley, G. J., Kolpate, F. J., Crawford, J. L., van Boom, J. H., van der Marel, G., and Rich, A. (1979). Molecular structure of a left-handed double helical DNA fragment at atomic resolution. *Nature (London)* **282**, 680–686.

Wang, A. H.-J., Quigley, G. J., Kolpak, F. J., van der Marel, G., van Boom, J. A., and Rich, A. (1981). Left-handed double helical DNA: Variations in backbone conformation. *Science* **211**, 171–176.

Wang, A. H.-J., Fujii, S., van Book, J. H., and Rich, A. (1982). Molecular structure of the octamer d(GGCCGGCC): Modified A-DNA. *Proc. Natl. Acad. Sci. U.S.A.* **79**, 3968–3972.

Wang, A. H.-J., Fujii, S., van Boom, J. H., van der Marel, G. A., van Boeckel, S. A. A., and Rich, A. (1983). Molecular structure of r(GCG)d(TATACGC): A DNA–RNA hybrid helix joined to double helical DNA. *Nature (London)* **299**, 601–604.

Wang, J. C. (1980). Superhelical DNA. *Trends Biochem. Sci.* **5**, 219–221.

Wartell, R. M., Klysik, J., Hillen, W., and Wells, R. D. (1982). Junction between Z and B conformations in a DNA restriction fragment: Evaluation by Raman spectroscopy. *Proc. Natl. Acad. Sci. U.S.A.* **79**, 2549–2553.

Watson, J. D., and Crick, F. H. C. (1953). A structure for desoxyribose nucleic acid. *Nature (London)* **171**, 737–738.

Wells, R. D., Goodman, T. C., Hillen, W., Horn, G. T., Klein, R. D., Larson, J. E., Müller, U. R., Neuendorf, S. K., Panayotatos, N., and Stirdivant, S. M. (1980). DNA structure and gene regulation. *Prog. Nucleic Acid Res. Mol. Biol.* **24**, 167–267.

Wells, R. D., Miglietta, J. J., Klysik, J., Larson, J. E., Stirdivant, S. M., and Zacharias, W. (1982). Spectroscopic studies on acetylaminofluorene-modified $(dT-dG)_n \cdot (dC-dA)_n$ suggest a left-handed conformation. *J. Biol. Chem.* **257**, 10166–10171.

Wing, R., Drew, H., Takano, T., Broka, C., Tanaka, S., Itakura, K. and Dickerson, R. E. (1980). Crystal structure analysis of a complete turn of B-DNA. *Nature (London)* **287**, 755–758.

Wu, H. M., Dattagupta, N., and Crowthers, D. M. (1981). Solution structural studies of the A and Z forms of DNA. *Proc. Natl. Acad. Sci. U.S.A.* **78**, 6808–6811.

Zacharias, W., Larson, J. E., Klysik, J., Stirdivant, S. M., and Wells, R. D. (1982). Conditions which cause the right-handed to left-handed DNA conformational transition: Evidence for several types of left-handed DNA structures in solution. *J. Biol. Chem.* **257**, 2775–2782.

Zimmer, C., Tymen, S., March, C., and Guschlbauer, W. (1982). Conformational transitions of poly(dA-dC)·poly(dG-dT) induced by high salt or in ethanolic solution. *Nucleic Acids Res.* **10**, 1081–1091.

Zimmerman, S. B. (1982). The three-dimensional structure of DNA. *Annu. Rev. Biochem.* **51**, 395–427.

15

Initiation of DNA Replication in Eukaryotes

ROBERT M. BENBOW, MICHELLE F. GAUDETTE,
PAMELA J. HINES,[1] AND MASAKI SHIODA

Department of Biology
The Johns Hopkins University
Baltimore, Maryland
and
Department of Zoology
Iowa State University
Ames, Iowa

I. INTRODUCTION

DNA replication is one of the most fundamental cellular processes. The molecular structures of DNA and DNA replication intermediates were described, more or less correctly, more than 30 years ago (Watson and Crick, 1953a,b; see Chapter 14). It is somewhat surprising, therefore, that little progress has been made in defining the molecular mechanisms of chromosomal DNA replication in higher eukaryotes. By contrast, the mechanisms used for replication of prokaryotic DNA (reviewed by Kornberg, 1980, 1982), mitochondrial DNA (reviewed by

---

[1]Present address: Zoology Department, University of Washington, Seattle, Washington 98195.

449

Clayton, 1982), viral DNA (reviewed by DePamphilis and Wasserman, 1980, and Challberg and Kelly, 1982), and the chromosomal DNA of lower eukaryotes (reviewed by Campbell, 1983) have been described in considerable detail.

In the 15 years since the pioneering work of Huberman and Riggs (1968), DNA replication in eukaryotes generally has been described using the concepts and vocabulary of prokaryotic DNA replication (e.g., see Edenberg and Huberman, 1975; Hand, 1978). Replicons, defined origins of DNA replication, replication forks, Okazaki fragments, these and numerous other concepts of prokaryotic DNA replication have been appropriated, often without convincing evidence, to describe eukaryotic DNA replication.

Recently, it has become apparent that some of these concepts must be modified to describe chromosomal DNA replication in higher eukaryotes. The replicon, for example, originally designated a continuous, usually circular sequence of DNA under the positive control of a single origin of DNA replication (Jacob *et al.*, 1963). Replicons were explicitly defined as units which could only replicate as a whole, such as bacterial chromosomes, F factors, and bacteriophage genomes. The concept of a replicon necessarily becomes ambiguous when applied to the contiguous replicating units of eukaryotic chromosomal DNA. A replicon is particularly difficult to define during embryogenesis in higher eukaryotes when sequences of DNA which may function as independent replicons early in development no longer do so later in development (Callan, 1972; Blumenthal *et al.*, 1974).

Similarly, the idea of mandatory sequence-specific initiation at a unique origin of replication is difficult to reconcile with the observation that apparently any piece of DNA can be replicated after microinjection into *Xenopus laevis* eggs (Harland and Laskey, 1980). Origins of replication in prokaryotes were defined as sequence-specific elements of recognition at which DNA replication nearly always begins. By contrast, Harland and Laskey (1980; Harland, 1981) have suggested that origins of replication are not required in *X. laevis* embryos, a view which is in apparent contradiction with reports that initiation of DNA replication commences at specific sites (Hines and Benbow, 1982; Hiraga *et al.*, 1982).

Moreover, the idea that chromosomal DNA replication in higher eukaryotes occurs exclusively at replication forks is difficult to reconcile with the apparent lack of sufficient forks to account for replication in rapidly dividing embryonic cells (Baldari *et al.*, 1978; Micheli *et al.*, 1982; Gaudette and Benbow, 1985). Even in cases where strong selection procedures to isolate replication forks have been applied, few if any convincing micrographs of replication forks have been obtained in DNA isolated from enterocoelomates (Burks and Stambrook, 1978). By contrast, *in vitro* DNA replication in extracts of *X. laevis* embryos resulted more in efficient formation of microbubble structures rather than traditional replication forks (Benbow and Ford, 1975). These microbubble structures have been observed to be correlated with rapid DNA replication in sea urchin embryos and numerous other eukaryotic organisms (Baldari *et al.*, 1978; Micheli *et al.*, 1982). Some authors have viewed microbubbles as artifacts generated during the isolation of replication forks (Hardman and Gillespie, 1980; Micheli *et al.*, 1982). Alternatively, we have suggested (Benbow, 1984; Gaudette and Benbow, 1985) that microbubbles (separons in our nomenclature) reflect normal nonfork intermediate structures in embryonic DNA replication. In addition, the unexpected abundance of extensive regions of single-stranded DNA in rapidly dividing cells of higher eukaryotes (reviewed by Micheli *et al.*, 1982; Benbow, 1984) lends credence to the possibility that alternative replication mechanisms not involving replication forks may exist.

One of the most significant recent developments in eukaryotic DNA replication is the independent discovery of self-priming DNA polymerases (DNA polymerase–primase complexes) in *Drosphila* (Conaway and Lehman, 1982a), mouse (Kozu *et al.*, 1982), and *X. laevis* (Riedel *et al.*, 1982; Shioda *et al.*, 1982). The tight association of DNA polymerase and primase activities suggests that eukaryotic DNA synthesis may not necessarily require assembly of the elaborate multienzyme replication complexes used in *Escherichia coli* (reviewed by Kornberg, 1980, 1982). Tight DNA polymerase–primase association in a relatively small soluble macromolecule also makes possible an alternative to the replication fork which might be useful during embryogenesis and other intervals when large amounts of chromosomal DNA require replication in a very short period of time.

In this review, we would like to suggest that there is a feasible alternative for replicating chromosomal DNA which does not involve replication forks. This hypothetical mechanism has certain features which might prove highly advantageous during periods of extremely rapid chromosomal DNA replication. The essential feature of this mechanism is that the strands of the double helix effectively separate over large regions to form "single strands" prior to replication. Mechanisms of this type were originally proposed for replication of *E. coli* DNA (Rosenberg and Cavalieri, 1964), but have been ruled out in prokaryotes (reviewed by Kornberg, 1980, 1982). We propose to call mechanisms of this type strand separation mechanisms to distinguish them from replication fork (DePamphilis and Wasserman, 1980; Edenberg and Huberman, 1975) and displacement synthesis (Challberg and Kelly, 1982; Clayton, 1982; Krimer and Hof, 1983) mechanisms.

We emphasize that at present there is no proof that strand separation occurs in higher eukaryotes. Nevertheless, considerable evidence has accumulated over the past 10 years which is consistent with this hypothesis. In this chapter we have concentrated on reviewing the extensive but generally uncited literature consistent with strand separation. We believe this emphasis is justified, since replication fork mechanisms have exclusively dominated discussion of eukaryotic DNA replication for the past decade.

II. ORIGINS OF DNA REPLICATION

Huberman and Riggs (1968) interpreted their now classic fiber autoradiography experiments in HeLa or Chinese hamster cells in terms of bidirectional movement of forklike growing points from unique origins of DNA replication. It is of interest that very few if any autoradiograms shown by Huberman and Riggs actually exhibited the appearance of labeled replication forks (see Section VI). In spite of this, the Huberman and Riggs bidirectional model for DNA replication, in which forklike growing points proceed outward from fixed origins of DNA replication, has dominated all discussions of chromosomal DNA replication in eukaryotes since its publication.

The tandem arrays of grain tracks described by Huberman and Riggs (1968) have been confirmed many times. For example, Amaldi *et al.* (1973) similarly used fiber autoradiography to demonstrate reproducible fixed sites of initiation in synchronized Chinese hamster ovary cells. Until recently, most researchers have accepted the existence of unique origins of eukaryotic chromosomal DNA without question (reviewed by Edenberg and Huberman, 1975; Hand, 1978; Taylor, 1984).

Direct evidence in favor of specific sites for initiation of DNA replication *in vivo* in eukaryotic cells is compelling. Botchan and Dayton (1982) convincingly showed by electron microscopy at least one specific locus of small replicated regions in the nontranscribed spacer region of sea urchin ribosomal DNA. McKnight *et al.* (1977) showed by electron microscopy of chromatin spreads that replication was initiated in the nontranscribed spacer regions in *Drosophila melanogaster* ribosomal DNA. Bozzoni *et al.* (1981) showed by electron microscopy of DNA an apparently nonrandom distribution of "microbubbles" in *X. laevis* ribosomal DNA; again these "origins" were predominantly localized in the nontranscribed spacer region. Bozzoni *et al.* (1981) also showed preferential labeling of restriction fragments located in the nontranscribed spacer region of ribosomal DNA in synchronized cultured cells of *X. laevis*.

Replication of amplified genes also begins at unique sites *in vivo* in a wide range of eukaryotic organisms. In *Tetrahymena* (Truett and Gall, 1977) and in *Physarum polycephalum* (Vogt and Braun, 1977) replication begins at specific sites near the center of long palindromic repeats of amplified extrachromosomal ribosomal DNA. Heintz and Hamlin (1982) showed that the initial incorporation of DNA precursors in S phase was into specific restriction fragments of 1000-fold amplified dihydrofolate reductase genes in specially selected methotrexate-resistant Chinese hamster ovary cells. This initiation region has been cloned (Heintz *et al.*, 1983) and its nucleotide sequence is being determined.

Spradling (1981) has shown that amplification of the *Drosophila* chorion genes during oogenesis proceeds from specific sites on the X and third chromosomes. Moreover, an inversion which caused a portion of a chorion gene cluster on the X chromosome to be moved to a different locus caused genes near the new position to be amplified; the remaining previously amplified genes of the chorion cluster left in the normal locus were no longer amplified (Spradling and Mahowald, 1981). All of these data are consistent with site specificity for initiation of DNA replication in amplified genes *in vivo*. They do not, however, necessarily demonstrate sequence specificity.

Additional indirect evidence for specific sites of initiation of replication *in vivo* comes from numerous reports that specific genes are replicated at specific times in S phase (reviewed by Goldman *et al.*, 1984). Mouse satellite DNA and monkey component α DNA both replicate late in S phase (Parker *et al.*, 1975). By contrast, bovine satellite I DNA initiates replication early in S phase (Matsumoto and Gerbi, 1982). Condensed heterochromatin, including the inactivated X chromosome, replicates late in S phase (Balazs *et al.*, 1974). α-Globin sequences are located in the region of early-replicating DNA (Furst *et al.*, 1981; Epner *et al.*, 1981). Newly synthesized DNA isolated from the middle of S phase from Friend leukemia cells shows greater hybridization to globin cDNA than newly synthesized DNA isolated from very early or late S phase (Lo *et al.*, 1980). In murine erythroleukemia cells, the immunoglobulin heavy chain constant region genes are replicated in order of their location in the genome (Braunstein *et al.*, 1982). Stambrook (1974) showed the regulated temporal replication of ribosomal genes in Chinese hamster ovary cells. All of these data are consistent with initiation in specific regions of chromosomes at specific times in the cell cycle. Taylor (1984) has postulated that replication in early versus late S phase is a major factor controlling gene expression.

Early replicated regions of chromosomal DNA are frequently actively transcribed (Bello, 1983; reviewed by Goldman *et al.*, 1984; Taylor, 1984). In addition, the pattern of early- and late-replicating chromosomal regions can be tissue specific, as demonstrated by fluorescence staining of early and late bands in the X chromosomes of human lymphocyte and amniotic fluid cells where the order of replication of bands differs (Willard, 1977). Locations of late-replicating chromosomal regions mapped by autoradiography are stage and tissue dependent in developing *Rana* embryos (Flickinger *et al.*, 1965, 1966, 1967a,b; Stambrook and Flickinger, 1973).

Replication of specific subsets of genes in response to cellular signals is also observed in higher plants. Isolated soybean nuclei selectively replicate certain repetitive DNA sequences *in vitro* (Caboche and Lark, 1981). Treatment with a cytokinin causes an apparent change in the pattern of DNA replication *in vitro* due either to replication of different sequence families or modification of newly synthesized DNA.

Recently, however, an apparent paradox regarding the function and even existence of replication origins in higher eukaryotes has arisen. Harland and Laskey (1980) and Laskey and Harland (1981) have shown that apparently any DNA was replicated after microinjection into *X. laevis* eggs. Mechali *et al.* (1983a) and Mechali and Kearsey (1984) showed that the amount of replication was proportional to the length of microinjected circular template, at least for templates 4–12 kb in length. These data suggest that origins were not required for replication of microinjected DNA molecules *in vivtro*. In apparent contradiction, Hines and Benbow (1982; Hines *et al.*, 1985) have shown that replication of microinjected DNA in *X. laevis* eggs begins frequently at specific sites (presumptive specific origins). Etkin *et al.* (1984) have shown that microinjected DNA containing eukaryotic DNA sequences was replicated up to 50-fold more efficiently in *X. laevis* embyros than the prokaryotic cloning vector. This suggests that the cloned eukaryotic DNA sequences functioned as origins.

McTiernan and Stambrook (1984) have reported that replication is initiated randomly on SV40 microinjected into *X. laevis* eggs; however, only 11 replicated regions were mapped by electron microscopy and 8 of these overlapped one site on the genome. McTiernan and Stambrook (1980, 1984) have also shown that some plasmids were preferentially replicated relative to others after microinjection into *X. laevis* eggs, although all plasmids appeared to be replicated semiconservatively as monitored by density shift experiments. This confirms the differential replication of microinjected plasmids in *X. laevis* eggs reported by Hines and Benbow (1982) and Watanabe and Taylor (1980) and also the lack of a requirement for a specific origin reported by Harland and

Laskey (1980). Watanabe and Taylor (1980) have isolated a recombinant plasmid containing a sequence of X. *laevis* DNA which directs up to 17-fold more synthesis than the cloning vector when microinjected into X. *laevis* eggs (Chambers et al., 1982). Hiraga et al. (1982) demonstrated preferential replication *in vitro* of a plasmid containing an inserted repetitive sequence of X. *laevis* sperm DNA. The replicated regions mapped to nonrandom loci. We propose the term "replication enhancer sequences" to designate sequences of DNA which increase the efficiency of replication of the vector into which they are cloned.

Replication enhancer sequences which function in X. *laevis* eggs do not appear to function in lower eukaryotes. Zakian (1981) transformed plasmids containing X. *laevis* rDNA, 5 S rDNA, several human *Alu*II sequences, and X. *laevis* mitochondrial DNA into yeast cells. Only the mitochondrial DNA functioned as an "ars" (autonomously replicating sequence) in yeast; however, ars may require sequences with other functions (centromeric, etc.) in addition to those found in replication enhancer sequences. It also should be noted that Mechali et al. (1983b) and Mechali and Kearsey (1984) found no evidence for the preferential use of cloned DNA in any plasmid DNA microinjected into X. *laevis* eggs. The reason for the differences in results of these researchers and the results of the laboratories of Taylor, Etkin, Benbow, and Hiraga is not known.

The apparent contradiction between initiation of replication at specific sites and the lack of a requirement for specific "origins" can be resolved. If the minimal requirement for replication is strand separation (see Section VI), any DNA which is separated will be replicated. If strand separations occur initially in specific regions (Case and Baker, 1975; Wortzman and Baker, 1980, 1981) and/or priming occurs at somewhat specific sites (Hay et al., 1984; Tseng and Ahlem, 1984), then specific "origins" will exist as monitored by electron microscopy and fiber autoradiography. Initiation of DNA synthesis at specific sites, however, would not be rigidly determined by a specific sequence or secondary structure as in prokaryotes or eukaryotic viruses, but rather would be statistically determined.

An additional feature which may be important for the functioning of specific sites of initiation of DNA synthesis, at least in embryos, is the stage of development of the organism. Hines and Benbow (1982) did not observe significant differences in the replication of microinjected pXlr11 relative to ColE1 in X. *laevis* eggs at low DNA concentrations; this is in agreement with Mechali et al. (1983b). Newport and Kirschner (1982a,b), however, showed that microinjection of an amount of DNA equivalent to that found in a blastula induced a "midblastula" transition with striking consequences, including increased transcription and a variety of intracellular changes. The putative origin of Hines and Benbow (1982) appeared to function only when amounts of DNA greater than needed to induce the midblastula transition were microinjected. However, Mechali and Kearsey (1984) did not observe the increased replication efficiency reported by Hines and Benbow (1982) at high DNA concentrations. By contrast, saturation was observed at only 10 ng of DNA per egg in these experiments. The reasons for this discrepancy are not known. In conclusion, it is possible that unique initiation sites for DNA synthesis begin to function efficiently only after specific stages in development, but the existing experimental evidence is still contradictory.

III. REPLICATION FORKS

Synthesis of chromosomal DNA in prokaryotes takes place at structures called replication forks (reviewed by Kornberg, 1980, 1982). Morphologically similar replication fork structures have been observed in lower eukaryotes including *P. polycephalum* (Funderud *et al.,* 1979) and *Saccharomyces cerevisiae* (Newlon *et al.,* 1974; Petes and Newlon, 1974). In addition, numerous laboratories have reported replication forks and replication eyes in DNA isolated from *Dipteran* embryos (Wolstenholme, 1973; Kriegstein and Hogness, 1974; Blumenthal *et al.,* 1974; Lee and Pavan, 1974; Zakian, 1976; Baldari *et al.,* 1978).

McKnight and Miller (1979) have observed replication forks by electron microscopy in chromatin isolated from *D. melanogaster* embryos. Busby and Bakken (1980) also observed replication forks in chromatin isolated from sea urchin embryos. It should be emphasized, however, that it is not possible to distinguish single- from double-stranded DNA in replication forks observed in chromatin spreads.

In contrast to the abundant replication forks reported in prokaryotes, lower eukaryotes, and schizocoelomates, there are few reports of replication forks in enterocoelomates. Baldari *et al.* (1978) and Kurek *et al.* (1979) observed few replication forks and frequent microbubbles and single-stranded regions in DNA isolated from rapidly dividing sea urchin embryos. Micheli *et al.* (1982) observed only low levels of replication forks in DNA isolated from a variety of eukaryotes including chick embryos, *X. laevis* embryos, regenerating rat liver, and *Vicia faba* root tips; microbubbles and extensive single-stranded regions were far more abundant. Gaudette and Benbow (1985) did not find any replication forks in DNA isolated from rapidly dividing *X. laevis* cleavage embryos. A low level of possible forks was observed in DNA isolated from more slowly dividing X. laevis gastrulae and hatched embryos.

Even when extreme selective pressures were employed to isolate replication forks in DNA from enterocoelomate cells, only low levels of replication fork structures have been observed. Burks and Stambrook (1978) enriched for replicating structures in DNA isolated from cultured Chinese hamster cells by banding in Cs_2SO_4 gradients in the presence of $HgCl_2$; very few branched replicating structures were observed and only one was a convincing replication fork. Taylor et al. (1973) and Valenzuela et al. (1983), among others, have shown isolated examples of convincing replication forks in DNA isolated from mammalian cells. However, to our knowledge, no one has reported a sufficient level of replication forks to account for the rates of replication observed in rapidly dividing cells of any enterocoelomate.

It is to be emphasized that the autoradiographic experiments of Huberman and Riggs (1968) and others (reviewed by Edenberg and Huberman, 1975) did not show labeled replication forks. Instead, tandem arrays of grains were reported which were interpreted in terms of replication forks based on the measured grain densities. These arrays are equally consistent with the alternative replication mechanism presented in Section VI below.

In enterocoelomate organisms where replication forks apparently have been isolated, these structures could be accounted for by alternative replication mechanisms (see Section VI). It is important to emphasize that a structure which looks like a replication fork will be isolated if the replicative DNA polymerase synthesizes in close temporal and spatial proximity to a region of active strand separation. Forklike structures will also be generated if proteins are removed from DNA replicating by the strand separation mechanism (see Section VI) and the strands are subsequently allowed to renature.

Hines and Benbow (1982) and Benbow et al. (1978; Benbow and Krauss, 1977) showed that structures with the morphology of replication forks were formed after microinjection of circular DNA molecules into X. laevis eggs or incubation of DNA in extracts of X. laevis eggs. These results have been confirmed, both in vivtro (McTiernan and Stambrook, 1980, 1984) and in vitro (Hiraga et al., 1982). While some of the apparent late replicative intermediates could have been formed by the efficient X. laevis DNA recombinase (Benbow and Krauss, 1977), as suggested by Mechali and Harland (1982), this is unlikely for the majority of the putative early replicative intermediates. At least two explanations are possible for the observed structures. It is certainly possible that at least some microinjected DNA is replicated by a mechanism involving replication forks; Richter *et al.* (1981) have shown that X. *laevis* egg extracts catalyze apparently normal chain elongation of replicating SV40 minichromosomes. Presumably this elongation involves replication forks (DePamphilis and Wasserman, 1980), although this was not shown explicitly. Alternatively, the microinjected circular DNA could have been replicated by a strand separation mechanism (see Section VI) with subsequent reannealing of the topologically interlocked strands during extraction and incubation with restriction enzymes. Putative replicative intermediates which are consistent with the latter alternative have recently been described (Hines *et al.*, 1985).

IV. EXTENSIVE SINGLE-STRANDED REGIONS

The association of single strandedness with newly replicated DNA was reported over 15 years ago (Schandl and Taylor, 1969). This has been confirmed in numerous reports (among others, Case *et al.*, 1974; Wanka *et al.*, 1977; Henson, 1978; Wortzman and Baker, 1981; Carnevali and Filetici, 1981), and the single-strand regions have been shown to be associated with specific sequences in the genome (Case and Baker, 1975; Wortzman and Baker, 1980). The single-stranded regions usually were presumed to be associated with the lagging strands of replication forks, although there were indications that the single-stranded regions might be longer than expected (Wortzman and Baker, 1981). In one significant early study, Tan and Lerner (1972) correlated the appearance of single-stranded DNA with phases of the cell cycle in growing diploid lymphoblasts using rabbit antisera to single-stranded DNA which did not cross-react to double-stranded DNA. Single-stranded DNA was not detected in G_1 phase cells, but was abundantly present midway through S phase. In metaphase cells, staining for single-stranded DNA was observed in the central portion of some nuclei with radiations to the periphery.

More recently, however, it has become apparent that very extensive regions of single-stranded DNA are involved (10–75 kb in many cases). Electron micrographs of extended single-stranded regions are shown in Fig. 1. The extent of these regions is so great that a number of artifacts (branch migration, shear forces during isolation, etc.) which could result in the appearance of single-stranded regions become less plausible. Still unresolved is the question whether the DNA is single-stranded in the nucleus or is in a destabilized state such that the extraction conditions cause conversion to a fully single-stranded state (Habener *et al.*, 1970). From our point of view, it is only essential that the double helix be sufficiently destabilized to permit relatively free access by the DNA polymerase–primase complex at a number of sites (see Section VI). The essential point is that the parental DNA duplex structure of replicating DNA is perturbed during S phase: It is either fully single-stranded *in vivo* or readily converted to a fully single-stranded state during extraction of the DNA.

Collins (1977) showed that human diploid fibroblasts accumulated single-stranded (or destabilized) parental DNA as they entered S phase. DNA isolated from G_0 cells was native duplex DNA with essentially no "breaks" or "gaps." As the cells entered G_1 the amount of single-strandedness and gaps in the isolated DNA began to increase. During S phase a substantial fraction of the parental DNA was isolated in the single-stranded form as estimated from $Cs_2SO_4/AgClO_4$ density gradient centrifugation. This fraction was estimated as at least 15% by S1 nuclease sensitivity, and perhaps as much as 46% based on treatment with *Neurospora crassa* endonuclease. Similar results were shown for HeLa cells as they traversed the cell cycle (Collins *et al.*, 1977) and in regenerating rat liver (Hoffman and Collins, 1976). Bjursell *et al.* (1979) similarly showed that single-stranded DNA of 10–70 kb could be isolated from actively dividing human cells. The single-stranded DNA was far more abundant in early or late S-phase cells than in G_2-phase cells. In addition, Gaudette and Benbow (1985) showed that molecules longer than 70 kb which appeared to be entirely single stranded by electron microscopy could be isolated from rapidly dividing *X. laevis* embryos. It is of interest that extensively single-stranded DNA molecules can be isolated from *X. laevis* embryos prior to the midblastula transition (i.e., before any transcription takes place). The design of the experiments above was such that the extensive single-stranded regions were unlikely to have resulted simply from loss of small nascent DNA fragments.

Table I summarizes the frequencies (A) and extent (B) of single-stranded regions and their preferential association with S phase in a wide variety of eukaryotic organisms. Micheli *et al.* (1982) have published a similar compilation of the frequencies of single-stranded regions in eukaryotic DNA, which contains additional examples. It must therefore be regarded as an established fact that a strong correlation exists between the occurrence of extensive single-stranded (or destabilized) regions and chromosomal DNA replication. The reason for this correlation is not yet known.

It should be pointed out that the highly efficient linear concatenation systems found in a number of organisms (Bayne *et al.*, 1984; Etkin and Roberts, 1983; Bendig, 1981; Rusconi and Schaffner, 1981; Huttner *et*

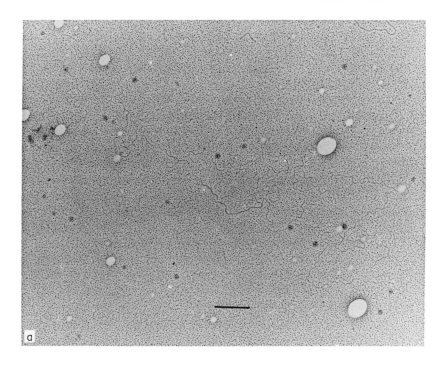

Fig. 1. Electron micrographs of extensive single-stranded regions. DNA was gently isolated from *Xenopus laevis* gastrulae by the procedure of Gaudette and Benbow (1985) and spread for electron microscopy by the formamide procedure of Davis *et al.* (1971). (a) A DNA molecule with a short double-stranded region separating into two extensive single-stranded regions. It is assumed that tightly bound proteins bind to the exposed

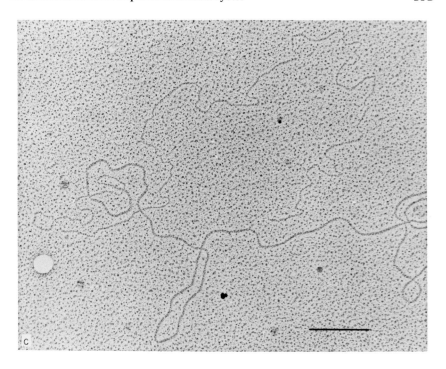

single-stranded regions and prevent renaturation. From Benbow (1984). (b) Schemetic diagram of the molecule shown in (a);—, double-stranded DNA; (· · ·), single-stranded DNA. (c) Enlargement of a DNA molecule isolated from gastrulae showing the progression of an extended double-stranded region into a extended single-stranded region. (d) Schematic diagram of the molecule shown in (c). Bars represent 1 kb.

TABLE I

Occurrence of Potential Replication Intermediates

A. Relative Frequencies of Separons, Replication Forks, and Extensive Single-Stranded Regions[a]

System	Separons	Forks	Partially or entirely single-stranded DNA	Reference
Vicia faba (germinated seed)				
Root tips	3.9	0.6	23.0	Micheli *et al.* (1982)
Drosophila melanogaster				
Preblastoderm embryos (stage 6–10)	2.6	0.7	5.5	Micheli *et al.* (1982)
Cellular blastoderm (stage 14a,b)	1.0	0.8	8.3	
Gastrulae	0.5	0.9	7.9	
Larvae: I instar	0.7	0.2	6.3	
Larvae: II instar	1.0	0.3	7.7	
Larvae: III instar	0.9	0.4	6.1	
Drosophila virilis				
Preblastoderm embryos: main band DNA	1.5	3.1	DNR	Zakian (1976)
Preblastoderm embryos: satellite DNA	0.5	0.7	DNR	
Adult flies	Very rare if at all		DNR	
Paracentrotus lividus				
Third S-phase embryo	16.6	Rare	23.5	Baldari *et al.* (1978)
Gastrulae	2.5	Rare	7.1	
Ovaries	2.1	Rare	8.1	
Ring canal	1.7	Rare	7.4	
Sperm	0	0	0	
Unfertilized eggs	0	0	0	
Arabacia punctulata				
Pregastrula embryos	<1.0	+	+	Kurek *et al.* (1979)
Sperm and unfertilized eggs	0	0	0	
Salmo irideus (18 weeks old)				
Gonads	16.0	0	29.9	Micheli *et al.* (1982)
Crassius auratus (adult)				
Testis, mid-October	3.9	1.0	15.6	Micheli *et al.* (1982)

(continued)

				Reference
Testis, mid-November	13.4	0.6	6.4	Gaudette and R. Benbow (1985)
Testis, mid-December	0.6	0.8	17.3	Micheli et al. (1982)
Ovary, mid-October	0.8	0.5	23.5	Micheli et al. (1982)
Spleen	0.9	1.4	5.7	Bozzoni et al. (1981)
Gills	1.0	0.5	12.1	Bozzoni et al. (1981)
Kidney	0.9	0.4	8.1	Micheli et al. (1982)
Xenopus laevis				
Gastrulae (stage 10–12)	91	3–5	75	Gaudette and R. Benbow (1985)
Embryos (stage 33–36)	3.8	0.4	2.4	Micheli et al. (1982)
Larvae (stage 40–50)	3.4	0.9	6.5	Micheli et al. (1982)
Larvae (stage 40–50) main peak DNA	+	+	+	Bozzoni et al. (1981)
Larvae (stage 40–50) rDNA	+	+	+	Bozzoni et al. (1981)
Unsynchronized culture fibroblast	+	Rare	+	Micheli et al. (1982)
Gallus domesticus				
Embryos (stage 19–21)	2.3	0.5	15.2	Micheli et al. (1982)
Rattus rattus (adult)				
Liver, normal	0.2	0.1	2.2	Micheli et al. (1982)
Liver, 18-hr posthepatectomy	3.3	0.2	13.6	Micheli et al. (1982)
Chinese hamster culture cells (CHEF 125)				
Asynchronous	1.1	0.5	9.1	Micheli et al. (1982)
Asynchronous	Very rare	Rare	2.1	Carnevali and P. Filetici (1981)
Synchronous at $t = 0$ (9 hr hydroxyurea block)	+	0	3.5	
$t = 0$ (15 hr hu block)	+	0	4.4	
$t = 0$ (19 hr hu block)	+	0	6.2	
$t = 20$ min	+	0	1.8	
$t = 180$ min	+	0	1.9	
Hu inhibited	0	0	0	
Bovine liver cell line	DNR	DNR	+	Wanka et al. (1977)
Asynchronous human melanoma cells	DNR	DNR	+	Lonn and Lonn (1983)
Synchronous human diploid fibroblasts				
$t = 4$ hr	DNR	DNR	25 (many sites)	Collins (1977)
$t = 24$ hr	DNR	DNR	67 (many sites)	
$t = 30$ hr	DNR	DNR	66 (few sites)	

TABLE I (Continued)

B. Extent of Single-Stranded Regions

Organism	Partially or entirely DNA single-stranded	Reference
Strongylocentrotus purpuratus		
Morulae	19	Wortzman and Baker (1981)
Human lymphoid (Raji) cells		Bjursell *et al.* (1979)
Early S phase	9.2	
Late S phase	9.1	
G phase	2.9	
Human diploid fibroblasts		Collins (1977)
$t = 0$ hr G_0	3	
2 $G_0 \rightarrow G_1$	3	
4 G_1	4	
6 G_1	7	
12 $G_1 \rightarrow S$	12	
16 S	13	
20 S	15	
24 S	11	
30 $S \rightarrow G_2$	3	
Xenopus laevis		Gaudette and Benbow (1985)
Blastulae (stage 7–9)	42[b]	
Gastrulae (stage 10–12)	7.1–12.0[c]	
Hatched embryos (34/35)	4[d]	
Adult (brain)	<1	

[a] +, observed but not quantitated; 0, not observed; DNR, data not reported.
[b] Includes 1.8% separons.
[c] Includes 1.3–2.1% separons.
[d] Includes 1.0% separons.

al., 1981) could reduce the risk to the continuity of genetic information posed by the extensive single-stranded regions.

Small denatured (i.e., single-stranded) regions in duplex DNA have also been shown to exist in the chromosomal DNA isolated from a number of organisms (Klein and Byers, 1978; reviewed by Conrad and Newlon, 1983). Electron micrographs of small strand *separated regions* (separons) in DNA isolated from *X. laevis* gastrulae are shown in Fig. 2. Although the median size of a separon is 0.2 kb, regions as long as 30 kb (i.e., extensive single-stranded regions) have also been seen (Fig. 1; also Gaudette and Benbow, 1985). Separons frequently appear in clusters as shown in Fig. 2c (see also Baldari *et al.*, 1978; Gaudette and Benbow, 1985). Microbubbles which presumably correspond to strand-separated regions have also been reported in *Drosophila virilis* (Zakian, 1976), in sea urchins (Baldari *et al.*, 1978), and in most eukaryotic organisms (reviewed by Micheli *et al.*, 1982). Since stably denatured regions have been found in the absence of DNA replication, Conrad and Newlon (1983), among others, have concluded they were not intermediates in DNA replication.

Almost a decade ago, however, Benbow and Ford (1975) showed that small structures with the morphology of "D loops" appeared in DNA isolated from nuclei that were induced to replicate by incubation in extracts of *X. laevis* embryos. These structures were formed at a much higher frequency than replication forks (Benbow and Ford, 1975). Because the term "D loop" implies that a specific mechanism generated the structures, we now propose to rename these structures 1,2 separons. Reexamination of our original electron micrographs suggest that some of the D loop structures should have been classified as stable denatured regions (1,1 separons). Nevertheless, structures with the morphology of genuine D loops (1,2 separons) predominated. Circumstantial evidence for the existence of stable *nonreplicating* D-loop structures (1,2 separons) was provided by Wanka *et al.* (1977) who showed that small, prelabeled, single-stranded DNA fragments were released by branch migration from nuclear DNA isolated from bovine liver cells. 1,1 and 1,2 separons were equally frequent in replicating DNA isolated from *X. laevis* embryos (Table I).

In Table I we have shown that there is a striking correlation between 1,1 and 1,2 separons and the occurrence (though not the rate) of DNA synthesis (see also Micheli *et al.*, 1982). This table also documents the relative lack of replication forks in DNA isolated from rapidly dividing cells in a wide range of organisms. There are two orders of magnitude too few replication forks in DNA isolated from cleaving *X. laevis* embryos to account for the rate of DNA synthesis (Benbow, 1984; Gaudette

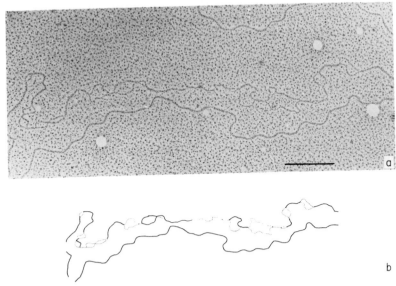

Fig. 2. Electron micrographs of strand-separated regions. (a) An enlargement of a small section of a long DNA molecule isolated from gastrulae showing strand separation regions (above) and fully double-stranded DNA (below). (b) Schematic diagram of the molecule shown in (a). (c) Enlarged section of a DNA molecule isolated from newly hatched embryos. The circular molecule is a *ColE1* double-stranded marker molecule added before spreading. (d) Schematic diagram of the molecule shown in (c). Bars represent 1 kb. (*continued*)

and Benbow, 1985). The only significant exception in higher eukaryotes seems to be *Dipteran* embryos such as *Drosophila* (Kriegstein and Hogness, 1974; Blumenthal *et al.*, 1974) and the screwworm fly (Lee and Pavan, 1974). In *D. virilis*, however, it should be noted that Zakian (1976) reported high levels of separons. By contrast, Wolstenholme (1973) and Kriegstein and Hogness (1974) did not report these structures in DNA isolated from *D. melanogaster* embryos. More recently, however, Micheli *et al.* (1982) have reported the occurrence of both separons and extensive single-stranded regions in DNA isolated from embryos of *D. melanogaster*.

The existence of stable denatured regions in duplex DNA implies the existence of a stabilization factor, possibly a DNA binding protein. The stabilization factor must be refractory to removal from the DNA, since stable denatured regions have been seen (often at reduced frequencies) in DNA extracted by a variety of protocols involving phenol, protease digestion, or treatment with detergents (Baldari *et al.*, 1978; Micheli *et*

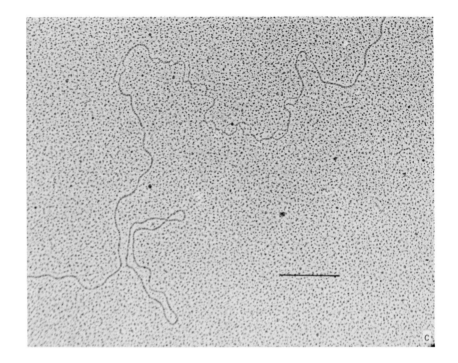

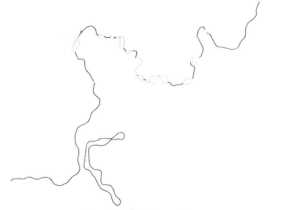

Fig. 2. (Continued.)

al., 1982; Gaudette and Benbow, 1985). The fact that stable denatured regions can be observed in nonreplicating DNA (Conrad and Newlon, 1983) does not, we believe, rule out a role for separons in DNA replication. Instead, we speculate that strand separation is a fundamental cellular mechanism in eukaryotes which may also play a role in transcription (Kohwi-Shigematsu *et al.*, 1983), recombination (Klein and Byers, 1978), replication (Gaudette and Benbow, 1985), and perhaps even gene amplification (Spradling, 1981; Spradling and Mahowald, 1981).

We also speculate that strand separation is not catalyzed by energetically "expensive" DNA helicases, but instead results from a shift in the normal equilibrium between denatured and double-stranded DNA. This shift may be promoted by a strand stabilization protein, DNA topoisomerase I (to remove the topological constraints imposed by plectonemic coiling), and a "crowbar" protein.

Our current hypothesis is that DNA in lower eukaryotes and schizocoelomates (such as *Dipterans*) is replicated by a mechanism involving replication forks. In contrast, we speculate that DNA in enterocoelomates, at least in embryos, is replicated by a mechanism which does not involve forks (see Section VI). We further speculate that this mechanism evolved to obviate the difficulties involved in the replication of the generally much larger genomes of enterocoelomates (13,500 kb for yeast and 165,000 kb for *Drosphila* versus ~3,000,000 kb for *human*, *Xenopus*, and chick).

V. DNA POLYMERASE–PRIMASE COMPLEXES

Eukaryotic DNA polymerases have recently been exhaustively reviewed (Fry, 1983), as have DNA polymerases isolated from embryos (Nelson *et al.*, 1983; Benbow, 1984). We will confine our discussion to the newly discovered DNA polymerase–primase complexes, which are confined to the nuclei (Fox *et al.*, 1980; Bensch *et al.*, 1982) and are the presumptive "replicative" polymerases.

Recently, it has become apparent that single-stranded DNA is an excellent template for DNA synthesis in extracts of eukaryotic embryos (Mechali and Harland, 1982; Riedel *et al.*, 1982; Yoda and Okazaki, 1983). The reason for this became clear when Conaway and Lehman (1982a), Kozu *et al.* (1982), and Shioda *et al.* (1982) reported that purified DNA polymerase activities contained an associated primase activity. In *X. laevis* (Benbow *et al.*, 1975; Shioda *et al.*, 1982; Riedel *et al.*, 1982; Nelson *et al.*, 1983) and in most other eukaryotic organisms (Yagura *et al.*, 1983b) except *Drosophila* (Conaway and Lehman, 1982a,b; Kaguni *et al.*, 1983a,b), DNA polymerase is found in two forms: one associated with primase and one devoid of primase activity. It is potentially interesting that the DNA polymerase–primase complex is apparently more resistant to aphidicolin than the form of DNA polymerase devoid of primase activity (Nelson et al., 1983; Wang et al., 1984). It is clear that the DNA polymerase and primase are tightly associated (Yagura et al., 1983a). The D. melanogaster (Conaway and Lehman, 1982a,b; Kaguni et al., 1983a,b) and X. laevis (Shioda et al., 1982) DNA polymerase and primase activities copurify through up to 10 purification steps, resulting in nearly homogeneous enzymes. Very highly purified KB cell DNA polymerase α, purified by an affinity procedure involving monoclonal antibodies (Tanaka et al., 1982; Bensch et al., 1982), also contains DNA primase activity (Wang et al., 1984).

The molecular basis of the tight association between DNA polymerase α and DNA primase is less clear. Hubscher (1983) and König et al. (1983) have reported that the DNA primase and polymerase activities are both associated with a large (100+ kilodaltons) polypeptide, whereas Lehman's group has reported that the D. melanogaster DNA polymerase activity is associated with the large polypeptide, but the primase is associated with a small polypeptide (50 or 60 kilodaltons) (Kaguni et al., 1983a,b). The X. laevis DNA polymerase α, lacking DNA primase activity, consists of two subunits of ~120 and 50 kilodaltons (Benbow et al., 1985), whereas the DNA polymerase–primase complex consists of a ~160-kilodalton polymerase subunit and a 60-kilodalton nonpolymerase subunit (Hiriyanna, personal communication), providing indirect support for primase association with a small polypeptide. With the exception of Tseng and Ahlem's report of a free DNA primase of low molecular weight from human lymphocytes (Tseng and Ahlem, 1982, 1983), most reports of DNA primase have consistently been associated with a DNA polymerase α-type activity.

DNA primase activities from a wide range of eukaryotes have exhibited remarkably similar properties. With natural single-stranded DNA templates, primase synthesized oligoribonucleotide primers of about 10 nucleotides (Conaway and Lehman, 1982b; Shioda et al., 1982). There was an absolute requirement for ATP, both in the priming reaction (Conaway and Lehman, 1982a; Shioda et al., 1982) and in the subsequent elongation reaction (König et al., 1983). The composition of the oligonucleotide primer was exquisitely sensitive to concentrations of dNTPs and NTPs: In the absence of dNTPs, oligoribonucleotides were synthesized; in the presence of low levels of dNTPs, mixed tandem arrays of oligoribo- and oligodeoxyribonucleotides were synthesized; above 5 μM dNTP only deoxyribonucleotides were incorporated after the initial 5'-terminal oligoribonucleotide (Hu et al., 1983).

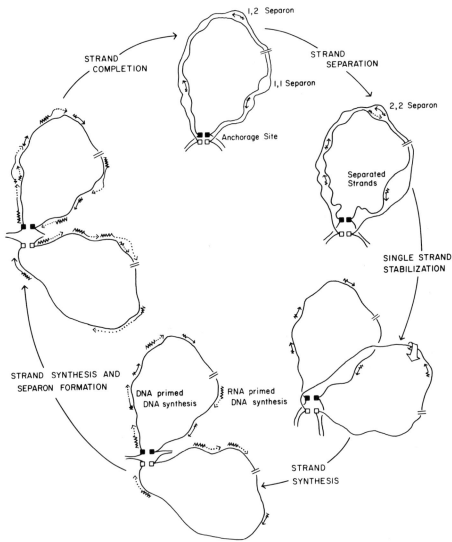

Fig. 3.   The strand separation mechanism. Chromosomal DNA is shown organized in double-stranded loops (domains) of 30–70 kb which are attached to a proteinaceous subnuclear structure at anchorage sites. The strands of the double helix appear as separated at nonrandom locations on the DNA. The separated regions (separons) are usually 0.2 kb median length. 1,1 separons are those in which both separated strands appear to be single-stranded in the electron microscope. 1,2 separons appear to consist of one single and one double strand. 2,2 separons in which both strands appeared to be double-stranded were extremely rare. Replication of a chromosomal domain begins with extensive strand separation. This may occur first near the anchorage sites, but in any event

VI. STRAND SEPARATION MECHANISM

An extreme form of the strand separation mechanism is drawn in Fig. 3 (Benbow, 1984). Following others, we have assumed that chromosomal DNA is organized in loops which are torsionally stressable (Paulson and Laemmli, 1977; Pardoll et al., 1980). The double-stranded loops (domains) are 30–70 kb and are attached to a proteinaceous subnuclear structure at anchorage sites. The size of the domains apparently changes during early embryogenesis. During cleavage in X. laevis the domains appear to be considerably smaller than in gastrulae, neurulae, or later embryonic stages (Buongiorno-Nardelli et al., 1982). It should be noted that organization into matrix-bound domains is not essential for the strand separation mechanism. Strand separation is not inconsistent with involvement of matrix attachment sites in initiation or termination of replication (see Cook and Lang, 1984); nor does it require matrix attachment sites.

The strands of the double helix are shown as separated (stably denatured) at nonrandom locations on the DNA. Essentially every chromosomal DNA molecule isolated from rapidly dividing X. laevis gastrulae contained these strand-separated regions (Gaudette and Benbow, 1985). These separated regions (separons) were usually 0.2 kb median length (Gaudette and Benbow, 1985), but were observed to be much longer on occasion. 1,1 separons are defined as those in which both separated strands appear to be single-stranded in the electron microscope. 1,2 separons appear to consist of one single and one double strand. 2,2 separons, in which both strands appeared to be double stranded, were extremely rare. We emphasize that these definitions are based on morphological criteria and do not imply that the DNA in a separon is exclusively single- or double-stranded.

Replication of a chromosomal domain begins with extensive strand separation. This may occur first near the anchorage or matrix attach-

---

rapidly traverses the entire domain. The separated single strands are stabilized by a DNA (tight) binding protein which also renders the exposed single-stranded regions less susceptible to nucleolytic attack. RNA-primed DNA synthesis catalyzed by the newly described DNA polymerase–primase complex commences at a number of sites throughout the domain. DNA-primed DNA synthesis may also occur at 1,2 and 2,2 separon sites. Additional separons are formed as the DNA becomes double-stranded. This occurs prior to strand completion. The completion of an intact fully double-stranded domain causes DNA to become subject to torsional stress. A conformational change is induced which triggers the subsequent round of strand separation and replication. Wavy loops, DNA; dotted lines, newly synthesized DNA; jagged lines, oligoribonucleotides primer; solid and open squares, nuclear attachment sites. From Benbow (1984).

ment sites (if they exist), but in any event rapidly traverses the entire domain. The separated single strands are stabilized by a DNA (tight?) binding protein which also renders the exposed single-stranded regions less susceptible to nucleolytic attack (Gargiulo *et al.*, 1982; Gargiulo and Worcel, 1983; M. F. Gaudette, P. J. Hines, and R. M. Benbow unpublished observations).

RNA-primed DNA synthesis catalyzed by the newly described DNA polymerase–primase complex (Conaway and Lehman, 1982a; Kozu *et al.*, 1982; Riedel *et al.*, 1982; Shioda *et al.*, 1982) commences at a number of sites throughout the domain. DNA-primed DNA synthesis may also occur at existing 1,2 and 2,2 separon sites. Additional separons are formed as soon as the DNA becomes double-stranded (Gaudette and Benbow, 1985). This occurs prior to strand completion.

We postulate that completion of an intact covalently closed double-stranded domain causes DNA within the domain to become subject to torsional stress. We further speculate that the torsional stress induces a conformational change in the DNA which serves as trigger for the subsequent round of strand separation and DNA synthesis.

We emphasize that Fig. 3 depicts an extreme version of the strand separation mechanism. Models which combine features of the strand separation, replication fork, and displacement synthesis mechanisms are readily constructed. The mechanism in Fig. 3 has one overriding virtue, however; it makes specific, testable predictions which clearly differentiate among the alternative mechanisms. Among these are the occurrence of extensive single-stranded parental DNA during S phase, an absolute requirement for a freely diffusible DNA polymerase–primase complex at high levels per domain, the occurrence of rapid strand separation, an absence of replication forks accompanied by the concomitant presence of separons, and, most importantly, the occurrence of DNA synthesis at a number of sites distant from the region of active strand separation. The latter prediction can be readily tested by electron microscope autoradiography (Rochaix *et al.*, 1974) or ultrastructural visualization of incorporated biotinylated deoxynucleotide precursors.

The most essential feature of the proposed mechanism is that the strands of the double helix separate over an extensive region prior to replication. This implies that replication forks will not be found. There will be no leading strand synthesis and no elaborate replication fork machinery. As a consequence of the strand separation mechanism, origins of replication in the conventional sense will not exist. The strand separation mechanism also has a number of unpalatable features, most of which also center around the formation of the extensive single-stranded regions. In this context it is of interest that kinetoplast DNA replication has recently been shown to proceed via fully single-stranded intermediate structures (Englund et al., 1982; Kitchin et al., 1984). Nevertheless, single strands which extend for 30–70 kb obviously place the genetic material at considerable risk. We would therefore like to point out that substantial literature exists (reviewed in Section III) which shows that very long single-stranded chromosomal DNA molecules can be isolated during S phase in most higher eukaryotes. This observation is valid whether or not the single-stranded DNA molecules are intermediates in DNA replication.

VII. REPLICATION AND THE NUCLEAR MATRIX

In this review we have focused exclusively on events occurring at the level of the structure of the DNA. We have omitted all studies examining the relationship between nucleosome assembly and DNA replication; in our view these have proved both conflicting and confusing. Another highly controversial area, the relationship of DNA replication to fixed sites in the nucleus, will be briefly discussed only with regard to the consequences for the strand separation mechanism. In our view, it is premature to conclude that chromosomal replication necessarily takes place on a nuclear matrix.

Whether or not DNA is replicated at fixed sites (variously termed nuclear matrix, nuclear cage, nuclear envelope, or pore complex) has a long and controversial history (reviewed by Dijkwel et al., 1979; Mc-Cready et al., 1980; Hunt and Vogelstein, 1981; Vogelstein et al., 1980; Cook and Lang, 1984). Although definitive experiments to rule out artifacts of association have not been carried out, it seems probable that at least some newly synthesized DNA is preferentially associated with a residual nuclear skeleton or scaffold (Pardoll et al., 1980; McCready et al., 1980; Vogelstein et al., 1980; Cook and Lang, 1984).

DNA replication by the strand separation mechanism can readily accommodate association with the matrix if one assumes that strand separation initiates at or near matrix attachment sites. DNA synthesis would then occur first (statistically) near the matrix. Nascent DNA would be attached to the matrix via a single strand, explaining the extreme sensitivity of nascent DNA to exogenously added DNase I or micrococcal nuclease (Berezney and Bucholtz, 1981). The linkage would be relatively resistant to the restriction endonuclease EcoRI, as reported by McCready et al. (1980). If matrix attachment sites corresponded to a subset of specific sequences or secondary structures of the

DNA (Small *et al.*, 1982; Goldberg *et al.*, 1983), DNA replication would *appear* to begin at unique sites. To replicate the DNA more rapidly during embryogenesis, additional matrix attachment sites on the DNA could be activated, resulting in an apparent shrinkage of domain size (Buongiorno-Nardelli *et al.*, 1982). Forbes *et al.* (1983) recently reported that nucleus-like structures were assembled around DNA microinjected into unfertilized X. *laevis* eggs or incubated in extracts of X. *laevis* eggs. This raises the exciting possibility that subnuclear structures can self-assemble with exogenously added DNA molecules to form "functional" nuclei.

VIII. CONCLUSION

We apologize for being unable to review a number of other areas of interest to eukaryotic DNA replication, such as the fascinating results of Noguchi *et al.* (1983), which suggest that ribonucleoside diphosphates rather than deoxyribonucleoside triphosphates may be the normal precursors to chromosomal DNA synthesis. We have not discussed the possible role of diadenosine tetraphosphate in the control of DNA replication, the relationship of nucleosome assembly to DNA synthesis, the rereplication which occurs during gene amplification, and new developments in the structural characterization of eukaryotic DNA replication proteins.

In this review we have chosen instead to emphasize an alternative view of eukaryotic DNA replication from that generally presented. We have suggested that DNA replication in rapidly dividing cells may occur by a strand separation mechanism. We wish to point out that all of the data presented can still be reconciled with a replication fork mechanism. In this interpretation, the failure to observe replication forks could result from their destruction during extraction, perhaps by DNA topoisomerase I (Liu and Wang, 1979; Gourlie and Pigiet, 1983) or II (Liu *et al.*, 1983) cleavage. The extensive single-stranded regions could be explained either by nucleolytic degradation by *rec* B/C-type nucleases (Burton *et al.*, 1977; Clough, 1980), by displacement synthesis (Krimer and Hof, 1983), or by strand shearing mechanisms (Rosenberg and Cavalieri, 1968). Strand-separated regions could be related to transcription (Kohwi-Shigematsu *et al.*, 1983), to recombination (Klein and Byers, 1978), or to artifacts (Hardman and Gillespie, 1980; Micheli *et al.*, 1982). Nevertheless, we believe the strand separation mechanism is a plausible alternative which is compatible with the existing literature in eukaryotes and which makes a number of specific testable predictions. It also suggests an obvious mechanism to coordinate replication, transcription, and gene amplification.

We also wish to emphasize that our view is derived predominantly from rapidly dividing cells, and we cannot at present generalize to cultured or other slowly dividing cells which may use other mechanisms. In addition, our hypothesis has been based only on data from enterocoelomates. Finally, we emphasize that the mechanism we suggest for chromosomal DNA replication in enterocoelomates is an unproved hypothesis at best. It is proposed in the spirit of stimulating new experiments and represents an attempt to reconcile a number of apparent contradictions in the literature. While we believe that some variant of the strand separation mechanism is likely to function in most embryonic cells, it is unlikely to be correct in all of its particulars or necessarily to be universal for all chromosomal DNA replication in higher eukaryotes.

ACKNOWLEDGMENTS

We thank K. T. Hiriyanna, Nicholas Marini, and Alayne Senior for discussion of the manuscript, Deborah J. Stowers for artwork and technical assistance, and Karen Kalbaugh Morris for her hard work and patience in preparing the final manuscript.

REFERENCES

Amaldi, F., Buongiorno-Nardelli, M., Carnevali, F., Leoni, L., Mariotti, D., and Pomponi, M. (1973). Replicon origins in Chinese hamster cell DNA. Exp. Cell Res. 80, 79–87.

Balazs, I., Brown, E. H., and Schildkraut, C. L. (1974). Temporal order of replication of some DNA cistron. Cold Spring Harbor Symp. Quant. Biol. 38, 239–245.

Baldari, C. T., Amaldi, F., and Buongiorno-Nardelli, M. (1978). Electron microscopic analysis of replicating DNA of sea urchin embryos. Cell (Cambridge, Mass.) 15, 1095–1107.

Bayne, M. L., Alexander, R. F., and Benbow, R. M. (1984). DNA binding protein from ovaries of the frog, Xenopus laevis which promotes concatenation of linear DNA. J. Mol. Biol. 172, 87–108.

Bello, L. J. (1983). Differential transcription of early and late replicating DNA in human cells. Exp. Cell Res. 146, 79–86.

Benbow, R. M. (1984). Activation of DNA synthesis during early embryogenesis. In "Biology of Fertilization" (C. B. Metz, and A. Monroy, eds.), Vol. 3, pp. 299–345. Academic Press, New York.

Benbow, R. M., and Ford, C. C. (1975). Cytoplasmic control of nuclear DNA synthesis during early development of the frog, Xenopus laevis: A cell-free assay. Proc. Natl. Acad. Sci. U.S.A. 72, 2437–2441.

Benbow, R. M., and Krauss, M. R. (1977). Recombinant DNA formation in a cell-free system from Xenopus laevis eggs. Cell (Cambridge, Mass.) 12, 191–204.

Benbow, R. M., Pestell, R. Q. W., and Ford, C. C. (1975). Appearance of DNA polymerase activities during early development of *Xenopus laevis*. *Dev. Biol.* **43**, 159–174.

Benbow, R. M., Krauss, M. R., and Reeder, R. H. (1978). DNA synthesis in a multienzyme system from *Xenopus laevis* eggs. *Cell (Cambridge, Mass.)* **13**, 307–318.

Benbow, R. M., Stowers, D. J., Hiriyanna, K. T., Dunne, V. C., and Ford, C. C. (1985). DNA polymerase-α_2 from ovaries of the frog, *Xenopus laevis*. Submitted for publication.

Bendig, M. M. (1981). Persistence and expression of histone genes injected into *Xenopus* eggs in early development. *Nature (London)* **292**, 65–67.

Bensch, K. G., Tanaka, S., Hu, S.-Z., Wang, T. S.-F., and Korn, D. (1982). Intracellular localization of human DNA polymerase with monoclonal antibodies. *J. Biol. Chem.* **257**, 8391–8396.

Berezney, R., and Bucholtz, L. A. (1981). Dynamic association of replicating DNA fragments with the nuclear matrix of regenerating rat liver. *Exp. Cell Res.* **132**, 1–13.

Bjursell, G., Gussander, E., and Lindahl, T. (1979). Long regions of single-stranded DNA in human cells. *Nature (London)* **280**, 420–423.

Blumenthal, A. B., Kriegstein, H. J., and Hogness, D. S. (1974). The units of DNA replication in *Drosophila melanogaster* chromosomes. *Cold Spring Harbor Symp. Quant. Biol.* **38**, 205–223.

Botchan, P. M., and Dayton, A. I. (1982). A specific replication origin in the chromosomal rDNA of *L. variegatus*. *Nature (London)* **299**, 453–456.

Bozzoni, I., Baldari, C. T., Amaldi, F., and Buongiorno-Nardelli, M. (1981). Replication of ribosomal DNA in *Xenopus laevis*. *Eur. J. Biochem.* **118**, 585–590.

Braunstein, J. D., Schulze, D., DelGuidice, T., Furst, A., and Schildkraut, C. L. (1982). The temporal order of replication of murine immunoglobulin heavy chain constant region sequences corresponds to their linear order in the genome. *Nucl. Acids Res.* **10**, 6887–6902.

Buongiorno-Nardelli, M., Micheli, G., Carri, M. T., and Marilley, M. (1982). A relationship between replicon size and supercoiled loop domains in the eukaryotic genome. *Nature (London)* **298**, 100–102.

Burks, D. J., and Stambrook, P. J. (1978). Enrichment and visualization of small replication units from cultured mammalian cells. *J. Cell Biol.* **77**, 762–773.

Burton, W. G., Roberts, R. J., Myers, P. A., and Sager, R. (1977). A site-specific single-strand endonuclease from the eukaryote *Chlamydomonas*. *Proc. Natl. Acad. Sci. U.S.A.* **74**, 2687–2691.

Busby, S., and Bakken, A. H. (1980). Transcription in developing sea urchins: Electron microscopic analysis of cleavage, gastrula, and prism stages. *Chromosoma* **79**, 85–104.

Callan, H. G. (1972). Replication of DNA in the chromosomes of eukaryotes. *Proc. R. Soc. London, Ser. B* **181**, 19–41.

Campbell, J. L. (1983). Yeast DNA replication. *Genet. Eng.* **5**, 109–156.

Caboche, M., and Lark, K. G. (1981). Preferential replication of repeated DNA sequences in nuclei isolated from soybean cells grown in suspension culture. *Proc. Natl. Acad. Sci. U.S.A.* **78**, 1731–1735.

Carnevali, F., and Filetici, P. (1981). Single-stranded molecules in DNA preparations from cultured mammalian cells at different moments of the cell cycle. *Chromosoma* **82**, 377–384.

Case, S. T., Mongeon, R. L., and Baker, R. F. (1974). Single-stranded regions in DNA isolated from different developmental stages of the sea urchin. *Biochim. Biophys. Acta* **349**, 1–12.

Case, S. T., and Baker, R. F. (1975). Detection of long eukaryote-specific pyrimidine runs in repetitive DNA sequences and their relation to single-stranded regions in DNA isolated from sea urchin embryos. *J. Mol. Biol.* **98**, 69–92.

Challberg, M. D., and Kelly, T. J., Jr. (1982). Eukaryotic DNA replication: Viral and plasmid model systems. *Annu. Rev. Biochem.* **51**, 901–934.

Chambers, J. C., Watanabe, S., and Taylor, J. H. (1982). Dissection of a replication origin of *Xenopus* DNA. *Proc. Natl. Acad. Sci. U.S.A.* **79**, 5572–5576.

Clough, W. (1980). An endonuclease isolated from Epstein-Barr virus-producing human lymphoblastoid cells. *Proc. Natl. Acad. Sci. U.S.A.* **77**, 6194–6198.

Collins, J. H., Berry, D. E., and Cobbs, C. S. (1977). Structure of parental DNA of synchronized HeLa cells. *Biochemistry* **16**, 5438–5444.

Collins, J. M. (1977). Deoxyribonucleic acid structure in human diploid fibroblasts stimulated to proliferate. *J. Biol. Chem.* **252**, 141–147.

Conaway, R. C., and Lehman, I. R. (1982a). A DNA primase activity associated with DNA polymerase alpha from *Drosphila melanogaster* embryos. *Proc. Natl. Acad. Sci. U.S.A.* **79**, 2523–2527.

Conaway, R. C., and Lehman, I. R. (1982b). Synthesis by the DNA primase of *Drosophila melanogaster* of a primer with a unique chain length. *Proc. Natl. Acad. Sci. U.S.A.* **79**, 4585–4588.

Conrad, M. N., and Newlon, C. S. (1983). Stably denatured regions in chromosomal DNA from the *cdc2 Saccharomyces cerevisiae* cell cycle mutant. *Mol. Cell. Biol.* **3**, 1665–1669.

Cook, P. R., and Lang, J. (1984). The spatial organization of sequences involved in initiation and termination of eukaryotic DNA replication. *Nucleic Acids Res.* **12**, 1069–1075.

Davis, R. W., Simon, M., and Davidson, N. (1971). Electron microscope heteroduplex methods for mapping regions of base sequence homology in nucleic acids. *Meth. Enzymol.* **21D**, 413–428.

DePamphilis, M. L., and Wasserman, P. M. (1980). Replication of eukaryotic chromosomes: A closeup of the replication fork. *Annu. Rev. Biochem.* **49**, 627–666.

Dijkwel, P. A., Mullenders, L. H. F., and Wanka, F. (1979). Analysis of the attachment of replicating DNA to a nuclear matrix in mammalian interphase nuclei. *Nucleic Acids Res.* **6**, 219–230.

Edenberg, H. J., and Huberman, J. A. (1975). Eukaryotic chromosome replication. *Annu. Rev. Genet.* **9**, 245–284.

Englund, P. T., Hajduk, S. L., Marini, J. C., and Plunkett, M. L. (1982). Replication of kinetoplast DNA. *Cold Spring Harbor Monogr. Ser.* **2**, 423–433.

Epner, E., Rifkind, R. A., and Marks, P. A. (1981). Replication of alpha and beta globin DNA sequences occurs during early S phase in murine erythroleukemia cells. *Proc. Natl. Acad. Sci. U.S.A.* **78**, 3058–3062.

Etkin, L. D., and Roberts, M. (1983). Transmission of integrated sea urchin histone genes by nuclear transplantation in *Xenopus laevis*. *Science* **221**, 67–69.

Etkin, L. D., Pearman, B., Roberts, M., and Bektesh, S. L. (1984). Replication, integration, and expression of exogenous DNA injected into fertilized eggs of *Xenopus laevis*. *Differentiation* **26**, 194–202.

Flickinger, R. A., Coward, S. J., Miyagi, M., Moser, C., and Rollins, E. (1965). The ability of DNA and chromatin of developing frog embryos to prime for RNA polymerase-dependent RNA synthesis. *Proc. Natl. Acad. Sci. U.S.A.* **53**, 783–790.

Flickinger, R. A., Green, R., Kohl, D. M., and Miyagi, M. (1966). Patterns of synthesis of DNA-like RNA in parts of developing frog embryos. *Proc. Natl. Acad. Sci. U.S.A.* **56**, 1712–1718.

Flickinger, R. A., Miyagi, M., Moser, C. R., and Rollins, E. (1967a). The relation of DNA synthesis to RNA synthesis in developing frog embryos. *Dev. Biol.* **15,** 414–431.

Flickinger, R. A., Freedman, M. L., and Stambrook, P. J. (1967b). Generation times and DNA replication patterns of cells of developing frog embryos. *Dev. Biol.* **16,** 457–473.

Forbes, D. J., Kirschner, M. W., and Newport, J. W. (1983). Spontaneous formation of nucleus-like structures around bacteriophage DNA microinjected into *Xenopus* eggs. *Cell (Cambridge, Mass.)* **34,** 13–23.

Fox, A. M., Breaux, C. B., and Benbow, R. M. (1980). Intracellular localization of DNA polymerase activities within large oocytes of the frog, *Xenopus laevis. Dev. Biol.* **80,** 79–95.

Fry, M. (1983). Eukaryotic DNA polymerases. *In* "Enzymes of DNA Synthesis and Processing" (S. T. Jacob, ed.), vol. 1, pp. 39–92. CRC Press, Cleveland, Ohio.

Funderud, S., Andreassen, R., and Haugli, F. (1979). DNA replication in *Physarum polycephalum:* Electron microscopic and autoradiographic analysis of replicating DNA from defined stages of S-period. *Nucleic Acids Res.* **6,** 1417–1431.

Furst, A., Brown, E. H., Braunstein, J. D., and Schildkraut, C. L. (1981). Alpha-globin sequences are located in a region of early replicating DNA. *Proc. Natl. Acad. Sci. U.S.A.* **78,** 1023–1027.

Gargiulo, G., and Worcel, A. (1983). Analysis of chromatin assembled in germinal vesicles of *Xenopus* oocytes. *J. Mol. Biol.* **170,** 699–722.

Gargiulo, G., Wasserman, W., and Worcel, A. (1982). Properties of the chromatin assembled on DNA microinjected into *Xenopus* oocytes and eggs. *Cold Spring Harbor Symp. Quant. Biol.* **47,** 549–556.

Gaudette, M. F., and Benbow, R. M. (1985). Chromosomal DNA replication in *Xenopus laevis* embryos: A strand separation mechanism. Submitted for publication.

Goldberg, G. I., Collier, I., and Cassel, A. (1983). Specific DNA sequences associated with the nuclear matrix in synchronized mouse 3T3 cells. *Proc. Natl. Acad. Sci. U.S.A.* **80,** 6887–6891.

Goldman, M. A., Holmquist, G. P., Gray, M. C., Caston, L. A., and Nag, A. (1984). Replication timing of genes and middle repetitive sequences. *Science* **224,** 686–692.

Gourlie, B. B., and Pigiet, V. P. (1983). Polyoma virus minichromosomes: Characterization of the products of *in vitro* DNA synthesis. *J. Virol.* **45,** 585–593.

Habener, J. F., Bynum, B. S., and Shack, J. (1970). Destabilized secondary structure of newly replicated HeLa DNA. *J. Mol. Biol.* **49,** 157–170.

Hand, R. (1978). Eukaryotic DNA: Organization of the genome for replication. *Cell (Cambridge, Mass.)* **15,** 317–325.

Hardman, N., and Gillespie, D. A. F. (1980). DNA replication in *Physarum polycephalum.* Analysis of replicating nuclear DNA using the electron microscope. *Eur. J. Biochem.* **106,** 161–167.

Harland, R. M. (1981). Initiation of DNA replication in eukaryotic chromosomes. *Trends. Biochem. Sci.* **6,** 71–74.

Harland, R. M., and Laskey, R. A. (1980). Regulated replication of DNA microinjected into eggs of *Xenopus laevis. Cell (Cambridge, Mass.)* **21,** 761–777.

Hay, R. T., Hendrickson, E. A., and DePamphilis, M. L. (1984). Sequence specificity for the initiation of RNA-primed simian virus 40 DNA synthesis *in vivo. J. Mol. Biol.* **175,** 131–157.

Heintz, N. H., and Hamlin, J. L. (1982). An amplified chromosomal sequence that includes the gene for dihydrofolate reductase initiates replication within specific restriction fragments. *Proc. Natl. Acad. Sci. U.S.A.* **79,** 4083–4087.

Heintz, N. H., Milbrandt, J. D., Greisen, K. S., and Hamlin, J. L. (1983). Cloning of the initiation region of a mammalian chromosomal replicon. *Nature (London)* 302, 439–441.

Henson, P. (1978). The presence of single-stranded regions in mammalian DNA. *J. Mol. Biol.* 119, 487–506.

Hines, P. J., and Benbow, R. M. (1982). Initiation of replication at specific origins in DNA molecules microinjected into unfertilized eggs of the frog, *Xenopus laevis. Cell (Cambridge, Mass.)* 30, 459–468.

Hines, P. J., Hoehn-Saric, E., and Benbow, R. M. (1985). Replication of DNA microinjected into unfertilized eggs of the frog, *Xenopus laevis.* I. Characterization of replicating molecules. Submitted for publication.

Hiraga, S., Sudo, T., Yoshida, M., Kubota, H., and Ueyama, H. (1982). In vitro replication of recombinant plasmids carrying chromosomal segments of *Xenopus laevis. Proc. Natl. Acad. Sci. U.S.A.* 79, 3697–3701.

Hoffman, L. M., and Collins, J. M. (1976). Single-stranded regions in regenerating rat liver DNA. *Nature (London)* 260, 642–643.

Hu, S.-Z., Wang, T. S.-F., and Korn, D. (1984). DNA primase from KB cells evidence for a novel model of primase catalysis by a highly purified primase/polymerase alpha complex. *J. Biol. Chem.* 259, 2602–2609.

Huberman, J. A., and Riggs, A. D. (1968). On the mechanism of DNA replication in mammalian chromosomes. *J. Mol. Biol.* 32, 327–341.

Hübscher, U. (1983). The mammalian primase is part of a high molecular weight DNA polymerase-α polypeptide. *EMBO J.* 2, 133–136.

Hunt, B. F., and Vogelstein, B. (1981). Association of newly replicated DNA with the nuclear matrix of *Physarum polycephalum. Nucleic Acids Res.* 9, 349–363.

Huttner, K. M., Barbosa, J. A., Scangos, G. A., Pratchea, D. D., and Ruddle, F. H. (1981). DNA-mediated gene transfer without carrier DNA. *J. Cell Biol.* 91, 153–156.

Jacob, F., Brenner, S., and Cuzin, F. (1963). On the regulation of DNA replication in bacteria. *Cold Spring Harbor Symp. Quant. Biol.* 28, 329–348.

Kaguni, L. S., Rossignol, J.-M., Conaway, R. C., Banks, G. R., and Lehman, I. R. (1983a). Association of DNA primase with the β/γ subunits of DNA polymerase alpha from *Drosophila* embryos. *J. Biol. Chem.* 258, 9037–9039.

Kaguni, L. S., Rossignol, J.-M., Conaway, R. C., and Lehman, I. R. (1983b). Isolation of an intact DNA polymerase-primase from embryos of *Drosophila melanogaster. Proc. Natl. Acad. Sci. U.S.A.* 80, 2221–2225.

Kitchin, P. A., Klein, V. A., Fein, B. I., and Englund, P. T. (1984). Gapped minicircles. A novel replication intermediate of kinetoplast DNA. *J. Biol. Chem.* 259, 15532–15539.

Klein, H. L., and Byers, B. (1978). Stable denaturation of chromosomal DNA from *Saccharomyces cerevisiae* during meiosis. *J. Bacteriol.* 134, 629–635.

Kohwi-Shigematsu, T., Gelinas, R., and Weintraub, H. (1983). Detection of an altered DNA conformation at specific sites in chromatin and supercoiled DNA. *Proc. Natl. Acad. Sci. U.S.A.* 80, 4389–4393.

König, H., Riedel, H. D., and Knippers, R. (1983). Reactions in vitro of the DNA polymerase-primase from *Xenopus laevis* eggs: A role for ATP in chain elongation. *Eur. J. Biochem.* 135, 435–442.

Kornberg, A. (1980). "DNA Replication." Freeman, San Francisco, California.

Kornberg, A. (1982). "DNA Replication," Suppl. Freeman, San Francisco, California.

Kozu, T., Yagura, T., and Seno, T. (1982). De novo DNA synthesis by a novel mouse DNA polymerase associated with primase activity. *Nature (London)* 298, 180–182.

Kriegstein, H. J., and Hogness, D. S. (1974). Mechanism of DNA replication in *Drosophila* chromosomes: Structure of replication forks and evidence for bidirectionality. *Proc. Natl. Acad. Sci. U.S.A.* **71**, 135–139.

Krimer, D. B., and Hof, J. V. (1983). Extrachromosomal DNA of pea (*Pisum sativum*) root-tip cell replicates by strand displacement. *Proc. Natl. Acad. Sci. U.S.A.* **80**, 1933–1937.

Kurek, M. P., Billig, D., and Stambrook, P. J. (1979). Size classes of replication units in DNA from sea urchin eggs. *J. Cell Biol.* **81**, 698–703.

Laskey, R. A., and Harland, R. M. (1981). Replication origins in the eukaryotic chromosome. *Cell (Cambridge, Mass.)* **24**, 283–284.

Lee, C. S., and Pavan, C. (1974). Replicating DNA molecules from fertilized eggs of *Cochliomyia homnivorax* (Diptera). *Chromosoma* **47**, 429–437.

Liu, L. F., and Wang, J. C. (1979). Interaction between DNA and *Escherichia coli* DNA topoisomerase. I. Formation of complexes between the protein and superhelical and non-superhelical duplex DNAs. *J. Biol. Chem.* **254**, 11082–11088.

Liu, L. F., Rowe, T. C., Yang, L., Tewey, K., and Chen, G. L. (1983). Cleavage of DNA by mammalian DNA topoisomerase. II. *J. Biol. Chem.* **258**, 15365–15370.

Lo, S. C., Ross, J., and Mueller, G. C. (1980). Localization of globin gene replication in Friend leukemia cells to a specific interval of the S-phase. *Biochim. Biophys. Acta* **608**, 103–111.

Lonn, U., and Lonn, S. (1983). Aphidicolin inhibits the synthesis and joining of short DNA fragments but not the union of 10-kilobase DNA replication intermediates. *Proc. Natl. Acad. Sci. U.S.A.* **80**, 3996–3999.

McKnight, S. L., and Miller, O. L. (1979). Electron microscopic analysis of chromatin replication in the cellular blastoderm *Drosophila melanogaster* embryo. *Cell (Cambridge, Mass.)* **12**, 795–804.

McKnight, S. L., Bustin, M., and Miller, O. L. (1977). Electron microscopic analysis of chromosome metabolism in the *Drosophila melanogaster* embryo. *Cold Spring Harbor Symp. Quant. Biol.* **82**, 741–754.

McCready, S. J., Goodwin, J., Mason, D. W., Brazell, A., and Cook, P. R. (1980). DNA is replicated at the nuclear cage. *J. Cell Sci.* **46**, 365–386.

McTiernan, C. F., and Stambrook, P. J. (1980). Replication of DNA templates injected in frog eggs. *J. Cell Biol.* **87**, 45a.

McTiernan, C. F., and Stambrook, P. J. (1984). Initiation of SV40 DNA replication after microinjection in *Xenopus* eggs. *Biochim. Biophys. Acta* **782**, 295–303.

Matsumoto, L. H., and Gerbi, S. A. (1982). Early initiation of bovine satellite I DNA replication. *Exp. Cell Res.* **140**, 47–54.

Mechali, M., and Harland, R. M. (1982). DNA synthesis in a cell-free system from *Xenopus* eggs: Priming and elongation on single-stranded DNA *in vitro*. *Cell (Cambridge, Mass.)* **30**, 93–101.

Mechali, M., and Kearsey, S. (1984). Lack of specific sequence requirement for DNA replication in *Xenopus* eggs compared with high sequence specificity in yeast. *Cell (Cambridge, Mass.)* **38**, 55–64.

Mechali, M., Harland, R., and Laskey, R. (1983a). DNA replication in Xenopus eggs: *In vitro* and *in vivo* studies. *J. Cell Biochem.* **7B**, 125.

Mechali, M., Mechali, F., and Laskey, R. A. (1983b). Tumor promotor TPA increases initiation of replication on DNA injected into *Xenopus* eggs. *Cell (Cambridge, Mass.)* **35**, 63–69.

Micheli, G., Baldari, C. T., Carri, M. T., Cello, G. D., and Buongiorno-Nardelli, M. (1982). An electron microscope study of chromosomal DNA replication in different eukaryotic systems. *Exp. Cell Res.* **137**, 127–140.

Nelson, E. M., Stowers, D. J., Bayne, M. L., and Benbow, R. M. (1983). Classification of DNA polymerase activities from ovaries of the frog, *Xenopus laevis*. *Dev. Biol.* **96,** 11–22.

Newlon, C. S., Petes, T. D., Hereford, L. H., and Fangman, W. L. (1974). Replication of yeast chromosomal DNA. *Nature (London)* **247,** 32–35.

Newport, J., and Kirschner, M. (1982a). A major developmental transition in early *Xenopus* embryos. I. Characterization and timing of cellular changes at the mid-blastula stage. *Cell (Cambridge, Mass.)* **30,** 675–686.

Newport, J., and Kirschner, M. (1982b). A major developmental transition in early *Xenopus* embryos. II. Control of the onset of transcription. *Cell (Cambridge, Mass.)* **30,** 687–696.

Noguchi, H., Reddy, G. P., and Pardee, A. B. (1983). Rapid incorporation of label from ribonucleoside diphosphates into DNA by a cell-free high molecular weight fraction from animal cell nuclei. *Cell (Cambridge, Mass.)* **32,** 443–451.

Pardoll, D. M., Vogelstein, B., and Coffey, D. S. (1980). A fixed site of DNA replication in eukaryotic cells. *Cell (Cambridge, Mass.)* **19,** 527–536.

Parker, R. J., Tobia, A. M., Baum, S. G., and Schildkraut, C. L. (1975). DNA replication in synchronized cultured mammalian cells: V. The temporal order of synthesis of component α DNA during monkey DNA synthesis induced by SV40 virus. *Virology* **66,** 82–93.

Paulson, J. R., and Laemmli, U. K. (1977). The structure of histone-depleted metaphase chromosomes. *Cell (Cambridge, Mass.)* **12,** 817–828.

Petes, T. D., and Newlon, C. S. (1974). Structure of DNA in DNA replication mutants of yeast. *Nature (London)* **251,** 637–639.

Richter, A., Otto, B., and Knippers, R. (1981). Replication of SV40 chromatin in extracts from eggs of *Xenopus laevis*. *Nucleic Acids Res.* **9,** 3793–3807.

Riedel, H-D., König, H., Stahl, H., and Knippers, R. (1982) Circular single-stranded phage M13-DNA as a template for DNA synthesis in protein extracts from *Xenopus laevis* eggs: Evidence for a eukaryotic DNA priming activity. *Nucleic Acids Res.* **10,** 5621–5635.

Rochaix, J-D., Bird, A., and Bakken, A. (1974). Ribosomal RNA gene amplification by rolling circles. *J. Mol. Biol.* **87,** 473–487.

Rosenberg, B. H., and Cavalieri, L. F. (1964). On the transient template for *in vivo* DNA synthesis. *Proc. Natl. Acad. Sci. U.S.A.* **51,** 826–834.

Rosenberg, B. H., and Cavalieri, L. F. (1968). Shear sensitivity of the *E. coli* genome: Multiple membrane attachment points of the *E. coli* DNA. *Cold Spring Harbor Symp. Quant. Biol.* **33,** 65–72.

Rusconi, S., and Schaffner, W. (1981). Transformation of frog embryos with a rabbit β globin gene. *Proc. Natl. Acad. Sci. U.S.A.* **78,** 5051–5055.

Schandl, E. K., and Taylor, J. H. (1969). Early events in the replication and integration of DNA into mammalian chromosomes. *Biochem. Biophys. Res. Commun.* **34,** 291–300.

Shioda, M., Nelson, E. M., Bayne, M. L., and Benbow, R. M. (1982). DNA primase activity associated with DNA polymerase alpha from *Xenopus laevis* ovaries. *Proc. Natl. Acad. Sci. U.S.A.* **79,** 7209–7213.

Small, D., Nelkin, B., and Vogelstein, B. (1982). Nonrandom distribution of repeated DNA sequences with respect to supercoiled loops and the nuclear matrix. *Proc. Natl. Acad. Sci. U.S.A.* **79,** 5911–5915.

Spradling, A. C. (1981). The organization and amplification of two chromosomal domains containing *Drosophila* chorion genes. *Cell (Cambridge, Mass.)* **27,** 193–201.

Spradling, A. C., and Mahowald, A. P. (1981). A chromosome inversion alters the pattern of specific DNA replication in *Drosophila* follicle cells. *Cell (Cambridge, Mass.).* **27,** 203–209.

Stambrook, P. J. (1974). Temporal replication of ribosomal gene in synchonized Chinese hamster cells. *J. Mol. Biol.* **82,** 303–313.

Stambrook, P. J. and Flickinger, R. A. (1973). Changes in chromosomal DNA replication patterns in developing frog embryos. *J. Exp. Zool.* **174,** 101–104.

Tan, E. M., and Lerner, R. A. (1972). An immunological study of the fates of nuclear and nucleolar macromolecules during the cell cycle. *J. Mol. Biol.* **68,** 107–114.

Tanaka, S., Hu, S.-Z., Wang, T. S.-F., and Korn, D. (1982). Preparation and preliminary characterization of monoclonal antibodies against human DNA polymerase α. *J. Biol. Chem.* **257,** 8386–8390.

Taylor, J. H. (1984). Origins of replication and gene regulation. *Mol. Cell. Biol.* **61,** 99–109.

Taylor, J. H., Wu, M., and Erikson, L. C. (1973). Functional subunits of chromosomal DNA from higher eukaryotes. *Cold Spring Harbor Symp. Quant. Biol.* **38,** 225–232.

Truett, M. A. and Gall, J. G. (1977). The replication of ribosomal DNA in the macronucleus of *Tetrahymena. Chromosoma* **64,** 295–303.

Tseng, B. Y., and Ahlem, C. N. (1982). DNA primase activity from human lymphocytes: Synthesis of oligoribonucleotides that prime DNA synthesis. *J. Biol. Chem.* **257,** 7280–7283.

Tseng, B. Y., and Ahlem, C. N. (1983). A DNA primase from mouse cells: Purification and partial characterization. *J. Biol. Chem.* **258,** 9845–9849.

Tseng, B. Y., and Ahlem, C. N. (1984). Mouse primase initiation sites in the origin region of simian virus 40. *Proc. Natl. Acad. Sci. U.S.A.* **81,** 2342–2346.

Valenzuela, M. S., Mueller, G. C., and Dasgupta, S. (1983). Nuclear matrix–DNA complex resulting from *Eco*RI digestion of HeLa nucleoids is enriched for DNA replicating forks. *Nucleic Acids Res.* **11,** 2155–2164.

Vogelstein, B., Pardoll, D. M., and Coffey, D. S. (1980). Supercoiled loops and eukaryotic DNA replication. *Cell (Cambridge, Mass.)* **22,** 79–85.

Vogt, V. M., and Braun, R. (1977). The replication of ribosomal DNA in *Physarum polycephalum. Eur. J. Biochem.* **80,** 557–566.

Wang, T. S.-F., Hu, S.-Z., and Korn, D. (1984). DNA primase from KB cells. *J. Biol. Chem.* **259,** 1854–1865.

Wanka, F., Brouns, R. M. G. M. E., Aelen, J. M. A., and Eygensteyn, J. (1977). The origin of nascent single-stranded DNA extracted from mammalian cells. *Nucleic Acids Res.* **4,** 2083–2097.

Watanabe, S., and Taylor, J. H. (1980). Cloning of an origin of DNA replication of *Xenopus laevis. Proc. Natl. Acad. Sci. U.S.A.* **77,** 5292–5296.

Watson, J. D., and Crick, F. H. C. (1953a). Molecular structure of nucleic acids: A structure for deoxyribose nucleic acid. *Nature (London)* **171,** 737–738.

Watson, J. D., and Crick, F. H. C. (1953b). Genetic implications of the structure of deoxyribonucleic acid. *Nature (London)* **171,** 964–967.

Willard, H. F. (1977). Tissue specific heterogeneity in DNA replication patterns of human X-chromosomes. *Chromosoma* **61,** 61–73.

Wolstenholme, D. R. (1973). Replication DNA molecules from eggs of *Drosophila melanogaster. Chromosoma* **43,** 1–18.

Wortzman, M. S., and Baker, R. F. (1980). Specific sequences within single-stranded regions in the sea urchin genome. *Biochim. Biophys. Acta* **609,** 84–96.

Wortzman, M. S., and Baker, R. F. (1981). Two classes of single-stranded regions in DNA from sea urchin embryos. *Science* **211,** 588–590.

Yagura, T., Tanaka, S., Kozu, T., Seno, T., and Korn, D. (1983a). Tight association of DNA primase with a subspecies of mouse DNA polymerase alpha. *J. Biol. Chem.* **258,** 6698–6700.

Yagura, T., Kozu, T., Seno, T., Saneyoshi, M., Hiraga, S., and Nagano, H. (1983b). Novel form of DNA polymerase alpha associated with DNA primase activity of vertebrates: Detection with mouse stimulating factor. *J. Biol. Chem.* **258,** 13070–13075.

Yoda, K.-Y., and Okazaki, T. (1983). Primer RNA for DNA synthesis on single-stranded DNA template in a cell-free system from *Drosophila melanogaster* embryos. *Nucleic Acids Res.* **11,** 3433–3450.

Zakian, V. A. (1976). Electron microscopic analysis of DNA replicating in main band and satellite DNAs of *Drosophila virilis. J. Mol. Biol.* **108,** 305–331.

Zakian, V. A. (1981). Origin of replication from *Xenopus laevis* mitochondrial DNA promotes high frequency transformation of yeast. *Proc. Natl. Acad. Sci. U.S.A.* **78,** 3128–3132.

16

Role of Phosphorylation of Nonhistone Proteins in the Regulation of Mitosis

RAMESH C. ADLAKHA, FRANCES M. DAVIS,
AND POTU N. RAO

Department of Chemotherapy Research
The University of Texas M. D. Anderson Hospital
and Tumor Institute
Houston, Texas

I. INTRODUCTION

Our understanding of the eukaryotic cell cycle has improved a great deal since Howard and Pelc (1953) subdivided it into four phases, i.e.,

485

pre-DNA synthesis (G$_1$) period, DNA synthesis (S) period, post-DNA synthesis (G$_2$) period, and mitosis (M) period (for reviews, see Mazia, 1974; Baserga, 1976; 1981; Prescott, 1976; Pardee *et al.*, 1978). The S phase is the period in which DNA replicates, but it is during mitosis that the replicated chromosomes are equally distributed between the two daughter cells. Although the four stages of mitosis, i.e., prophase, metaphase, anaphase, and telophase, are well recognized, the complex mechanisms which regulate the process of mitosis are not yet well understood.

Using the technique of cell fusion, we have previously shown that mitotic cells can induce nuclear envelope breakdown and chromosome condensation in interphase cells (Johnson and Rao, 1970). These experiments suggested the presence of inducers of mitosis in mitotic cells. Kinetic studies on the entry of cells into mitosis in the presence of inhibitors of protein synthesis have provided evidence that the commitment of a cell to enter mitosis is dependent on continued protein synthesis. The results showed that inducers of mitosis are probably proteins.

There is also persuasive evidence from our laboratory and others that postsynthetic modification, especially phosphorylation, of the proteins synthesized in G$_2$ is equally important for the entry of cells into mitosis. The modification of proteins via reversible phosphorylation–dephosphorylation by phosphoprotein kinases and phosphoprotein phosphatases has been reported to be a mechanism of paramount importance in the regulation of numerous intracellular events (for reviews, see Krebs and Beavo, 1979; Cohen, 1982; Fischer, 1983; Ingebritsen and Cohen, 1983; Nestler and Greengard, 1983). The importance of protein phosphorylation in various steps of mitosis has been suggested by a number of studies. This review summarizes some of the recent evidence for the role of protein phosphorylation in the regulation of mitosis. However, we shall review only briefly the studies of others in this area and will draw heavily from recent results obtained in our own laboratory.

II. ROLE OF PROTEIN AND RNA SYNTHESIS IN THE ENTRY OF CELLS INTO MITOSIS

A. Effect of Inhibitors of RNA and Protein Synthesis

As the cell proceeds through DNA synthesis, the biosynthetic activities related to the entry of cells into mitosis are initiated and reach a peak at the G_2–M transition. Continued protein synthesis is known to be essential for the cells to complete G_2 and enter mitosis. Taylor (1963) showed that in human carcinoma strain KB cells, if protein synthesis was blocked by puromycin or chloramphenicol, the mitotic index remained constant for about 1 hr after addition of the inhibitor, indicating that any protein synthesis necessary for mitosis is completed by prophase. These studies also showed that partial inhibition of protein synthesis prolonged the G_2 period. Inhibition of RNA synthesis in mouse cell line L-929 by actinomycin resulted in a cessation of the entry of cells into mitosis within 2 hr (Caspersson et al., 1963). Kishimoto and Lieberman (1964), working with primary cultures of rabbit kidney cortex, observed that for the entry of cells into mitosis, RNA synthesis was necessary for at least 1 hr during G_2, while protein synthesis was needed until the initiation of prophase. Similar observations were made by Tobey et al. (1966) using Chinese hamster ovary (CHO) cells. These indicated that division-related RNA and protein synthesis occurred until 1.9 hr and 1.0 hr, respectively, prior to cell division. Using the protein synthesis inhibitors emetine or cycloheximide, Wagenaar and Mazia (1978) and Wagenaar (1983a) have shown that the synthesis of proteins needed for mitosis occurs just before mitosis in sea urchin embryos and Allium root tip cells.

B. G_2-Specific Proteins

To determine whether specific proteins may be involved in the G_2–M transition, we compared the total cellular proteins of G_2-synchronized, G_2-arrested, and S-phase HeLa cells by two-dimensional polyacrylamide gel electrophoresis (Al-Bader et al., 1978). In this study, HeLa cells were synchronized in S phase by the excess thymidine double-block method (Rao and Engelberg, 1966). Cells in S and G_2 phases were obtained by harvesting at 1 hr and 7 hr, respectively, after the reversal of the second thymidine block. To induce G_2 arrest, synchronized S-phase cells were exposed to cis acid (cis-4-[[[(2-chloroethyl)nitrosoamino]carbonyl]amino] cyclohexanecarboxylic acid) (75 μg/ml) for 60 min, the drug was removed and the cells incubated in drug-free medium for 16 hr, at which time 90% of the cells were blocked in G_2, as revealed by the premature chromosome condensation (PCC) method of cell cycle analysis (Rao et al., 1977).

The total cellular proteins of G_2-arrested, G_2-synchronized, and S-phase populations, as analyzed by two-dimensional polyacrylamide gel electrophoresis, are shown in Figs. 1–3. The number of proteins present in each of the 14 clusters in the three different cell populations

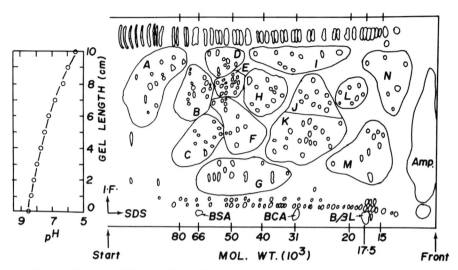

Fig. 1.   Tracings of the two-dimensional polyacrylamide gel of total cellular proteins of G_2-synchronized HeLa cells. The letters A–N indicate the various clusters of the protein spots. BSA, bovine serum albumin; BCA, bovine carbonic anhydrase; BBL, bovine β-lactoglobulin; Amp, ampholines; IF, isoelectric focusing. From Al-Bader *et al.* (1978).

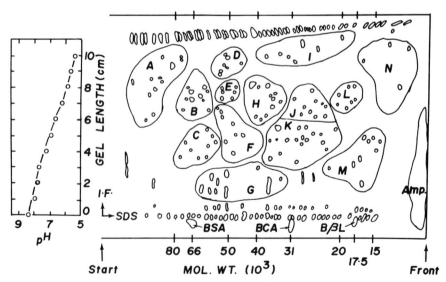

Fig. 2.   Tracings of the two-dimensional polyacrylamide gel of the total cellular proteins of G_2-arrested HeLa cells. Other details are the same as in Fig. 1. From Al-Bader *et al.* (1978).

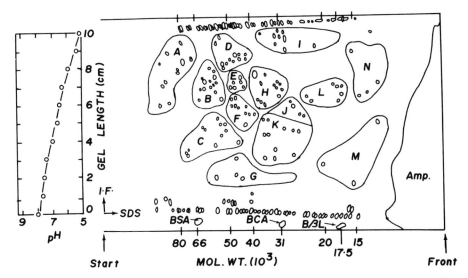

Fig. 3. Tracing of the two-dimensional polyacrylamide gel of total cellular proteins of synchronized S-phase HeLa cells. Other details are the same as in Fig. 1. From Al-Bader *et al.* (1978).

are summarized in Table I. A significant number of proteins present in clusters G, J, K, and M of both G_2-synchronized and G_2-arrested cells were absent from S-phase cells. These proteins are probably specific to the G_2 phase. A major difference between G_2-arrested and G_2-synchronized cells could be seen in cluster E. At least nine spots present in G_2-synchronized cells were missing in S-phase and G_2-arrested cells. These proteins may be required for the G_2–M transition. Interestingly, when these proteins were supplied to the G_2-arrested cells by fusion with synchronized G_2 cells, the nuclei of treated cells entered mitosis in synchrony with those of the untreated cells residing in the binucleate cells (Al-Bader *et al.*, 1978).

C. Cell Fusion Studies

The fusion of a mitotic cell with an interphase cell results in chromosome condensation and dissolution of the nuclear envelope in the interphase cell within 45 min following fusion. This phenomenon has been called "premature chromosome condensation" (Johnson and Rao, 1970). It has also been called "prophasing" by Matsui *et al.* (1972), since the changes observed mimic to a great extent the normal prophase condensation patterns. No effort will be made in this chapter to review

TABLE I

Comparison of the Protein Profiles in S-Phase, G_2-Synchronized, and G_2-Arrested HeLa Cells[a]

	Number of protein spots in			Major differences in number of spots between		
	S phase	G_2 arrested	G_2 synchronized	G_2 synchronized and S	G_2 arrested and S	G_2 synchronized and G_2 arrested
A	12	12	13			
B	14	10	14			
C	12	9	10			
D	12	7	9			
E	6	6	15	9	0	9
F	10	8	11			
G	4	10	11	7	6	1
H	12	11	11			
I	9	7	9			
J	4	10	10	6	6	0
K	11	18	18	7	7	0
L	7	7	7			
M	3	9	9	6	6	0
N	4	4	6			
	120	128	153	35	25	10

[a]From Al-Bader *et al.* (1978).

this phenomenon (for review, see Rao, 1982). However, two points of special relevance to this chapter will be briefly mentioned. First, experiments of Rao and Johnson (1974) showed that when mitotic cells prelabeled during G_2 with ^{3}H-labeled tryptophan and leucine were fused with unlabeled interphase cells in the presence of cycloheximide (to block protein synthesis during fusion), autoradiographs of the cells showed, without exception, that the label from mitotic cells had migrated onto the prematurely condensed chromosomes. These results clearly indicate that certain of the inducing factors in mitotic cells are proteins, and that these proteins bind to chromatin of the induced nucleus and are responsible either wholly or in part for the ordered condensation of chromosomes. The second point of interest stems from the studies of Matsui *et al.* (1971). Their results showed that mitotic cells that completed the G_2–M transition in the presence of either puromycin or cycloheximide failed to induce prophasing when subsequently fused with interphase cells. Detailed kinetic analysis of prophasing frequency revealed that the synthesis of mitotic factors was continuous, and the amount of mitotic factors necessary for a single nucleus to enter prophase was synthesized by 60–65 min before mitosis.

III. THE INDUCTION OF GERMINAL VESICLE BREAKDOWN AND CHROMOSOME CONDENSATION IN *XENOPUS* OOCYTES BY MITOTIC FACTORS

Sunkara *et al.* (1979a,b) have demonstrated that extracts from mitotic HeLa cells can induce germinal vesicle breakdown (GVBD) and chromosome condensation when injected into immature *Xenopus laevis* oocytes. Thus, a bioassay has been developed for the study of mitotic inducer proteins. These studies also revealed that the mitotic factors accumulate slowly in the beginning of G_2, increase rapidly during late G_2, and reach a threshold at the G_2–M transition when the chromatin is transformed into discrete chromosomes. Extracts from mitotic CHO cells (Nelkin *et al.*, 1980) and from cdc mutants of *Saccharomyces cerevisiae* (Weintraub *et al.*, 1983) have also been reported to induce GVBD in *Xenopus* oocytes. We have further demonstrated that the mitotic factors have a great affinity for chromatin and are localized on metaphase chromosomes as well as in cytoplasm. They are nondialyzable, heat- and Ca^{2+}-sensitive, and Mg^{2+}-dependent nonhistone proteins with an approximate molecular mass of 100 kilodaltons (Sunkara *et al.*, 1982; Adlakha *et al.*, 1982a,b).

These factors are inactivated at the end of mitosis by another set of factors, called the inhibitors of mitotic factors (IMF), that are activated at telophase (Adlakha *et al.*, 1983, 1984). However, it is not yet clear whether the binding of mitotic factors to chromatin induces the tight packaging of chromatin fibers into metaphase chromosomes directly or whether the mitotic factors become associated with chromatin and then modify (e.g., phosphorylate or acetylate) the structural proteins to induce chromosome condensation indirectly. Our studies have also revealed that these mitotic factors are greatly stabilized by phosphatase inhibitors (Adlakha *et al.*, 1982a), suggesting a possible role for phosphorylation and dephosphorylation in regulating the biological function of these factors. Recently, Wagenaar (1983b) has shown that microinjection of free Ca^{2+} into fertilized and artificially activated sea urchin eggs delayed the onset of mitosis. We have shown that mitotic factors are inactivated by Ca^{2+} (Adlakha *et al.*, 1982a). Since fluctuations in the Ca^{2+} concentration are known to alter the state of phosphorylation of a number of proteins, and hence their biological activity (Cohen, 1983), these observations suggest a possible role for protein phosphorylation in the regulation of mitosis (see Chapter 12).

IV. ROLE OF PROTEIN PHOSPHORYLATION
DURING MITOSIS AND MEIOSIS

A. Phosphorylation of Histones

The recognition of an association between histone (especially H1) phosphorylation and mitosis dates back to the early work of Lake (1973; Lake and Saltzman, 1972; Lake *et al.*, 1972) and of Bradbury's group (1973, 1974). The most detailed studies on the phosphorylation of histones during the cell cycle in mammalian cells have been carried out by Gurley *et al.* (1975, 1978a,b), although several other groups have also made important contributions (Marks *et al.*, 1973; Balhorn *et al.*, 1975; Sherod *et al.*, 1975; Hohmann *et al.*, 1976; Johnson and Allfrey, 1978; Fischer and Laemmli, 1980; Ajiro *et al.*, 1981a,b). These studies have shown that the superphosphorylation of histone H1 (and H3), which coincides with the G_2–M transition, plays an important role in chromosome condensation and the initiation of mitosis.

More recently, Krystal and Poccia (1981), Hanks *et al.* (1983), and Ajiro *et al.* (1983), working with three different systems, i.e., sea urchin embryos, HeLa cells, and temperature-sensitive baby hamster kidney cells ($tsBN_2$), respectively, showed that H1 is also phosphorylated during premature chromosome condensation. However, a large body of evidence indicates that the superphosphorylation of histone H1 by itself is not sufficient for chromosome condensation (Gorovsky and Keevert, 1975; Tanphaichitr *et al.*, 1976; Krystal and Poccia, 1981; Allis and Gorovsky, 1981; Hanks *et al.*, 1983). Therefore, the molecular mechanisms by which the phosphorylation of H1 modulates the structure of nuclear chromatin during the cell cycle remain unclear.

B. Phosphorylation of Intermediate Filament Proteins

A major reorganization of the microfilaments and microtubules takes place during mitosis (Sanger and Sanger, 1976; Brinkley *et al.*, 1976). Aubin *et al.* (1980), working with PTK_2 cells, have shown that vimentin and cytokeratin filaments are distributed differently during mitosis and have suggested a role for the vimentin filaments in orienting the spindle and/or the chromosomes during mitosis. The involvement of phosphorylation in filament formation or aggregation was suggested by Lazarides (1980). Recently, several groups have shown that vimentin and keratin polypeptides are phosphorylated at an increased rate during mitosis as compared to interphase (Robinson *et al.*, 1981; Evans and Fink, 1982; Bravo *et al.*, 1982). These investigators speculated that the phosphorylation–dephosphorylation of these polypeptides might be responsible for the changes in filament organization observed during mitosis.

C. Phosphorylation of Ribosomal Protein S6 and Other Proteins during Meiotic Maturation of X. laevis Oocytes

In immature X. laevis oocytes, meiosis can be induced to resume by stimulation in vitro with progesterone or by microinjection of maturation-promoting factor (MPF) from matured oocytes or of mitotic factors from mammalian cells. Progesterone induces the production of MPF through a complex pathway that requires dephosphorylation. Maller and Krebs (1977) first demonstrated that the microinjection of the catalytic subunit of a cyclic AMP-dependent protein kinase inhibited progesterone-induced maturation, while microinjection of the regulatory subunit or a specific heat-stable protein inhibitor of the catalytic subunit of cyclic AMP-dependent kinase was able to induce maturation, even in the absence of progesterone. More recently, it was reported that the microinjection of the inhibitor of protein phosphatase-1 blocked progesterone-induced maturation of oocytes and also the maturation induced by microinjection of the heat-stable protein kinase inhibitor, but not the maturation of oocytes induced by microinjection of MPF (Huchon et al., 1981). These results clearly demonstrate that protein phosphatase-1 is responsible for the dephosphorylation event known to be involved in the triggering of maturation (Maller and Krebs, 1980). Once MPF has become active, protein phosphorylation has been correlated with the breakdown of the germinal vesicle and the condensation of the meiotic chromosomes. Boyer et al. (1983) reported an increase in the phosphorylation of a 105-kilodalton protein during meiotic maturation. Phosphorylation of the ribosomal protein S6 during maturation, as determined by two-dimensional gel electrophoresis, has also been reported by Hanocq-Quetier and Baltus (1981), Kalthoff et al. (1982), Nielsen et al. (1982), and Kruppa et al. (1983). Maller et al. (1977), Wu and Gerhart (1980), and Nielsen et al. (1982) have reported that microinjection of crude or partially purified MPF into oocytes stimulated protein phosphorylation (including ribosomal protein S6) and that these changes were evident within minutes of MPF injection. These results clearly demonstrate that protein phosphorylation occurs during meiotic maturation as a result of MPF action. However, it has not yet been established that the activity of MPF is due to a kinase, since only partially purified preparations of MPF were used in most of these studies.

D. Phosphorylation of High Mobility Group (HMG), Nuclear Lamina, Nuclear Matrix, and Nucleolar Proteins during Mitosis

Among the four major HMG proteins identified, only HMG 14 and 17 have been shown to be phosphorylated *in vivo* (Bhorjee, 1981; Levy-Wilson, 1981; Saffer and Glazer, 1982). Bhorjee (1981) reported a sevenfold increase in the phosphorylation of HMG 14 in the G_2 phase as compared to the G_1 phase in HeLa cells. In agreement with these results Paulson and Taylor (1982) found that phosphorylation of HMG 14 was greater in metaphase chromosomes than in interphase chromatin. Walton and Gill (1983) were able to phosphorylate HMG 14 with casein kinase II *in vitro* at the site identical to the *in vivo* phosphorylation site in HeLa cells. These authors suggested that since HMG 14 and 17 bind specifically to the DNase I-sensitive sites correlated with the transcriptionally active or potentially active gene sequences (for reviews, see Allfrey, 1982; Weisbrod, 1982), their extensive phosphorylation at mitosis may be responsible for the decrease in transcription that has been observed during mitosis.

Gerace and Blobel (1980) and Jost and Johnson (1981) suggested that the phosphorylation of laminar proteins A, B, and C, the three major polypeptides of the nuclear envelope lamina, is related to the depolymerization of the lamina and hence the breakdown of the nuclear envelope. Dephosphorylation of these proteins at the end of mitosis could allow their repolymerization and the re-formation of the nuclear envelope.

Henry and Hodge (1983) have recently shown both *in vivo* and *in vitro* that several nuclear matrix polypeptides are extensively phosphorylated during premitosis and dephosphorylated during postmitosis. These studies have implied a relationship between nuclear protein biochemistry and function during mitosis. Working with *Physarum polycephalum*, Shibayama *et al.* (1982) observed cell cycle-dependent changes in the phosphorylation of several nucleolar proteins *in vivo*. A fivefold increase in the specific activity as well as incorporation of ^{32}P in some nucleolar proteins occurred about 2 hr before mitosis. The period of phosphorylation coincided with the swelling of the nucleoli and increased ribosomal RNA synthesis.

E. Phosphorylation of Nonhistone Proteins Extractable by 0.2 *M* NaCl

The phosphorylation of nonhistone proteins (NHP) has also been implicated in the regulation of transcription. An enhancement in NHP phosphorylation represents an early event coinciding with gene activity and has been well documented (Stein *et al.*, 1974; Kleinsmith, 1975; Jungmann and Kranias, 1977; Ahmed and Wilson, 1978). The suggestion that phosphorylation and dephosphorylation of NHP extractable in 0.2 *M* NaCl may play a role in mitosis stems mainly from two independent studies from our laboratory and will be discussed in detail here.

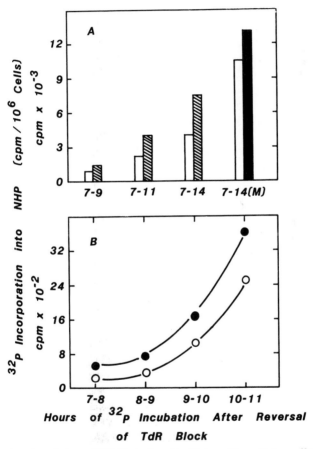

Fig. 4. Phosphorylation of NHP during G_2–M transition. HeLa cells were synchronized in S phase by the excess thymidine (2.5 m*M*) double-block method. At the end of S phase (i.e., 7 hr after reversal), the cells were incubated with ^{32}P continuously (A) or for 1 hr (B). Cells were collected and extracts prepared and processed for scintillation counting. Symbols in (A): open bar, cytoplasmic extracts; cross-hatched bar, nuclear extracts; solid bar, chromosomal extracts. Symbols in (B): open circle, cytoplasmic extracts; filled circle, nuclear extracts. 7–14 (M) represents mitotic cells, whereas the other time points refer to G_2 cells. From Sahasrabuddhe *et al.* (1984).

1. Phosphorylation of NHP during the G_2–M Transition

Since the mitotic factors can be completely extracted in 0.2 M NaCl and the maturation-promoting activity (MPA) of these factors is greatly stabilized by phosphatase inhibitors (Adlakha *et al.*, 1982a), we investigated the role of phosphorylation of NHP (extractable in 0.2 M NaCl) in the initiation of mitosis. Cytoplasmic and nuclear or chromosomal NHP were extracted as previously described (Adlakha *et al.*, 1982a,c; Sahasrabuddhe *et al.*, 1984). For these studies, HeLa cells synchronized in S phase by double-thymidine block were labeled with ^{32}P beginning at the end of S phase. Cells were collected at different points in G_2 and mitosis, NHP extracted, and radioactivity in the NHP measured. The level of phosphorylation of NHP increased gradually during early to mid G_2 and rapidly during late G_2, reaching a maximum at mitosis (about 10-fold above early G_2; Fig. 4A). Pulse labeling experiments gave identical results (Fig. 4B).

2. Continuous Phosphorylation of NHP in Cells Held in Mitosis

We studied the effect of holding cells in mitosis for a long period, using Colcemid, N_2O (Rao, 1968), and Taxol (P. N. Rao and B. R. Brinkley, unpublished data), on the level of phosphorylation of NHP. We observed a continuous increase in the level of phosphorylation of NHP over a period of 8–10 hr (Table II). To determine whether this

TABLE II

Increase in Uptake of ^{32}P into NHP
during Prolonged Mitotic Arrest[a]

	Incorporation of ^{32}P into NHP (cpm/10^6 cells)		
Hours held in mitosis	N_2O	Colcemid	Taxol
1	572	511	459
3	1893	1564	1427
6	3478	3316	3197
8	4394	3948	3913
10	5405	4826	4709

[a]HeLa cells were synchronized in mitosis by the N_2O block method. ^{32}P was added, cells were divided into three groups and exposed to three different mitotic blocking agents, and incubation continued. Cell samples taken at various times to incubation were processed for the incorporation of label into NHP. From Sahasrabuddhe *et al.* (1984).

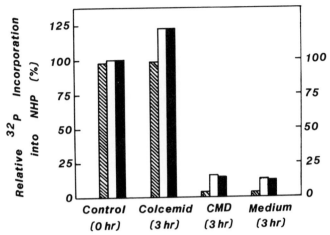

Fig. 5. Dephosphorylation of NHP during M–G_1 transition. HeLa cells were labeled with [32P] during G_2 and were synchronized in mitosis by N_2O block method. Mitotic cells were allowed to divide in the presence or absence of cycloheximide (CMD). Three hours after reversal of the N_2O block, cells were collected and extracts prepared and processed for scintillation counting. Symbols: Open bar, cytoplasmic extracts; solid bar, nuclear or chromosomal extracts; cross-hatched bars, mitotic index (%). From Sahasrabuddhe *et al.* (1984).

increased incorporation of [32P] was due to superphosphorylation of NHP or to turnover of the existing phosphate, prelabeled ([32P]) mitotic HeLa cells were incubated for 9 hr in a medium containing Colcemid and unlabeled pyrophosphate. Cell samples were taken at regular intervals and the amount of radioactivity in NHP determined. We observed a linear decrease in [32P] label with time, reaching the 50% mark in 6 hr. These results indicate that the phosphate groups on proteins turn over *in vivo* when cells are held in mitosis for prolonged periods of time.

3. Effect of Cell Division on NHP Phosphorylation

[32P]-labeled, N_2O-blocked mitotic cells were allowed to divide in the presence or absence of cycloheximide, and G_1 cells were collected 3 hr after the reversal of the N_2O block. The level of phosphorylation of NHP in G_1 cells was reduced to 10–15% of that in mitotic cells (Fig. 5). The presence of cycloheximide during the M–G_1 transition had no effect either on cell division or on the levels of phosphorylation of NHP. These results show that there is a strong correlation between the phosphorylation of NHP and the entry of cells into mitosis, on the one hand, and the dephosphorylation of NHP at the time of the exit of cells from mitosis, on the other.

4. Cyclical Pattern of NHP Phosphorylation during the HeLa Cell Cycle

The phosphorylation of NHP at various points during G_1 and S was studied. We observed a gradual decrease in the level of NHP phosphorylation during G_1, with the minimum level occurring at early to mid S phase (3–5% of that of mitotic cells). The phosphorylation of NHP began to increase in late S to early G_2 and reached a peak at mitosis (Adlakha *et al.*, 1982c; Sahasrabuddhe *et al.*, 1984). Proteins from extracts of mid G_2, mitotic, and early G_1 cells labeled with ^{32}P-orthophosphate were separated by SDS–polyacrylamide gel electrophoresis. Autoradiographs showed that eight major proteins (with apparent molecular masses of 100, 92, 70, 55, 43, 36, 30, and 27.5

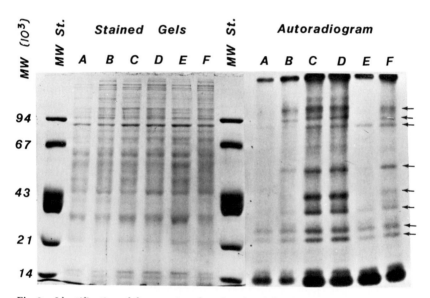

Fig. 6. Identification of the proteins phosphorylated during G_2–M and dephosphorylated during M–G_1 transitions. HeLa cells synchronized in S phase by double-thymidine block were labeled with ^{32}P continuously beginning at the end of S phase. Cells were collected in mid G_2, mitosis, and early G_1 and extracts prepared as described in the legend to Fig. 5. Proteins were separated by SDS–polyacrylamide gel electrophoresis and the gels stained by Coomassie Blue. For radioautography Kodak XAR-5 film was exposed to dried gels. Lanes: A, G_2 cytoplasmic; B, G_2 nuclear; C, mitotic cytoplasmic; D, mitotic chromosomal; E, G_1 cytoplasmic; F, G_1 nuclear. The migration of molecular weight standards (MW St.) is shown. Arrows indicate the eight major protein bands phosphorylated during mitosis. Note the decrease in the intensity of labeling of these eight bands in early G_1 (lanes E and F) as compared to mitotic (lanes C and D) extracts. From Sahasrabuddhe *et al.* (1984).

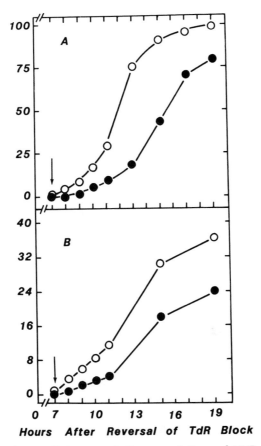

Fig. 7. Effect of X irradiation on mitotic accumulation and NHP phosphorylation. HeLa cells were synchronized in S phase by double-TdR block method. At 7 hr after reversal of the second TdR block, an aliquot of cells was exposed to 500 rad of X rays (indicated by arrow), whereas another aliquot of unirradiated cells served as control. ^{32}P-orthophosphate and Colcemid were added to both the control and treated cultures and incubation continued. Cell samples were taken at different times for mitotic index and the preparation of extracts to determine the incorporation of ^{32}P into NHP. (A) Mitotic index (%); (B) ^{32}P incorporation into NHP (cpm $\times$ $10^{-3}/10^{6}$ cells). Open circle, control; filled circle, X irradiated. From Sahasrabuddhe *et al.* (1984).

kilodaltons were extensively phosphorylated only in mitosis (Fig. 6). The relative amount of ^{32}P incorporation into these NHP in mid-G_2 and early-G_1 cells was only 10–20% of that in mitotic cells. These studies suggest that phosphorylation–dephosphorylation of these NHP may play a role in mitosis.

5. Effect of Radiation-Induced Mitotic Delay on the Phosphorylation of NHP

We have also studied the effects of X ray-induced mitotic delay on the levels of phosphorylation of NHP. Figure 7 shows that the mitotic delay induced by 500 rad of X rays resulted in a corresponding delay in the incorporation of ^{32}P into NHP, again demonstrating a strong correlation between the phosphorylation of NHP and the entry of cells into mitosis.

6. Effect of Cycloheximide on G_2 Traverse and the Phosphorylation of NHP

As described in Section II,A, the synthesis of new proteins until about 1 hr before mitosis is necessary for G_2 cells to enter into mitosis. Our own studies have shown that mitotic factors become available only during G_2 and that they reach a critical level by the G_2–M transition. We wished to determine whether these newly synthesized proteins were phosphorylated at the G_2–M transition. HeLa cells synchronized in S phase were labeled with ^{32}P from the end of S phase, and aliquots of cells were pulse treated with cycloheximide for 1 hr at different times during G_2. While the level of phosphorylation of NHP increased with time in the control cells, it ceased in the treated cells as soon as cycloheximide was added (Fig. 8).

Why should cycloheximide block phosphorylation? If cycloheximide were blocking the synthesis of enzymes necessary for phosphorylation, i.e., phosphokinases, the enzymes existing prior to the addition of cycloheximide should be active and continue to phosphorylate NHP unless they have an extremely short half-life. An alternative explanation would be that most of the NHP phosphorylation observed during G_2 occurs in the newly synthesized NHP. Therefore, it appears that continued synthesis and immediate phosphorylation of NHP may be equally important for the G_2–M transition. Westwood *et al.* (1983) have recently reported results very similar to ours on the phosphorylation of NHP during the cell cycle of CHO cells. In contrast, Song and Adolph (1983), also working with HeLa cells, reported that NHP of isolated metaphase chromosomes are strikingly dephosphorylated in comparison to those of S-phase chromatin. The differences between these two studies most likely stem from the different experimental protocols used for the extraction of NHP. In our studies, we are dealing with only a small subset of NHP extractable with 0.2 M NaCl from whole cells, a major portion of these proteins being present in the cytoplasm (Adlakha *et al.*, 1982a,b). Song and Adolph, on the other hand, dealt exclusively with NHP extracted from isolated chromatin

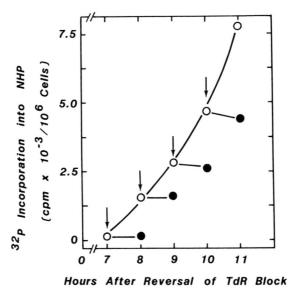

Fig. 8.   Effect of cycloheximide on the phosphorylation of NHP during G_2–M transition. HeLa cells were synchronized in S phase as monolayer cultures. ^{32}P was added to all the dishes when the cells were entering G_2 (i.e., 7 hr after reversal of second TdR block) and incubation continued in the presence of Colcemid. Dishes were grouped into two sets. In the first set, one of the dishes was taken every hour and pulse treated with cycloheximide for 1 hr. The other set served as control. Cell samples taken at 1-hr intervals from the control and treated culture was processed for determination of the incorporation of ^{32}P into NHP. Symbols: Open circle, control; filled circle, cycloheximide treated. Arrows indicate the time of addition of cycloheximide. From Sahasrabuddhe *et al.* (1984).

obtained after digestion of the nuclei with micrococcal nuclease. The two studies, therefore, investigated phosphorylation in altogether different subsets of NHP. Identification of these mitosis-specific proteins, the kinases, and the phosphorylated products will be crucial to the elucidation of the molecular mechanisms involved in the regulation of mitosis as well as meiosis.

F. Monoclonal Antibodies that Recognize Mitosis-Specific Phosphoproteins

1. Characterization of the Mitosis-Specific Phosphoprotein Antigens

Our laboratory has recently isolated two monoclonal antibodies, MPM-1 and MPM-2, that react specifically with mitotic cells, but not

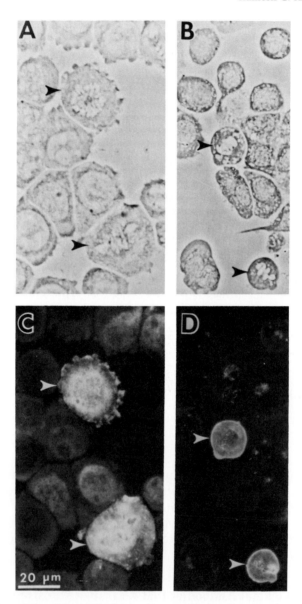

Fig. 9. Specificity of the antibodies to mitotic cells. Random populations of HeLa (A and C) and mosquito (B and D) cells were stained by indirect immunofluorescence with monoclonal antibody MPM-1. (A and B) Phase-contrast photomicrographs. (C and D) Fluorescent photomicrographs. Fluorescence is present both on the chromosomes and in the cytoplasm of the mitotic cells, which are indicated by arrows. From Davis *et al.* (1983).

with interphase cells (Davis *et al.*, 1983). MPM-1 is an IgM antibody with κ light chains, and MPM-2 is an IgG antibody with κ light chains. The two antibodies bind to both chromosomes and cytoplasm in the mitotic cells as determined by indirect immunofluorescence (Fig. 9). Mitotic and meiotic cells of all animal and plant species tested, including human, mouse, hamster, chicken, frog, mosquito, nematode, and carrot, react specifically with these antibodies. Since the specificity of the antibodies for mitotic cells extends over a broad range of species, it is likely that the antigenic determinant is highly conserved during evolution and must be of functional value to the process of cell division. Electroimmunoblots of proteins of synchronized cells separated by

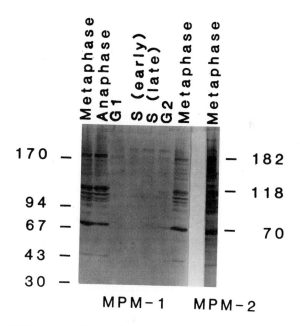

Fig. 10. Cell cycle specificity of polypeptides recognized by the antibodies. Polypeptides from whole HeLa cells synchronized in various cell cycle phases were separated in SDS–gradient polyacrylamide gels, electrophoretically transferred to nitrocellulose sheets, and stained with monoclonal antibodies MPM-1 and MPM-2 by an indirect immunoperoxidase procedure. Approximately 25 μg of protein per slot was loaded. Metaphase cells and Anaphase cells 1 hr after reversal of N_2O blockade; G_1, cells in G_1 phase 3.5 hr after reversal of N_2O blockade; S (early), cells in early S phase 2 hr after reversal of second thymidine blockade; S (late), cells in late S phase 6.5 hr after reversal of second thymidine blockade; G_2, cells in G_2 phase 10 hr after reversal of second thymidine blockade. Sizes shown in kilodaltons are of marker proteins (on the left) and of the major antigens (on the right). From Davis *et al.* (1983).

electrophoresis in SDS–polyacrylamide gels revealed that the two monoclonal antibodies recognized the same family of polypeptide bands in mitotic HeLa cells (Fig. 10). Three major reactive bands of 182, 118, and 70 kilodaltons and at least 13 other bands ranging in apparent molecular mass from 40 to >200 kilodaltons exhibited affinity for the antibodies. No differences in relative antibody affinity among the major bands were observed, and no evidence for the proteolysis of a single antigen into multiple antigens was found.

Since the experiments described above suggested that the phosphorylation of NHP proteins might play a role in mitosis, the phosphorylation of these mitosis-specific antigens was investigated. Phosphorylation of the antigens was studied by using synchronized HeLa cells labeled with ^{32}P during the G_2 phase and arrested in mitosis by using Colcemid. An extract from these cells was immunoprecipitated, and the proteins were separated on polyacrylamide gels, transferred to nitrocellulose paper, and processed either for autoradiography or for antibody reactivity by indirect immunoperoxidase staining.

Figure 11 shows that the 182-, 118-, and 70-kilodalton polypeptide antigens were ^{32}P labeled. Other ^{32}P-labeled NHP proteins did not react with the antibody. Treatment of a mitotic cell extract with alkaline phosphatase in solution removed detectable antigenic sites (Fig. 11B, lanes 5 and 6). No significant proteolysis of the phosphatase-digested samples was detected on amido black-stained immunoblots, nor did phosphatase digestion have any effect on immunostained bands recognized by an unrelated antibody specific for nucleoli.

The susceptibility of the immunoreactivity to phosphatase digestion was further tested by the following experiment. Proteins from mitotic cells were separated on polyacrylamide gels and transferred to nitrocellulose paper. Then the immunoblot was incubated in the presence or absence of bacterial alkaline phosphatase (10 units/ml). The immunoblots were stained by the indirect immunoperoxidase method. Phosphatase treatment completely removed the antigenic reactivity of the 182-kilodalton polypeptide and reduced the reactivity of other bands, as shown in Fig. 11. In control experiments, a similar treatment of immunoblots with phosphatase had no effect on indirect immunoperoxidase staining by an antibody specific for nucleoli. These results have recently been confirmed in the nematode *Caenorhabditis elegans* (Hecht *et al.*, 1983), although the apparent molecular masses of the major phosphorylated polypeptide bands that bind the antibodies are somewhat different in this organism than in mammalian cells.

Because the antigens recognized are all phosphoproteins and the antigenic reactivity is lost when the phosphate groups are removed by

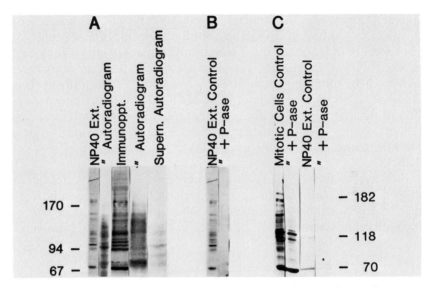

Fig. 11. Antibodies recognize a family of phosphorylated polypeptides. Synchronized HeLa cells were labeled with ^{32}P-orthophosphate during the G_2 phase, and cells were arrested in mitosis with 0.05 μg of Colcemid per milliliter. An extract was prepared from one aliquot of the cells by sonication in a buffer consisting of 0.5% Nonidet P-40, 0.15 M NaCl, 10 mM NaH$_2$PO$_4$, and 1 mM phenylmethylsulfonyl fluoride, pH 8. After immunoprecipitation or digestion with alkaline phosphatase, the polypeptides were separated by electrophoresis and stained by the indirect immunoperoxidase procedure with antibody MPM-1, as described in the legend to Fig. 10. Proteins of whole mitotic HeLa cells were stained by the indirect immunoperoxidase procedure (lane 1; lane 2 is an autoradiograph of lane 1). Antigenic proteins from an extract of mitotic HeLa cells were precipitated with antibody MPM-2 and then separated on the polyacrylamide gel and stained by the indirect immunoperoxidase procedure (lane 3; lane 4 is an autoradiograph of lane 3). The sensitivity of the antigens to alkaline phosphatase was investigated by incubating an extract from mitotic HeLa cells in the absence (lane 6) or the presence (lane 7) of alkaline phosphatase before gel electrophoresis and staining by the indirect immunoperoxidase procedure. Nitrocellulose transfers of separated proteins of whole mitotic cells were incubated also in the absence (lane 8) or presence (lane 9) of alkaline phosphatase before staining by the indirect immunoperoxidase method. Approximately 25 μg of protein per lane was loaded. Molecular sizes are shown in kilodaltons for marker proteins (on the left) and major antigens (on the right). From Davis et al. (1983).

alkaline phosphatase digestion, it seems likely that the polypeptides recognized share a common or similar antigenic phosphorylated site.

2. Biological Activity of the Antibodies

Experiments using microinjection of antibodies revealed that they were unable to block entry of cells into mitosis or meiosis. The maturation of *X. laevis* oocytes induced by progesterone was not prevented or delayed by antibody injection. Antibodies introduced into synchronized S- or G_2-phase HeLa cells by either glass needle (Diacumakos *et al.*, 1970) or red blood cell ghost-mediated (Schlegel and Rechsteiner, 1978) microinjection failed to block entry into mitosis. These studies suggest that the antibodies do not block either the phosphorylation of these mitosis-specific phosphoproteins or the biological function of the proteins as cells enter mitosis. In addition, the antibodies are not able to immunoprecipitate the MPA from an extract of mitotic HeLa cells.

However, in these experiments the percentage of mitotic cells among those injected with mitosis-specific antibodies was frequently higher than that among those cells injected with control antibodies. Therefore, the antibodies were injected into mitotic HeLa cells, synchronized by

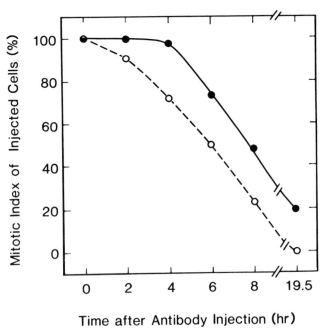

Fig. 12. Monoclonal antibodies delay M–G_1 transition. MPM-2 and control (SP2/0-Ag14) ascites fluids were separately loaded into human red blood cell ghosts. Antibody-loaded ghosts were fused with synchronized, mitotic HeLa cells using polyethylene glycol. Cells injected with antibodies were detected using fluorescein isothiocyanate-conjugated rabbit anti-mouse IgG. The mitotic index of cells injected with MPM-2 (solid circles) or control mouse immunoglobulins (open circles) was scored at various times after antibody injection. From F. M. Davis and P. N. Rao, unpublished data.

N_2O blockade, and the ability of these cells to complete cell division and enter the G_1 period upon reversal of the blockade was determined. Completion of mitosis was delayed by 2–3 hr, and 20% of the cells were irreversibly blocked in mitosis by antibody injection (Fig. 12). In addition, binding the antibodies to the antigens in vitro reduced the dephosphorylation of the antigens by alkaline phosphatase by 85%. These data suggest that the microinjected antibodies bind to the phosphoproteins in mitotic cells and interfere with or delay the dephosphorylation of these proteins that usually occurs as the cells complete mitosis and enter G_1 phase.

Thus, the phosphorylation of a set of polypeptides has been shown to occur only at mitosis, and the dephosphorylation of these same polypeptides seems to be necessary for the exit from mitosis.

V. CONCLUDING REMARKS

When cells approach mitosis, various dramatic events occur in the nucleus: the nuclear envelope breaks down, chromatin condenses into discrete chromosomes, and gene transcription is shut off. Synthesis of RNA and proteins has long been known to be necessary for the G_2–M transition. Certain G_2-specific polypeptides have been detected using two-dimensional polyacrylamide gel electrophoresis. The results summarized in this chapter strongly suggest that not only new protein synthesis, but also phosphorylation of the newly synthesized proteins may be involved in carrying out or regulating the various processes that occur at mitosis. Although most of the evidence has come from studies of a correlative nature, some direct evidence for the involvement of protein phosphorylation in the regulation of mitosis and meiosis has been provided recently by the microinjection of protein kinase and phosphatase inhibitors into amphibian oocytes. Mitosis-specific monoclonal antibodies that recognize proteins phosphorylated only at mitosis have been isolated. With the availability of biological assays for both the mitotic proteins that induce germinal vesicle breakdown and chromosome condensation in Xenopus oocytes, and the inhibitors of the mitotic factors that may be involved in re-formation of the nuclear envelope and the decondensation of chromosomes, the means to isolate and characterize these factors appear to be at hand. Characterization of these factors will increase our understanding of the biochemical events associated with mitosis. It is tempting to speculate that protein phosphorylation and dephosphorylation, either by or of the mitotic factors and the inhibitors of the mitotic factors, will represent a crucial regulatory mechanism for mitosis.

ACKNOWLEDGMENTS

We thank Twee Y. Tsao, Susan K. Fowler, and Hélène Bigo for their excellent technical assistance, and Josephine Neicheril for her superb secretarial assistance in the preparation of this manuscript. We also thank Drs. Abdullatif A. Al-Bader and Chintaman G. Sahasrabuddhe for useful discussions during the course of these studies. This work was supported in part by research grants CA-27544 and CA-34783 from the National Cancer Institute, Department of Health and Human Services, and CH-205 from the American Cancer Society.

REFERENCES

Adlakha, R. C., Sahasrabuddhe, C. G., Wright, D. A., Lindsey, W. F., and Rao, P. N. (1982a). Localization of mitotic factors on metaphase chromosomes. *J. Cell Sci.* **54**, 193–206.

Adlakha, R. C., Sahasrabuddhe, C. G., Wright, D. A., Lindsey, W. F., Smith, M. L., and Rao, P. N. (1982b). Chromosome bound mitotic factors: Release by endonucleases. *Nucleic Acids Res.* **10**, 4107–4117.

Adlakha, R. C., Sahasrabuddhe, C. G., and Rao, P. N. (1982c). Role of phosphorylation of nonhistone proteins in the condensation and decondensation of chromosomes during G_2–M and M–G_1 transitions. *J. Cell Biol.* **95**, 75a.

Adlakha, R. C., Sahasrabuddhe, C. G., Wright, D. A., and Rao, P. N. (1983). Evidence for the presence of inhibitors of mitotic factors during G_1 period in mammalian cells. *J. Cell Biol.* **97**, 1707–1713.

Adlakha, R. C., Wang, Y. C., Wright, D. A., Sahasrabuddhe, C. G., Bigo, H., and Rao, P. N. (1984). Inactivation of mitotic factors by ultraviolet irradiation of HeLa cells in mitosis. *J. Cell Sci.* **65**, 279–295.

Ahmed, K., and Wilson, M. J. (1978). Chromatin controls in the prostate. *In* "The Cell Nucleus" (H. Busch, ed.), Vol. 6, pp. 409–459. Academic Press, New York.

Ajiro, K., Broun, T. W., and Cohen, L. H. (1981a). Phosphorylation states of different histone$_1$ subtypes and their relationship to chromatin functions during the HeLa S-3 cell cycle. *Biochemistry* **20**, 1445–1454.

Ajiro, K., Borun, T. W., Shulman, S. D., McFadden, G. M., and Cohen, L. H. (1981b). Comparison of the structures of human histones 1A and 1B and their intramolecular phosphorylation sites during the HeLa S-3 cell cycle. *Biochemistry* **20**, 1454–1464.

Ajiro, K., Nishimoto, T., and Takhashi, T. (1983). Histone H1 and H3 phosphorylation during premature chromosome condensation in a temperature-sensitive mutant (tsBN$_2$) of baby hamster kidney cells. *J. Biol. Chem.* **258**, 4534–4538.

Al-Bader, A. A., Orengo, A., and Rao, P. N. (1978). G_2 phase-specific proteins of HeLa cells. *Proc. Natl. Acad. Sci. U.S.A.* **75**, 6064–6068.

Allfrey, V. G. (1982). Postsynthetic modifications. *In* "The HMG Chromosomal Proteins" (E. W. Johns ed.), pp. 123–148. Academic Press, New York.

Allis, C. D., and Gorovsky, M. A. (1981). Histone phosphorylation in macro- and micronuclei of *Tetrahymena thermophila*. *Biochemistry* **20**, 3828–3833.

Aubin, J. E., Osborn, M., Franke, W. W., and Weber, K. (1980). Intermediate filaments of the vimentin and cytokeratin-type are differently distributed during mitosis. *Exp. Cell Res.* **129**, 149–165.

Balhorn, R., Jackson, V., Granner, D., and Chalkley, R. (1975). Phosphorylation of the lysine-rich histones throughout the cell cycle. *Biochemistry* **14**, 2504–2511.

Baserga, R. (1976). "Multiplication and Division in Mammalian Cells." Dekker, New York.

Baserga, R. (1981). The cell cycle. *N. Engl. J. Med.* **304,** 453–459.

Bhorjee, J. S. (1981). Differential phosphorylation of nuclear nonhistone high mobility group proteins HMG 14 and HMG 17 during the cell cycle. *Proc. Natl. Acad. Sci. U.S.A.* **78,** 6944–6948.

Boyer, J., Belle, R., Capony, J.-P., and Ozon, R. (1983). Early increase of a 105,000-dalton phosphoprotein during meiotic maturation of *Xenopus laevis* oocytes. *Biochimie* **65,** 15–23.

Bradbury, E. M., Inglis, R. J., Matthews, H. R., and Sarner, N. (1973). Histone H1 phosphorylation in *Physarum polycephalum.* Correlation with chromosome condensation. *Eur. J. Biochem.* **33,** 131–13⁰.

Bradbury, E. M., Inglis, R. J., Matthews, H. R., and Langan, T. A. (1974). Molecular basis of control of mitotic cell division in eukaryotes. *Nature (London)* **249,** 553–556.

Bravo, R., Fey, S. J., Mose Larsen, P., and Celis, J. E. (1982). Modification of vimentin polypeptides during mitosis. *Cold Spring Harbor Symp. Quant. Biol. Vol.* **46,** 379–385.

Brinkley, B. R., Fuller, G. M., and Highfield, D. P. (1976). Tubulin antibodies as probes for microtubules in dividing and nondividing mammalian cells. *Cold Spring Harbor Conf. Cell Proliferation* **3,** 435–455.

Caspersson, T., Farber, S., Foley, G. E., and Killander, D. (1963). Cytochemical observations on the nucleolus-ribosome system. Effects of actinomycin D and nitrogen mustard. *Exp. Cell Res.* **32,** 529–552.

Cohen, P. (1982). The role of protein phosphorylation in neural and hormonal control of cellular activity. *Nature (London)* **296,** 613–620.

Cohen, P. (1983). Protein phosphorylation and hormone action—An overview. *Cell Biol. Int. Rep.* **7,** 479–480.

Davis, F. M., Tsao, T. Y., Fowler, S. K., and Rao, P. N. (1983). Monoclonal antibodies to mitotic cells. *Proc. Natl. Acad. Sci. U.S.A.* **80,** 2926–2930.

Diacumakos, E. G., Holland, S., and Pecora, P. (1970). A microsurgical methodology for human cells *in vitro:* Evolution and applications. *Proc. Natl. Acad. Sci. U.S.A.* **65,** 911–918.

Evans, R. M., and Fink, L. M. (1982). An alteration in the phosphorylation of vimentin-type intermediate filaments is associated with mitosis in cultured mammalian cells. *Cell (Cambridge, Mass.)* **29,** 43–52.

Fischer, E. H. (1983). Cellular regulation by protein phosphorylation. *Bull. Inst. Pasteur (Paris)* **81,** 7–31.

Fischer, S. G., and Laemmli, U. K. (1980). Cell cycle changes in *Physarum polycephalum* histone H1 phosphate: Relationship to deoxyribonucleic acid binding and chromosome condensation. *Biochemistry* **19,** 2240–2246.

Gerace, L., and Blobel, G. (1980). The nuclear envelope lamina is reversibly depolymerized during mitosis. *Cell (Cambridge, Mass.)* **19,** 277–287.

Gorovsky, M. A., and Keevert, J. B. (1975). Absence of histone F1 in a mitotically dividing genetically inactive nucleus. *Proc. Natl. Acad. Sci. U.S.A.* **72,** 2672–2676.

Gurley, L. R., Walters, R. A., and Tobey, R. A. (1975). Sequential phosphorylation of histone sub-fractions in the Chinese hamster cell cycle, *J. Biol. Chem.* **250,** 3936–3944.

Gurley, L. R., D'Anna, J. A., Barham, S. S., Deaven, L. L., and Tobey, R. A. (1978a). Histone phosphorylation and chromatin structure during mitosis in Chinese hamster cells. *Eur. J. Biochem.* **84,** 1–15.

Gurley, L. R., Tobey, R. A., Walters, R. A., Hildebrand, C. E., Hohmann, P. D., D'Anna, J. A., Barham, S. S., and Deaven, L. L. (1978b). Histone phosphorylation and chromatin structure in synchronized mammalian cells. *In* "Cell Cycle Regulation" (J. R. Jeter, I. L. Cameron, G. M. Padilla, and A. M. Zimmerman, eds.), pp. 37–60. Academic Press, New York.

Hanks, S. K., Rodriguez, L. V., and Rao, P. N. (1983). Relationship between histone phosphorylation and premature chromosome condensation. *Exp. Cell Res.* **148,** 293–302.

Hanocq-Quetier, J., and Baltus, E. (1981). Phosphorylation of ribosomal proteins during maturation of *Xenopous laevis* oocytes. *Eur. J. Biochem.* **120,** 351–355.

Hecht, R. M., Berg-Zabelshansky, M., and Davis, F. M. (1983). Embryonic-arrest mutant of *C. elegans* fails to phosphorylate mitosis-specific phosphoproteins. *J. Cell Biol.* **97,** 254a.

Henry, S. M., and Hodge, L. D. (1983). Nuclear matrix: A cell cycle-dependent site of increased intranuclear protein phosphorylation. *Eur. J. Biochem.* **133,** 23–29.

Hohmann, P., Tobey, R. A., and Gurley, L. R. (1976). Phosphorylation of distinct regions of F_1 histone: Relationship to cell cycle. *J. Biol. Chem.* **251,** 3685–3692.

Howard, A., and Pelc, S. R. (1953). Synthesis of deoxyribonucleic acid in normal and irradiated cells and its relation to chromosome breakage. *Heredity Suppl.* **6,** 261–273.

Huchon, D., Ozon, R., and Demaille, J. G. (1981). Protein phosphatase-1 is involved in *Xenopus* oocyte maturation. *Nature (London)* **294,** 358–359.

Ingebritsen, J. S., and Cohen, P. (1983). Protein phosphatases: Properties and role in cellular regulation. *Science* **221,** 331–338.

Johnson, E. M., and Allfrey, V. G. (1978). Post-synthetic modifications of histone primary structure. Phosphorylation and acetylation as related to chromatin conformation and function. *In* "Biochemical Action of Hormones" (G. Litwack, ed.), Vol. 5, pp. 1–51. Academic Press, New York.

Johnson, R. T., and Rao, P. N. (1970). Mammalian cell fusion: Induction of premature chromosome condensation in interphase nuclei. *Nature (London)* **226,** 717–722.

Jost, E., and Johnson, R. T. (1981). Nuclear lamina assembly, synthesis and disaggregation during the cell cycle in synchronized HeLa cells. *J. Cell Sci.* **47,** 25–53.

Jungmann, R. A., and Kranias, E. G. (1977). Nuclear phosphoprotein kinases and the regulation of gene transcription. *Int. J. Biochem.* **8,** 819–830.

Kalthoff, H., Darmer, D., Towbin, H., Gordon, J., Amons, R., Moller, W., and Richter, D. (1982). Ribosomal proteins S_6 from *Xenopus laevis* oocytes. *Eur. J. Biochem.* **122,** 439–443.

Kishimoto, S., and Lieberman, I. (1964). Synthesis of RNA and protein required for the mitosis of mammalian cells. *Exp. Cell Res.* **36,** 92–101.

Kleinsmith, L. J. (1975). Phosphorylation of non-histone proteins in the regulation of chromosome structure and function. *J. Cell. Physiol.* **85,** 459–475.

Krebs, E. G., and Beavo, J. A. (1979). Phosphorylation–dephosphorylation of enzymes. *Annu. Rev. Biochem.* **48,** 923–959.

Kruppa, J., Darmer, D., Kalthoff, H., and Richter, D. (1983). The phosphorylation of ribosomal protein S_6 from progesterone-stimulated *Xenopus laevis* oocytes. Kinetic studies and phosphopeptide analysis. *Eur. J. Biochem.* **129,** 537–542.

Krystal, G. W., and Poccia, D. L. (1981). Phosphorylation of cleavage stage histone H1 in mitotic and prematurely condensed chromosomes. *Exp. Cell Res.* **134,** 41–48.

Lake, R. S. (1973). Further characterization of the F1 histone phosphokinase of metaphase arrested animal cells. *J. Cell Biol.* **58,** 317–333.

Lake, R. S., and Saltzman, N. P. (1972). Occurrence and properties of a chromatin-associated F-1 histone phosphokinase in mitotic Chinese hamster cells. *Biochemistry* **11**, 4817–4826.

Lake, R. S., Goidl, J. A., and Salzman, N. P. (1972). F1 histone modification at the metaphase in Chinese hamster cells. *Exp. Cell Res.* **73**, 113–121.

Lazarides, E. (1980). Intermediate filaments as mechanical integrators of cellular space. *Nature (London)* **283**, 249–256.

Levy-Wilson, B. (1981). Enhanced phosphorylation of high-mobility group proteins in nuclease-sensitive mononucleosomes from butyrate-treated HeLa cells. *Proc. Natl. Acad. Sci. U.S.A.* **78**, 2189–2193.

Maller, J. L., and Krebs, E. G. (1977). Progesterone-stimulated meiotic cell division in *Xenopus* oocytes: Induction by regulatory subunit and inhibition by catalytic subunit of adenosine-3′,5′-monophosphate-dependent protein kinase. *J. Biol. Chem.* **252**, 1712–1718.

Maller, J. L., and Krebs, E. G. (1980). Regulation of oocyte maturation. *Curr. Top. Cell. Regul.* **16**, 271–311.

Maller, J. L., Wu, M., and Gerhart, J. C. (1977). Changes in protein phosphorylation accompanying maturation of *Xenopus laevis* oocytes. *Dev. Biol.* **58**, 295–312.

Marks, D. B., Paik, W. K., and Borun, T. W. (1973). The relationship of histone phosphorylation to DNA replication and mitosis during the HeLa cell cycle. *J. Biol. Chem.* **248**, 5660–5667.

Matsui, S., Weinfeld, H., and Sandberg, A. A. (1971). Dependence of chromosome pulverization in virus-fused cells on events in the G_2 period. *J. Natl. Cancer Inst. (U.S.)* **47**, 401–411.

Matsui, S., Yoshida, H., Weinfeld, H., and Sandberg, A. A. (1972). Induction of prophase in interphase nuclei by fusions with metaphase cells. *J. Cell Biol.* **54**, 120–132.

Mazia, D. (1974). The cell cycle. *Sci. Am.* **1**, 55–64.

Nelkin, B., Nichols, C., and Vogelstein, B. (1980). Protein factor(s) from mitotic CHO cells induce meiotic maturation in *Xenopus laevis* oocytes. *FEBS Lett.* **109**, 233–238.

Nestler, E. J., and Greengard, P. (1983). Protein phosphorylation in the brain. *Nature (London)* **305**, 583–588.

Nielsen, P. J., Thomas, G., and Maller, J. L. (1982). Increased phosphorylation of ribosomal protein S_6 during meiotic maturation of *Xenopus*. *Proc. Natl. Acad. Sci. U.S.A.* **79**, 2937–2941.

Pardee, A. B., Bubrow, R., Hamlin, J. L., and Kletzien, R. F. (1978). Animal cell cycle. *Annu. Rev. Biochem.* **47**, 715–750.

Paulson, J. R., and Taylor, S. S. (1982). Phosphorylation of histones 1 and 3 and non-histone high mobility group 14 by an endogenous kinase in HeLa metaphase chromosomes. *J. Biol. Chem.* **257**, 6064–6072.

Prescott, D. M. (1976). "Reproduction of Eukaryotic Cells." Academic Press, New York.

Rao, P. N. (1968). Mitotic synchrony in mammalian cells treated with nitrous oxide at high pressure. *Science* **160**, 774–776.

Rao, P. N. (1982). The phenomenon of premature chromosome condensation. *In* "Premature Chromosome Condensation: Application in Basic, Clinical, and Mutation Research" (P. N. Rao, R. T. Johnson, and K. Sperling, eds.), pp. 1–41. Academic Press, New York.

Rao, P. N., and Engelberg, J. (1966). Effect of temperature on the mitotic cycle of normal and synchronized mammalian cells. *In* "Cell Synchrony—Studies in Biosynthetic Regulation" (I. L. Cameron and G. M. Padilla, eds.), pp. 332–352. Academic Press, New York.

Rao, P. N., and Johnson, R. T. (1974). Regulation of cell cycle in hybrid cells. *Cold Spring Harbor Conf. Cell Proliferation* **1**, 785–800.

Rao, P. N., Wilson, B. A., and Puck, T. T. (1977). Premature chromosome condensation and cell cycle analysis. *J. Cell. Physiol.* **91**, 131–142.

Robinson, S. I., Nelkin, B., Kaufmann, S., and Vogelstein, B. (1981). Increased phosphorylation rate of intermediate filaments during mitotic arrest. *Exp. Cell Res.* **133**, 445–449.

Saffer, J. D., and Glazer, R. I. (1982). The phosphorylation of high mobility group proteins 14 and 17 and their distribution in chromatin. *J. Biol. Chem.* **257**, 4655–4660.

Sahasrabuddhe, C. G., Adlakha, R. C., and Rao, P. N. (1984). Phosphorylation of nonhistone proteins associated with mitosis in HeLa cells. *Exp. Cell Res.* **153**, 439–450.

Sanger, J. W., and Sanger, J. M. (1976). Actin localization during cell division. *Cold Spring Harbor Conf. Cell Proliferation* **3**, 1295–1316.

Schlegel, R. A., and Rechsteiner, M. C. (1978). Red cell-mediated microinjection of macromolecules into mammalian cells. *Methods Cell Biol.* **20**, 341–354.

Sherod, D., Johnson, G., Balhorn, R., Jackson, V., Chalkley, R., and Granner, D. (1975). The phosphorylation region of very lysine-rich histone in dividing cells. *Biochim. Biophys. Acta* **381**, 337–347.

Shibayama, T., Nakaya, K., Matsumoto, S., and Nakamura, Y. (1982). Cell cycle-dependent change in the phosphorylation of the nucleolar proteins of *Physarum polycepholum in vivo*. *FEBS Lett.* **139**, 214–216.

Song, M. K. H., and Adolph, K. W. (1983). Phosphorylation of nonhistone proteins during the HeLa cell cycle. Relationship to DNA synthesis and mitotic chromosome condensation. *J. Biol. Chem.* **258**, 3309–3318.

Stein, G. S., Spelsberg, T. C., and Kleinsmith, L. J. (1974). Nonhistone chromosomal proteins and gene regulation. *Science* **183**, 817–824.

Sunkara, P. S., Wright, D. A., and Rao, P. N. (1979a). Mitotic factors from mammalian cells induce germinal vesical breakdown and chromosome condensation in amphibian oocytes. *Proc. Natl. Acad. Sci. U.S.A.* **76**, 2799–2802.

Sunkara, P. S., Wright, D. A., and Rao, P. N. (1979b). Mitotic factors from mammalian cells: A preliminary characterization. *J. Supramol. Struct.* **11**, 189–195.

Sunkara, P. S., Wright, D. A., Adlakha, R. C., Sahasrabuddhe, C. G., and Rao, P. N. (1982). Characterization of chromosome condensation factors of mammalian cells. *In* "Premature Chromosome Condensation: Application in Basic, Clinical, and Mutation Research" (P. N. Rao, R. T. Johnson, and K. Sperling, eds.), pp. 233–251. Academic Press, New York.

Tanphaichitr, N., Moore, K. C., Granner, D. K., and Chalkley, R. (1976). Relationship between chromosome condensation and metaphase lysine-rich histone phosphorylation. *J. Cell Biol.* **69**, 43–50.

Taylor, E. W. (1963). Relationship of protein synthesis to the division cycle in mammalian cell cultures. *J. Cell Biol.* **19**, 1–18.

Tobey, R. A., Petersen, D. F., Anderson, E. C., and Puck, T. T. (1966). Life cycle analysis of mammalian cells. III. The inhibition of division in Chinese hamster cells by puromycin and actinomycin. *Biophys. J.* **6**, 567–581.

Wagenaar, E. B. (1983a). Inhibition of protein synthesis during G_2 induces irregular metaphase and anaphase in mitosis of root cells of *Allium sativum L. Cell Biol. Int. Rep.* **7**, 827–833.

Wagenaar, E. B. (1983b). Increased free Ca^{2+} levels delay the onset of mitosis in fertilized and artificially activated eggs of the sea urchin. *Exp. Cell Res.* **148**, 73–82.

Wagenaar, E. B., and Mazia, D. (1978). The effect of emetine on first cleavage division in sea urchin, *Strongylocentrotus purpuratus* In: "Cell Reproduction: Essays in Honor of Daniel Mazia" (E. R. Dirksen, D. M. Prescott, and C. F. Fox, eds.), pp. 539–545. Academic Press, New York.

Walton, G. M., and Gill, G. N. (1983). Identity of the *in vivo* phosphorylation site in high mobility group 14 protein in HeLa cells with the site phosphorylated by casein kinase II *in vitro*. *J. Biol. Chem.* **258,** 4440–4446.

Weintraub, H., Buscaglia, M., Ferrez, M., Weiller, S., Boulet, A., Fabre, F., and Baulieu, E. E. (1983). "MPF" activity in *Saccharomyces cerevisiae. C. R. Hebd. Seances Acad. Sci.,* Ser. 3 **295,** 787–790.

Weisbrod, S. (1982). Active chromatin. *Nature (London)* **297,** 289–295.

Westwood, J. T., Wagenaar, E. B., and Church, R. B. (1983). Changes in protein phosphorylation during the cell cycle of CHO cells. *J. Cell Biol.* **97,** 147a.

Wu, M., and Gerhart, J. C. (1980). Partial purification and characterization of the maturation-promoting factor from eggs of *Xenopus laevis. Dev. Biol.* **79,** 465–477.

17

Translational Regulation of Eukaryotic Protein Synthesis by Phosphorylation of eIF-2α

IRVING M. LONDON, DANIEL H. LEVIN, ROBERT L. MATTS,
N. SHAUN B. THOMAS, RAYMOND PETRYSHYN,
AND JANE-JANE CHEN

Harvard-MIT Division of Health Sciences and Technology
and
Department of Biology
Massachusetts Institute of Technology
Cambridge, Massachusetts

I. INTRODUCTION

The focus of this chapter is on the regulation of initiation of protein synthesis in eukaryotic cells and, more specifically, on the phosphorylation of the initiation factor, eIF-2, as a key element in the mechanism of regulation. We shall not attempt a review of this subject; in the past few years several reviews have been published to which the interested reader may refer (Kramer and Hardesty, 1980; Jagus *et al.*,

515

1981; London *et al.*, 1981; Ochoa, 1983). We prefer to present a brief historical account followed by a description of recent studies in our laboratory on the mechanism of regulation by phosphorylation.

Most of these studies have been carried out in rabbit reticulocytes and in lysates of these cells. The reticulocyte possesses mitochondria required for the biosynthesis of heme, and ribosomes and messenger RNA for the synthesis of protein. At the same time, the absence of a nucleus serves to simplify the elucidation of the protein-synthesizing system by eliminating transcriptional regulation.

Following the observations in the 1950s that inorganic iron promotes protein synthesis in immature erythroid cells (Kruh and Borsook, 1956, Kassenaar *et al.*, 1957; Nizet, 1957; Morell *et al.*, 1958), we observed in 1964 that heme enhances the synthesis of protein in rabbit reticulocytes (Bruns and London, 1965). In heme deficiency, protein synthesis is diminished and polyribosome disaggregation occurs; on addition of hemin, polyribosomes reform and protein synthesis is restored (Grayzel *et al.*, 1966). These findings, which occur at the level of translation, pointed to a defect in initiation in heme deficiency. We found further that iron exerts its effects on the synthesis of protein in reticulocytes by stimulating the synthesis of heme.

The development of a reticulocyte lysate for protein synthesis (Zucker and Schulman, 1968) facilitated progress in understanding the mechanisms involved in these effects of heme. Mitochondria required for heme synthesis are removed and the system is dependent on exogenous hemin. In the presence of added hemin (10–30 μM), protein synthesis and polyribosomes are well maintained for 60–90 min; when hemin is not added, several rounds of initiation occur during the first 5 min, and then there is an abrupt decline in the rate of protein synthesis ("shutoff") associated with disaggregation of polyribosomes (Hunt *et al.*, 1972). With shutoff there is also a marked diminution in 40 S methionyl-tRNA$_f$ initiation complexes (Legon *et al.*, 1973). These complexes comprise the ternary complex of eukaryotic initiation factor 2 (eIF-2), Met-tRNA$_f$, and GTP bound to the 40 S ribosomal subunit. It was, therefore, of particular interest to observe that the addition of eIF-2, to be sure in relatively high and nonphysiologic concentrations, could prevent shutoff or restore the synthesis of protein in a shutoff lysate (Beuzard and London, 1974; Kaempfer, 1974; Clemens *et al.*, 1975).

These findings clearly implicated eIF-2 in the regulatory mechanism affected by heme deficiency. That a mere depletion of eIF-2 was not an adequate explanation of shutoff was evident from the major earlier finding that an inhibitor was formed in these heme-deficient lysates

(Maxwell and Rabinovitz, 1969; Adamson et al., 1968; Howard et al., 1970). On purification, the inhibitor was shown to be a cAMP-independent protein kinase that specifically phosphorylates the α subunit, 38,000 daltons, of eIF-2 (Kramer et al., 1976; Levin et al., 1976; Farrell et al., 1977; Gross and Mendelewski, 1977). In hemin-supplemented lysates, the kinase is present in an inactive form, whereas it is activated in the absence of hemin or on treatment with sulfhydryl reagents such as N-ethylmaleimide (Gross and Rabinovitz, 1972; Ranu and London, 1976; Farrell et al., 1977).

Initiation of protein synthesis is inhibited not only in heme deficiency, but also on treatment with double-stranded RNA (dsRNA, 1–50 ng/ml) (Ehrenfeld and Hunt, 1971) or oxidized glutathione (GSSG, 50–500 mM) (Kosower et al., 1972). Inhibitions of initiation by heme deficiency, dsRNA, and GSSG share several common features: biphasic kinetics (a normal rate of synthesis for about 5 min followed by abrupt shutoff), prevention or reversal of inhibition by addition of eIF-2, and activation of cAMP-independent protein kinases that phosphorylate eIF-2α. The heme-regulated inhibitor (HRI) and the dsRNA-activated inhibitor (dsI) phosphorylate the same site on eIF-2α (Ranu, 1979; Ernst et al., 1980; Levin et al., 1980), but they are readily distinguished immunologically by a chicken antiserum to the rabbit reticulocyte HRI. The eIF-2α kinase activity induced by GSSG, like that of HRI, is inhibited by the antiserum to HRI, whereas dsI is unaffected (Petryshyn et al., 1979).

HRI has been purified in a heme-reversible form (Trachsel et al., 1978). Phosphorylation appears to be required for its eIF-2α kinase activity. In heme-deficiency, phosphorylation of HRI and of eIF-2α is concomitant with inhibition of protein synthesis (Ernst et al., 1979). With increased phosphorylation of HRI in vitro, eIF-2α kinase activity is increased (Fagard and London, 1981). The rate of phosphorylation of purified HRI is not affected by dilution, a finding indicative of autophosphorylation rather than heterophosphorylation (Gross and Mendelewski, 1978; Hunt, 1979). But the possibility remains that under physiologic conditions HRI may be activated by heterophosphorylation.

The claim of Datta et al. (1978) that HRI is activated by a cAMP-dependent protein kinase has been disproved by Levin et al. (1979), Gross (1979), Hunt (1979), and Grankowski et al. (1980). Nevertheless, the activation of HRI by heterophosphorylation is a possible mechanism that may explain the inhibition of protein synthesis that occurs in heme-supplemented lysates on the addition of GSSG (Ernst et al., 1978) or Ca^{2+} (deHaro et al., 1983). Both inhibitions are accompanied by the phosphorylation and activation of HRI and the phosphorylation of eIF-2α. Neither GSSG nor Ca^{2+} has been found to have a direct effect on the activation and phosphorylation of purified HRI *in vitro*. An inhibitor of protein synthesis has been partially purified from GSSG-inhibited reticulocyte lysates with properties distinct from those of HRI (Kan *et al.*, 1983); it is a heat-stable protein with an apparent MW of 23,000, and its activation requires both GSSG and ATP. It may be identical with the 24,000-dalton sulfhydryl-sensitive protein observed by Jackson *et al.* (1983) in gel-filtered reticulocyte lysates. The inhibition of protein synthesis and the phosphorylation of eIF-2α induced by the addition of Ca^{2+} are apparently effected by a Ca^{2+}–phospholipid-activated protein kinase (deHaro *et al.*, 1983). Further exploration of these intriguing findings may help to illuminate the mechanism of activation of HRI.

In investigating the activation of HRI, we have employed reticulocyte lysates that were filtered through dextran gels to remove low-molecular-weight components. Protein synthesis was sustained by the addition of both hemin (20 μM) and glucose 6-phosphate (50–200 μM); the addition of either one alone only partially restored protein synthesis. In the absence of both hemin and glucose 6-phosphate, HRI was rapidly activated with phosphorylation of HRI and eIF-2α and inhibition of protein synthesis. Activation was partially blocked by hemin or glucose 6-phosphate (500 μM); deoxyglucose 6-phosphate but not fructose 1-6-diphosphate could replace glucose 6-phospate. We conclude that the activation of HRI is controlled by both hemin and glucose 6-phosphate (Michelson *et al.*, 1984).

The regulation of HRI by hemin has been studied by determining whether hemin is bound by HRI (Fagard and London, 1981). On addition of HRI to a solution of hemin, the spectrum of hemin is modified: The maximum is shifted from 380–390 μm to 418 μm, the peak is sharpened and increased in intensity, and the difference spectrum shows a single peak at 418 μm. Treatment with NEM and phosphorylation of HRI diminish the binding of hemin. These findings are consistent with a direct effect of heme on HRI: In the heme-replete cell or lysate, heme is bound to the enzyme and maintains it in an inactive form; in heme deficiency, activation can occur. Still to be determined is whether the activation occurs by autophosphorylation, by heterophosphorylation, or by both.

Evidence of a functional role for the phosphorylation of eIF-2α in the regulation of initiation is provided by studies in reticulocyte lysates *in situ* and in intact reticulocytes. The phosphorylation of eIF-2α in heme-deficient lysates occurs rapidly and well before the onset of inhibition. When hemin is added to an inhibited heme-deficient lysate, dephosphorylation of eIF-2α precedes the restoration of protein synthesis. In lysates inhibited by heme deficiency, GSSG, or added HRI, the phosphorylation of eIF-2α is accompanied by the phosphorylation of HRI, which migrates as an 80,000-dalton polypeptide in SDS–PAGE. The inhibition of initiation induced by dsRNA is accompanied by the phosphorylation of eIF-2α and a 67,000-dalton polypeptide, which is dsI. These observations *in situ* indicated that the phosphorylation of eIF-2α is a critical factor in the inhibition of initiation of protein synthesis and that the phosphorylation of HRI, like that of dsI, is associated with its activation (Ernst et al., 1979).

Further evidence of a functional role for the phosphorylation of eIF-2α in the regulation of initiation was provided by studies in intact reticulocytes (Leroux and London, 1982). On treatment of intact reticulocytes with isonicotinic acid hydrazide to inhibit heme synthesis, the phosphorylation of eIF-2α was significantly greater (30–40%) than in control cells (10%). The narrow range of phosphorylated eIF-2α observed in heme-deficient as compared to control reticulocytes indicated that a relatively modest concentration of phosphorylated eIF-2α can suffice to inhibit the cycle of initiation. These findings further suggested that phosphorylation of a portion of eIF-2α could suffice if a factor with which eIF-2 must interact becomes limiting when eIF-2α is phosphorylated. A likely candidate to serve as a rate-limiting factor was reversing factor (RF). RF is a multipolypeptide factor, consisting of five subunits, that can prevent or reverse inhibition of protein synthesis in heme-deficient or HRI-inhibited reticulocyte lysates. It has been variously called anti-HRI, RF, rRF, sRF, eIF-2B, and GEF, and has been purified from postribosomal supernatants and ribosomal salt washes, either in a free form or in a 1 : 1 complex with eIF-2 (Amesz et al., 1979; Siekierka et al., 1981; Matts et al., 1983; Ralston et al., 1978, 1979; Konieczny and Safer, 1983; Panniers and Henshaw, 1983). We refer to it as RF.

II. FUNCTION OF THE REVERSING FACTOR

Restoration of linear synthesis in inhibited lysates can be achieved by the addition of high nonphysiological levels of eIF-2 or by the addition of ~ 10-fold lower levels of RF. The composition of RF, purified and isolated as a 1 : 1 complex with eIF-2, is shown in Fig. 1. We

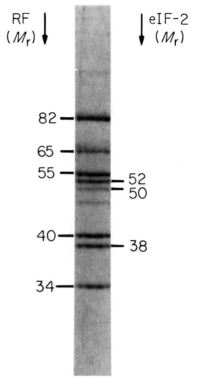

Fig. 1. SDS–polyacrylamide gel of purified RF·eIF-2 complex. From Matts *et al.*, (1983).

estimate that RF is present in the reticulocyte lysate at a 10- to 20-fold lower concentration than eIF-2, based on yields of the purified factors. Addition of RF has been found to stimulate protein synthesis catalytically in inhibited lysates, with 10–20 pmol of globin synthesized in 30 min for each picomole of RF added. Addition of eIF-2 to heme-deficient lysates has been previously found to result at best in a stoichiometric synthesis of globin chains.

At physiological Mg^{2+} concentrations, the presence of RF is essential for the dissociation of GDP from the eIF-2·GDP binary complex (Siekierka *et al.*, 1982; Matts *et al.*, 1983; Panniers and Henshaw, 1983). The primary function of RF in the initiation cycle then appears to be the catalysis of the dissociation of the eIF-2·GDP complex, which permits the exchange of GTP for GDP and, in the presence of Met-tRNA$_f$, promotes the formation of the ternary complex eIF-2·GTP·Met-tRNA$_f$. The eIF-2·GDP complex is efficiently phosphorylated by HRI with the result that RF-catalyzed dissociation of the eIF-2·GDP complex is in-hibited (Matts et al., 1983). In the absence of RF-catalyzed dissociation of GDP from eIF-2, the conversion of the eIF-2·GDP complex to the eIF-2·GTP·Met-tRNA$_f$ ternary complex is inhibited (Siekierka et al., 1982; Matts et al., 1983; Panniers and Henshaw, 1983; Pain and Clem-ens, 1983), since GDP will not spontaneously dissociate from eIF-2 under physiological conditions. Inasmuch as the function of RF in the formation of the ternary complex may be due solely to its dissociation of eIF-2·GDP, we have used eIF-2·GDP dissociation as an assay for specific RF activity.

III. MECHANISM OF DIMINUTION OF RF ACTIVITY BY eIF-2α PHOSPHORYLATION

Phosphorylation of eIF-2α in heme-deficient reticulocytes or their lysates was found to reach a maximum level at approximately the time of shutoff of protein synthesis (Farrell et al., 1978; Leroux and London, 1982). Only 30–40% of the eIF-2α present was phosphorylated when translation was inhibited by greater than 90%, a finding that prompted

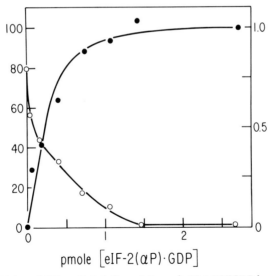

Fig. 2. Inhibition of RF-mediated dissociation of eIF-2·[³H]GDP by eIF-2(α P)·GDP. Values are plotted as percentage inhibition of dissociation (●) and as picomoles of eIF-2·[³H]GDP dissociated (○) in 7 min at 30°C as described in Matts et al. (1983). From Matts et al. (1983).

us to suggest that RF might become rate limiting in the initiation cycle (Leroux and London, 1982). There are two mechanisms by which eIF-2α phosphorylation could inhibit the ability of RF to function: (1) eIF-2α phosphorylation could inhibit the interaction of RF and eIF-2, and (2) RF could bind to phosphorylated eIF-2 to form a non-dissociable, complex and thus make RF unavailable. The first mechanism, however, would not explain why RF would not be free to interact with the 60–70% of the eIF-2 that remains unphosphorylated in heme-deficient reticulocytes. If this mechanism were correct, in a reaction mixture containing phosphorylated eIF-2·GDP complex, free RF should catalyze the dissociation of subsequently added unphosphorylated eIF-2·[³H]GDP complexes. This does not occur, however. The data in Fig. 2 indicate that, in the presence of GDP and RF, the phosphorylation of increasing levels of eIF-2·GDP leads to the complete inhibition

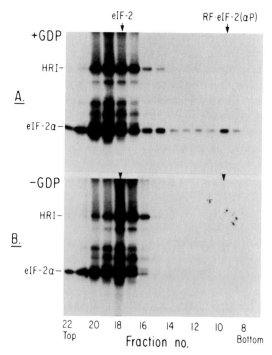

Fig. 3. Formation of an RF·eIF-2(α P) complex in the presence of GDP. Autoradiogram of SDS–polyacrylamide gels run on fractions from glycerol gradient analysis of reaction mixtures. eIF-2 was phosphorylated in reaction mixtures by addition of HRI and [γ-³²P]ATP in the presence of RF and 40 μM GDP (A) or no GDP (B). Experimental conditions are described in Matts *et al.* (1983). From Matts *et al.* (1983).

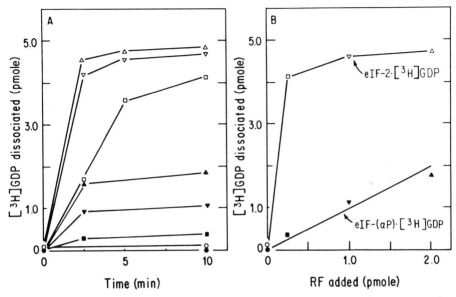

Fig. 4. Effect of eIF-2α phosphorylation on the kinetics (A) and stoichiometry (B) of RF-catalyzed dissociation of GDP. (A) Rate of dissociation of GDP from eIF-2·[³H]GDP (○, □, ▽, △) and eIF-2(α P)·[³H]GDP (●, ■, ▼, ▲) in the absence of RF (○, ●) and upon addition of 0.25 pmol RF (□, ■), 1.0 pmol RF (▽, ▼) and 2.0 pmol RF (△, ▲). (B) Picomoles of [³H]GDP dissociated in 10 min from unphosphorylated binary complex (○, □, ▽, △) and phosphorylated binary complex (●, ■, ▼, ▲) plotted against the amount of RF added. From Thomas et al. (1984).

of RF-catalyzed dissociation of unphosphorylated eIF-2·[³H]GDP. In vitro studies in which reaction mixtures containing [γ-³²P]ATP were analyzed by glycerol gradient centrifugation (Fig. 3A) show that in the presence of GDP, RF reacts with phosphorylated eIF-2·GDP complex to form an RF·eIF-2(α P) complex. These findings have been subsequently confirmed (Siekierka et al., 1984). In accord with previous reports, RF was found not to bind phosphorylated eIF-2 in the absence of GDP (Fig. 3B) (Siekierka et al., 1982; Voorma and Amesz, 1982).

To study further the mechanism by which the RF·eIF-2(α P) complex is formed, we examined the effect of the prephosphorylation of the eIF-2·GDP complex on the kinetics of RF-stimulated GDP dissociation in vitro. RF catalytically stimulated the complete exchange of [³H]GDP from unphosphorylated eIF-2·[³H]GDP complexes at a rate dependent upon the concentration of RF present (Fig. 4). However, if the eIF-2·GDP complex is phosphorylated prior to the addition of RF, there is only a limited, rapid dissociation of [³H]GDP from the complex. The extent of [³H]GDP exchanged from the phosphorylated binary complex was proportional to the amount of RF present in the assay, with a stoichiometry of 1 pmol of [³H]GDP dissociated per picomole of RF added. The data indicate that the mechanism by which RF activity is diminished by eIF-2α phosphorylation involves the direct binding of RF to the phosphorylated binary complex eIF-2(α P)·GDP from which the GDP becomes dissociable; the RF·eIF-2(α P) complex formed is nondissociable and, as a result, the RF is sequestered. Since the concentration of eIF-2 in reticulocyte lysates is estimated to be 10- to 20-fold higher than that of RF, the sequestering of RF in an RF·eIF-2(α P) complex explains how the phosphorylation of as little as 30–40% of the eIF-2 can inhibit protein synthesis in reticulocyte lysates.

IV. LOCALIZATION AND COMPOSITION OF THE SEQUESTERED RF COMPLEX IN RETICULOCYTE LYSATES

The presence of RF in initiation complexes was determined under physiological conditions in the reticulocyte lysate cell-free system. Lysates were incubated and then analyzed immediately on sucrose gradients. RF was localized in gradient fractions by the functional GDP exchange assay (Matts *et al.*, 1983). Under all conditions studied, RF

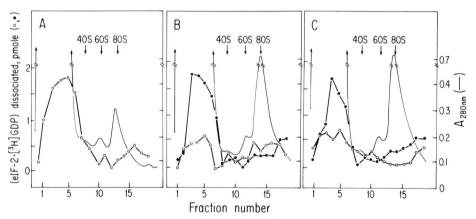

Fig. 5. Localization of RF in whole lysates. Standard reticulocyte lysate protein synthesis mixes were incubated for 15 min at 30°C (A) with 20 μM hemin Cl; (B) without hemin Cl; (C) with 20 μM hemin Cl and 15 ng/ml reoviral dsRNA. These samples immediately fractionated on sucrose gradients, and fractions were either assayed for RF activity after treatment with bovine intestinal alkaline phosphatase for 30 min at 30°C (●), or after incubation in the absence of alkaline phosphatase (○), as described in Thomas *et al.* (1984). From Thomas *et al.* (1984).

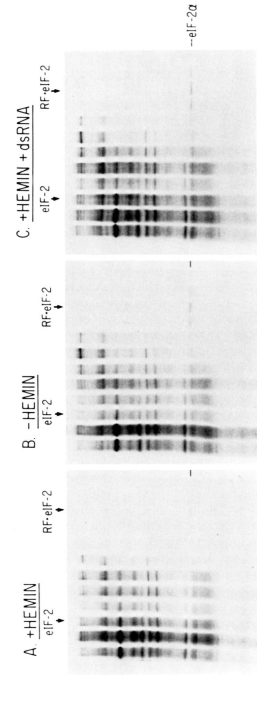

Fig. 6. Nonribosomal RF complex. Reticulocyte lysate cell-free protein synthesis mixes were incubated as described in the legend to Fig. 5 (A–C) except that [γ-^{32}P]ATP (100 μCi) was added at 10 min. Samples taken at 15 min were immediately fractionated on glycerol gradients as described in Thomas et al. (1984). The position of purified, phosphorylated eIF-2α is noted. The figure is an autoradiogram. From Thomas et al. (1984).

was detectable as a nonribosomal complex. In heme-deficient lysate (Fig. 5B), or lysates inhibited by dsRNA (Fig. 5C), the RF sedimented as a nonfunctional complex, but was liberated by the action of alkaline phosphatase. Treatment with alkaline phosphatase led to the de-phosphorylation of many phosphoproteins, including 80–90% of the eIF-2(α P) in the gradient (data not shown).

To determine whether RF forms a nonribosomal complex with phos-phorylated eIF-2α, reticulocyte lysates were incubated with [γ-^{32}P]ATP, and nonribosomal complexes were fractionated by sedi-mentation through glycerol gradients. The phosphoprotein profiles are shown in Fig. 6. At a time when protein synthesis has ceased in a heme-deficient lysate or in a lysate incubated with dsRNA, phosphorylated eIF-2α sediments at exactly the same position as purified RF·eIF-2 (Fig. 6B and C, respectively).

The results presented here show that RF is sequestered predomi-nantly as a nonribosomal complex in lysates inhibited by heme depri-vation or by dsRNA (Fig. 5). The complex contains phosphorylated eIF-2α (Fig. 6). We have also determined that there is no unphosphory-lated eIF-2α present in this complex and that when purified RF is added to the heme-deficient lysate, the amount of this eIF-2(α P)-con-taining complex increases (data not shown). Further, all nonribosomal eIF-2(α P), comprising ∼ 5% of the cellular total, is complexed with RF. Since there is only ∼ 5–10% as much RF in reticulocyte lysates as there is eIF-2, the data are consistent with the proposal that RF is sequestered as a nonribosomal complex with a small proportion of the total eIF-2(α P). We have also determined that the lysate complex does not contain tightly bound guanine nucleotide, in full agreement with *in vitro* stud-ies described previously (Fig. 4).

Treatment with a phosphatase which dephosphorylates many phos-phoproteins, including 80–90% of the eIF-2(α P) in the gradient, is required to liberate functional RF from the complex (Fig. 5B,C). Since a concomitant dephosphorylation of eIF-2(α P) is observed when protein synthesis is resumed, for example, on addition of hemin to a heme-deficient lysate, it is probable that the primary site of phosphatase action is the nonribosomal RF·eIF-2(α P) complex. This dephosphory-lation of eIF-2α would liberate functional RF to catalyze the recycling of eIF-2 and the resumption of protein synthesis.

V. CORRELATION BETWEEN RF ACTIVITY AND PROTEIN SYNTHESIS IN RETICULOCYTE LYSATES

The dissociation of eIF-2·GDP appears to be catalyzed specifically by RF in the reticulocyte lysate. We have examined the capacity of whole lysates to catalyze the dissociation of exogenously added eIF-2·[³H]GDP as a measure of their RF activity under various conditions of protein synthesis. In heme-deficient lysates, when protein synthesis is inhibited by over 90%, the rate of eIF-2·[³H]GDP dissociation was 5% of that found in hemin-supplemented lysates. To examine the relationship between RF activity and protein synthesis, we have used the ability of lysates to dissociate eIF-2·[³H]GDP in 2 min as a measure of RF function. Under these conditions, the rate of dissociation of eIF-2·GDP was linear, so that changes in the amount of eIF-2·[³H]GDP dissociated reflect changes in the amount of active RF present (Matts and London, 1984).

Incubation of heme-deficient lysates for various periods of time results in a progressive loss of the lysate's ability to catalyze the dissociation of eIF-2·[³H]GDP. At 10 min the loss of the ability of heme-defi-

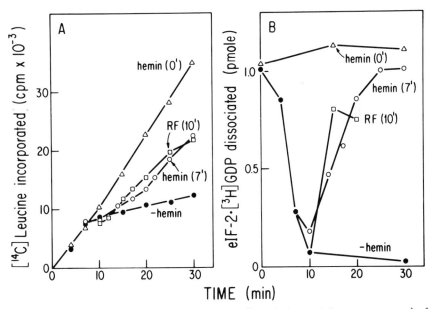

Fig. 7. Restoration of lysate eIF-2·[³H]GDP dissociation activity upon reversal of protein synthesis inhibition by the addition of hemin or RF. Experimental procedures are described in Matts and London (1984). (A) Incorporation of [¹⁴C]leucine into protein was measured with 20 μM hemin Cl added at 0 min, △; without hemin Cl, ●; with 20 μM hemin Cl added at 7 min, ○; without hemin Cl, but with 1 pmol of RF added at 10 min, □. (B) RF activity was determined by measuring the ability of protein synthesis mixture to catalyze the dissociation of [³H]GDP from preformed eIF-2·[³H]GDP added to the lysate; △, with 20 μM hemin Cl added at 0 min; ●, without hemin Cl; ○, with 20 μM hemin Cl added at 7 min; □, without hemin Cl, but with 2 pmol of RF added per 100 μl of protein synthesis mix at 10 min. From Matts and London (1984).

cient lysates to stimulate eIF-2·[³H]GDP dissociation corresponds to the time at which protein synthesis becomes maximally inhibited (Fig. 7). At 7 min, when the rate of protein synthesis is beginning to decline rapidly, heme-deficient lysates show a substantial decrease in eIF-2·GDP dissociation activity. Addition of hemin to a heme-deficient lysate after 7 min of incubation results in a restoration of eIF-2·GDP dissociation activity in the lysate that correlates with the restoration of protein synthesis (Fig. 7). Addition of an RF preparation to a heme-deficient lysate at 10 min predictably restores protein synthesis and eIF-2·[³H]GDP dissociation activity.

As noted, inhibition of protein synthesis in reticulocyte lysates by dsRNA or GSSG is also associated with the phosphorylation of eIF-2α. On addition of dsRNA or GSSG there is a rapid loss of RF activity in the reticulocyte lysate that corresponds to the time of shutoff of protein synthesis (Fig. 8).

The loss of RF activity in the reticulocyte lysate, measured as

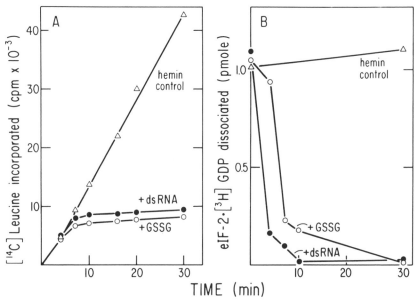

Fig. 8.   Effect of double-stranded RNA and oxidized glutathione on protein synthesis and lysate RF activity. (A) Protein synthesis assay was carried out as in Fig. 7; △, with 20 μM hemin Cl added at 0 min; ○, with 20 μM hemin Cl and 500 μM oxidized glutathione (GSSG) added at 0 min; ●, with hemin Cl 20 μM and dsRNA, 20 ng/ml, added at 0 min. (B) RF activity was determined in lysates incubated as above, as described in Fig. 7. From Matts and London (1984).

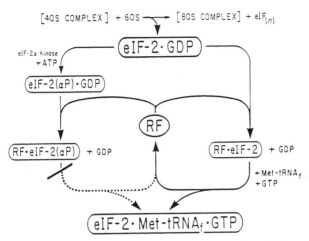

Fig. 9.   Schema of interactions of RF and eIF-2·GDP and effect of phosphorylation of eIF-2α. From Matts *et al.* (1983).

eIF-2·[³H]GDP dissociation activity, correlates with the known activation of eIF-2α kinases, the phosphorylation of eIF-2α, and the inhibition of protein synthesis. The correlation of shutoff of protein synthesis with the lack of significant eIF-2·GDP dissociation activity demonstrates that this function is specific for RF; RF has become sequestered in a complex with eIF-2(α P). These relationships are presented schematically in Fig. 9.

VI. DOUBLE-STRANDED RNA-DEPENDENT eIF-2α KINASE

We have noted that protein synthesis in reticulocyte lysates is inhibited by the addition of low levels of dsRNA (1–50 ng/ml). As in the case of heme deficiency, biphasic kinetics are observed: a brief period of normal synthesis followed by an abrupt decline in the rate of synthesis (Fig. 10A). The inhibition in response to dsRNA is due to the activation of a cAMP-independent protein kinase (dsI) that phosphorylates eIF-2α (Farrell *et al.*, 1977; Levin and London, 1978). The phosphoprotein profile in Fig. 10B is an autoradiogram which demonstrates that the addition of dsRNA to partially purified dsI and [γ-³²P]ATP results in the phosphorylation of a 70,000/72,000-dalton protein doublet (70K/72K)(dsI) and the 38,000-dalton α subunit of added eIF-2 (track 2). In the absence of dsRNA, no phosphorylation of dsI or eIF-2α is observed (track 1). The inhibitions of protein synthesis by dsRNA and

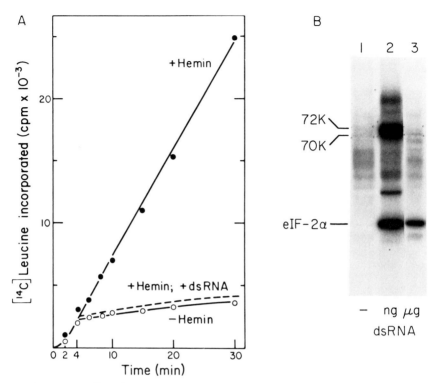

Fig. 10. (A) Effect of dsRNA on protein synthesis. Protein synthesis reactions (25 μl) using reticulocyte lysates were incubated in the presence and absence of hemin (20 μM) and in the presence of dsRNA (20 ng/ml), as indicated. Aliquots were assayed for the extent of protein synthesis at the indicated times. For experimental details, see Petryshyn *et al.* (1980). (B) Effect of dsRNA on the eIF-2α kinase activity of dsI. Protein kinase assays (20 μl) containing partially purified latent dsI were carried out in the presence of [γ-^{32}P]ATP, as described (Petryshyn *et al.*, 1980). All assays were supplemented with purified eIF-2 (0.5 μg). Lane 1 contained no dsRNA; lane 2 contained low levels of dsRNA (20 ng/ml); lane 3 contained high levels of dsRNA (25 μg/ml). All assays were analyzed by SDS–PAGE and autoradiography. The figure is an autoradiogram.

by heme deficiency share many characteristics (for review, see London *et al.*, 1981): the phosphorylation of proximal or identical site(s) of eIF-2α; polyribosome disaggregation; 40 S-Met-tRNA$_f$ depletion; potentiation of inhibition by ATP; and reversal of inhibition by GTP (2 mM), glucose 6-phosphate (1 mM), eIF-2, or RF. In spite of these similarities, the two eIF-2α kinases are distinct molecular entities by several criteria: They differ in intracellular localization and modes of activation (Farrell *et al.*, 1977; Levin and London, 1978), immunological properties (Petryshyn *et al.*, 1979), phosphoprotein profiles (Ernst *et al.*, 1979; Levin *et al.*, 1980), and molecular weights and chromatographic properties (Levin *et al.*, 1980); Trachsel *et al.*, 1978; Petryshyn *et al.*, 1980).

VII. ACTIVATION OF dsI

Activation of dsI from its latent precursor takes place on ribosomes in lysates (Farrell *et al.*, 1977; Levin and London, 1978) and requires both ATP and low levels of dsRNA. dsI can also be activated in purified preparations of the latent precursor of dsI (Petryshyn *et al.*, 1980). In all preparations activation is accompanied by the dsRNA-dependent phosphorylation of a polypeptide doublet that migrates as a 70K/72K doublet (previously denoted as 67K/68.5K) in SDS–PAGE (Petryshyn *et al.*, 1980, 1982; Levin *et al.*, 1981). It is an interesting paradox that the addition of high levels of dsRNA (10–20 μg/ml) to reticulocyte lysates results in no inhibition of protein synthesis (Hunter *et al.*, 1975), and that high levels of dsRNA prevent the phosphorylation of the 70K/72K doublet of dsI so that little or no phosphorylation of eIF-2α occurs (Fig. 10B, track 3). Although the activation of dsI is prevented by high levels of dsRNA, the activity of dsI, once activated by low levels of dsRNA, is not affected (Hunter *et al.*, 1975; Levin and London, 1978). During activation the phosphorylation of dsI occurs in an asymmetric manner in which a 70K polypeptide component is phosphorylated first followed by the rapid phosphorylation of a 72K component (Levin *et al.*, 1981). The kinetics of the dsRNA-dependent phosphorylation of the 70K/72K polypeptide doublet is shown in the autoradiogram (Fig. 11B). The 70K and 72K components represent different phosphorylated states of the same polypeptide (dsI) (Levin *et al.*, 1981). The functional significance of multiphosphorylation of dsI is unknown. However, the phosphorylation of the 70K/72K doublet results in an active eIF-2α kinase and inhibition of protein synthesis. Furthermore, maximal eIF-2α phosphorylation is achieved when the phosphate level is significantly higher in the 72K polypeptide than in the 70K polypeptide (Levin *et al.*, 1981; Petryshyn *et al.*, 1982). dsI has been purified from rabbit reticulocytes in its active (Levin *et al.*, 1980; Grosfeld and Ochoa, 1980) and latent form (Petryshyn *et al.*, 1980; Levin *et al.*, 1981). The activation of highly purified latent dsI proceeds through an autokinase mechanism similar to that observed for HRI (Petryshyn *et al.*, 1984). The possibility that activation of dsI occurs through a heterokinase mechanism or may be induced by agents other than dsRNA remains to be explored.

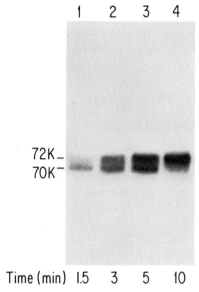

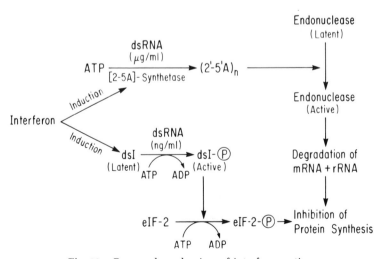

Fig. 11. Kinetics of dsI autophosphorylation. Protein kinase assays containing dsRNA (20 ng/ml), [γ-^{32}P]ATP, and highly purified latent dsI (see Petryshyn *et al.*, 1980, for details) were incubated at 10°C for the indicated times. All assays were analyzed by SDS–PAGE and autoradiography. The figure is an autoradiogram.

Fig. 12. Proposed mechanism of interferon action.

VIII. BIOLOGICAL SIGNIFICANCE OF dsI

dsI is not restricted to rabbit reticulocytes. A similar dsRNA-dependent eIF-2α kinase that inhibits protein synthesis is observed in normal human reticulocytes, but not erythrocytes (Petryshyn et al., 1984). Furthermore, the results of several studies indicate the presence of dsRNA-dependent eIF-2α kinase activity which is inhibitory of protein synthesis in extracts of interferon (IFN)-treated nonerythroid cells (Lebleu et al., 1976; Zilberstein et al., 1978). The antiviral property of IFNs is

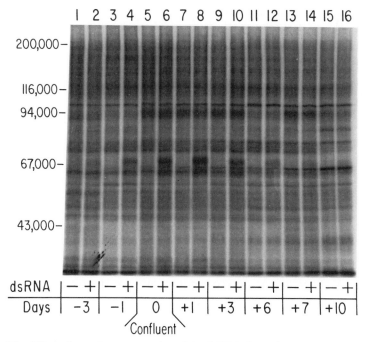

Fig. 13. Effect of growth on expression of the dsRNA-dependent protein kinase. 3T3-F442A cells were grown in the Dulbecco–Vogt modified Eagle's minimal essential medium supplemented with 10% calf serum. For experiments 3×10^4 cells/100 mm plate were inoculated and ~ 7 days after inoculations confluence was attained. Cell extracts were prepared at the indicated days of growth by resuspending cell pellets in buffer containing 10 mM Tris-HCl (pH 7.7), 6 mM MgCl$_2$, 80 mM KCl, 2 mM DTT, 250 mM sucrose, and 0.1 mM EDTA. Cells were lysed by the addition of Triton X-100 to a final concentration of 0.1% (w/v). Homogenates were centrifuged at 10,000 g for 10 min at 4°C, and the soluble fraction (S10) was used for protein kinase assays. Protein kinase assays (20 μl) contained 10 mM Tris-HCl (pH 7.7), 20–50 mM KCl, 5 mM Mg(OAc)$_2$, 20–50 μM [γ-^{32}P]ATP (5–10 μCi), dsRNA (20 ng/ml) where indicated, and 4 μg of S10 protein. Incubations were for 15 min at 30°C. All assays were analyzed by SDS–PAGE and autoradiography. The figure is an autoradiogram. From Petryshyn et al. (1984).

attributable in part to the activation of two cellular enzymes which inhibit protein synthesis. The first enzyme, upon activation by dsRNA, polymerizes ATP into $(2'-5'A)_n$-linked oligoadenylates (Kerr and Brown, 1978). These oligonucleotides activate a cellular endonuclease which degrades RNA in virus-infected, IFN-treated cells and *in vitro* (Clemens and Williams, 1978; Nilsen *et al.*, 1983). The second effect of IFN is to increase the activity of a dsRNA-dependent eIF-2α protein kinase (Lebleu *et al.*, 1976; Zilberstein *et al.*, 1978). The effect of IFN in enhancing the activity of these enzyme systems is shown in Fig. 12.

Recent evidence suggests that IFN and its associated enzyme, the $(2'-5'A)_n$ synthetase, may also play a role in regulating growth and differentiation of cells (Lieberman *et al.*, 1974; Rossi *et al.*, 1977; Krishnan and Baglioni, 1980; Creasey *et al.*, 1980). We have observed in a clone of mouse fibroblasts (3T3-F442A) a pattern of dsRNA-dependent phosphorylation of a 67K protein that is related to the various stages of growth of these cells (Petryshyn *et al.*, 1983). 3T3-F442A cells have a high rate of spontaneous differentiation to adipocytes after the cells become confluent. The dsRNA-dependent phosphorylation of the 67K protein (3T3-dsI) increases with cell growth until the cultures become confluent and protein synthesis is reduced. After confluence there is a rapid and marked decrease in 3T3-dsI phosphorylation, and the cells begin to differentiate to adipocytes. A representative experiment of the phosphoprotein profile in dsRNA-dependent phosphorylation of 3T3-F442A cells is shown in Fig. 13. These results raise the question of a possible role of dsI in regulating the growth and differentiation of cells.

IX. REGULATION OF PROTEN SYNTHESIS BY PROTEIN PHOSPHATASES

It is evident from many studies that protein phosphatases play a significant role in the control of translation. The *in situ* phosphorylation of eIF-2α in heme-deficient or dsRNA-inhibited reticulocyte lysates is a reversible event due to the action of protein phosphatase(s). Hence, the extent of *in situ* phosphorylation of eIF-2α in inhibited lysates is determined by an eIF-2α kinase–eIF-2α phosphatase equilibrium (Ernst *et al.*, 1979; Leroux and London, 1982; Petryshyn *et al.*, 1982). In heme-deficient lysates, phosphorylation of eIF-2α is rapid and achieves a steady-state level within the first 5 min of incubation. In dsRNA-inhibited lysates, the rate of phosphorylation of eIF-2α is initially slower, but then increases, reaching a steady-state level of phosphorylation at 10–15 min (Leroux and London, 1982). In both inhibited lysates the turnover of the phospate on eIF-2α is rapid, with an estimated half-life of 2–3 min. The eIF-2α phosphatase(s) appears to be at least partially active at all times and may be presumed to be responsible for maintaining eIF-2α in dephosphorylated form in normal lysates and in catalyzing the rapid dephosphorylation of eIF-2(α P) upon reversal of inhibition in heme-deficient or dsRNA-inhibited lysates. As described, phosphorylated eIF-2α appears to be present in situ as an eIF-2(α P)·GDP complex which binds tightly to RF to form an RF·eIF-2(α P) complex which is inactive in protein synthesis (Matts et al., 1983). Recent data support the conclusion that this complex is the primary target for eIF-2α phosphatase activity in inhibited lysates (Matts and London, 1984). The dephosphorylation of eIF-2α therefore restores RF activity and permits the resumption of protein synthesis.

The in situ activation of both the heme-regulated eIF-2α kinase (HRI) and the dsRNA-activated eIF-2α kinase (dsI) in lysates is accompanied by phosphorylation of the two kinase molecules themselves. The reversal of heme deficiency by the delayed addition of hemin or of dsRNA-induced inhibition by high levels of poly(I) · poly(C) (20 μg/ml) is accompanied by a reduction in the level of phosphorylation of the respective kinase (Ernst et al., 1979; Petryshyn et al., 1982). Since the phosphorylation of the two eIF-2α kinases is essential for activity, the dephosphorylation is concomitant with a reduced capacity to phosphorylate eIF-2(α P). These observations indicate that protein phosphatases also regulate the activation of the eIF-2α kinases.

The mechanisms of these specific dephosphorylations and the nature of the protein phosphatases involved are not clear. In a preliminary analysis of the protein phosphatase species present in reticulocytes (Foulkes et al., 1983), type 1 (PP-1) and type 2 (PP-2) activities were identifed and characterized based on criteria previously established for similar activities in rabbit skeletal muscle and rabbit liver (Cohen et al., 1981). These include (a) chromatographic separation on DEAE-cellulose, (b) substrate specificity toward glycogen phosphorylase a and the α and β subunits of phosphorylase kinase, (c) differential sensitivity to the heat-stable protein phoshatase inhibitors 1 and 2, and (d) sensitivity to MgATP. When total lysate phosphatases are assayed in the presence of 1 mM MnCl$_2$, protein phosphatase type 2 represents 84% of lysate phosphorylase phosphatase activity. However, when phosphatase assays are carried out with MgATP concentrations similar to those in the lysate, type 2 activity is diminished, and the levels of type 1 (41%) and type 2 (59%) protein phosphatases are comparable. A small proportion (6%) of total lysate phosphatase is tightly bound to the ribosomes, where type 1 phosphatase predominates.

At least five species of protein phosphatases can be identified in lysates. These constitute two forms of protein phosphatase type 1, one of which (designated F_C) is dependent on MgATP and a lysate activator protein F_A; both F_C and F_A have been identified previously in skeletal muscle. Three species of protein phosphatase type 2 have been identified and designated PP-2B, PP-$2A_1$, and PP-$2A_2$ in accord with designations for rabbit skeletal muscle and rabbit liver phosphatases, which display similar phosphatase profiles (Cohen *et al.*, 1981). Lysate protein phosphatases types 1, F_C, $2A_1$, and $2A_2$ can all act on phosphorylase *a* and the α (type 2) or β (type 1) subunit of phosphorylase kinase. PP-2B, a Ca^{2+}/calmodulin-dependent phosphatase, specifically dephosphorylates the α subunit of phosphorylase kinase, but does not act on phosphorylase *a*. The heat-stable protein phosphatase inhibitor 2 from skeletal muscle completely blocks the activity of the type 1 phosphatases (PP-1, F_C), but has no effect on the three species of type 2 protein phosphatase. A preliminary assay of the two heat-stable phosphatase inhibitors 1 and 2 in lysates indicates significant levels of inhibitor 2, but little or no detectable inhibitor 1 (Foulkes *et al.*, 1983; Reddy and Ernst, 1983).

In a related study on the function and specificity of lysate protein phosphatase activities involved in the regulation of protein synthesis, the effects of protein phosphatase inhibitor 2 on reticulocyte protein synthesis and protein phosphorylation were examined (Ernst *et al.*, 1982). It was found that inhibitor 2 inhibited protein chain initiation in normal hemin-supplemented lysates. As in heme deficiency, inhibition was characterized by biphasic kinetics and the phosphorylation of eIF-2α, and was reversed by the delayed addition of purified eIF-2. The inhibition of protein synthesis in lysates by inhibitor 2 is presumed to be due to the inhibition of a type 1 eIF-2α phosphatase, which permits a basal eIF-2α kinase activity to be expressed resulting in the accumulation of eIF-2(α P). Inhibitor 2 itself (29,000 daltons) was partially phosphorylated in the lysate by a cAMP-independent protein kinase. This accords with recent findings that inhibitor 2 is associated in skeletal muscle and in reticulocytes with F_C phosphatase, and that the phosphorylation of inhibitor 2 by F_A inactivates the inhibitor and permits expression of F_C (Resnik *et al.*, 1983; Reddy and Ernst, 1983).

However, as reflected in other studies on reticulocyte phosphatases, assignment of phosphatase specificity is premature. Crouch and Safer (1980) have purified a type 2 phosphatase ($2A_2$) with eIF-2α phosphatase activity against purified eIF-2(α P). A related study, however, demonstrated that purified PP-1 from rabbit skeletal muscle dephosphorylated purified eIF-2(α P) at the same relative rate as PP-$2A_2$

(Stewart *et al.*, 1982). Grankowski *et al.* (1980) partially purified a protein phosphatase with activity toward both the α and β subunits of eIF-2 and a putative heme-regulated eIF-2α kinase. The same study characterized two small heat-stable proteins which stimulate phosphatase activity toward different substrates, thereby imposing some degree of specificity. Mumby and Traugh (1980a,b) partially purified several protein phosphatases which were characterized in part, based on chromatographic properties, cation sensitivities, and substrate specificities. From these and other studies, it is clear that the nature and regulation of the protein phosphatases which act on eIF-2α and the eIF-2α kinases remain to be resolved.

A crucial target of phosphatase action is the RF·eIF-2(α P) complex in which the RF is tightly bound. Dephosphorylation of eIF-2α in this complex is essential for the release of RF which is required for the dissociation of GDP from eIF-2, and thus for the recycling of eIF-2 by permitting the formation of the eIF-2·GTP·Met-tRNA$_f$ ternary complex.

X. CONCLUDING REMARKS

The regulation of initiation of eukaryotic protein synthesis by phosphorylation of eIF-2α is now far better understood than was true only a few years ago. The heme-regulated eIF-2α kinase and the dsRNA-dependent eIF-2α kinase, although distinctly different molecular entities, appear to phosphorylate the same site on eIF-2α. These kinases are activated by phosphorylation, but the issue of the mechanism of physiologic activation, by auto- or heterophosphorylation, remains to be resolved.

The generality of eIF-2α phosphorylation as a major mechanism of regulation of initiation also remains to be determined. The presence of these eIF-2α kinases in normal human reticulocytes provides a basis for investigating the role of eIF-2α phosphorylation in various human hematologic disorders such as iron deficiency anemia, sideroblastic anemias, plumbism, and thalassemic syndromes with secondary impairment of heme synthesis. And still to be explored is the activity of these, and perhaps other, eIF-2α kinases in a wide variety of nonerythroid eukaryotic cells.

The apparent paradox of the occurrence of shutoff of protein synthesis when only 30–40% of the eIF-2α is phosphorylated is now readily explained by the finding that RF present in limiting concentration is sequestered in a tight association with eF-2(α P) and is unavailable for its function, the dissociation of GDP from eIF-2. This dissociation is essential for the recycling of eIF-2 which involves the dissociation of the (eIF-2·GDP) complex, the exchange of GTP for GDP, and the formation of the (Met-tRNA$_f$·eIF-2·GTP) ternary complex. When shutoff of initiation occurs because of the sequestration and unavailability of RF, the RF·eIF-2(α P) complex is a prime target for phosphatase activity which is necessary for the liberation of RF and the resumption of protein synthesis.

The action of RF in promoting the exchange of GTP for GDP in the complex eIF-2·GDP is very similar to the mechanism by which the prokaryotic elongation factor, Ts, promotes the exchange of GTP for GDP in the Tu·GDP complex. These guanine nucleotide-binding regulatory mechanisms bear a resemblance to those observed with the G proteins of adenylate cyclase and transducin. In the light of the findings with eIF-2 and RF, study of the possible effects of phosphorylation on the regulatory mechanisms of guanine nucleotide binding and exchange may prove fruitful.

REFERENCES

Adamson, S. D., Herbert, E., and Godchaux, W. (1968). Factors affecting the rate of protein synthesis in lysate systems from reticulocytes. *Arch. Biochem. Biophys.* **125**, 671–683.

Amesz, H., Goumans, H., Hambrich-Morree, T., Voomra, H. O., and Benne, R. (1979). Purification and characterization of a protein factor that reverses the inhibition of protein synthesis by the heme-regulated translational inhibitor in rabbit reticulocyte lysates. *Eur. J. Biochem.* **98**, 513–520.

Beuzard, Y., and London, I. M. (1974). The effects of hemin and double-stranded RNA on α- and β-globin synthesis in reticulocyte and Krebs II ascites cell-free systems and the reversal of these effects by an initiation factor preparation. *Proc. Natl. Acad. Sci. U.S.A.* **71**, 2863–2866.

Bruns, G. P., and London, I. M. (1965). The effect of hemin on the synthesis of globin. *Biochem. Biophys. Res. Commun.* **18**, 236–242.

Clemens, M. J., and Williams, B. R. G. (1978). Inhibition of cell free protein synthesis by pppA2′p5′A: A novel oligonucleotide synthesized by interferon-treated L cell extracts. *Cell (Cambridge, Mass.)* **13**, 565–572.

Clemens, M. J., Safer, B., Merrick, W. C., Anderson, W. F., and London, I. M. (1975). Initiation of protein synthesis in rabbit reticulocyte lysates by double-stranded RNA and oxidized glutathione: Indirect mode of action on polypeptide chain initiation. *Proc. Natl. Acad. Sci. U.S.A.* **72**, 1286–1290.

Cohen, P., Foulkes, J. G., Goris, J., Hemmings, B. A., Ingebritsen, T. S., Stewart, A. A., and Strada, S. T. (1981). Classification of protein phosphatases involved in cellular regulation. In "Metabolic Interconversion of Enzymes" (H. Holzer, ed.), pp. 28–43. Springer-Verlag, Berlin and New York.

Creasey, A., Bartholomew, J. C., and Merigan, T. C. (1980). Role of G_0-G_1 arrest in the inhibition of tumor cell growth by interferon,. *Proc. Natl. Acad. Sci. U.S.A.* **77**, 1471–1475.

Crouch, D., and Safer, B. (1980). Purification and properties of eIF-2 phosphatase. *J. Biol. Chem.* **255**, 7918–7924.

Datta, A., deHaro, C., and Ochoa, S. (1978). Translational control by hemin is due to binding to cyclic-AMP-dependent protein kinase. *Proc. Natl. Acad. Sci. U.S.A.* **75**, 1148–1152.

deHaro, C., DeHerreros, A. G., and Ochoa, S. (1983). Activation of the heme-stabilized translational inhibitor of reticulocyte lysates by calcium ions and phospholipids. *Proc. Natl. Acad. Sci. U.S.A.* **80**, 6843–6847.

Ehrenfeld, E., and Hunt, T. (1971). Double-stranded polio virus RNA inhibits initiation of protein synthesis by reticulocyte lysates. *Proc. Natl. Acad. Sci. U.S.A.* **68**, 1075–1078.

Ernst, V., Levin, D. H., Leroux, A., and London, I. M. (1980). Site-specific phosphorylation of the α subunit of eukaryotic initiation factor by the heme-regulated and double-stranded RNA-activated eIF-2α kinases from rabbit reticulocyte lysates. *Proc. Natl. Acad. Sci. U.S.A.* **77**, 1286–1290.

Ernst, V., Levin, D. H., and London, I. M. (1979). *In situ* phosphorylation of the α subunit of eukaryotic initiation factor 2 in reticulocyte lysates inhibited by heme deficiency, double-stranded RNA, oxidized glutathione, or heme-regulated protein kinase. *Proc. Natl. Acad. Sci. U.S.A.* **76**, 2118–2122.

Ernst, V., Levin, D. H., Leroux, A., and London, I. M. (1980). Site-specific phosphorylation of the α subunit of eukaryotic initiation factor eIF-2α kinase from rabbit reticulocyte lysates. *Proc. Natl. Acad. Sci. U.S.A.* **77**, 1286–1290.

Ernst, V., Levin, D. H., Foulkes, J. G., and London, I. M. (1982). Effects of skeletal muscle protein phosphatase inhibitor-2 on protein synthesis and protein phosphorylation in rabbit reticulocyte lysates. *Proc. Natl. Acad. Sci. U.S.A.* **79**, 7092–7096.

Fagard, R., and London, I. M. (1981). Relationship between the phosphorylation and activity of the heme-regulated eIF-2α kinases. *Proc. Natl. Acad. Sci. U.S.A.* **78**, 866–870.

Farrell, P. J., Balkow, K., Hunt, T., Jackson, R. J., and Trachsel, H. (1977). Phosphorylation of initiation factor eIF-2 and the control of reticulocyte protein synthesis. *Cell (Cambridge, Mass.)* **11**, 187–200.

Farrell, P. J., Hunt, T., and Jackson, R. J. (1978). Analysis of phosphorylation of protein synthesis initiation factor eIF-2 by two-dimensional gel electrophoresis. *Eur. J. Biochem.* **89**, 517–521.

Foulkes, J. G., Ernst, V., and Levin, D. H. (1983). Separation and identification of type 1 and type 2 protein phosphatases from rabbit reticulocyte lysates. *J. Biol. Chem.* **258**, 1439–1443.

Grankowski, N., Lehmusvirta, D., Kramer, G., and Hardesty, B. (1980). Partial purification and characterization of reticulocyte phosphatase with activity for phosphorylated peptide initiation factor 2. *J. Biol. Chem.* **255**, 310–317.

Grayzel, A. I., Horchner, P., and London, I. M. (1966). The stimulation of globin synthesis by heme. *Proc. Natl. Acad. Sci. U.S.A.* **55**, 650–655.

Grosfeld, H., and Ochoa, S. (1980). Purification and properties of double-stranded RNA-activated eIF-2α kinase from reticulocyte lysates. *Proc. Natl. Acad. Sci. U.S.A.* **77**, 6526–6530.

Gross, M. (1979). Control of protein synthesis by hemin. *J. Biol. Chem.* **254**, 2370–2377.

Gross, M., and Mendelewski, J. (1977). Additional evidence that the hemin-controlled translational repressor from rabbit reticulocytes is a protein kinase. *Biochim. Biophys. Res. Commun.* **74**, 559–569.

Gross, M., and Mendelewski, J. (1978). Control of protein synthesis by hemin. An association between the formation of the hemin-controlled translational repressor and the phosphorylation of a 100,000 molecular weight protein. *Biochim. Biophys. Acta* **520,** 650–663.

Gross, M., and Rabinovitz, M. (1972). Control of globin synthesis by hemin: Factors influencing formation of an inhibitor of globin chain initiation in reticulocyte lysates. *Biochim. Biophys. Acta* **287,** 340–352.

Howard, G. A., Adamson, S. D., and Herbert, E. (1970). Studies on cessation of protein synthesis in a reticulocyte lysate cell-free system. *Biochim. Biophys. Acta* **213,** 237–240.

Hunt, T. (1979). Control of initiation of mammalian protein synthesis. From gene to protein. *Miami Winter Symp.* **16,** 321–345.

Hunt, T., Vanderhoff, G., and London, I. M. (1972). Control of globin synthesis: The role of heme. *J. Mol. Biol.* **66,** 471–481.

Hunter, T., Hunt, T., Jackson, R. J., and Robertson, H. D. (1975). The characteristics of inhibition of protein synthesis by double-stranded ribonucleic acid in reticulocyte lysates. *J. Biol. Chem.* **250,** 409–417.

Jackson, R. J., Herbert, P., Campbell, E. A., and Hunt, T. (1983). The roles of sugar phosphates and thiol reducing systems in the control of reticulocyte protein synthesis. *Eur. J. Biochem.* **131,** 313–324.

Jagus, R., Anderson, W. F., and Safer, B. (1981). The regulation of initiation of mammalian protein synthesis. *Prog. Nucleic Acid Res. Mol. Biol.* **25,** 127–185.

Kaempfer, R. (1974). Identification and RNA-binding properties of an initiation factor capable of relieving translational inhibition induced by heme deprivation or double-stranded RNA. *Biochem. Biophys. Res. Commun.* **61,** 591–597.

Kan, B., Kanigur, G., Tiryaki, D., Gokhan, N., and Bermek, E. (1983). High pO_2-promoted activation of a translational inhibitor in rabbit reticulocytes: Role of the oxidation state of sulfhydryl groups. *In* "Molecular Mechanisms of Protein Synthesis" (E. Bermek, ed.), pp. 169–177. Springer-Verlag, Berlin and New York.

Kassenaar, A., Morell, H., and London, I. M. (1957). The incorporation of glycine into globin and the synthesis of heme *in vitro* in duck erythrocytes. *J. Biol. Chem.* **229,** 423–435.

Kerr, I. M., and Brown, R. E. (1978). pppA2'p5'A2'p5'A: An inhibitor of protein synthesis synthesized with an enzyme fraction from interferon-treated cells. *Proc. Natl. Acad. Sci. U.S.A.* **75,** 256–260.

Konieczny, A., and Safer, B. (1983). Purification of the eukaryotic initiation factor 2–eukaryotic initiation factor 2B complex and characterization of its guanine nucleotide exchange activity during protein synthesis initiation. *J. Biol. Chem.* **258,** 3402–3408.

Kosower, N. S., Vanderhoff, G. A., and Kosower, E. M. (1972). The effect of glutathine disulfide on initiation of protein synthesis. *Biochim. Biophys. Acta* **272,** 623–627.

Kramer, G., and Hardesty, B. (1980). Regulation of eukaryotic protein synthesis. *In* "Cell Biology: A Comprehensive Treatise" (L. Goldstein and D. H. Prescott, eds.), Vol. 4, pp. 69–105. Academic Press, New York.

Kramer, G., Cimadevilla, M., and Hardesty, B. (1976). Specificity of the protein kinase activity associated with the hemin-controlled repressor of rabbit reticulocytes. *Proc. Natl. Acad. Sci. U.S.A.* **73,** 3078–3082.

Krishnam, I., and Baglioni, C. (1980). Increased levels of (2'–5')oligo(A) polymerase activity in human lymphoblastoid cells treated with glucocortoids. *Proc. Natl. Acad. Sci. U.S.A.* **77,** 6506–6510.

Kruh, J., and Borsook, G. (1956). Hemoglobin synthesis in rabbit reticulocytes *in vitro. J. Biol. Chem.* **220,** 905–915.

Lebleu, B., Sen, G. C., Shaila, S., Cabrer, B., and Lengyel, P. (1976). Interferon, double-stranded RNA, and protein phosphorylation. *Proc. Natl. Acad. Sci. U.S.A.* **73,** 3107–3111.

Legon, S., Jackson, R., and Hunt, T. (1973). Control of protein synthesis in reticulocyte lysates by haemin. *Nature (London), New Biol.* **241,** 150–152.

Leroux, A., and London, I. M. (1982). Regulation of protein synthesis by phosphorylation of eukaryotic initiation factor 2α in intact reticulocytes and reticulocyte lysates. *Proc. Natl. Acad. Sci. U.S.A.* **79,** 2147–2151.

Levin, D. H., and London, I. M. (1978). Regulation of protein synthesis: Activation by double-stranded RNA of a protein kinase that phosphorylates eukaryotic initiation factor 2. *Proc. Natl. Acad. Sci. U.S.A.* **75,** 1121–1125.

Levin, D. H., Ranu, R. S., Ernst, V., and London, I. M. (1976). Regulation of protein synthesis in reticulocyte lysates. Phosphorylation of methionyl-tRNA$_f$ binding factor by protein kinase activity of the translational inhibitor isolated from heme-deficient lysates. *Proc. Natl. Acad. Sci. U.S.A.* **73,** 3112–3116.

Levin, D. H., Ernst, V., and London, I. M. (1979). Effects of the catalytic subunit of cAMP-dependent protein kinase (type II) from reticulocytes and bovine heart muscle on protein phosphorylation and protein synthesis in reticulocyte lysates. *J. Biol. Chem.* **254,** 7935–7941.

Levin, D. H., Petryshyn, R., and London, I. M. (1980). Characterization of double-stranded RNA-activated kinase that phosphorylates α subunit of eukaryotic initiation factor 2(eIF-2α) in reticulocyte lysates. *Proc. Natl. Acad. Sci. U.S.A.* **77,** 832–836.

Levin, D. H., Petryshyn, R., and London, I. M. (1981). Characterization of purified double-stranded RNA-activated eIF-2α kinase from rabbit reticulocyte. *J. Biol. Chem.* **256,** 7638–7641.

Lieberman, D., Voloch, Z., Aviv, H., Nudel, U., and Revel, M. (1974). Effects of interferon on hemoglobin synthesis and leukemia virus production in Friend cells. *Mol. Biol. Rep.* **1,** 447–451.

London, I. M., Ernst, V., Fagard, R., Levin, D. H., and Petryshyn, R. (1981). Regulation of protein synthesis by phosphorylation and heme. *Cold Spring Harbor Conf. Cell Proliferation* **8,** 931–1012.

Matts, R. L., and London, I. M. (1984). The regulation of initiation of protein synthesis by phosphorylation of eIF-2α and the role of reversing factor in recycling of eIF-2. *J. Biol. Chem.* **259,** 6708–6711.

Matts, R. L., Levin, D. H., and London, I. M. (1983). Effect of phosphorylation of the α subunit of eukaryotic initiation factor eIF-2 on the function of reversing factor (RF) in the initiation of protein synthesis. *Proc. Natl. Acad. Sci. U.S.A.* **80,** 2559–2563.

Maxwell, C. R., and Rabinovitz, M. (1969). Evidence for an inhibitor in the control of globin synthesis by hemin in reticulocyte lysates. *Biochem. Biophys. Res. Commun.* **35,** 79–85.

Michelson, A. M., Ernst, V., Levin, D. H., and London, I. M. (1984). Effects of glucose 6-phosphate and hemin on activation of heme-regulated eIF-2α kinase in gel-filtered reticulocyte lysates. *J. Biol. Chem.* **259,** 8529–8533.

Morell, H., Savoie, J. C., and London, I. M. (1958). The biosynthesis of heme and the incorporation of glycine into globin in rabbit bone marrow *in vitro. J. Biol. Chem.* **233,** 923–929.

Mumby, M., and Traugh, J. A. (1980a). Multiple forms of phosphoprotein phosphatase from rabbit reticulocyte. *Biochim. Biophys. Acta* **611,** 342–350.

Mumby, M., and Traugh, J. A. (1980b). Dephosphorylation of translational initiation factors and 40 S ribosomal subunits by phosphoprotein phosphatases from rabbit reticulocytes. *Biochemistry* **18,** 4548–4556.

Nilsen, T. W., Maroney, P. A., and Baglioni, C. (1983). Maintenance of protein synthesis in spite of mRNA breakdown in interferon-treated HeLa cells infected with reovirus. *Mol. Cell. Biol.* **3**, 64–69.

Nizet, A. (1957). Recherches sur les rélations entre les biosynthésis de l'heme et de la globine. *Bull. Soc. Chim. Biol.* **39**, 265.

Ochoa, S. (1983). Regulation of protein synthesis in eukaryotes. *Arch. Biochem. Biophys.* **223**, 325–349.

Pain, V. M., and Clemens, M. J. (1983). Assembly and breakdown of mammalian protein synthesis initiation complexes: Regulation by guanine nucleotides and by phosphorylation of initiation factor eIF-2. *Biochemistry* **22**, 726–733.

Panniers, R., and Henshaw, E. C. (1983). A GDP/GTP exchange factor essential for eukaryotic initiation factor 2 cycling in Ehrlich ascites tumor cells and its regulation by eukaryotic initiation factor 2 phosphorylation. *J. Biol. Chem.* **258**, 7928–7934.

Petryshyn, R., Trachsel, H., and London, I. M. (1979). Regulation of protein synthesis in reticulocyte lysates: Immune serum inhibits heme-regulated protein kinase activity and differentiates heme-regulated protein kinase from double-stranded RNA-induced protein kinase. *Proc. Natl. Acad. Sci. U.S.A.* **76**, 1575–1579.

Petryshyn, R., Levin, D. H., and London, I. M. (1980). Purification and characterization of a latent precursor of a double-stranded RNA dependent protein kinase from reticulocyte lysates. *Biochem. Biophys. Res. Commun.* **94**, 1190–1198.

Petryshyn, R., Levin, D, H., and London, I. M. (1982). Regulation of double-stranded RNA-activated eukaryotic initiation factor 2 α kinase by type 2 protein phosphatase in reticulocyte lysates. *Proc. Natl. Acad. Sci. U.S.A.* **79**, 6512–6516.

Petryshyn, R., Chen, J.-J., Levin, D. H., and London, I. M. (1983). Growth-dependent expression of a dsRNA-dependent protein kinase in non-interferon treated 3T3 fibroblasts. *Fed Proc., Fed. Am. Soc. Exp. Biol.* **42**, 2248.

Petryshyn, R., Rosa, F., Fagard, R., Levin, D. H., and London, I. M. (1984). Control of protein synthesis in human reticulocytes by heme-regulated and double-stranded RNA dependent eIF-2α kinases. *Biochem. Biophys. Res. Commun.* **119**, 891–899.

Ralston, R. O., Das, A., Dasgupta, N., Roy, R., Palmieri, S., and Gupta, N. K. (1978). Protein synthesis in rabbit reticulocytes: Characteristics of a ribosomal factor that reverses inhibition of protein synthesis in heme-deficient lysates. *Proc. Natl. Acad. Sci. U.S.A.* **75**, 4858–4862.

Ralston, R. O., Das, A., Grace, M., Das, H., and Gupta, N. K. (1979). Protein synthesis in rabbit reticulocytes: Characteristics of a postribosomal supernatant factor that reverses inhibition of protein synthesis in heme-deficient lysates and inhibition of ternary complex (Met-tRNA_f·eIF-2·GTP) formation by heme-regulated inhibitor. *Proc. Natl. Acad. Sci. U.S.A.* **76**, 5490–5494.

Ranu, R. S. (1979). Regulation of protein synthesis in rabbit reticulocyte lysates: The heme-regulated protein kinase (HRI) and double-stranded RNA-induced protein kinase (dRI) phosphorylate the same site(s) on initiation factor eIF-2. *Biochem. Biophys. Res. Commun.* **91**, 1437–1444.

Ranu, R. S., and London, I. M. (1976). Regulation of protein synthesis in rabbit reticulocytes: Purification and initial characterization of the cyclic 3' : 5'-AMP-independent protein kinase of the heme-regulated translational inhibitor. *Proc. Natl. Acad. Sci. U.S.A.* **73**, 4349–4363.

Reddy, P., and Ernst, V. G. (1983). Partial purification and characterization of heat-stable protein phosphatase inhibitor-2 from rabbit reticulocytes. *Biochem. Biophys. Res. Commun.* **114**, 1089–1096.

Resnik, T., Hemmings, B. A., Lim Tuny, H. Y., and Cohen, P. (1983). Characterisation of a reconstituted MgATP dependent protein phosphatase. *Eur. J. Biochem.* **114**, 1089–1096.

Rossi, G. B., Dolei, A., Cioe, L., Benadetto, A., Matarese, G. P., and Belardelli, F. (1977). Inhibition of transcription and translation of globin messenger RNA in dimethyl sulfoxide-stimulated Friend erythroleukemic cells treated with interferon. *Proc. Natl. Acad. Sci. U.S.A.* **74**, 2036–2040.

Siekierka, J., Mitsui, K., and Ochoa, S. (1981). Mode of action of the heme-controlled translational inhibitor: Relationship of eukaryotic initiation factor-2-stimulating protein to translation restoring factor. *Proc. Natl. Acad. Sci. U.S.A.* **78**, 220–223.

Siekierka, J., Mauser, L., and Ochoa, S. (1982). Mechanism of polypeptide chain initiation in eukaryotes and its control by phosphorylation of the α subunit of initiation factor 2. *Proc. Natl. Acad. Sci. U.S.A.* **79**, 2537–2540.

Siekierka, J., Manne, V., and Ochoa, S. (1984). Mechanism of translational control by partial phosphorylation of the α subunit of eukaryotic initiation factor 2. *Proc. Natl. Acad. Sci. U.S.A.* **81**, 352–356.

Stewart, A. A., Ingebritsen, T. S., Manalan, A., Klee, C. B., and Cohen, P. (1982). Discovery of a Ca^{2+} and calmodulin-dependent protein phosphatase. *FEBS Lett.* **137**, 80–84.

Thomas, N. S. B., Matts, R. L., and London, I. M. (1984). Distribution of reversing factor (RF) in reticulocyte lysates during active protein synthesis and on inhibition by heme deprivation or double-stranded RNA. *Proc. Natl. Acad. Sci. U.S.A.* **81**, 6998–7002.

Trachsel, H., Ranu, R. S., and London, I. M. (1978). Regulation of protein synthesis in rabbit reticulocyte lysates: Purification and characterization of heme-reversible translational inhibitor. *Proc. Natl. Acad. Sci. U.S.A.* **75**, 3654–3658.

Voorma, H. O., and Amesz, H. (1982). The control of the rate of protein synthesis initiation. *Dev. Biochem.* **24**, 297–309.

Zilberstein, A., Kimchi, A., Schmidt, A., and Revel, M. (1978). Isolation of two interferon-induced translational inhibitors: A protein kinase and an oligo-isoadenylate synthetase. *Proc. Natl. Acad. Sci. U.S.A.* **75**, 4734–4738.

Zucker, W. V., and Schulman, H. M. (1968). Stimulation of globin-chain initiation by hemin in the reticulocyte cell-free system. *Proc. Natl. Acad. Sci. U.S.A.* **59**, 582–589.

Index

R

S